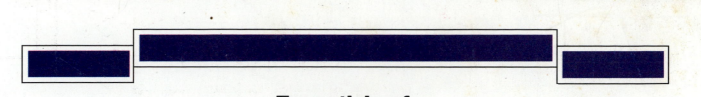

Essentials of
Physical Geography

Essentials of
Physical Geography

Ralph C. Scott

Towson State University
Towson, Maryland

West Publishing Company

St. Paul   New York   Los Angeles   San Francisco

COPYRIGHT © 1991 By WEST PUBLISHING
COMPANY
50 W. Kellogg Boulevard
P.O. Box 64526
St. Paul, MN 55164-0526

Printed in the United States of America

98 97 96 95 94 93 92 91   8 7 6 5 4 3 2 1

Library of Congress Cataloging-in-Publication Data

Scott, Ralph C. (Ralph Carter), 1944-
    Essentials of physical geography / Ralph C. Scott.
    p.   cm.
    Includes index.
    ISBN 0-314-79264-3
    1. Physical geography.   I. Title.
GB54.5.S37   1991
910'.02—dc20
                                        90-44892
                                            CIP

PRODUCTION CREDITS

Interior Design: David Farr,
                IMAGESMYTHE, INC.
Cover Design: Katherine Townes
Cover Image: Mike Surowiak, @ Tony Stone Worldwide
Copyeditor: Kim Kaliszewski
Cartographer: Alice B. Thiede, Carto-Graphics
Artwork: Rolin Graphics
Compositor: Carlisle Communications, Ltd.

The map on page 394 is supplied by Marie Tharp, 1 Washington Ave., South Nyack, NY 10960.

CHAPTER OPENER CREDITS

Introduction   The Upper Snake River and Grand Teton Mountains of northwestern Wyoming. (Clark T. Heglar, The Image Bank)   **Chapter One**   The Earth, as seen from space, is dominated by clouds and water. (Department of the Interior, U.S. Geological Survey)   **Chapter Two**   A cloud-filled sky is reflected in a peaceful lake setting. (H. Armstrong Roberts)   **Chapter Three**   A recent snow up in the mountains. (Ross M. Horowitz, The Image Bank)   **Chapter Four**   Wonderful color is reflected out on the sea with these sailboats. (Ken Fraser, FPG International)   **Chapter Five**   A beautiful sunset. (Don King, The Image Bank)   **Chapter Six**   Weather satellite view of an Atlantic hurricane and a frontal cyclone over the northeastern United States. (NOAA, JLM Visuals)   **Chapter Seven**   Summer showers near Mt. Ruddle, Banff National Park, Alberta (T. Algire/H. Armstrong Roberts)   **Chapter Eight**   A splendid view of the beach. (M. Thonig/H. Armstrong Roberts)   **Chapter Nine**   The Apalachicola River, in Florida, is bordered by a forest of water-tolerant cypress trees. (W. Metzen/H. Armstrong Roberts)   **Chapter Ten**   Winter view of rows of corn stubble in central Missouri. (Marbut Memorial Slide Collection, American Society of Agronomy)   **Chapter Eleven**   Hawaii's Mt. Kilauea in eruption. (Joanna McCarthy, The Image Bank)   **Chapter Twelve**   Zigzag ridge patterns produced by the tilting and folding of resistant rock strata in the Central Appalachians. (U.S. Geological Survey)   **Chapter Thirteen**   An eroded sandstone rock formation near Laramie, Wyoming. (David Butler)   **Chapter Fourteen**   Entrenched meanders of the Upper Colorado River in San Juan County, Utah. (U.S. Geological Survey)   **Chapter Fifteen**   View of glacially sculptured Yosemite Valley, California. (B.F. Molnia)   **Chapter Sixteen**   Saharan landscape in southern Algeria. (H. Armstrong Roberts)   **Chapter Seventeen**   A volcanic island surrounded by a barrier reef in the Society Islands of French Polynesia. (Marcel Isy-Schwart, The Image Bank)

Contents

Chapter Three

Energy Flow and Air Temperature 36

Chapter Four

Air Pressure and Wind 52

Chapter Five

Atmospheric Moisture 74

Chapter Six

Weather Systems   97

Chapter Seven

Climates of the World   123

Chapter Eight

Water on and Beneath the Earth's Surface   152

Chapter Nine

Natural Vegetation 175

Chapter Ten

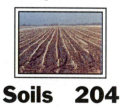

Soils 204

Chapter Eleven

Landforms and the Earth's Interior 233

Chapter Twelve

Diastrophic and Volcanic Landforms 258

Chapter Thirteen

Weathering, Mass Wasting, and Karst Topography 282

Chapter Fourteen

Fluvial Processes and Landforms 302

Chapter Fifteen

Glacial Processes and Landforms 324

Chapter Sixteen

Eolian Processes and Desert Landscapes   348

Chapter Seventeen

Coastal Processes and Landforms   368

Preface

As we enter the 1990s, geographic education in North America seems to be experiencing a long-overdue revival. Students, educators, and the general public are once again coming to realize that other places, and the events that occur in them, have an important bearing on our own lives. Especially crucial to our future well-being is the recent growth in public environmental consciousness. Our planet is a finely balanced and fragile one, and humanity now has a greater potential than ever before to alter permanently its physical characteristics. It is essential that the next generation of leaders gain an understanding and appreciation of the environment greater than that of the leaders of the past.

My primary goal in writing this book is to provide its readers with some understanding and appreciation of the world around them. The book is intended chiefly for college undergraduate students with limited physical science and geography backgrounds. Most of these students will likely be taking a geography course to help meet general university requirements; others, I hope will be taking it as an early course within a geography major program. This book is approximately 20 percent shorter than my previous textbook, *Physical Geography,* also published by West Publishing Company. Its somewhat reduced scope should make it easier to cover within the time constraints of a one-semester or one-quarter course.

The optimal sequence of topics in a physical geography course has long been a popular subject of debate among those who teach in this field. Before I began writing this book, we surveyed a large group of physical geography instructors to assess their feelings on this important matter. A clear-cut majority believed that the first of the "big four" subjects to be covered should be weather and climate, followed, in turn, by treatment of natural vegetation, soils, and landforms. I have followed this suggested order in this book. For those who prefer other sequences, the material in each of these sections is, I believe, sufficiently self-contained that there should not be any major problems of comprehension or continuity if the chapters are covered out of numerical order.

The approach I have used for each topic was called by one reviewer the "process/distribution approach." By this he meant that a certain amount of general science background material appears first to explain the origin and characteristics of the topic. This is followed by a discussion of the topic's geographical distribution and, finally, by the reasons for its distribution. Throughout the book, I have tried to stress the causative and distributional linkages among the various earth phenomena.

I have treated the subject of human interaction with the natural environment lightly in the main body of the text itself. This is not because this factor is unimportant; rather, it is because there is simply not enough space to systematically cover this crucial subject in an introductory physical geography textbook. Important environmental concerns are, however, discussed in the Case Studies found at the end of most chapters and in many of the Focus boxes scattered throughout the text. I hope that these studies will promote student interest, will illustrate the relevance of the material to contemporary issues, and will foster an increased environmental consciousness.

A number of additional learning aids have been provided in the text. At the beginning of each chapter is a subject outline and several Focus Questions that pose broad questions addressing the core subjects of the chapter material. Within the body of the chapters, key terms appear in boldface type for easy recognition, while other important terms are italicized. At the end of each chapter is a summary, a number of review questions, and a listing of the key terms. An index and extensive glossary appear at the end of the text, and appendix sections deal with the topics of scale conversions, maps and remote sensing, and weather map interpretation. In addition, you have probably already noted that the book is extensively illustrated with full-color photographs, maps, and diagrams.

West has also provided a comprehensive package of ancillaries that I believe will greatly aid adopters of this book. These include an instructor's manual, a laboratory manual, a student study guide, a computerized study guide available for IBM and Macintosh, and a computerized test bank for IBM, Macintosh, and Apple computers. A set of nearly 60 color slides or overhead transparencies of the more important maps and diagrams in the text is also available for teachers using the book. In addition, a free *PC Globe 3.0* program is available for adopters. Please contact the publisher for further information on these materials.

Acknowledgments

The writing of a book such as this is a massive and time-consuming undertaking, especially for a single author. Its preparation involves a great many individuals, all of whom are essential to its eventual successful completion.

I would first of all like to thank the editorial and production staff at West, some of whom I have never met, who enabled this book to become a reality. Foremost among these individuals is my Acquiring Editor, Clark Baxter, who persuaded me to write a second book just when I was beginning to recover from the first one. In addition, Nancy Crochiere, our Developmental Editor, has provided expert analyses of the numerous manuscript reviews and has kept me on the right track with regard to proper procedures. Beth Kennedy, our Promotion Manager, has been responsible for the development of advertising brochures and for making sure that they reached potential users of the book. Lastly, the individual who has almost been single-handedly turning my manuscripts into books is my good friend Nancy Roth, our Production Editor.

Another group of individuals whose long hours of work with the manuscript have been vital to both its content and quality have been the reviewers. The work of many of the reviewers of my book *Physical Geography* is also reflected in the content of this book, and I would again like to thank them all for their generous assistance. I wish particularly to thank reviewers who worked with the manuscript for *Essentials of Physical Geography*. They are William D. Brooks, Indiana State University; Anthony Orr Clarke, University of Louisville; Robert Cullison, Essex Community College; Donald W. Duckson Jr., Frostburg State University; Dennis Edgell, Kent State University; David Fitzgerald, St. Mary's University; Roland L. Grant, Eastern Montana College; Clark Hilden, Blue Mountain Community College; Solomon A. Isiorho, Indiana University-Purdue University, Fort Wayne; Rudi Kiefer, University of North Carolina-Wilmington; Steve LaDochy, California State University-Los Angeles; Francis Magilligan, Georgia State University; Bob Phillips, University of Wisconsin-Platteville; David R. Privette, Central Piedmont Community College; Robert Quinn, Eastern Washington University; Michael Sady, Western Nevada Community College; Brent R. Skeeter, Salisbury State University; Jim Switzer, Southwestern College; Paul Weser, Scottsdale Community College. I also want to thank John Morgan, of Towson State University, for his careful review of the map appendix material, and for supplying information on geographic information systems for use in the appendix.

Still another group of people who played a key role in this project were the artists and photo suppliers. I wish especially to thank Alice Thiede, who drew most of the maps appearing in the book, and the artists at Rolin Graphics, who produced most of the figures and graphs. The majority of the photos were supplied by John S. Shelton, the U.S. Geological Survey, JLM Visuals and H. Armstrong Roberts, Inc. Others were provided by David Butler of the University of Georgia, by my Towson State University colleagues Wayne McKim and John Morgan, and by my mother, Jerry Scott.

Lastly, I again owe a huge debt of gratitude to my wife Judi, who has spent countless hours typing, and on occasion editing, the manuscript. I wish also to thank my daughter Kelli, who provided valuable typing service during the latter, hectic stages of the manuscript preparation period.

Ralph C. Scott

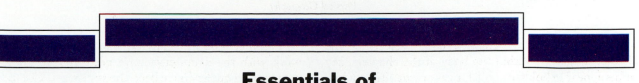

Essentials of
Physical Geography

Introduction

THE DISCIPLINE OF GEOGRAPHY

Most fields of knowledge today are systematically defined; that is, they study everything about one *topic*. An example of a topically oriented discipline is biology, which is the study of living organisms. All lifeforms as well as the environmental factors that directly influence them fall within the bounds of accepted biological studies.

An alternate approach in defining a field of knowledge is to study everything from a single *perspective*. In this case, the topics that can be examined are not restricted, but each topic is properly examined only from the single perspective of the discipline. History, which studies how things change through time, is organized in this fashion. Another such discipline is **geography,** which is concerned with the locational aspects of phenomena. Because virtually all topics change through time and vary in location, neither history nor geography is topically constrained.

The phenomena that can be studied geographically include not only physical objects like people, rivers, or types of vegetation, but also less tangible things such as religious beliefs, military alliances, or tastes in music or clothing. The distributions of these phenomena are not random; everything has a reason for being located where it is. In nearly all cases, the distribution of one type of thing is influenced by the presence or distribution of other things. For example, factors such as climate, landforms, patterns of transportation routes, and the availability of housing and employment influence human population distributions. Factors such as availability of solar energy, proximity to water bodies, and elevation control world climate patterns. In examining the causes for the distributions of earthly phenomena, geographers cannot avoid noting that interrelationships exist among them. The systematic study of these interrelationships has therefore also become a focus of geographic research.

Geographers feel that most fields of knowledge, especially those with rigidly defined topical boundaries, do not adequately emphasize the broader relationships that exist between their topics and other phenomena. The familiar expression of being "unable to see the forest for the trees" is particularly applicable in this context. While many fields of study, by analogy, may provide their practitioners with detailed knowledge of individual trees, the geographer is more concerned with the characteristics of the forest as a whole. Geography is an integrative discipline, then, with a wide breadth of coverage (see Figure I.1). Like a person assembling a jigsaw puzzle, the geographer's goal is to put enough pieces of knowledge about the earth and the universe in their proper locational settings to see the "big picture." The following definition summarizes the concepts presented in the preceding paragraphs: *Geography is the study of the distributions and interrelationships of phenomena.* The geographer is therefore concerned both with the locations of things and with the causes and consequences of those locations. A basic objective of geographic research is to understand better the nature of places, and especially how distributions of earth phenomena interact to cause similarities and differences between places. The ultimate goal of geography is to develop, through a sufficient understanding of spatial interaction, a comprehension of the entire earth as a functioning system.

Subdivisions of Geography

As the examination of a college textbook will show, the major academic disciplines are typically subdivided into

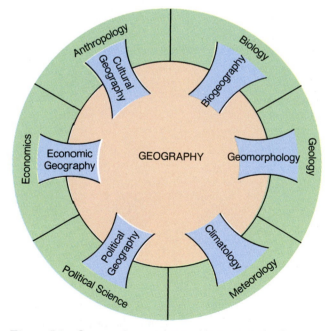

Figure I.1 Geography overlaps with most topically defined fields of knowledge, because each contains spatial components.
(*Source: Reprinted with permission from Julia A. Tuason, "Reconciling the Unity and Diversity of Geography."* Journal of Geography *vol. 86, no. 5, p. 193.*)

units of material. The field of geography is no exception. Although geography as a whole is defined by its method of approach rather than by topic, its subdivisions *are* topically defined. While the basic spatial emphasis remains, each branch of geography deals with a different category of earth phenomena.

A hierarchy of geographical subdivisions exists, and in general, the more advanced a study is, the more narrowly its topical borders are defined. At the most general level, the field of geography is partitioned into two major branches: physical geography and human geography. **Physical geography** is concerned with locational aspects of natural earth phenomena—that is, with those not produced or primarily controlled by human beings. The four major divisions of physical geography are weather and climate, the earth's surface features (including landforms and water bodies), natural vegetation, and soils. Most of this textbook is devoted to the geographical analysis of these four subjects.

Human geography, conversely, deals with those subjects whose distributional patterns are largely or entirely controlled by people. Included are such topics as patterns of population, agricultural and industrial activities, urban areas, religion, transportation routes, political regions, and recreational facilities. A basic understanding of physical patterns, however, is important to the human geographer, as well as to any other informed person, because the earth's physical environment affects nearly all human activities.

In reality, the interaction of earth phenomena, both physical and human, produces a great deal of topical overlap, both in geography and in other disciplines. Geographers, because of their interests and breadth of training, are especially able to appreciate these interrelationships. Geography, then, is both a physical and a social science, and one of the geographer's chief goals is to emphasize the interconnections among the two groups of phenomena.

FUNDAMENTAL GEOGRAPHICAL CONCEPTS

The geographic distribution of the earth's physical features, as already noted, is not the result of chance, but exists in response to natural laws, which provide reasons for the characteristics and locations of terrestrial phenomena. A basic understanding of these laws makes the study of physical geography, and, indeed, of any physical science, much more logical and intellectually satisfying. Explanations for the distribution of the earth's features will be an essential component of our geographic study throughout the book.

An important method of geographical analysis is the formulation of **regions** that display relative similarity in selected attributes. Regionalization simplifies and organizes patterns of earth phenomena and aids geographers in understanding why these patterns exist. Throughout this book, regional patterns of physical earth phenomena are displayed on maps and diagrams and their causes and characteristics are discussed.

The fact that all earthly phenomena are to some extent interrelated means that any action that affects one component will ultimately influence everything else. We can therefore view the earth as a single system of enormous complexity. A **system** is a set of interrelated components through which energy flows to produce orderly changes. In contrast to a **closed system,** which contains a finite and confined supply of energy, an **open system** has access to an unlimited supply of energy from one or more external sources. The earth system can be considered as an open system powered by two different "unlimited" energy sources. These consist of solar energy and of heat from the decay of radioactive elements inside the earth. The total quantity of energy reaching the earth system from these two sources is relatively constant but is capable of powering a great number of earth processes because of the many routes it can take as it permeates the earth system.

The earth system is composed of a number of interconnected subsystems, often described as "environmental spheres" (see Figure I.2). The four major subsystems are the *atmosphere,* the ocean of air that overlies the entire earth's surface; the *hydrosphere,* the water of the surface and near-surface regions of the earth; the *lithosphere,* the massive accumulation of rock and metal that forms the solid body of the planet itself; and the *biosphere,* the layer of living organisms of which we are a part. All four respond in various ways to the flow of energy and materials through the earth system. The resultant distributional patterns and movements of these subsystems form the basis for the material content of this book.

Human activities are increasingly altering or disrupting natural terrestrial processes. This disruption produces rapid environmental changes, often of an undesirable nature. Although this textbook deals primarily with the natural environment, Case Studies at the end of each chapter and Focus Boxes throughout the text examine human influences on the environment.

(a)

(b)

(c)

Figure I.2 The four major terrestrial subsystems, or "environmental spheres," are the (a) atmosphere, (b) hydrosphere, (c) lithosphere, and (d) biosphere. *(a: Jerry Scott; b & c: H. Armstrong Roberts; d: M. Thonig/H. Armstrong Roberts)*

(d)

Selected References

The following books are recommended for those seeking a more comprehensive discussion of geographic methodology, philosophy, and history:

Broek, Jan O. M. *Geography: Its Scope and Spirit.* Columbus, Ohio: Merrill Publishing Co., 1965.

Gaile, Gary, and Cort Wilmott, eds. *Geography in America.* Columbus, Ohio: Merrill Publishing Co., 1989.

Holt-Jenson, Arild. *Geography: Its History and Concepts.* Totowa, N. J.: Barnes & Noble Books, 1980.

James, Preston E., and Geoffrey J. Martin. *All Possible Worlds: A History of Geographical Ideas.* 2d ed. New York: John Wiley & Sons, Inc., 1981.

Key Terms

Geography	**System**
Physical geography	**Closed system**
Human geography	**Open system**
Region	

The Planetary Setting

Outline

Focus Questions

1. What is the geographical setting of the earth in space, and what movements is it making through space?
2. How is the global system of time zones organized?
3. What causes the changing seasons?

THE PLACE OF THE EARTH IN THE UNIVERSE

The earth is the third of nine known planets orbiting the star we call the sun. Sharing our solar system with the sun and planets are a number of planetary satellites or moons as well as asteroids, comets, dust, gases, and a great deal of empty space. Two opposing factors are responsible for the maintenance of stability in the solar system. One is the outward thrust of centrifugal "force" caused by the inertia of the planets as they orbit the sun; the second is the inward pull of gravity. The revolution of the planets around the sun, and of the moons around the planets, is made possible by the precise balance between these forces.

The sun is only one relatively undistinguished member of a giant assemblage of stars called the Milky Way galaxy. Galaxies are large groups of stars held together by their mutual gravitational attraction. They vary tremendously in size, shape, and density and contain anywhere from a few thousand to hundreds of billions of stars. Most galaxies also contain extensive clouds of gases and dust out of which new stars constantly form. The Milky Way is a large, discus-shaped galaxy consisting of more than 100 billion (or 10^{11}) stars. Despite its enormous stellar population, the great preponderance of the galaxy consists of virtually empty space containing only the most rarefied concentrations of gas and dust.

Our solar system is located near the outer fringe of the Milky Way on a spiral arm an estimated 30,000 light years from the galactic center. The center of the galaxy is rendered invisible to us by vast clouds of interstellar gas and dust. The stars that so liberally sprinkle the sky on a clear night are our stellar neighbors, occupying nearby portions of our arm of the galaxy.

PLANETARY MOTIONS

The earth is engaged in several motions through space, each at a differing level of magnitude. These movements can be divided into two categories—large-scale movements and small-scale movements. The large-scale movements are of limited direct significance to the earth because millions or even billions of years are needed for them to produce major changes. The earth's small-scale movements, however, critically affect nearly every aspect of our environment and will be discussed in greater detail. For clarity, the various earth motions will be discussed in order of decreasing magnitude.

Large-scale Motions

The largest-scale motion of all is the movement of the earth, along with the rest of the solar system and galaxy, away from the center of the universe—that is, from the site of the Big Bang (see the Focus box on page 8). The universe is still expanding at great speed.

The second large-scale motion is the revolution of our solar system around the center of the Milky Way galaxy. The time required for one full revolution is estimated to be 230 million years.

Small-scale Motions

The small-scale motions cause the earth to change its position constantly with respect to the sun. Because the sun is our only major source of planetary heat and light, and because it powers external earth processes such as the weather, plant growth, and erosion, these movements are extremely crucial to the earth and its life-forms. The small-scale motions are also responsible for the changing seasons and the alternation of day and night and form the basis of our system of keeping time.

Revolution

The larger of the two primary small-scale motions of the earth is its revolution around the sun. This motion is made possible by the combined effect of centrifugal force generated by the inertia of the earth as it travels in a curved orbit around the sun and the inward pull of the sun's gravitational force.

The earth's revolution is the basis of our calendar year. To be precise, the period of time needed for one complete revolution around the sun is 365 days, 5 hours, and 49 minutes. For the sake of convenience, though, calendar years are defined in whole day increments. Thus, most years are 365 days long. Because the extra 5 hours and 49 minutes required for a full revolution is very close to an extra quarter of a day, every fourth year has an added day (February 29) and is termed a *leap year*.

The earth orbits the sun on an angular plane termed the *plane of the ecliptic* in the same direction as all the other planets (see Figure 1.1). This direction appears as counterclockwise when earth is viewed from a point in space above the northern hemisphere. The vantage point in space is crucial because the direction of revolution becomes clockwise when earth is viewed from above the southern hemisphere.

Formation of the Universe, Solar System, and Earth

Astronomical observations have revealed that the universe is composed of vast expanses of nearly perfectly empty space, as well as the widely separated clouds of gases, dust, and stars we call galaxies. Within the known universe are many billions of galaxies, each containing, on the average, tens of billions of stars. The number of stars in the universe is, for all practical purposes, limitless. It has been said that there are more stars in the universe than grains of sand on all the beaches of the world, and this is apparently a great understatement. Moreover, a large proportion of these stars may have planetary systems similar to our own solar system.

The origin of the universe is one of the great mysteries facing humanity, and numerous theories concerning its formation have been developed over the centuries. Any modern scientific theory, however, must take into account a startling fact revealed through the spectrographic analysis of the light from distant galaxies. This fact is that the galaxies are rushing away from one another. In general, the farther apart two galaxies are, the faster their rate of recession seems to be. The only way astronomers can explain this phenomenon is by theorizing that the occupied universe is expanding in all directions, much like an inflating balloon, as the result of a titanic explosion.

The Big Bang Theory

Over the past few decades a theory of the origin of the universe—the so-called "Big Bang Theory"—has been developed and refined. It seems to fit most known facts, and is now generally accepted by the scientific community. According to this theory, some 15 or so billion years ago, all matter in the known universe was concentrated into one tiny and inconceivably dense "cosmic egg." How it formed, and what events occurred prior to this time, nobody knows. The universe as we know it was initiated by the explosion of this egg. The explosion, the "big bang," destroyed the egg and hurled all the matter it contained into space in all directions, much like shrapnel from a detonating bomb. The violence of the explosion was so great that the super-dense matter of the cosmic egg was reduced to its constituent atomic particles, which soon reassembled to form only two elements, hydrogen and helium. Shortly after the explosion, then, the universe consisted of an expanding cloud of gas composed of approximately 73 percent hydrogen and 27 percent helium.

Gravity, however, began to alter the characteristics of the cloud. The expanding cloud gradually lost its homogeneity as it was drawn into many smaller gas clouds, each held together by internal gravitational attraction. Within these gas clouds, or protogalaxies, gravity produced much denser and more localized concentrations of gas. As these gas clouds con-

continued on next page

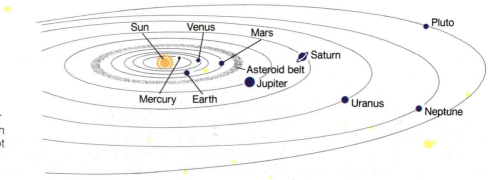

Figure 1.1 The planets of our solar system all orbit the sun in the same direction and, except for Pluto, on nearly the same orbital plane.

tracted, their central cores were gradually heated by compression until thermonuclear fusion reactions were spontaneously initiated; and stars were born. The basic thermonuclear reaction that powers the stars, including the sun, involves the conversion of hydrogen to helium. A series of other stellar fusion reactions are believed to have produced all the elements heavier than helium.

The earliest stars formed under highly favorable conditions and were large, extremely hot, and short-lived. These massive stars became increasingly unstable as their fuel supplies were exhausted and eventually exploded as supernovas, spewing throughout the galaxy a portion of the heavy elements they had produced. Succeeding generations of stars, including our sun, thus formed from concentrations of gas containing not only hydrogen and helium, but the heavier elements as well. It is believed that the elements that constitute most of the earth, and even our own bodies, originated in the cores of these supernova stars!

Formation of the Solar System and Earth

When a cloud of gas contracts to form a star, any initial rotational motion is magnified because of the conservation of angular momentum. (This is the same principle that causes ice skaters to spin faster when their arms are drawn in.) The generated centrifugal force draws the contracting gases into a disk shape. The major concentration of gases at the center of the disk is under enough gravitational pressure to initiate thermonuclear reactions and become a star, while it is believed that smaller nodes of gases form on the rotating rim of the disk to produce planets. This would explain why the planets of our solar system all orbit the sun on roughly the same angular plane and why they revolve around the sun in the same direction. If this system

Figure 1.2 Planet earth as viewed by the Apollo astronauts. (*Department of the Interior, U.S. Geological Survey*)

operates everywhere, most stars would have planetary systems, and, of course, the chances of earth-like planets and the existence of life elsewhere in the universe would greatly increase.

In the case of our solar system, it is believed that the weaker gravity of the earth and the other inner planets, coupled with intense radiation emanating from the nearby sun (the so-called "solar wind"), drove off the hydrogen and helium that formed the initial bulk of the masses of these planets. The inner planets therefore consist primarily of the heavier residual elements that originated in supernova stars. The massive outer planets, though, were able to retain most of their light gases. They still consist largely of hydrogen and helium, which probably surround small solid cores of heavier elements.

The earth's orbital path is roughly circular, but actually forms a broad ellipse. The mean radius of this ellipse, which represents the average distance between the earth and sun, is approximately 93 million miles (150 million km). At its closest approach to the sun, the earth is said to be at *perihelion* (from the Greek *peri*, or "near," and *helios*, or "sun"). At this point in its orbit, which occurs about January 3 each year, the earth is only 91.5 million miles (147.5 million km) from the sun (see Figure 1.3). Six months later, on or about July 4, the earth is at its farthest point from the sun, or at *aphelion*

(from the Greek *ap*, or "away from"). Our distance from the sun is then approximately 94.5 million miles (152.5 million km). Therefore, the earth's distance from the sun varies by about 3 percent during the course of a year.

Rotation

Rotation, the most small-scale motion of the earth, involves the spinning of our planet on its *axis*. The rotation of the earth on its axis is similar to the rotation

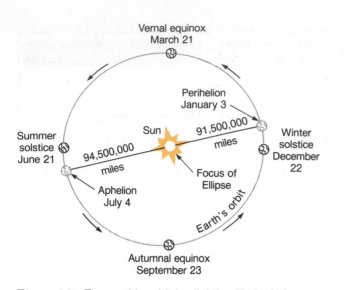

Figure 1.3 The earth's orbit is slightly elliptical. As a consequence, we are approximately three million miles nearer the sun in early January than in early July.

of a model globe on its axis bar except that, of course, the real earth's axis is not a physical entity and provides no support. The rotational motion of the earth allows for the definition of three important and very basic locations. The **North Pole** and **South Pole**, located on opposite sides of the planet, are the two points where the axis intersects the earth's surface. Midway between the poles, and exactly equidistant from both along its entire length, lies the **equator.** This is the line of the fastest rotational speed on the earth.

Rotation is the basis of our calendar day and is the primary reason for the apparent westward motion of the sun, moon, and stars through the sky. (The situation is analogous to riding eastward in an automobile and watching the outside scenery apparently rushing by to the west.) The earth rotates in the same direction as it revolves; this rotational direction can be described as counterclockwise from north, clockwise from south, or eastward if the earth is viewed from a point in space above the equator.

Because all places on earth complete one 360° rotation in a 24-hour period, they rotate through an angle of 1° every 4 minutes. The linear speed of rotation, however, is highly variable and is dependent upon distance from the equator. The speed of rotation is at its maximum value of approximately 1040 miles per hour (465 meters/sec) at the equator, but diminishes to 0 miles per hour at the poles.

Earth's axis is not oriented at a perpendicular angle to the plane of the ecliptic, but instead is inclined from the perpendicular at an angle of approximately $23\frac{1}{2}°$ (see Figure 1.4). Because this value does not change significantly over the short term as the earth revolves around the sun, the earth's axis is described as exhibiting **parallelism**, and the poles remain directed toward essentially the same points in space for extended periods of time. This is why the North Star, or Polaris, occupies a fixed position above the North Pole.

SIZE AND SHAPE OF THE EARTH

It is common knowledge that the earth is approximately spherical in shape. The shape of the earth, however, is not readily apparent from observation and was the

Figure 1.4 The earth's axis is tilted at a constant angle with respect to the plane of the ecliptic as it revolves around the sun during the course of the year.

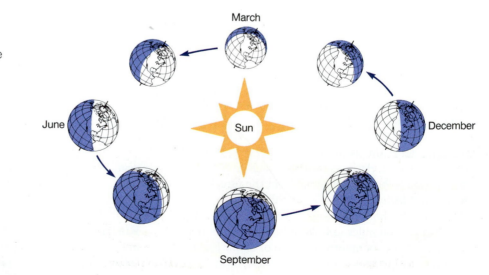

subject of speculation and debate for centuries. Most early scientists and philosophers believed that the earth was flat. This view was based on the physical appearance of the earth's surface, just as the early belief that the rest of the universe revolved around the earth was based on the apparent motion of the sun, moon, and stars. Modern science, however, has produced many proofs of the sphericity of the earth. These include the virtual equality of gravitational attraction at all points on the earth's surface, numerous precise surface and astronomical measurements of distances and directions, and the visual evidence provided by high altitude and space photography.

Earth's spherical shape is produced by gravity. This incompletely understood force causes all matter to have an attraction for, and to be attracted by, all other matter. The mutual gravitational attraction of the earth's matter has caused the planet to assume a spherical form—the most compact form that exists. Gravity has also produced a layering of the earth by density, with lighter materials tending to overlie heavier ones. (Fortunately for us, this layering is far from perfect.) All other known astronomical objects of large size are also essentially spherical, indicating that gravity is apparently a universal force that operates wherever a sufficient mass of matter has assembled.

The three basic measurements of the size of a sphere are its radius, diameter, and circumference. For a given sphere, if any one of the measurements is known, the other two can be readily calculated. The radius of the earth, or the distance from its surface to its center, is approximately 3950 miles (6350 km). Earth's diameter, the linear distance from any point on its surface through the center of the earth to its opposite (antipodal) point,

is equal to two radii, or 7900 miles (12,700 km). Finally, earth's circumference, the length of a line drawn on the surface that bisects the planet, is 24,900 miles (40,000 km).

Departures from Perfect Sphericity

Although gravity causes the earth to be approximately spherical, it is not perfectly so because other forces also influence the shape of our planet. The effects of two of these forces are sufficiently important to merit discussion. They are the outward thrust of centrifugal force generated by the earth's rotation on its axis and tectonic forces produced deep within the earth.

The centrifugal force of rotation is much too weak to overcome the binding force of gravity and cause the earth to fly apart. It is, however, sufficient to slightly distort our planet's shape. In the vicinity of the equator, where the speed of rotation is greatest, the earth is bulged out from its spherical form. In contrast, the slowly rotating polar areas are somewhat flattened as the earth's rotational inertia displaces their mass equatorward (Figure 1.5). The earth thus assumes the shape of an oblate ellipsoid, a three-dimensional ellipse that is flattened at the poles. The amount of bulging and flattening is too small to be visually perceptible, and from space the earth looks perfectly spherical. However, this departure from perfect sphericity affects the precise calculations required for surveying and mapping and also influences the force of gravity on different parts of the earth.

A second, more important departure from perfect sphericity (discussed at length later in this book) results from the earth's variable surface features. If our planet

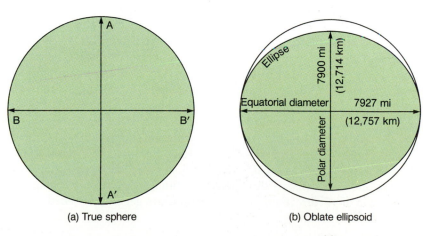

Figure 1.5 The earth's shape is somewhat distorted from that of a true sphere (Figure a) because of the centrifugal force of the earth's rotation on its axis. This motion causes our planet to assume the shape of an oblate ellipsoid (Figure b), with its equatorial diameter slightly larger than its polar diameter.

(a) True sphere

(b) Oblate ellipsoid

were a perfect sphere, or even a perfect oblate ellipsoid, its surface would be smooth and would appear flat to an observer at ground level. Instead, the surface is highly wrinkled and contorted in places, exhibiting a great variety of landforms of many sizes and shapes. Total surface irregularity increases if we consider the sea floor, rather than the ocean surface, as comprising more than 71 percent of our planet's surface. Thus the earth, when examined in sufficient detail, has a unique shape, undoubtedly duplicated by no other planet in the universe.

How great a departure from perfect sphericity do earth's surface features represent? One way of judging is by determining the total vertical distance between the world's highest mountain and its greatest ocean depth. Mt. Everest in the Himalayas is the highest peak on earth, reaching an elevation of 29,028 feet (8847 m) above sea level. Challenger Deep in the western Pacific is the deepest known point of any ocean, with a maximum depth of 36,198 feet (11,033 m) below sea level. The total vertical difference between these two extreme points is 65,226 feet (19,880 m), or about $12\frac{1}{3}$ miles. This value is less than one-third of 1 percent of the earth's total radius of 3950 miles.

Despite the relatively diminutive size of the earth's surface features when compared to the size of the entire planet, they are extremely important to humankind because we happen to live just where these features are located—at the surface. As a result, the earth's surface features have strongly influenced most major aspects of human society, including population distributions, transportation routes, and agricultural patterns. In addition, earth's landforms profoundly influence global patterns of climate, vegetation, and soils.

DIRECTIONS

The concept of direction is vital to the field of geography because of the discipline's emphasis on location. Unfortunately, due to a lack of reference lines or surfaces, no natural "universal" directions exist in space. In outer space, in fact, even the words "up" and "down" are meaningless.

The basis for the horizontal directional systems used on earth is our planet's rotational orientation. Rotation has established the positions of the North and South Poles and the equator. These locations enable us to construct reference lines for stating horizontal directions. Vertical directions are based on the orientation of the earth's gravitational force, which is considered to pull directly downward toward the center of the earth.

North, often considered the most basic of all directions, is by definition the straight-line direction from any point toward the North Pole. Likewise, *south* is the direction from any point toward the South Pole. These two directions, of course, are always 180° apart with respect to a given starting point because the poles are on opposite sides of the earth. *East* and *west* may either be defined as those directions lying perpendicular to north and south, or as those lying parallel to the orientation of the equator. The earth rotates in an eastward direction, and when an observer stands facing northward, east is on his right and west on his left. Any other direction can be determined by reference to these four cardinal directions.

It is important to note that the directions north, south, east, and west are all equally horizontal. Expressions such as "up North" and "down South" are strictly figures of speech and are probably based on the fact that many maps showing these areas are displayed vertically, as on classroom walls.

Magnetic North and South

The exact determination of direction is difficult or impossible without reference to an accurate map or the availability of surveying equipment, although estimates can often be made from the position of the sun, the North Star, or other celestial objects. The most commonly employed method for determining direction in the field is the use of a magnetic compass. A compass does not point toward the geographic (rotational) North Pole, however, but instead aligns itself with the magnetic field extending outward from the magnetic North Pole.

The magnetic North and South Poles, unlike the geographic poles, are not fixed in position, but drift slowly about. The magnetic North Pole is currently located in the Canadian Arctic, north of the Hudson Bay, at a point some 1000 miles (1600 km) south of the geographic North Pole. The magnetic South Pole is located near the coast of Antarctica south of Sydney, Australia, at a point about 1800 miles (2900 km) from the geographic South Pole. Because the magnetic poles are at considerable distances from the geographic poles, compass directions can differ greatly from geographical directions, and corrections are generally needed.

LATITUDE AND LONGITUDE

Geographers are concerned with understanding spatial phenomena and interactions over the entire earth. It is therefore necessary for them to employ locational systems that can accurately describe the geographic position of any point on the earth's surface.

A number of global locational systems have been devised, but the most widely used, by far, is the system of **latitude** and **longitude**. This system is based on the locations of the North and South Poles and the equator. Related to these locations are the directions established by the earth's rotational orientation. All lines in the system of latitude and longitude are oriented either due north-south or east-west. The system will be examined by discussing first the characteristics of lines of latitude, then lines of longitude. Finally, the two will be put together to show how these lines form a system for describing locations.

Latitude

Lines of latitude have the following characteristics, as illustrated in Figure 1.6.

1. They extend in an east-west direction but are used to state locations in a north-south reference frame.
2. Lines of latitude are really circles of latitude if viewed in their entirety, because they form full circles on the globe.
3. Lines of latitude are parallel, or equidistant to one another, along their entire lengths. For this reason, they are commonly termed *parallels*.
4. Lines (or circles) of latitude vary greatly in length (circumference), but do so in an orderly fashion. The longest circle of latitude is the equator. Progressing poleward, the circumferences of the circles of latitude become progressively shorter until at the poles they have shrunk to a point.
5. An unlimited number of parallels exist, but on a map they are usually spaced at set intervals, decided upon by the cartographer (mapmaker) in view of the map's intended purpose.

By international agreement the equator is used as the starting line for the determination of latitude values. Parallels of latitude are identified by their angular distance, in degrees, north or south of the equator. As the starting line, the equator is assigned a value of 0° latitude. The half of the earth's surface located north of

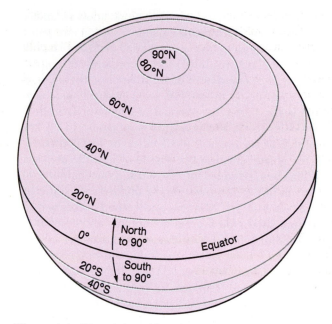

Figure 1.6 The globe with selected lines of latitude. These lines actually take the form of concentric circles, numbered from the equator.

the equator is termed the *northern hemisphere,* and all lines of latitude here contain an "N" after their value. Likewise, the half of the earth's surface located south of the equator is termed the *southern hemisphere;* and all latitude values here end with an "S." If we took a large globe and drew a parallel for each degree of curvature north of the equator, we would find that the fortieth line we drew would pass through Philadelphia, Pennsylvania. Philadelphia therefore has a latitude of 40° N. So do Columbus, Ohio; Boulder, Colorado; Ankara, Turkey; Beijing (Peking), China; and many other cities and towns. It is therefore evident that a statement of latitude is insufficient by itself to pinpoint a location.

The higher the numerical value of a line of latitude, the farther it is from the equator. The point farthest north is the North Pole, which is 90 degrees north of any point on the equator. Its latitude is therefore 90° N. By the same token, the latitude of the South Pole is 90° S. Because all low-numbered parallels are near the equator, areas within about 30° of the equator are frequently referred to as the *low latitudes.* Areas between about 30° and 60° N and S latitude have intermediate values and are referred to as the *middle latitudes.* Most of the United States and Europe lie within this zone. Areas poleward of about 60° N and S are considered to be in the *high latitudes.*

Lines of latitude are parallel to one another, and, if the earth is considered to be a sphere with a circumference of 24,900 miles, the length of one degree of latitude can be calculated readily. A circle contains 360° of arc. The length of one degree of latitude must therefore be 24,900/360 or 69 miles (111 km). Actually, since the earth is not perfectly spherical, a degree of latitude is slightly longer at the poles (69.4 miles, or 111.7 km) than it is at the equator (68.7 miles, or 110.6 km).

For greater precision, a degree of latitude or longitude can be subdivided into sixtieths, termed *minutes* ('), and each minute, in turn, can be divided into sixtieths, termed *seconds* ("). Minutes and seconds of curvature therefore subdivide degrees as minutes and seconds of time subdivide hours. One minute of latitude is approximately 1.15 miles (2.85 km) long, and one second of latitude is about 100 feet (31 m) long. A latitudinal reading is given from left to right in progressively smaller units, as in the following example:

39° 07′ 15″ S.

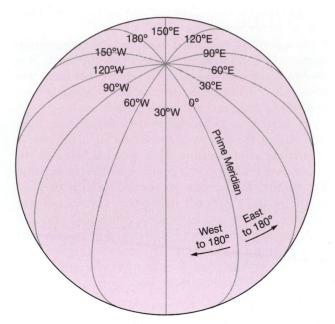

Figure 1.7 Lines of longitude are half circles with their end points at the poles. They are spaced farthest apart at the equator and are numbered in degrees east and west of the prime meridian.

Longitude

Lines of longitude, or *meridians*, are similar to lines of latitude but they also possess a number of unique characteristics (see Figure 1.7). The term "meridian" means "noon" or "midday," and, as we shall see, longitude is the basis of our system of time.

Lines of longitude have the following characteristics:

1. They extend in a north-south direction but are used to state locations in an east-west reference frame.
2. Each meridian has its endpoints at the North and South Poles and its midpoint at the equator.
3. Meridians are *not* parallel to one another; they are spaced farthest apart at the equator and converge northward and southward to meet at the poles. Because meridians converge, the surface length of a degree, minute, or second of longitude depends on its latitude. The length of one degree of longitude at the equator is approximately the same as that of a degree of latitude: 69.2 miles (111.1 km). This length decreases to 0 at the poles.
4. Meridians always intersect parallels at 90° angles.
5. No limitation exists as to the number of meridians that may be placed on a map, but, like parallels, they are usually spaced at set intervals.

The numbering of meridians is accomplished in the same way as is the numbering of parallels—by choosing a starting 0° line and measuring subsequent lines in degrees of surface curvature, this time to the east and west. A problem long existed, though, in choosing an appropriate starting meridian because no one meridian differs from the rest in any geometrical manner. As a result, many countries that made maps established 0° meridians through their own capital cities or other points of national importance. This practice resulted in a great deal of confusion because longitude values for the entire world depended upon the country of origin of the map or book being consulted. It was finally decided at an international meeting in 1884 to adopt the English system, in which the starting or **prime meridian** passed through Greenwich Observatory on the outskirts of London.

From the prime meridian (also called the *Greenwich meridian*) westward, longitude values increase from 0° W to a maximum of 180° W halfway around the world from Greenwich. Similarly, going eastward, values of longitude increase from 0° E to 180° E. The numerical values of meridians therefore go twice as high as values for parallels. (The two 180° lines are actually one and the same, and thus two meridians, 0° and 180°, can be labeled either E or W.)

Longitude divides the world into the *eastern hemisphere* and the *western hemisphere*. These terms are not as frequently employed as the terms northern hemisphere and southern hemisphere, but such longitudinally derived expressions as the Western World and the Middle East are in common usage.

Determining Geographical Coordinates

Taken together, the latitude and longitude of any location describe its geographical position. A given parallel intersects a given meridian at only one point on the earth, so if the correct value of each line is given, the position of any place can be pinpointed. The accuracy of placement of any point depends on the accuracy with which its position is stated. Locations stated to the nearest whole degree contain a potential margin of error of many miles, while those stated to the nearest second of latitude and longitude are accurate to within 50 feet.

When the geographical coordinates of a site are given, latitude is always given first. An example of a location stated to the nearest second is as follows:

39° 22′ 18″ N 76° 38′ 05″ W

It must be remembered that each number is always followed by the letter, N, S, E, or W, and that minutes and seconds are sixtieths, not hundredths. Most maps have selected lines of latitude and longitude either drawn on them or indicated by tick marks on their margins. The spacing of lines of latitude is not necessarily the same as that for lines of longitude.

TIME

Our system of keeping time is based on the east-west (or longitudinal) position of the sun. The sun's position and apparent motion during the course of a day are associated primarily with the earth's axial rotation but are also slightly affected by the earth's revolution around the sun. The basic unit of time is the *solar day*, which is the period of time needed for the sun to make one apparent 360° circuit of the sky. The solar day is, of course, divided into units of hours, minutes, and seconds.

Local Time

The event historically used for establishing the time of day has been *solar noon*. Solar noon occurs the instant the sun reaches its daily high point in the sky. For centuries, until the late 1800s, communities made the time of the mean occurrence of solar noon the official instant of noon and then simply subdivided the periods between solar noons into the appropriate hours, minutes, and seconds. This system of **local time** was highly accurate with respect to the position of the sun, because the time was individually adjusted to each locality. During the nineteenth century, however, a major shortcoming in the local time system became increasingly evident and eventually led to its discontinuance over most of the world. The problem was that different places, even those located near one another, inevitably had somewhat different times. Because the local time system used the precise east-west position of the sun in the sky, only places on precisely the same meridian had exactly the same time at any one instant.

Prior to the nineteenth century, time differences between nearby localities using the local time system were only a minor inconvenience. Timepieces were rare and rather inaccurate, appointments were not generally made for exact times, and means of transport were so slow that it took a long period to reach a location with a significantly different time. In the nineteenth century, however, the advent of the railroads provided a rapid and long-distance means of transportation, while the telegraph offered a means of instantaneous long-distance communication. The railroad and telegraph companies, as might be imagined, experienced serious difficulties because of the multiplicity of times existing throughout the world. As a result, they increasingly pressured government agencies to develop standardized time zones.

Standard Time

Our present **standard time** system was officially established for the United States in November 1883 and was soon adopted by most other countries. The standard time system, like the local time system, is based on the position of the noon sun, but only at selected meridians of longitude rather than at each specific site. The areas surrounding these standard meridians all have the same official time. Because the earth rotates 15° in one hour, the standard time zones differ by exact one-hour intervals. Thus, 24 time zones cover the earth (see Figure

1.8). In changing from local time to standard time, a decision was made to sacrifice accuracy of time determination for much greater convenience.

The meridians used as the centers of their respective time zones are multiples of 15°. The meridian that provides the basis for all the time zones is the Greenwich meridian, and the time zone established by this meridian is called Greenwich mean time (GMT). Many people employed in fields such as transportation and communications base their activities directly on Greenwich mean time.

If the standard time system were rigorously followed, each time zone would be exactly centered on its standard meridian and would extend 7° 30′ both east and west from the meridian to the borders of the adjacent time zones. For example, the 75° W meridian is used as the basis for the Eastern Time Zone of the United States. In theory, Eastern Time should extend westward to 82° 30′ W, changing to Central Time from that line westward to 97° 30′ W. In reality, though, the time zone boundaries are generally drawn along state or county lines and are sometimes quite distant from the theoretical 7° 30′ boundaries (see Figure 1.9). The rationale for this, of course, is to avoid the problems that arise when time lines pass through the middle of a county, town, or other political unit. Internationally, time zone boundaries have also been adjusted considerably, as is evident from Figure 1.8. In many cases, entire countries use the same time, and the time changes as the international boundary is crossed. Most large countries, especially those of great longitudinal extent, have multiple time zones. The United States, for example, has 7

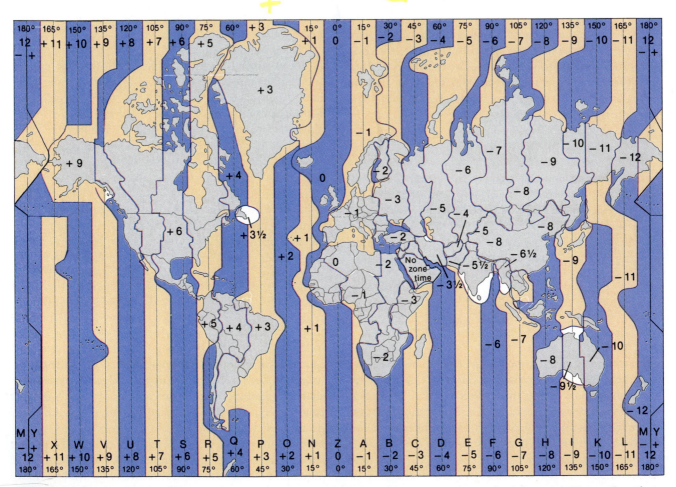

Figure 1.8 The global system of time zones. The International Date Line appears near the left and right margins of the map.

Figure 1.9 The standard time zones of the coterminous United States. The central Meridian of each time zone is indicated.

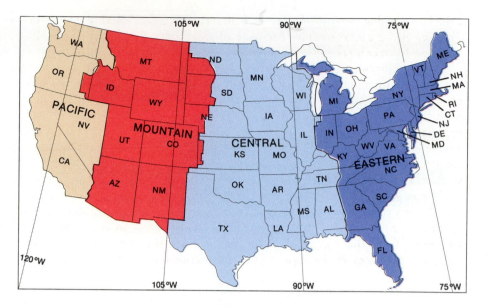

different time zones (4 in the coterminous United States) and the Soviet Union has 11! Over the open oceans, the official time boundaries are generally used because no practical advantage is gained from shifting them.

The International Date Line

The International Date Line is a unique line in the world standard time system. When it is crossed, time changes by a full 24 hours. Therefore, although the hour of the day remains the same, the calendar date and day of the week both change. In addition, the direction of the time change is the opposite of all those established by the 24 one-hour time zone boundaries. Crossing the Date Line from east to west causes the time to become one day later, and crossing it from west to east causes the time to become one day earlier.

The International Date Line is closely associated with the 180th meridian, and the two lines run together along much of their lengths through the Pacific. For the sake of convenience, though, the Date Line contains three major longitudinal digressions that allow portions of continents or island groups crossed by the 180th meridian to remain entirely on one side or the other. The 1883 decision to extend the 0° longitude line through England was at least partially made so as to place the International Date Line, with its potentially confusing

date changes, in as remote a portion of the earth as possible.

Daylight Saving Time

For most people, the daylight hours are those around which most activities, especially work-related ones, are centered. Certainly, major advantages exist in concentrating our waking periods and activities during the day; these include the presence of light to see by and the economic benefits derived from decreased fuel needs for lighting and heating. Yet, in many societies, especially in more technologically developed countries, the waking hours of most individuals are considerably offset from the daylight period. To illustrate this point, consider that on a typical day the sun is above the horizon from 6 A.M. to 6 P.M. The average adult in Europe or North America, however, is likely to awaken at perhaps 7 A.M. or later and to retire at 11 P.M. or later. This results in an hour or more of daylight being wasted in the morning, while nearly half the night is spent in wakefulness. Daylight saving time represents an attempt to achieve a greater correlation between the daylight hours and the human activity period.

To convert from standard time to daylight saving time, the official time is simply moved forward by one hour. This obliges individuals with set time schedules to do everything an hour earlier with respect to solar time. The implementation of daylight saving time changes no

17

Determining Times

Because time is based on the apparent westward movement of the sun in the sky, time zones to one's east are progressively later in hour while those to one's west are progressively earlier in hour regardless of the longitude. For example, the time in California is three hours earlier than on the East Coast. Announcements concerning the times of television programs have familiarized much of the American public with this fact. Conversely, standard time is five hours later in Great Britain than in the eastern United States.

The simplest method of calculating the difference in time between any two points is to determine their differences in longitude and then to divide this difference by 15 to determine the difference in hours. The time in the location farthest east in longitude is always the latest. When determining the difference in longitude between two places in opposite longitudinal hemispheres for the purpose of comparing their times, you should measure by way of the prime meridian in order to avoid the potential confusion encountered when crossing the International Date Line. The following two time problems should serve to clarify the methodology involved:

■ *Problem 1:* If it is 10 P.M., Sunday, at 135° W, what is the time and day at that same instant at 15° W?
■ *Analysis:* In going from 135° W to 15° W, we are traveling eastward; therefore, the time gets later. Since both points are in the same longitudinal hemisphere, we subtract their numerical values to determine their longitudinal separation. Since 135 − 15 = 120, the time must be 120/15 or 8 hours later at 15° W. Eight hours later than 10 P.M., Sunday, is 6 A.M., Monday.

■ *Problem 2:* If it is 9 A.M., Tuesday, at 105° E, what is the time and day at that same instant at 150° W?
■ *Analysis:* In going from 105° E to 150° W, we are traveling far west in longitude, so the time must become much earlier. (Actually, a much shorter distance would be involved if we traveled *eastward* past 180° longitude to reach 150° W. This direction, however, includes the complication of crossing the International Date Line.) Since the two given longitudes are in opposite longitudinal hemispheres, their total longitudinal separation is calculated by adding their numbers. Thus, we must go 105° + 150°, or 255°, westward from 105° E to reach 150° W. The time at 150° W must therefore be 255/15 or 17 hours earlier than 9 A.M., Tuesday. Subtracting, our new time and day becomes 4 P.M. Monday.

time zone boundaries; only the time within each time zone is changed.

Although it had long been in use in some parts of the United States, daylight saving time was adopted nationwide in 1966 for six months of the year, and this period was increased slightly in 1986. (A few localities have elected not to use the system.) Daylight saving time begins at 2 A.M. on the first Sunday in April (when the time is moved ahead to 3 A.M.) and ends at 2 A.M. on the last Sunday in October (when the time is moved back to 1 A.M. and standard time is resumed). A phrase sometimes employed to help keep track of the necessary clock adjustments is "spring forward and fall back." The reason for using daylight saving time only during what is essentially the summer half of the year is that the sun rises earliest during this period and, correspondingly, the greatest loss of daylight hours is likely to occur.

THE SEASONS

Over most of the earth's land surface, including nearly all areas lying outside the tropics, the seasonal climatic changes are one of the most important aspects of the natural environment. All lifeforms, both plant and animal, have had to adjust to the seasonal rhythm of the climate in these areas, and most species have evolved biological cycles that cause them to depend on the orderly seasonal progression of climatic changes.

Cause of the Seasons

The seasonal climatic changes are caused by the inclination of the earth's axis. This results in a yearly cycle of change in both the angle and duration of sunlight for all places on the earth (see Figure 1.4). The resulting changes in solar energy produce annual cycles of higher and lower temperatures over most of our planet.

Before the earth-sun relationships associated with each of the four seasons are examined, it is necessary to discuss the key factor responsible for the relationship between the sun's angle and its heating ability. The angle of the sun in degrees above the horizon is termed its *altitude*. The altitude of the sun is the most important factor controlling the amount of solar heating received at the earth's surface in all but the high latitudes. When the sun is directly overhead, or at an altitude of 90°, its heating power is greatest; as its altitude decreases, its rays gradually lose intensity. This loss of intensity occurs because a decrease in the sun's altitude causes an equivalent amount of solar energy to strike a larger earth surface area, spreading and reducing the energy received per unit area (see Figure 1.10). A further reduction in solar energy reaching the earth's surface occurs when the sun is at a low altitude because of increased scattering and absorption of sunlight resulting from its oblique passage through the atmosphere.

The tilt of the earth's axis causes the noon sun to vary in altitude at any place during the course of the year. This results in a considerable variation in the concentration and heating ability of the solar energy received. As the earth revolves around the sun, the North and South Poles constantly point toward the same two locations in space. The inclination of the poles, or of any other location on earth, is *not* constant, though, with respect to the sun. In fact, the combination of rotation and revolution constantly shifts both the latitude and longitude of the point on the earth's surface that has the sun directly overhead. Tracing the position of this point through the course of a year would show that the point shifts alternately northward and southward because of the earth's axial inclination (Figure 1.11) and moves constantly westward because of the earth's rotation.

The four seasons are based on the changing orientation of the earth's axis with respect to the sun during the course of a year. The official beginning of each season marks the instant that a key earth-sun positional relationship occurs.

Summer officially begins in the northern hemisphere when our hemisphere reaches its maximum inclination of $23\frac{1}{2}°$ toward the sun (see Figure 1.11). When this occurs, the sun is at its *summer* **solstice** position, and at solar noon is directly overhead at a point $23\frac{1}{2}°$ north of the equator. The northern hemisphere at this time receives a larger proportion of the earth's total allotment of solar energy than at any other time of the year. Because the southern hemisphere is on the opposite side of the earth, it receives a smaller amount of solar energy than at any other time during the year. Therefore, the northern hemisphere's summer solstice marks the beginning of winter in the southern hemisphere, where the sun is at its *winter solstice* position. All the seasons are exactly reversed between the two hemispheres, so that when any seasonal event occurs in one, the opposite event occurs in the other.

In the northern hemisphere, the summer solstice occurs on, or within two days of, June 21. In fact, each of the four seasonal positions occurs between the nineteenth and twenty-third day of its respective month. The exact time and date vary in different years because of leap years and slight variations in the orbital path of the earth from year to year. The decision to begin each of the four seasons at the four "quarter positions" of the

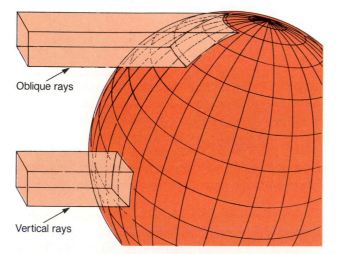

Figure 1.10 The intensity and heating power of the sun's rays are directly related to the sun's altitude. The same amount of energy is contained in both "boxes" of incoming radiation above, but the oblique rays striking the higher latitudes are spread over a larger area, resulting in lower temperatures.

Oblique rays

Vertical rays

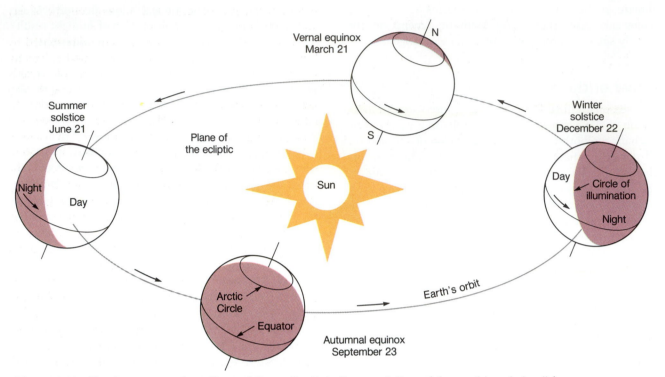

Figure 1.11 The four seasonal positions of the earth. Note the parallelism of the earth's axis in all four seasons.

earth's orbit around the sun was an arbitrary one; no natural law mandates this. As a consequence, the change of seasons often does not lead immediately to the changes in weather that some people seem to expect.

During the three months of summer that follow the summer solstice, the earth revolves so that the inclination of the North Pole toward the sun is progressively reduced. Consequently, the sun's vertical rays gradually shift southward from $23\frac{1}{2}°$ N until they reach the equator. When this occurs, between September 19 and 23, the sun is at its *autumnal* **equinox** position for the northern hemisphere, and the fall season begins. In the southern hemisphere, where spring is beginning, the sun is in its *spring* (or *vernal*) *equinox* position. At this time the axial tilt of the earth is precisely perpendicular to the sun's rays, and the amount of solar energy reaching both hemispheres is equal.

As the fall season progresses, the continued revolution of the earth causes the northern hemisphere to be inclined more and more away from the sun. Finally, between December 19 and 23, the entire $23\frac{1}{2}°$ tilt of the axis is directed away from the sun in the northern hemisphere and toward the sun in the southern hemisphere (see Figure 1.12). Winter now begins in the

northern hemisphere as it began six months earlier in the southern hemisphere. At this time, the sun follows its lowest course through the southern sky and provides less energy to the northern hemisphere than at any other time of the year.

During the three months of the winter season, the sun's vertical rays gradually progress northward from their initial position at $23\frac{1}{2}°$ S to reach the equator as the earth's inclination again becomes perpendicular to the sun's rays. Between March 19 and 23, the northern hemisphere spring equinox occurs as the sun crosses the equator; and both hemispheres again receive equal energy. The following three months of spring witness the increasing tilt of the northern hemisphere toward the sun until once again the summer solstice position is reached.

Two lines of latitude achieve importance as a result of the earth-sun relationships involved in the changing of the seasons. The **Tropic of Cancer** ($23\frac{1}{2}°$ N) and the **Tropic of Capricorn** ($23\frac{1}{2}°$ S) mark, respectively, the highest latitudes north and south that the sun's vertical rays ever reach. Stated differently, the sun is always directly overhead somewhere between $23\frac{1}{2}°$ N and $23\frac{1}{2}°$ S. Places between these two parallels have the sun directly over-

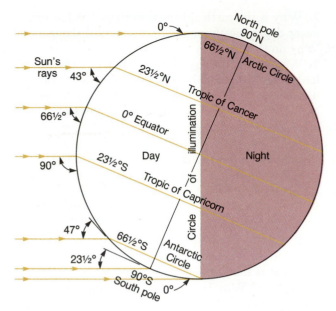

Figure 1.12 The northern hemisphere winter solstice. The sun is directly overhead at the Tropic of Capricorn. Days are longer than nights in the southern hemisphere, while nights are longer than days in the northern hemisphere.

head on two occasions during the year—once while the sun shifts northward toward the Tropic of Cancer and once while it shifts southward toward the Tropic of Capricorn. More than half the world is poleward of these lines (or outside the *tropics*) and never has the sun directly overhead. For nontropical locations, the sun reaches its highest altitude at solar noon on the date of the summer solstice.

Lengths of Day and Night

The changing of the seasons involves not only constant changes in the sun's noon altitude, but also variations in the lengths of day and night for all places on earth except the equator. All locations in the northern hemisphere experience longer days than nights during the spring and summer seasons, when the northern hemisphere is inclined toward the sun. Days are shorter than nights during the fall and winter seasons, when the northern hemisphere is tilted away from the sun. The lengths of day and night at equivalent latitudes in the southern hemisphere are exactly reversed from those in the northern hemisphere because the seasons are reversed. Only the equator, the line separating the two

hemispheres, experiences equal 12-hour lengths of day and night throughout the year.

The length of daylight is not evenly distributed by latitude, but becomes progressively less equal poleward. Figure 1.12 shows that more of each line of latitude north of the equator is on the night side than on the day side of the earth. North of the Arctic Circle, at $66\frac{1}{2}°$ N, all parallels are entirely in darkness. The sun therefore does not rise in this area at the time of the winter solstice. Conversely, increasing proportions of parallels progressively south of the equator (and experiencing the summer solstice) are in daylight until south of the Antarctic Circle, at $66\frac{1}{2}°$ S, continuous daylight prevails. Of course, the situation is exactly reversed six months later, so that both hemispheres (indeed, all places on earth) receive six months of day and six months of night over the course of a year.

The Arctic and Antarctic Circles therefore mark the equatorward limits of the two portions of our planet in which day and night periods exceed 24 hours at the times of the summer and winter solstices. All other parts of the world experience both day and night each calendar day. The situation is most extreme at the North and South Poles, where day and night each last for six months. The sun at these two locations remains constantly below the horizon during the fall and winter seasons, but is continuously above the horizon during the spring and summer seasons. On the dates of the equinoxes, all places on earth experience 12 hours of day and 12 hours of night.

Summary

This chapter has examined the earth's origin, position, and movements in space, as well as the size and shape of our planet. These factors have enabled us to devise systems for describing locations and determining times on the earth's surface; they are also responsible for the changing of the seasons.

The earth is the third planet from the sun in a solar system that contains nine known planets. The solar system is located in the outer portion of a spiral arm of the Milky Way galaxy—a rotating collection of many billions of stars. The earth is involved in several movements through space, occurring at greatly differing orders of magnitude. At the largest scale, we, along with the entire galaxy, are moving away from the center of the universe as an apparent consequence of the "big bang." We are also revolving around the center of the

Milky Way galaxy. On a smaller scale, earth revolves around the sun once each year and rotates on its axis once each day.

The earth is roughly spherical because of the inward pull of gravity and has a diameter of approximately 7900 miles (12,700 km). The centrifugal force of earth's axial rotation, however, has caused the equatorial regions to bulge outward and the polar regions to become flattened, so that the earth is more accurately described as an oblate ellipsoid. Forces above and below the earth's surface have also produced landforms of many sizes and shapes, so that in detail the shape of the earth is complex and undoubtedly unique.

The system of latitude and longitude is the most important locational system currently in use. Degrees of latitude are measured north and south of the equator, and degrees of longitude are measured east and west of the prime meridian.

The standard time zones are swaths of longitude 15 degrees wide, each centered on a meridian divisible by 15. Each of the 24 time zones differs in time by one hour from adjacent zones. Over land areas, the zones are usually adjusted to coincide with existing political boundaries. Across the International Date Line, located in the central Pacific, a 24-hour time difference exists.

The changing seasons occur because the earth's axis is inclined at $23\frac{1}{2}°$ from a perpendicular angle to the plane of the ecliptic. As the earth revolves around the sun during the course of the year, the northern and southern hemispheres alternately incline at varying angles toward and away from the sun. Spring and summer occur when a hemisphere is inclined toward the sun; the sun is therefore higher in the sky, periods of daylight longer, and the hemisphere's total energy receipt greater. Conversely, fall and winter occur when the hemisphere is inclined away from the sun.

Review Questions

1. Draw diagrams illustrating the revolution of the earth around the sun and the rotation of the earth on its axis. Be sure to indicate the correct directions of motion.

2. Why is the earth spherical in shape? What two major departures from a perfect spherical shape exist? Why do they occur?
3. What is the purpose of the system of latitude and longitude? What similarities and differences exist between parallels and meridians? How are degrees of latitude and longitude subdivided for greater precision?
4. Why was the system of local time discontinued? What advantages and disadvantages does the standard time system offer? Explain how this system is organized.
5. If it is 3 A.M., Thursday, at 30° E, what are the time and day at that same instant at 120° E?
6. If it is noon, Monday, at 135° W, what are the time and day at that same instant at 165° E?
7. What is the justification for the use of daylight saving time? Why is it used in the United States during only part of the year?
8. Draw a single diagram illustrating the parallelism of the earth's axis and the position of the earth with respect to the sun at the times of the solstices and the equinoxes. Label each position.
9. Describe the locations and explain the significance of the Arctic and Antarctic Circles and of the Tropics of Cancer and Capricorn.

Key Terms

Solar system
Milky Way galaxy
Revolution
Rotation
Geographic North and South Poles
Equator
Parallelism
Oblate ellipsoid
Latitude
Longitude
Prime meridian
Local time
Standard time
International Date Line
Daylight saving time
Solstice
Equinox
Tropics of Cancer and Capricorn
Arctic and Antarctic Circles

Chapter Two

Earth's Atmospheric Envelope

Focus Questions

1. What basic influences does the atmosphere have on the earth's surface environment?
2. Of what substances is the atmosphere composed?
3. How do the characteristics of the atmosphere change with altitude?
4. How are human activities influencing the composition of the atmosphere?

The *atmosphere* is the gaseous envelope that surrounds the earth. Although relatively transparent and seemingly insubstantial, it nonetheless profoundly affects nearly every aspect of our physical environment. Life on earth could not exist without it.

The atmosphere is the outermost of the four environmental spheres that interface at the surface of the earth. It essentially forms a canopy over the others: the hydrosphere, lithosphere, and biosphere. The cycles of development and change that characterize the four environmental spheres require tremendous amounts of energy. Most of this energy is derived from the sun. Because the atmosphere is the outermost of the four spheres, it is the first to receive the solar energy on which it and the others depend. As solar energy passes downward through the atmosphere, it is partially absorbed and redirected before it reaches the earth's surface and is converted to the heat energy that largely powers the other spheres.

ATMOSPHERIC ORIGIN AND FUNCTIONS

Earth's atmosphere is believed to have been derived primarily from volcanic gases released from our planet's interior. Volcanic eruptions not only produce large quantities of lava and ash, they also liberate great volumes of gases. Scientists believe that four billion years of volcanic eruptions have released not only the gases that have formed our atmosphere, but also the water vapor that has formed all the earth's water bodies and glaciers. Present-day eruptions continue to liberate gases originally dissolved under great pressure in rock deep beneath the surface. Evidence for the volcanic origin of the atmosphere is provided by measuring the composition of gases currently being released from active volcanos. These gases consist of approximately 85 percent water vapor, 10 percent carbon dioxide, and 1 to 2 percent nitrogen. These three gases and their derivatives, including oxygen, comprise about 99 percent of our atmosphere.

The atmosphere performs a number of vital functions that profoundly affect the nature of the earth's surface and its ability to support life. The most obvious function is the supplying of oxygen to animals and carbon dioxide to plants. These gases are essential to their respiratory processes.

Earth's atmosphere also protects our planet from the impact of small meteors, which are incinerated by friction before reaching the surface. The large meteorites that do reach the surface produce craters similar to those on the moon. Unlike lunar craters, meteorite craters on earth have relatively short spans of existence because the forces of erosion associated with atmospheric phenomena, especially precipitation and wind, gradually wear them away. In fact, all the earth's landforms exist for limited periods of geologic time because of erosional and depositional processes related to the atmosphere.

The atmosphere has the capacity to absorb large quantities of the radiation that reaches the earth both from the sun and from elsewhere in space. As a consequence, we are protected from lethal doses of short wave radiation and from extremes of temperature, and are substantially warmer than we otherwise would be at our distance from the sun.

Finally, the atmosphere makes possible the cycle of evaporation and precipitation that provides water to the land areas of the earth. Without this cycle (discussed in Chapter 5), there would be no clouds, no rain or snow, no rivers or lakes, little or no land vegetation or soil, and little animal life either on land or in the sea. In addition, many important minerals could not form, and most landforms with which we are familiar would not develop.

Figure 2.1 The gray, pockmarked surface of the moon stands in stark contrast to the brilliance of the distant earth in this photograph taken by the Apollo astronauts. If the earth had no atmosphere, its surface would closely resemble the moon's surface. *(Department of the Interior, U.S. Geological Survey)*

COMPOSITION OF THE ATMOSPHERE

The atmosphere is a physical mixture of gaseous elements and compounds. Most of these gases exist in constant proportions in all but the atmosphere's outermost layers (above 50 miles altitude). The reason for the atmosphere's homogeneity is the mixing action of the wind. If it were not for the turbulent transfer of air molecules from one level to another by wind currents, the various atmospheric gases would stratify into horizontal layers on the basis of density. A few gases, however, have characteristics that prevent the wind from mixing them evenly, and consequently they vary in their atmospheric concentrations. In addition, small quantities of liquids and solids are suspended in the atmosphere. All liquids and solids vary greatly in quantity because they are much denser than the gases and have a tendency to settle to the surface. The distribution of these substances depends on the size and location of their sources, the rate at which they enter the atmosphere, and the speed and direction of wind currents. In the following discussion, the components of the atmosphere are divided into three categories: nonvariable gases, variable gases, and particulates (liquids and solids).

Nonvariable Gases

The **nonvariable gases** comprise over 98 percent of the total atmosphere by volume (see Figure 2.2). These gases are sufficiently stable, both physically and chemically, to remain in the atmosphere long enough to be dispersed evenly by wind currents.

As Table 2.1 indicates, the most abundant gas by far is *nitrogen*, which by volume comprises 78 percent of the atmospheric total of nonvariable gases. Although it is only the third most abundant gas released by the volcanic eruptions that formed the atmosphere, nitrogen has achieved its position of dominance because of its physical and chemical stability. Gaseous nitrogen is diatomic; that is, two atoms of nitrogen combine to form one nitrogen molecule (N_2). In this form, nitrogen is relatively inactive and does not participate directly in the respiratory processes of either plants or animals.

Most of the remainder of the atmosphere consists of *oxygen* (O_2), another diatomic molecule, which is essential to the respiratory processes of plants and animals. Oxygen also enters into many chemical reactions with rock materials at or near the surface of the earth to form

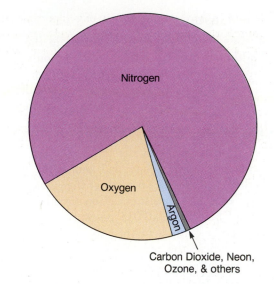

Nitrogen

Oxygen

Argon

Carbon Dioxide, Neon, Ozone, & others

Figure 2.2 The relative proportions of the major atmospheric gases.

a variety of products, including soil. Free oxygen is not liberated in quantity by volcanic eruptions. Instead, these eruptions release carbon dioxide (CO_2), which plants later separate into carbon and oxygen during photosynthesis. Plants have therefore been responsible for placing nearly all the free oxygen in the atmosphere. Nitrogen and oxygen together comprise slightly over 99 percent, by volume, of dry air.

Table 2.1
Composition of the Atmosphere

Name	Symbol	Percentage by Volume*
Nonvariable Gases		
Nitrogen	N_2	78.08
Oxygen	O_2	20.95
Argon	Ar	0.93
Neon	Ne	0.002
Others		0.001
Variable Gases and Particulates		
Water Vapor	H_2O	0.1 to 4.0
Carbon Dioxide	CO_2	0.035
Ozone	O_3	0.00006
Other Gases		Trace
Particulates		Normally trace

*Percentages, except for water vapor, are for dry air.

Variable Gases

Several gases exist in variable quantities in the atmosphere. Although they collectively comprise only a very small proportion of the atmosphere's total mass and volume, three of these variable gases—carbon dioxide, water vapor, and ozone—are crucial to the existence of life.

Carbon dioxide is essential to the photosynthetic processes of plants and provides the carbon that is the basic element of earth's plant and animal life. It was noted earlier that about 10 percent of the gases emitted by volcanic eruptions consist of carbon dioxide, yet only a very small percentage (0.035 percent) of the atmosphere currently consists of this gas. This invites the question of where the remainder has gone. A small portion of the carbon dioxide forms the bodies of plants, especially trees, that are living or have recently died. A considerably larger amount is stored beneath the earth's surface in the form of peat, coal, oil, and gas deposits. These deposits represent the remains of plants and animals, most of which lived hundreds of millions of years ago. The bulk of the carbon dioxide, though, has dissolved into the ocean waters. Carbon dioxide is highly soluble in water (a fact that the carbonated beverage industry uses to its advantage), and, over the course of geologic time, most atmospheric CO_2 has dissolved into the oceans, later to be precipitated in the form of carbonate rocks such as limestone and dolomite.

The atmosphere's carbon dioxide content has been slowly increasing, especially during the twentieth century, as a result of human activities. This phenomenon and its potential environmental consequences are discussed in the Case Study at the end of Chapter 3.

Water vapor (H_2O), or water in an invisible gaseous state, is one of the four most abundant gases in the atmosphere, but is highly variable in amount. Cold, dry air may contain less than 0.1 percent water vapor by volume, whereas very warm, moist air in the tropics may contain as much as 4 percent. Worldwide, air just above the earth's surface contains an estimated average of 1.4 percent water vapor. Because this proportion declines rapidly with altitude, approximately half the atmosphere's water vapor content exists within a mile (1.6 km) of sea level.

The major reason that the atmosphere's water vapor content varies so greatly is that the water vapor holding capacity of the air is highly dependent upon its temperature. Warm air can hold much more water vapor than can cold air. In addition, the availability of water vapor differs greatly in different areas because of the nature of the surface. For example, air over land areas is normally drier than air over oceans. As a result of the continuous temperature and locational changes to which the air is subjected, its water vapor content alternately increases through the process of evaporation and decreases through the processes of condensation (conversion to liquid water) and sublimation (conversion to ice).

Water vapor is critically important because it is the source of all the earth's precipitation. If the atmosphere could not alternately gain and lose huge quantities of this gas, the land areas of our planet would be sterile deserts. Atmospheric moisture is of sufficient importance to be treated in its own chapter (Chapter 5), and its effects are also discussed in subsequent chapters covering climate, the earth's surface waters, fluvial landforms, and glacial landforms.

The third important variable gas is ozone. Ozone (O_3), a triatomic form of oxygen, has three chemically bonded oxygen atoms instead of the normal two. It is produced by the bombardment of normal oxygen (O_2) molecules by cosmic radiation, which causes some to split and to recombine as ozone. The ozone molecule is not particularly stable chemically and tends to revert to the diatomic oxygen (O_2) form.

Most ozone exists between an altitude of about 10 miles (15 km) and 40 miles (65 km), a portion of the atmosphere sometimes referred to as the *ozone layer*. Concentrations are quite small; even at an altitude of 15 miles (24 km), where ozone is most abundant, it typically comprises only about 0.0012 percent of the total gases present.

The minute amounts of ozone produced in the upper atmosphere are important because ozone absorbs the potentially lethal quantities of incoming cosmic and ultraviolet radiation. Considerable concern has developed in recent years that human activities may be depleting the ozone layer. This subject is explored further in the Focus box on page 27.

Particulates

Atmospheric particulates consist of liquids and solids (with the exception of water droplets or ice crystals) derived primarily from the earth's surface and suspended temporarily in the atmosphere. They vary considerably in size, and the smallest are capable of remaining aloft almost indefinitely. Most particulates

Are Human Activities Depleting the Ozone Layer?

Considerable evidence has accumulated in recent years that human activities are depleting the ozone layer. It has been found that the average global concentration of stratospheric ozone declined by about 2 percent between 1969 and 1986. A continuation of this trend will eventually allow substantially more ultraviolet radiation to penetrate the atmosphere, leading to a major increase in the frequency of skin cancer. Plant and animal life could also be affected adversely, and many plastics would degrade more rapidly as a result of their sensitivity to ultraviolet radiation.

The scientific concern regarding possible human effects on the ozone layer increased dramatically in 1985, when the British Antarctic Survey discovered a hole in the ozone layer over Antarctica (see Figure 2.4). Apparently, this hole first appeared in 1975, and has been returning each Antarctic spring. The ozone-depleted zone extends from 7 to 12 miles (12 to 20 km) in altitude and exhibits an overall ozone loss of about 35 percent. In the center of this zone, between 8 and 11 miles in altitude, maximum ozone losses of 70 to 90 percent occur. Within the past few years, a second, weaker ozone hole has begun to form over the North Pole. Early in 1989, a scientific team measured 25 percent reductions in ozone concentrations at altitudes between 13 and 16 miles (22 to 26 km) over the Arctic Ocean.

The chief culprit responsible for the loss of ozone is believed by most researchers to be chlorofluorocarbon gases (CFCs), which currently are widely employed as refrigerants and aerosol propellants. When CFCs break down in the presence of ozone and sunlight, they release chlorine atoms that act as catalysts to trigger the conversion of ozone molecules into normal oxygen molecules. The chlorine atoms, unaffected by this transformation, act over and over again, eventually destroying great quantities of ozone. Although both the United States and Canada banned the nonessential use of CFCs as aerosol propellants in 1978, the resulting decline in emissions has been largely offset by an increase in their use in foreign countries.

CFCs are relatively stable gases, with chemical half-lives of 75 to 100 years. They also take several years to diffuse upward into the ozone layer following their release. Because these factors mean that there is a considerable period between the time CFCs are produced and the time they begin to affect the ozone layer; even a worldwide ban on their usage would be slow to show results.

A significant international agreement to limit future releases of CFCs was finally reached in September 1987, when 47 nations signed an accord pledging to reduce CFC production by 50 percent over a ten-year period. More recently, in June 1990, over 70 countries agreed to totally end CFC use by the year 2000. This agreement was especially significant because it included several less-developed countries, such as India, that had, until recently, been rapidly expanding their production of CFCs.

A NASA study published in the *American Journal of Science* (March 1988) compounded the mystery of the ozone hole by suggesting that methane gas (CH_4) may be even more important than CFCs as a destroyer of atmospheric ozone. Methane is produced by a variety of activities including the decay of wood, the decomposition of rice, and the digestive processes of cattle and termites. Atmospheric concentrations of this gas are currently increasing by 1 percent a year, and the total quantity of methane has doubled in the past century. The study concludes that the Antarctic ozone hole is likely to enlarge considerably and intensify in the coming years.

are solids, rather than liquids, and are collectively referred to as *dust*.

A considerable variety of sources of atmospheric dust particles exist, both as a result of natural processes and, increasingly, of human-related activities. Natural sources include ash from volcanic eruptions, wind-blown surface materials, smoke from fires, salt crystals released by the evaporation of sea spray, and various particles of biological origin, including pollen, spores, and bacteria (see Figure 2.3). Particulates released by human activities are derived primarily from urban activities such as factory and vehicle exhausts and smoke

(b)

(a)

Figure 2.3 Among the chief sources of atmospheric dust particles are (a) erupting volcanoes (in this case, Washington's Mount St. Helens), (b) urban and industrial activities (here, a paper mill near Jacksonville, Florida) and (c) forest fires. (*a: U.S. Geological Survey, JLM Visuals; b: W. Metzen/H. Armstrong Roberts; c: Charles Phelps Cushing/H. Armstrong Roberts*)

(c)

from heating and cooking, as well as from agricultural areas unprotected by vegetation (see the Case Study at the end of this chapter). The well-known hazy conditions that prevail in summer over much of the United States and Europe result largely from a combination of dust-generating human activities and relatively light summer winds.

Excessive quantities of airborne particulates can produce unpleasant and occasionally unhealthy atmospheric pollution. On the other hand, dust particles play a vital role in the formation of clouds and precipitation by serving as *condensation nuclei*. Water vapor under normal conditions requires solid or liquid surfaces on which to condense or sublime in order to produce water droplets or ice crystals, and airborne particulates provide these surfaces. Every raindrop is composed of many thousands of cloud droplets, each of which has formed around a mote of dust.

VERTICAL CHARACTERISTICS OF THE ATMOSPHERE

In the next few chapters, we examine the horizontal distribution of such atmospheric phenomena as temperature, air pressure, winds, and moisture. The horizontal (or at least relatively horizontal) level in which we are most interested is the base of the atmosphere; that is, the interface between the atmosphere and the earth's surface. It is in the vertical direction, however, that the greatest variations in the atmosphere occur.

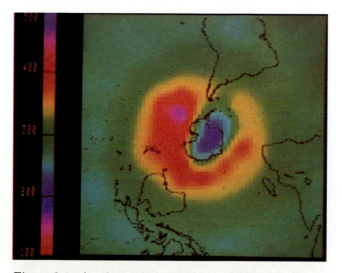

Figure 2.4 A color-coded satellite view of the ozone hole over Antarctica. (*National Oceanic and Atmospheric Administration/National Environmental Satellite, Data, and Information Service, Washington, D.C.*)

Weight, Density, and Vertical Extent

The air may appear weightless, but in reality it has mass and weight, just as does any other substance. The total weight of the earth's atmosphere is enormous. Estimated to be nearly 12 quintillion pounds, (normally written as 12×10^{18} lbs), the weight of the overlying air subjects each square inch of the earth's surface at sea level to an average pressure of 14.7 lbs (6.7 kg).

This weight is very unevenly distributed throughout the vertical extent of the atmosphere because the air is highly compacted by gravity. The great vertical extent of the atmosphere and the imperceptibility with which it merges with the scattered molecules of interplanetary space make it impossible to locate precisely the top of the atmosphere, but it extends several thousand miles above the earth's surface. Half of the atmosphere's total mass, however, lies below an altitude of only 3.5 miles, or 18,500 feet (5.6 km). At this altitude, the air pressure averages just half its value at sea level. Many of the world's higher mountain peaks extend well beyond this altitude; Mt. Everest, for example, has a summit elevation of 5.5 miles (8.8 km). Because the air pressure is so low, many mountain climbers employ oxygen masks when climbing above the 20,000 foot (6200 m) level.

Thermal Layers

The atmosphere not only contains layers of varying composition, it is also layered with respect to temperature (see Figure 2.5). The lowest and most important of these thermal layers is the troposphere, which contains nearly 80 percent of the atmosphere's total mass. This is the zone in which most observable weather phenomena occur and in which we live. The troposphere contains the great majority of the atmosphere's water vapor. Consequently, it contains nearly all clouds and is the origin of all precipitation reaching the surface. Winds in the troposphere are highly variable in both speed and direction, and these variations are directly responsible for many of the changing weather conditions at the surface.

The dominant vertical characteristic of the troposphere is a rather rapid decrease in temperature with increasing altitude. The actual rate of temperature change at any given time and place is quite variable, but over long periods a mean rate of decline of approximately 3.5 F° per thousand feet (6.4 C°/1000 m) has been found to exist. This value holds true as a global statistical mean at all levels in the troposphere.

Figure 2.5 The average vertical temperature profile of the atmosphere, and its relationship to the four thermal atmospheric layers.

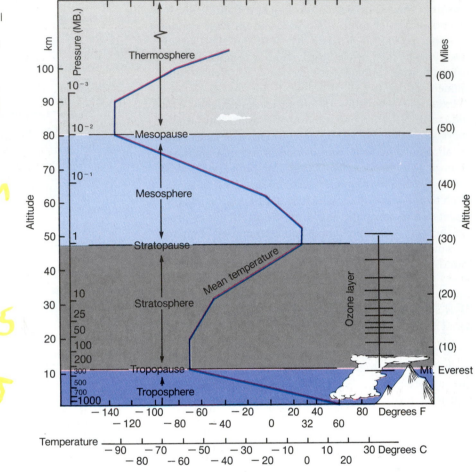

The rather sharply defined upper boundary of the troposphere, the **tropopause**, is the level at which the temperature stops falling with increasing altitude. The tropopause bulges outward above the tropics because of their high surface temperatures; over the equator it exists at an altitude of about 11 miles (18 km). It gradually lowers poleward, until at the North and South Poles it is only about 5 miles (8 km) above sea level. Its altitude also varies somewhat with the season of the year, being about a mile higher in summer than in winter over the middle and high latitudes. Temperatures at the tropopause usually have values between −60° F and −110° F (−50° C to −80° C).

The tropopause can be visualized as a boundary surface separating the troposphere from the **stratosphere**. This second thermal layer extends from the tropopause to an altitude of 30 miles (50 km). The lower stratosphere, up to about 20 miles (32 km) in altitude,

remains fairly constant in temperature; but above this level a gradual temperature rise occurs. This is caused by the absorption of solar ultraviolet radiation by ozone, which converts this energy into heat. Because temperatures increase with altitude in the stratosphere, the vertical air movements that form clouds have much less tendency to develop. The stratosphere is therefore characterized by cloudless, thin, dry air and is often subject to strong horizontal winds.

The upper edge of the stratosphere is marked by the **stratopause**, another well-defined thermal boundary. Crossing the stratopause, the temperature stops rising and begins falling as the atmosphere's third thermal layer, the **mesosphere**, is entered. The mesosphere is located between 30 and 50 miles (50 and 80 km) above the earth's surface. Because little solar energy is absorbed at this level, temperatures within the upper mesosphere fall to the lowest values found anywhere in the atmosphere. At the upper boundary of the mesosphere, termed the **mesopause**, temperatures are in the vicinity of -130° F (-90° C).

Above the mesosphere, a final reversal in temperature occurs in the **thermosphere**. In this zone, located at altitudes above 50 miles (80 km), the temperature rises gradually to reach the highest readings occurring anywhere in the atmosphere. In the upper thermosphere, temperatures may exceed 2000° F (1100° C) because of the absorption of ultraviolet radiation from the sun. Although this is a very high temperature, the air is so rarefied at these great heights that very little thermal energy is actually represented.

THE ATMOSPHERE, WEATHER, AND CLIMATE

The variability of the atmosphere is made evident by the constantly changing array of weather conditions with which we are all familiar. This chapter, in fact, serves as a preface to Chapters 3 through 7, which deal with global weather and climate characteristics.

Weather is the short-term condition of the atmosphere. Variations in the weather occur both over time, at any given location, and between different locations, at any one time. As a result, the weather becomes more complex when examined over progressively larger segments of time and space.

When discussing long-term weather conditions, it becomes necessary to generalize. These long-term conditions constitute an area's climate. The *climate* of a region is sometimes defined as its average weather because mean (or statistically average) conditions are usually stressed. Also of great importance, though, are extreme conditions because they can produce disastrous consequences. People should therefore expect conditions frequently near the climatic means, but should be prepared to cope occasionally with the extremes.

Weather and climate are therefore closely related but not identical concepts. The branch of science having the weather as its subject area is termed *meteorology* (from the Greek *meteora*, or "atmospheric phenomena," and *logos*, or "description"). Meteorology today is a highly sophisticated field that is considered a branch of applied physics. Forecasting meteorology involves the acquisition, rapid transmittal, and computer analysis of large volumes of surface and upper air data. *Climatology, the science that deals with the climate, is concerned with the causes and global distributions of climatic characteristics.

The atmospheric phenomena studied in weather and climate can be divided into six discrete segments or elements: temperature, air pressure, wind, humidity, clouds, and precipitation. Each element varies almost continuously over time and space, but, when analyzed in terms of its average condition, each displays an organized and logical geographical pattern. The next several chapters examine each weather element in turn in order to provide an understanding of what it is, how and why it operates, and how it is distributed around the world. Temperature is discussed first because it is not only one of the most important weather elements, but because, in a sense, it is also a measure of the energy that powers all the rest.

Summary

This chapter has examined the physical and chemical properties of the earth's atmosphere and has briefly discussed its influence on the surface environment. The atmosphere was formed as volcanic eruptions liberated dissolved gases that were originally locked inside the earth. The mixture of atmospheric gases has been altered gradually by physical, chemical, and biological activities.

The atmosphere performs several vital functions that profoundly affect the nature of the earth's surface and enable plant and animal life to exist. First, it provides the oxygen and carbon dioxide needed for animal and plant respiratory processes. Second, its frictional resistance destroys all but the largest incoming meteors before they reach the surface. Third, it protects the surface from temperature extremes and from potentially lethal cosmic and ultraviolet radiation. Finally, it makes possible the weather phenomena that water the earth's surface and produce a great variety of distinctive landforms and surface materials.

The atmosphere is composed of a physical mixture of gases. The most abundant are nitrogen (N_2), oxygen (O_2), argon (Ar), water vapor (H_2O), and carbon dioxide (CO_2). The proportions of the first three of these gases remain relatively constant, while those of the last two are subject to variation. In addition, small quantities of a number of liquids and solids exist within the atmosphere.

The most important vertical changes in the atmosphere are in density and temperature. The atmosphere rapidly becomes less dense with altitude, and half of the atmospheric gases are concentrated within 3.5 miles of sea level. Four distinctive thermal layers exist, each with vertical temperature characteristics reversed from adjacent layers. From the earth's surface to the top of the atmosphere, these layers are the troposphere, the stratosphere, the mesosphere, and the thermosphere.

Review Questions

1. How does the atmosphere affect the earth's surface environment? List several factors.
2. Explain how the atmosphere is believed to have originated.
3. What five gases comprise the majority of the atmosphere? What is the direct source of atmospheric oxygen?
4. What are the major natural and human sources of atmospheric dust particles? What vital functions do atmospheric dust particles perform?
5. What are the four thermal layers of the atmosphere, and what are the altitudes and temperature characteristics of each layer?

Key Terms

Nonvariable gases	Stratosphere
Variable gases	Stratopause
Ozone	Mesosphere
Troposphere	Mesopause
Tropopause	Thermosphere

CASE STUDY

Air Pollution: A Global Environmental Problem

The problem of air pollution has faced humanity for thousands of years. Within the past century, though, air pollution has increased from largely a local annoyance to an important global environmental problem. While air pollution is produced by both natural and human sources, the focus here is on the human sources.

Two general categories of human activity are major producers of polluted air: urban-related activities and agricultural activities. Urban air pollution is primarily associated with motor vehicles, factory emissions, and home heating. The leading agricultural air pollutant is wind-blown soil resulting from the removal of protective vegetation, plowing, and other soil-loosening activities. Airborne pesticides and agricultural burning also cause problems in some areas.

Urban Air Pollution

A combination of gaseous, liquid, and solid pollutants lower the quality of life in many large urban areas today. These substances produce unsightly deposits, reduce visibilities, irritate eyes and respiratory systems, and under some conditions can cause serious illnesses and even deaths. For example, the most severe urban air pollution episode so far recorded occurred from December 5 to December 10, 1952, in London, England, when 4000 deaths were attributed to air pollution resulting primarily from the household burning of low-quality soft coal. London has since enacted legislation which has curtailed this practice and brought about a noteworthy improvement in air quality.

Atmospheric conditions most associated with severe urban air pollution are a lack of precipitation, light winds, and a low-level temperature inversion. A *temperature inversion* occurs when a shallow layer of dense, cool air at the surface is trapped beneath lighter, warm air (see Figures 2.6 and 2.7). It is termed an inversion because it represents a localized reversal of the normal tropospheric characteristic of declining temperatures with increasing altitude. Short-lived temperature inversions are common occurrences. They form frequently in the late night and morning hours as a result of the nightly radiation of surface heat into space, only to dissipate by midday because of solar heating. Occasionally, though, foggy weather conditions within a stagnant air mass help maintain a temperature inversion for several days. This enables pollutants to accumulate to potentially dangerous levels in the trapped layer of cool air. Conditions favoring air pollution occur most frequently in late summer and autumn in the United States and western Europe. In the United States, the mountain-rimmed Los Angeles Basin traps pollutants produced by a major urban area, while the cold waters of the California Current and persistent high pressure systems produce low-level temperature inversions and frequent fog. As a consequence, Los Angeles experiences more days with moderate to severe air pollution than any other large city in the nation.

Fortunately, a combination of increased concern and technological advances has brought about an improvement in the urban air quality of most developed countries during the past decade. In the United States, especially, marked improvements have been recorded recently as a result of a number of factors. They include the desulfurization of coal and oil, the increased use of solar and nuclear power, the trend toward smaller cars with better-designed engines, the use of vehicle emissions control systems, and a substantial increase in the quantity and quality of industrial gaseous effluent treatment equipment. These changes have resulted largely from a growing public concern over urban air pollution and the consequent enactment of a great deal of anti-pollution legislation.

Improved air pollution control technology, however, has been largely offset by the rapid population growth and industrial development of many less-developed countries of the world, which are for the first time producing significant quantities of air pollution. The geographic consequence is that urban air pollution is becoming less restricted to established industrial areas such as North America, Europe, and Japan, and is increasingly becoming a global concern.

continued on next page

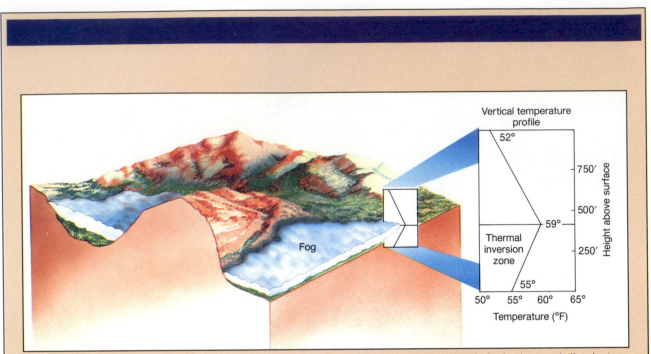

Figure 2.6   A low-level temperature inversion exists when a shallow layer of cool, relatively dense air lies just above the ground, with warmer air aloft. These inversions are frequently associated with fog that hinders the ability of sunlight to heat the surface and destroy the inversion.

Agricultural Air Pollution

While the quantity of urban air pollution produced around the world is probably remaining nearly steady or even declining slightly, agricultural air pollution continues to increase rapidly. The chief air pollutant of agricultural origin is windblown dust from unprotected soil. Dust storms are experienced in many parts of the world, especially where semiarid conditions prevail. These areas, despite their dryness, contain some of the world's most fertile soils, and they have been cleared and developed extensively for agriculture.

As in the case of urban air pollution, the erosion of soil by the wind is gradually being reduced in more technologically advanced countries such as the United States and Canada. During the Dust Bowl era of the 1930s, tremendous dust storms carried off millions of tons of fertile soil from the American Great Plains, reducing

Figure 2.7   Urban smog, produced in the Los Angeles Basin, is trapped near the surface by a thermal inversion and has spread eastward into the California desert near Twentynine Palms. (*B.F. Molnia*)

continued on next page

visibilities in the cities of the East Coast and producing spectacular sunsets in Europe. Improved agricultural techniques and changes in land use patterns have reduced the frequency of dust storms in this area, although continuing soil erosion remains a serious problem.

In contrast, high population growth rates in many less-developed countries of the subtropics, especially in Africa and Asia, necessitate a rapid expansion in agricultural land use. Much of this land is marginal in quality, and the removal of protective vegetation, plowing, and frequent droughts augment the potential for major dust storms. Dust storms not only temporarily add great quantities of dust to the air, they also remove vast amounts of irreplaceable soil. Much of this soil is deposited in the oceans, where it is irretrievably lost. The quantities of dust involved are so large that some scientists theorize it may be reducing the influx of solar energy sufficiently to account for the estimated worldwide drop in temperature of about 0.5 F° (0.3 C°) over the northern hemisphere between 1940 and the mid-1980s. Vegetation removal and soil erosion are leading to the formation of desertlike conditions over large areas. This process, called *desertification*, is discussed further in the Case Study at the end of Chapter 16. It represents one of the more serious long-range environmental problems faced by humanity.

Chapter Three

Energy Flow and Air Temperature

Focus Questions

1. How does solar energy get to the earth, and what happens to it once it reaches the earth?
2. What are the major factors controlling air temperatures at the earth's surface?
3. What is the global air temperature pattern that results from the action of these controls?

TERRESTRIAL ENERGY SOURCES

The *temperature* of any object or substance is a measure of its thermal energy. As energy is absorbed (gained from another energy source and converted to molecular kinetic energy), the temperature rises. If enough heating takes place the substance involved may undergo phase changes from a solid, to a liquid, and finally to a gaseous (vapor) form. Because no limiting maximum temperature exists, the temperature attained by any substance is restricted only by the limits of its ability to absorb more energy than it loses. The earth's air temperature therefore is a measure of the rate of motion of the air molecules that results from the absorption of energy from all sources.

Of the several energy sources that heat the atmosphere, the sun is overwhelmingly dominant. It is estimated to supply at least 99.97 percent of the atmosphere's heat. A second source, heat from the interior of the earth, provides nearly all the remainder.[1] The sun's energy is produced by the process of atomic fusion: within the sun's core, atoms of hydrogen are fused into helium under conditions of tremendous heat and pressure (see Figure 3.1). The mass of the helium produced is slightly less than that of the original hydrogen, and this lost mass is converted to the energy that is radiated in all directions from the sun. The earth is such a small target and is so far away from the sun that it intercepts only about one/two billionth of the sun's total energy output. This small proportion of energy, however, is the dominant power source for the earth system.[2]

ENERGY TRANSFER PROCESSES

How does solar energy cross the great void separating the earth from the sun? Once it has reached the earth system, how is this energy distributed to all parts of the earth? Three energy transport processes—radiation, conduction, and convection—collectively perform these tasks.

Radiation is the process of energy transmission by electromagnetic waves. These waves, so-named because

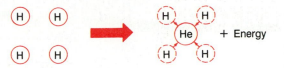

Figure 3.1 Hydrogen fusion involves the fusing of four hydrogen atoms into one helium atom. The attendant release of energy powers the sun and most other stars.

they display both electric and magnetic properties, have the ability to pass through a vacuum such as the 93 million miles (1.5×10^8 km) of space separating the earth from the sun. Radiation is divided into several major categories on the basis of wavelength to form an *electromagnetic spectrum* (see Figure 3.2). As the next section explains, the absorption of solar radiation converts this radiation to the heat energy that warms the earth's surface and atmosphere.

Once solar radiation reaches the earth system, the processes of conduction and convection play major roles in its subsequent distribution. Conduction involves heat transfer by physical contact between objects or substances of different temperatures. Heat always flows from the warmer body to the colder one and, if contact is maintained, will continue to do so until their temperatures have equalized. Conduction is the basis of our ability to sense heat and cold. For example, when you hold an ice cube, heat flows from your hand to the much colder ice. The loss of heat makes your hand cold, while the ice may gain enough energy from your hand to melt.

With respect to weather and climate, most conduction occurs between the earth's surface (both land and water) and the atmosphere. Because air, water, and land are all rather poor conductors, the direct effects of this energy transfer process are restricted largely to the lowest few feet of the atmosphere.

Convection is the process of energy transfer through the physical movement of a fluid (a liquid or gas) from one place to another. At the earth's surface, the winds and the ocean currents are the major convectors. A given quantity of air has a lower capacity for transporting heat than does the same quantity of water, but wind typically travels at a much greater speed than do ocean currents. As a result, approximately twice as much energy is convected by the atmosphere as by the ocean waters. On a global basis, convection acts to produce a poleward net transport of heat. If it were not for this exchange of energy between the lower and higher latitudes, the tropics would be significantly hotter, and the polar regions substantially colder, than

---

1. Heat from the earth's interior, termed *geothermal heat,* sometimes manifests itself spectacularly in the form of earthquakes, geysers, and volcanic eruptions. This energy source, although unimportant in providing atmospheric heat, plays a crucial role in the development of the earth's surface features, and will be examined in detail later in this book.

2. The earth and its atmosphere, taken together, will be referred to as the *earth system.*

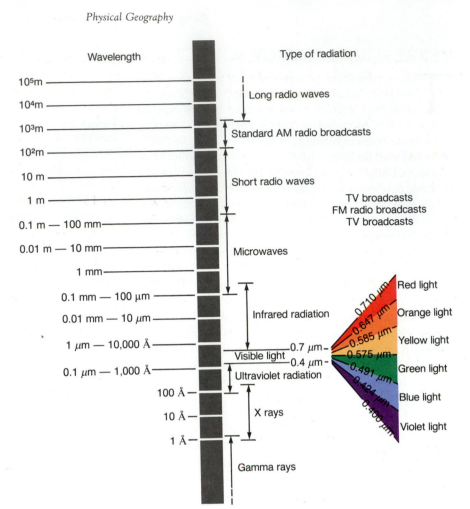

Figure 3.2 The electromagnetic spectrum.

they already are (see Figure 3.3). Such a situation would greatly reduce the habitability of our planet.

In addition, vast quantities of *latent heat* are transferred from the surface to the atmosphere through evaporation from surface water sources, including the oceans, lakes, rivers, damp soil, and even vegetation. The energy used to evaporate the water is released later as heat when the water vapor condenses back to liquid water or sublimes into ice crystals during the formation of clouds and precipitation (see Chapter 5).

The Global Energy Balance

Solar shortwave radiation passes through the 93 million miles (1.5×10^8 km) of space between the sun and the earth system without being appreciably altered in any way. During its comparatively short passage through

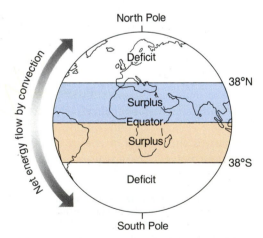

Figure 3.3 Latitudinal variation in the terrestrial energy budget. The earth receives more energy from the sun than it returns to space equatorward of 38°; this situation is reversed poleward of 38°.

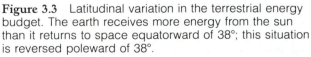

the atmosphere to the earth's surface, however, solar radiation is subjected to a number of redirections and transformations of great importance to life on earth.

Solar energy can be visualized as reaching the earth system in a steady stream of shortwave radiation. As this **insolation** (a shortening of the term *incoming solar radiation*) enters the atmosphere, it is effectively divided into three segments, as illustrated in Figure 3.4. Approximately 30 percent of the energy is redirected back to space. While most of this redirected energy is reflected from the upper surfaces of clouds, some is scattered by air molecules and dust particles, and some passes through the atmosphere to be reflected from the earth's surface. Scattered energy is broken up and reflected in all directions by air molecules and dust particles. Because the wavelengths that correspond to the color blue are scattered more effectively than those of the longer visual wavelengths, a clear sky in the presence of sunlight takes on a bright blue color.

The ability of different materials to reflect or scatter solar energy varies greatly; this ability is indicated by their visual brightness in sunlight. For example, the almost blinding brightness of ice and snow or of a cloud deck when viewed from above is produced by reflected sunlight and shows that these objects are excellent reflectors. In contrast, the dark green color of a dense forest or the black color of an asphalt road indicates that they are poor reflectors, and therefore good absorbers, of insolation. The reflection ratio, or *albedo*, of the earth system as a whole is 30 percent, but local albedos display considerable variation, both temporally and spatially (see Table 3.1). In general, albedos are higher in the high latitudes than in the lower latitudes because of the presence of ice and snow and the oblique passage of sunlight through the atmosphere in the high latitudes. For the same reasons, albedos are higher, on average, in winter than in summer outside the tropics.

The 30 percent of insolation that is reflected or scattered upward is not used in the terrestrial heating process because it is redirected back into space. The remaining 70 percent is absorbed to heat the earth's surface and atmosphere. The actual absorption of this energy occurs in two places: the atmosphere and the earth's surface. Approximately 19 percent of the total insolation is absorbed in the atmosphere. The remaining 51 percent passes through the atmosphere to be absorbed at the surface.

Energy is neither gained nor lost during the absorption process. What does occur is a major shift in the wavelengths of the radiation. Most insolation arrives as visible light and as near-infrared radiation, with an

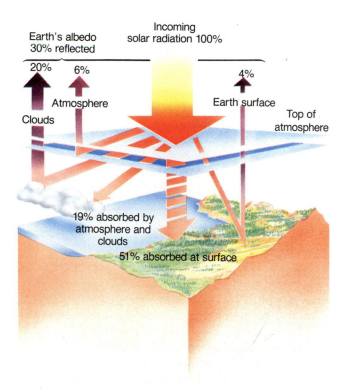

Figure 3.4 Approximately 30 percent of the solar radiation reaching the earth system is reflected or scattered back to space. The remaining 70 percent is absorbed within the atmosphere or at the surface.

Table 3.1
Typical Albedos of Terrestrial Materials*

Material	Albedo
Earth system as a whole	30
Sand	25
Dense forest	8
Field crops	25
Water (sun near horizon)	40–75
Water (sun overhead)	6
Fresh snow	85–95
Concrete	20–30

*The albedo is the short wave energy reflection rate of an object or substance, stated in percent.

Characteristics of Electromagnetic Radiation

Although electromagnetic radiation constantly surrounds us, and all objects or substances absorb and emit radiation, we still have a very imperfect understanding of this phenomenon. The radiation spectrum is commonly divided into categories based on wavelength, but all forms of electromagnetic radiation are essentially similar.

The hotter an object is, the greater the quantity of radiation it emits per unit area, and the shorter the wavelength of that energy. The sun, with a surface temperature of approximately 11,000° F (6000° C), radiates maximum energy in the vicinity of 0.5 micrometers, which is in the visible portion of the electromagnetic spectrum (see Figure 3.5). Because it is so much hotter than the earth, a given area on the surface of the sun emits about 160,000 times the radiation of an area of similar size on the earth's surface. The maximum radiation for the earth occurs at a wavelength of about 10 micrometers, well within the infrared range. Because wavelengths in the zone of maximum terrestrial radiation are roughly 20 times longer than wavelengths in the zone of maximum solar radiation, terrestrial radiation is generally referred to as longwave radiation and solar radiation as shortwave radiation.

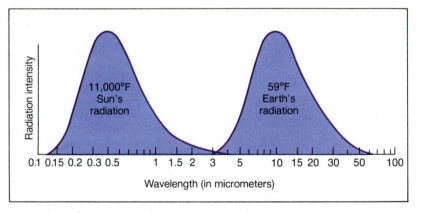

Figure 3.5 Comparison of solar and terrestrial radiation spectra. The peak radiation wavelength for the earth is approximately 10 micrometers, a value about twenty times longer than that of the sun's peak radiation wavelength.

average wavelength of about 0.5 μm (see the Focus box above).[3] The much cooler earth system, though, reradiates the absorbed energy in the form of longwave *terrestrial radiation,* which we sense as heat. This radiation has an average wavelength of about 10 μm. All insolation absorbed by the earth system is eventually returned to space. This process results in the maintenance of a global energy balance. A similar energy balance, however, does not exist at the local level over short periods of time. Whenever the local influx of energy exceeds its rate of loss, as during the day, the temperature rises. When a net loss of energy occurs, the temperature falls.

Most insolation that passes through the atmosphere is absorbed by the earth's surface and converted to heat. This explains, for example, why an asphalt driveway or a closed automobile that has been in sunlight is so much hotter than the surrounding air. The heated surface subsequently transfers the energy by reradiation, conduction, and evaporation to the surrounding atmosphere, which can then transport it over great distances. The atmosphere is therefore heated primarily from below; it absorbs much more heat from the underlying surface than from direct insolation. This is why the

3. Microscopic dimensions are commonly measured in micrometers, commonly written μm. One micrometer (1 μm) equals 10^{-6} meters.

warmest portion of the atmosphere is the lower troposphere and why tropospheric temperatures decline with increasing altitude.

The wavelengths of most of the terrestrial radiation that warms the atmosphere are between 3 and 4 μm. Most wavelengths in this range are absorbed by atmospheric water vapor and carbon dioxide. As a result, it is more difficult for longwave terrestrial radiation to leave the atmosphere than it was for the shortwave insolation to enter it. Heat is temporarily trapped in the atmosphere, and atmospheric temperatures become much warmer than they would otherwise be.

The heat-trapping ability of the atmosphere, which is analogous to the heating of the inside of a greenhouse or a closed car on a sunny day, is called the **greenhouse effect**. Glass, like our atmosphere, is relatively transparent to shortwave solar radiation, but not to longwave heat. As a result, solar energy easily enters a greenhouse, where it is converted to heat energy. This causes the greenhouse to heat up until it becomes so much warmer than the outside air that eventually as much heat energy escapes through the glass (by conduction) as shortwave energy enters, and a raised equilibrium temperature is established. Without the greenhouse effect, it is estimated that the earth's surface temperature would average about $-4°$ F ($-20°$ C) rather than its present value of 59° F (15° C). Because longwave terrestrial radiation escapes so slowly, the earth's surface and lower atmosphere cool only gradually during the nighttime hours, when no new insolation is available to offset radiational losses. Cooling is especially slow when the air is moist or a cloud cover is present to reradiate energy back to the surface.

WORLD TEMPERATURE PATTERN AND CONTROLS

Geographers are interested in terrestrial energy sources and energy transfer processes primarily because of the information they provide about the global temperature pattern. This pattern, in turn, strongly influences vegetative patterns, soil types, the distribution of human and animal populations, and many other factors. This section discusses the influence of the most important physical factors responsible for the earth's surface air temperature pattern. Before proceeding, a few basic terms used in geographic temperature analysis must be introduced.

Temperature patterns are usually mapped by the use of *isotherms*, lines connecting places with the same temperature. Figures 3.6 and 3.7 depict the mean temperatures for the world in January and July, the two months that represent the extremes of summer heat and winter cold for most land areas.

Weather reporting stations record a variety of temperature data. The daily high and low temperatures provide the basis for the calculation of several basic temperature figures. The *daily mean temperature* is calculated by adding the high and low temperatures and dividing the total by two. The *monthly mean temperature* is determined by averaging the daily means for the entire month.

The range, or variation, of temperature is also important. The *daily temperature range* is the number of degrees between the day's high and low temperatures. Last, and probably most important in terms of world temperature analysis, is the *annual temperature range*, defined as the number of degrees between the warmest and coldest monthly mean temperatures. For example, the warmest month in Baltimore, Maryland, is July, with a long-term mean of 77° F (25° C). Baltimore's coldest month is January, with a monthly mean of 33° F (1° C). Its annual temperature range is thus 44 Fahrenheit degrees (24 Celsius degrees). This figure gives a good indication of the "seasonality" of an area's climate, since higher values indicate a climate with a more pronounced summer-winter temperature variation. The global distribution of annual temperature ranges is displayed in Figure 3.8.

Primary Temperature Controls

Although a number of factors influence the earth's temperature pattern, five stand out as primary controls. These are the altitude of the sun in the sky, the duration of daylight, the distribution of land and water, elevation, and the pattern of ocean currents. After examining each of these primary controls, we will briefly discuss a number of secondary temperature influences.

Altitude of the Sun and Duration of Daylight

It has already been noted that nearly all atmospheric heating is derived directly or indirectly from the sun. When the sun is high in the sky, the radiant energy received per unit area of surface is maximized (see Figure 1.10). When the sun's altitude is low, radiation received per unit area is greatly reduced, and the energy that does arrive is more likely to be reflected and scattered by the atmosphere before reaching the surface.

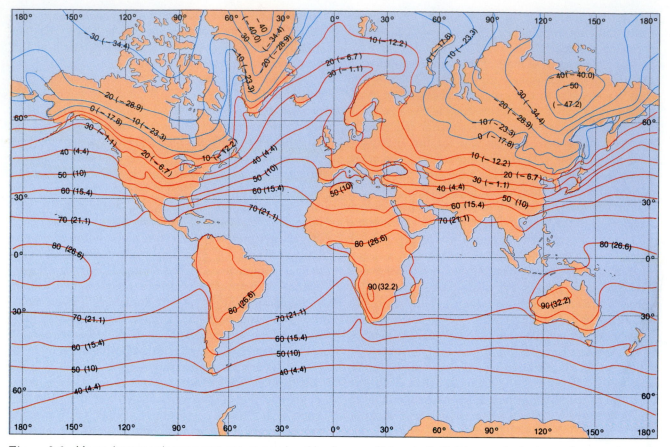

Figure 3.6   Mean January air temperatures at sea level in degrees Fahrenheit. Celsius equivalents are in parentheses.

For these reasons, *the altitude of the sun is the most important factor controlling air temperatures at the earth's surface*. This control affects the air temperature pattern over three differing timespans: long-term, annual, and daily.

The long-term pattern, the most basic and important, can be seen on both the January and July isotherm maps (Figures 3.6 and 3.7). The most fundamental long-term characteristic of the world temperature distribution is that temperatures are warmest in the tropics and decrease poleward. Put another way, as a general rule, temperatures decline with increasing latitude because of decreasing mean solar altitudes. Because isotherms are lines of equal temperature, they display an overall zonal, or east-west trend, paralleling lines of latitude.

Although the sun's average altitude at solar noon becomes progressively lower with increasing latitude, it is subject to seasonal variations in all parts of the world because of the changing inclination of the earth's axis with respect to the sun. For all locations from the Tropics of Cancer and Capricorn poleward, the sun gets progressively higher in the noon sky during the six-month period from the winter solstice to the summer solstice. It then decreases in altitude during the following six months. Locations within the tropics have two periods of increasing and of decreasing solar noon altitudes during the course of a year. At all locations except the equator itself, however, the sun's noon altitude averages higher in the spring and summer seasons than it does in the fall and winter seasons. In addition, all places on earth (again excepting the equator) have longer days than nights in spring and summer and longer nights than days in fall and winter. These two related factors are responsible for the seasonal variations in temperature distributions that can be observed between the January and July isotherm maps. For example, the combination of vertical sun angles,

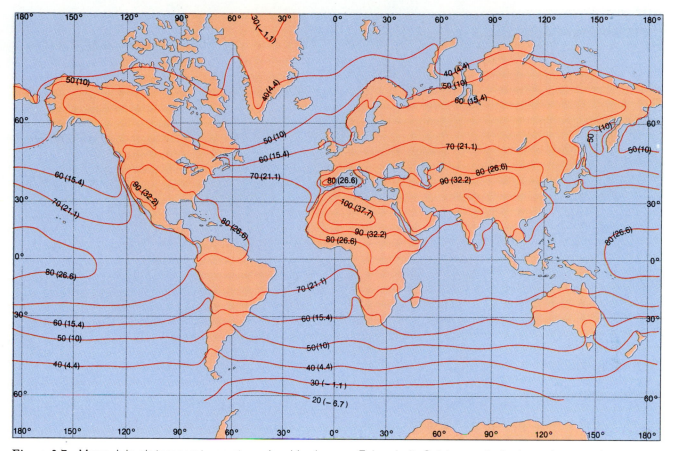

Figure 3.7 Mean July air temperatures at sea level in degrees Fahrenheit. Celsius equivalents are in parentheses.

limited cloud cover, and 13-to-14-hour periods of sunlight contribute to the fact that during their summer period the subtropics, rather than the equatorial regions, experience the hottest temperatures in the world.

The greater intensity and duration of insolation during the spring and summer typically cause these two seasons to be warmest. This tendency becomes progressively stronger poleward, because seasonal differences in the receipt of insolation increase with latitude. In the high latitudes, poleward of the Arctic and Antarctic Circles, seasonal differences in the daily duration of sunlight reach extremes, with 24 hours of daylight at the time of the summer solstice and 0 hours at the time of the winter solstice. Annual temperature ranges are therefore very large.

Because the earth's surface heats and cools rather slowly, seasonal temperature lags exist. This causes the hottest and coldest times of the year to occur a month or more after the dates of the summer and winter solstices.

For example, July and August are the two warmest months in most of North America, even though the greatest insolation receipts occur during the summer solstice month of June. Likewise, January and February are typically colder than the winter solstice month of December. As a result of the seasonal temperature lag, summer is warmer than spring, and winter colder than fall, in most places.

The familiar day/night cycle of air temperature is also caused by differing sun altitudes. This cycle is produced by the rotation of the earth on its axis. Daily heating and cooling lags also occur. Although the altitude of the sun begins to decrease after solar noon, the sun normally remains high enough in the sky for total insolation amounts to exceed outgoing terrestrial radiation totals until mid-afternoon. This causes maximum temperatures to typically occur between 2 P.M. and 4 P.M. By the same token, minimum temperatures most commonly occur near sunrise rather than near midnight.

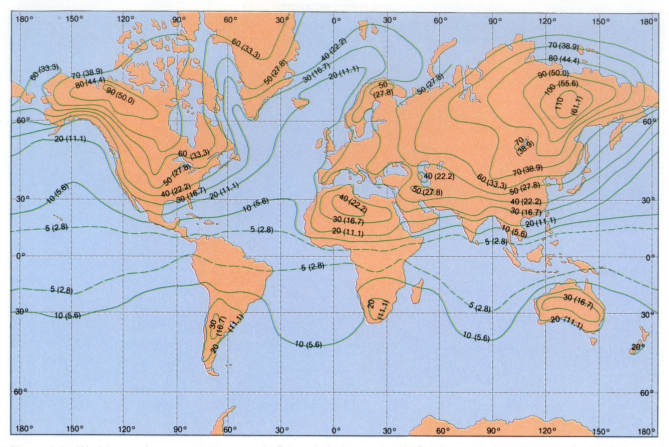

Figure 3.8 World annual temperature ranges in Fahrenheit degrees, with Celsius equivalents in parentheses. Values indicate the difference between the warmest and coldest monthly mean temperatures.

Land and Water Distribution

The complex distribution of land and water bodies over the earth's surface acts as a highly important world temperature control. Whereas the altitude of the sun largely affects the *mean* temperatures of different parts of the world, the distribution of land and water most greatly influences temperature *ranges*, both on a daily and an annual basis.

This factor's importance stems from the ability of land to change temperature far more readily than water can. As a result, land areas tend to have larger daily and annual temperature ranges and shorter seasonal and daily heating and cooling lags than do water bodies at similar latitudes. The influence of the surrounding land or water surface is so important that climates are often described as being either continental or maritime (see Figure 3.10). Continental climate characteristics include large daily and annual temperature ranges, one-month seasonal heating and cooling lags, and a tendency

toward relatively low amounts of precipitation, fog, cloudiness, and humidity. Continental climates are best developed in the interiors of large continents such as Asia and North America.

Maritime climate characteristics, conversely, include relatively small daily and annual temperature ranges, two-month seasonal heating and cooling lags, and a greater quantity of atmospheric moisture. Maritime climates are best developed over the oceans, but also occur in many coastal areas that have a predominantly onshore windflow (see Figure 3.11). They even develop to some extent along the shores of large lakes such as the Great Lakes.

The influence of land and water distribution on temperature ranges is clearly apparent from the world map of annual temperature ranges (Figure 3.8). A striking variation can be seen between the temperature ranges of large land and water masses at equivalent latitudes. Land areas have much larger ranges than do adjacent water bodies, especially in the higher latitudes.

Temperature Scales

Two temperature scales are in common use in the United States. Both have values based on the boiling and freezing points of water under standard atmospheric pressure conditions. The *Fahrenheit* scale, developed in 1714 by the German physicist Gabriel Daniel Fahrenheit, has been losing international popularity but is still the primary scale used in the United States. This scale has values of 32° F and 212° F, respectively, for the freezing and boiling points of water. The metric system temperature scale is the *Celsius,* or centigrade, scale, named in honor of Anders Celsius, a Swedish astronomer who devised it in 1742. The freezing and boiling points for water on the Celsius scale are 0° C and 100° C, respectively.

Where both temperature scales are commonly employed, as in the United States, it is often necessary to convert from one to the other. The conversion formulas are as follows:

$$C = \frac{5}{9}(F\text{-}32) \quad F = \frac{9}{5}C + 32$$

A third temperature scale, used primarily in the laboratory sciences, is the *Kelvin* (or Absolute) scale. Like the Celsius scale, it is calibrated to have 100 equal temperature increments between the freezing and boiling points of water. Numerical values, however, begin at absolute zero, the temperature at which all molecular vibration ceases. The freezing point of water is 273° K, and the boiling point is 373° K. Any Kelvin temperature has a numerical value 273 degrees higher than its Celsius equivalent. Stated as a formula, this relationship is as follows:

$$K = C + 273$$

The Fahrenheit, Celsius, and Kelvin scales are illustrated in Figure 3.9.

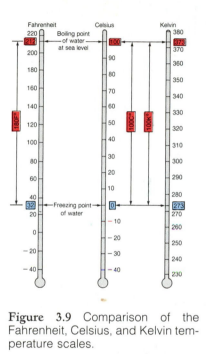

Figure 3.9 Comparison of the Fahrenheit, Celsius, and Kelvin temperature scales.

The world's largest annual temperature ranges occur on large continents in the high latitudes, where the influences of large seasonal variations of insolation and continentality are combined. In northeastern Asia, the monthly means of January and July differ by as much as 115 F° (64 C°). Ranges of over 80 F° (45 C°) occur in northwestern Canada and northeastern Alaska on the somewhat less massive continent of North America.

Elevation

Another important influence on the global temperature pattern is elevation. This factor primarily affects temperature means (Figure 3.12). For locations above approximately 7000 feet (2100 m), elevation becomes a dominant temperature influence. Peaks above 18,000 feet (5500 m) are perpetually covered with snow and ice, even at the equator. Only a small proportion of the world's surface area, however, is situated at such high elevations. Thus the importance of this control is somewhat reduced by the restricted areal extent of the earth's highland surfaces.

Elevation is more capable of producing significant differences in both temperature and precipitation means over small distances than is any other major climatic control. Climatic characteristics in mountainous areas therefore vary greatly; a hike or a short drive often

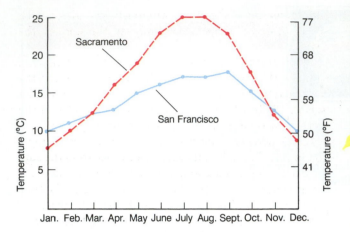

Figure 3.10 The annual temperature graphs of the coastal city of San Francisco and of Sacramento, located about 90 miles (145 km) inland, illustrate the influence of continentality.

brings travelers to places with climatic conditions very different from those they left. In lowland areas, in contrast, it is often necessary to travel hundreds of miles to reach locations with substantially different climates. Travel to both higher elevations and latitudes results in colder temperatures. Depending on the location and season of the year, however, temperatures decrease some 500 to 1000 times faster with elevation than with latitude, so that an increase in elevation of 1000 feet (300 m) may be equivalent, in thermal terms, to traveling 100 to 200 miles (160 to 320 km) poleward.

The cause of temperature decline with increased altitude in the troposphere has already been discussed. High-elevation locations are simply farther away from sea level, the level at which most insolation is converted to heat. As mentioned in Chapter 2, the worldwide average rate of vertical temperature decrease is approximately 3.5 F° per thousand feet (6.4 C° /1000 m) in the troposphere. This value is termed the **average environmental lapse rate.** It should be stressed that this is a mean value, and that the actual lapse rate for a location at any given time will probably differ from this figure.

Ocean Currents

The final major world temperature control is ocean currents, which are often classified as being warm or cold. Warm currents form when water which has absorbed heat in the low latitudes is transported poleward. Cold currents, conversely, have spent an ex-

tended period in the higher latitudes before flowing equatorward.

Ocean currents are capable of exerting a strong effect on air temperatures but are reduced in overall geographical influence by the limited areal extent of well-developed warm and cold currents. This is especially true with respect to land areas. Many areas are far inland, out of reach of the effects of ocean currents. *Leeward coasts*, which have a predominantly land-to-water wind flow, do not often experience the inland penetration of maritime air masses. Those land areas with temperatures most strongly influenced by ocean currents are associated with *windward coasts*, which have a predominantly water-to-land wind flow.

The ocean currents that most strongly influence world temperature patterns flow northward and southward along the east and west margins of the ocean basins in the middle latitudes and subtropics. In general, the east coasts of the continents are paralleled by warm ocean currents, while the west coasts are paralleled by cold currents (see Figures 3.13 and 8.12). The thermal effects of these currents, as might be expected, are most pronounced along the immediate coasts and diminish inland.

The January and July isotherm maps (Figures 3.6 and 3.7) illustrate, in places, the thermal effects of warm and cold currents. For example, the pronounced poleward bulge of isotherms over the North Atlantic on the

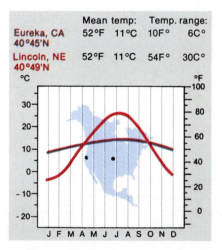

Figure 3.11 Eureka, California, and Lincoln, Nebraska, are at the same latitude and have identical annual mean temperatures. The difference in their continentality, however, causes their annual temperature ranges to differ greatly.

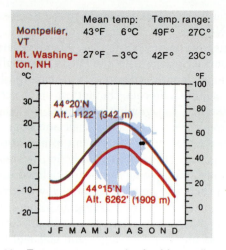

	Mean temp:		Temp. range:	
Montpelier, VT	43°F	6°C	49F°	27C°
Mt. Washington, NH	27°F	−3°C	42F°	23C°

Figure 3.12 Temperature graphs for Montpelier, Vermont, and Mt. Washington, New Hampshire, illustrate the influence of elevation on air temperature.

January map is produced by the warm North Atlantic Drift, a continuation of the Gulf Stream, which flows along the southeast coast of the United States. Mild air from the North Atlantic makes the climate of Europe much more comfortable than would be expected on the basis of latitude alone. The climate of London, for example, is often considered to be rather similar to that of Boston, Massachusetts. London, however, has a latitude of 52° N, which is 10 degrees farther north than Boston. In fact, London's latitude is the same as that of southern Labrador, where snow lies on the ground from September through May. Cold ocean currents show up best on summer isotherm maps. The pronounced equatorward bulge of isotherms off the California coast on the July map, and off the west coasts of South America, South Africa, and Australia on the January map, is quite evident.

Secondary Temperature Controls

In addition to the five primary controls, a number of secondary influences affect the global distribution of temperature. These factors are not dominant enough to be obvious on world temperature maps, but collectively they serve to modify global temperature patterns. The more important of these influences are the seasonally changing distance between the earth and the sun, geographical variations in cloud cover, variations in prevailing wind-flow patterns, topographic influences,

and, increasingly, the effects of human activities, which are discussed in the Case Study at the end of the chapter.

The earth's orbit around the sun is somewhat elliptical (Chapter 1), causing the distance between the earth and the sun to vary by about three million miles (4,800,000 km) during the course of a year. When the earth is at perihelion in January, it intercepts approximately 7 percent more insolation than it does at aphelion in July. Because of its varying distance from the sun, the earth might be expected to be slightly warmer as a whole near the time of perihelion than it is near the time of aphelion. With some time lag, this may be true; the difficulty in assessing the overall significance of this influence is that it is masked by two much more powerful temperature controls: seasonal differences in sun angle and the effect of land and water distribution.

Satellite photographs have shown that, on the average, slightly over half the earth's surface is overlain by clouds at any one time and that some areas are much cloudier than others. Because clouds are excellent reflectors of insolation, it might be expected that the temperature means of these cloudy zones would be substantially lowered. A cloud cover, however, is nearly as effective at keeping terrestrial radiation from escaping the atmosphere as it is at keeping solar radiation from entering. As a consequence, the daily and annual temperature ranges, much more than the means, are lowered by cloudy conditions. For this reason, not only are

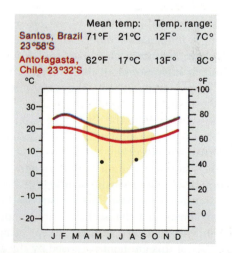

	Mean temp:		Temp. range:	
Santos, Brazil 23°58'S	71°F	21°C	12F°	7C°
Antofagasta, Chile 23°32'S	62°F	17°C	13F°	8C°

Figure 3.13 Ocean current effects on temperature. Santos, Brazil, is on the east coast of South America and is warmed by the Brazil Current. Antofagasta, Chile, on the west coast, is chilled by the Humboldt (Peru) Current.

daytime temperatures held down by a cloud cover, but nighttime temperatures are raised.

Wind currents are another secondary temperature influence. The area from which the wind comes largely determines day-to-day temperatures throughout most of the middle and higher latitudes. Daily variations can be quite striking, especially in the winter half of the year. The United States is frequently invaded by cold air masses from Canada and by warm air masses from Mexico and the Gulf of Mexico. In general, though, the effects of warm and cold air masses largely offset one another when it comes to computing long-term temperature means. This is especially true since the prevailing global wind belts are zonal (west to east, or east to west) in direction.

Many people believe that the speed of the wind, as well as its direction, strongly influences temperature. It is well known that a windy day normally feels much colder than a day when the air is calm. Windchill values, which indicate just how cold it feels due to the combined effect of air temperature and wind, have become popular statistics with radio and television weather broadcasters. In actuality, the wind does not lower the air temperature, but merely blows away the insulating layer of warm air next to our bodies. With sufficiently thick or impermeable clothing, this insulating layer stays intact and people are not chilled by a strong wind. An important influence of wind on air temperatures is to reduce daily temperature ranges by mixing air at different atmospheric levels.

The topography is the spatial distribution of land surface features. As is true of elevation, topography most strongly influences temperatures in mountainous regions, where mountain barriers block or redirect air masses. Just as high mountain ranges restrict the movement of people across them, they also impede the passage of air masses. Especially affected are cold, dense air masses, which resist the lifting necessary to cross mountain barriers. The most effective mountain barriers are those that are exceptionally high and continuous.

Probably the best example of the topographic influence is provided by the Himalayas along the India-Tibet border in south-central Asia. The January isotherm map (Figure 3.6) shows that the isotherms are compressed in this area, indicating a very large or "steep" temperature gradient. Part of this gradient is caused by the ability of the Himalayas to block the southward penetration of cold central Asian air masses into India. Similarly, the west coast of the United States, especially California, is normally spared incursions of cold Canadian air masses

in winter by the blocking effects of the Rockies and the Sierra Nevadas. The mountains of California are also very effective in restricting the inland penetration of cool, damp air masses from the Pacific, especially during the summer months (Figure 3.14), making places like Death Valley, California, the hottest on the continent.

Additional topographic influences on temperature result from the effects of slope angle and slope *aspect*, or compass orientation, on the receipt of insolation. In the northern hemisphere, particularly in the higher latitudes, steep south-facing slopes are inclined toward the sun and therefore have substantially warmer daytime temperatures than slopes facing in other directions, especially northward. Slope aspect can produce marked differences in vegetation types and soil moisture characteristics over small distances and is an important microclimatic influence in mountainous regions. In the southern hemisphere, of course, the situation is reversed; north-facing slopes receive the most insolation and have the warmest temperatures.

Summary

This chapter has discussed both the causes and the geographic distribution of the world temperature pattern. Temperature is the measurement of heat energy, and the heat energy that drives physical processes at the earth's surface comes primarily from the sun, which is a

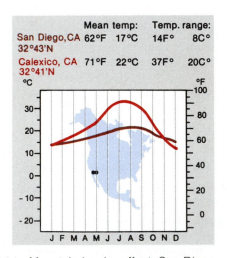

Figure 3.14 Mountain barrier effect. San Diego, California, is situated on the coast and frequently experiences onshore winds from the cool Pacific. Calexico is located about 90 miles (145 km) inland over several low mountain ranges.

giant hydrogen fusion reactor. Most earth processes are ultimately powered by this energy source.

Solar energy is radiated to earth in the form of electromagnetic waves that fall mostly within the visible light and near-infrared portions of the electromagnetic spectrum. Once this energy reaches the earth system, much is converted to longwave terrestrial radiation, which we sense as heat. The heat energy is distributed throughout the earth system by the combined processes of radiation, conduction, and convection before finally being reradiated to space.

A variety of factors is responsible for the global pattern of air temperatures. The most important factors are the altitude of the noon sun, the duration of daylight, the distribution of land and water, surface elevation, and ocean currents. Additional factors of more limited importance are the varying distance between the earth and sun, the amount of cloudiness, wind direction, topographic influences, and the influence of human activities.

Because each of the world temperature controls operates with differing intensities in different regions, the global pattern of temperatures is complex. The sun's altitude, the most dominant of all the controls, causes the overall temperature pattern to be latitudinal in nature, with mean temperatures generally decreasing as latitude increases. The duration of daylight is a seasonally varying factor which tends to cause progressively greater seasonal temperature contrasts with increasing latitude. Much of the thermal influence of the earth's land and water distribution is also seasonal in nature. Areas with continental climates, which are dominated by winds from the land, tend to have large temperature ranges on both a daily and an annual basis. Maritime climate areas, which receive the majority of their air from large water bodies, tend to have small daily and annual temperature ranges. Elevation has the effect of reducing mean temperature values at an average worldwide rate of 3.5 F° per 1000 feet (6.4 C°/1000m), so that high mountain and plateau regions are noted for their cold temperatures. Ocean currents may be either warm or cold, depending on their areas of origin, and correspondingly act to warm or cool regions in their vicinities.

Review Questions

1. Explain what happens to the insolation reaching the earth system. What proportion is absorbed, and where? What happens to the energy when it is absorbed? What proportion is reflected, and from where?

2. What is the greenhouse effect, and how does it influence world temperatures? Explain the origin of the term.

3. What are the five primary world temperature controls? Briefly explain how each is able to influence temperatures.

4. Explain the relationship between mean sun altitudes and the long-term global temperature pattern. What more short-term temperature characteristics are also controlled by the angle of the sun?

5. Does land and water distribution have a greater effect on temperature means or on temperature ranges? Why? What are the basic characteristics of a continental climate and of a maritime climate?

6. Which sides of continents are most commonly paralleled by warm ocean currents? By cold currents? Explain why windward coasts are more strongly influenced by ocean currents than are leeward coasts.

7. Briefly discuss the influence of each of the world temperature controls in your home town or college campus. How does your geographic location influence the effects or importance of each control?

8. Explain how human activities which involve the burning of fossil fuels might have an effect on global temperatures. What other types of human activities may be countering this tendency?

Key Terms

Radiation

Conduction

Convection

Insolation

Greenhouse effect

Continental climate

Maritime climate

Environmental lapse
   rate

CASE STUDY

Will Human Activities Alter the World Temperature Pattern?

Until recently, the direct human impact on global temperatures has been minor. Human-derived sources of energy, primarily from the burning of fossil fuels, produce only about 0.006 percent of the earth's total heat supply. This addition of heat is negligible on a global basis, but, unlike natural heat sources, it is highly concentrated in urban areas, which occupy a very small proportion of the earth's total area (see Chapter 7 for further discussion of urban climates). Of increasing concern, though, is the potential human influence on the global heat balance through the addition of pollutants to the air.

The threat to world temperatures is uncertain largely because some of humankind's activities have the potential to raise world temperatures, while others have the potential to reduce them. The most widespread concern in recent years has been over activities that may bring about climatic warming. Most controversy in this regard has involved the rapid increase in atmospheric carbon dioxide. Between 1850 and the present, atmospheric concentrations of CO_2 have risen from approximately 294 parts per million to 350 parts per million largely as a result of human activities. The rate of increase is accelerating, and it has been estimated that the amount of atmospheric carbon dioxide could double from its present value within the next half century (Figure 3.15). The increase has historically been attributed to the large-scale burning of fossil fuels and the consequent production of large quantities of carbon dioxide gas as a by-product. Recent studies, however, have indicated that the rapid worldwide removal of vegetation, especially of tropical forests, may also be of great importance. The decrease in global vegetation cover is reducing the amount of carbon dioxide withdrawn from the air by plants. In addition, although carbon dioxide is highly soluble in water and is being absorbed in large quantities by the oceans, the rate of oceanic assimilation is not keeping pace with the rate of CO_2 production. As a result, the amount of atmospheric CO_2 is increasing at nearly half the rate at which it is currently being added to the air through the burning of fossil fuels.

Carbon dioxide is the primary gas responsible for the greenhouse effect. It is estimated that doubling the atmosphere's CO_2 content could produce a mean global warming of as much as 5.5 F° (3 C°). A temperature rise of this magnitude would enlarge the arid zones of the subtropics and cause them to expand into the middle latitudes, reducing the total world area suitable for agriculture. In addition, computer models indicate that the warming would likely be most concentrated in the high latitudes, causing a partial melting of the Greenland and Antarctic ice sheets. This would raise sea levels worldwide and flood coastal cities. Moreover, it has even been suggested that a feedback cycle could become established, producing a runaway greenhouse effect. This concern stems from the fact that the capacity of water to hold carbon dioxide in solution decreases as its temperature rises. A major global warming trend, then, could result in the partial release of carbon dioxide already dissolved in the ocean waters. This would produce more warming, resulting in an additional release of oceanic CO_2, and so forth. A total melting of the polar ice sheets could ensue, perhaps over a period of several millennia, raising sea levels approximately 150 feet (45 m) and inundating the most populous ten percent of the world's land area.

The increase in carbon dioxide is enhancing the ability of the atmosphere to absorb terrestrial radiation. At the same time, however, several human activities seem to be working in the opposite direction by raising the reflectivity, or albedo, of the earth system. The activities of greatest concern are urban and agricultural operations that increase the amount of dust in the atmosphere. While much dust comes from natural sources, such as volcanic eruptions, it is estimated that human sources now produce over a fifth of the atmospheric dust total. Dust in sufficient quantities can reduce air temperatures by reflecting

continued on next page

and scattering large amounts of insolation. This has been proven during historical times by the effects of several major volcanic eruptions that have temporarily lowered worldwide temperatures by infusing massive quantities of dust into the stratosphere (see the Case Study at the end of Chapter 12). In addition, dust particles from human sources such as high-altitude aircraft serve as condensation nuclei, resulting in increased cloudiness in some areas.

Following a period of relatively cool temperatures in the early 1980s, a series of record-setting hot summers in the latter half of the decade has raised concerns that the long-heralded greenhouse effect global warming trend may finally be beginning. The rapid melting of glaciers in Alaska and Antarctica has provided further evidence of a recent trend toward increased warmth.

Humanity has a tremendous investment in the present world pattern of climates and coastlines. Virtually all economic development has taken place under the present set of conditions, and a significant alteration in world temperature patterns would doubtless produce a major worldwide calamity. More studies of the potential effects of human activities on global climatic patterns are needed before these effects become only too painfully evident.

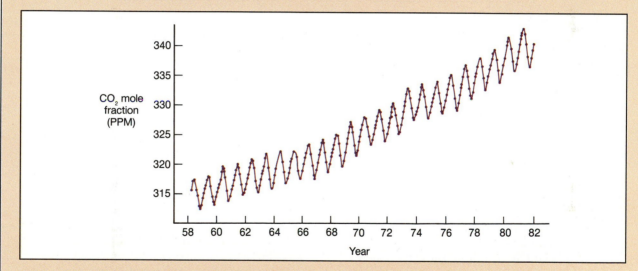

Figure 3.15 The average rise in global concentrations of atmospheric carbon dioxide is well illustrated by this graph of CO_2 levels monitored at the isolated measuring station atop Mauna Loa, Hawaii. CO_2 levels fall each northern hemisphere summer due to vegetative absorption, but then rise each winter to new record highs.

Chapter Four

Air Pressure and Wind

Outline

Focus Questions

1. What causes variations in air pressure, and how do these variations affect the weather?
2. What is the basic pattern of global air pressure belts?
3. What factors control wind speed and direction?
4. What is the basic pattern of global wind belts?
5. What are the jet streams, and how do they influence surface weather conditions?

This chapter examines the causes and global distributions of two additional important weather elements: air pressure and wind. Day-to-day variations in air pressure and wind, like those of temperature, are complex and subject to constant change; but both elements are organized into relatively simple patterns when examined at the global scale. Air pressure and wind are covered in a single chapter because they are more highly interrelated than any other pair of weather elements. They exist in a cause-and-effect relationship, with variations in pressure causing changes in wind direction and speed.

AIR PRESSURE

Air pressure is the force air exerts on its surroundings because of its weight. The amount of pressure exerted on any area at the surface or in the atmosphere is proportional to the weight of the column of air between that area and the outer edge of the atmosphere. Because the atmosphere is gaseous, this pressure is applied equally in all directions. The air pressure at sea level averages 14.7 pounds per square inch (1013.2 millibars). The pressure at any given site, however, fluctuates constantly and rarely equals this value exactly.

Analyzed at the molecular level, air pressure is produced by air molecule collisions. Air molecules at room temperature travel in random directions at speeds of about 1000 miles per hour (450 m/sec) and, near sea level, collide elastically with one another or with surrounding objects billions of times each second. The air pressure is actually determined by the number and force of these molecular collisions.

Differences in air pressure occur both horizontally and vertically. Because of the dominating influence of gravity on the vertical distribution of the atmosphere, however, the vertical pressure gradient, or the vertical variation in air pressure per unit of distance, is typically several thousand times greater than the horizontal pressure gradient.

Measured through time at any one place, air pressure variations are generally very small and can be neither seen nor felt. The practical importance of this weather element results largely from the fact that air pressure variations greatly influence, directly or indirectly, the global distributions of each of the other five weather elements. Air pressure patterns control both the speed and direction of the wind, and the wind, in turn, strongly influences global patterns of temperature and moisture.

Causes of Air Pressure Variations

Why do changes in air pressure occur? The constant variation in atmospheric pressure patterns is largely related to two factors: temperature changes and the rotation of the earth. When the air is heated, the air molecules increase their rate of vibration, resulting in more energetic collisions that push the molecules farther apart. This causes the heated air to expand in all directions, reducing the total mass of air molecules within any given area and causing a reduction in air pressure. Similarly, when air cools, its rate of molecular activity decreases. This allows the air molecules to crowd more closely together, increasing the surface air pressure. In summary, heating causes the air pressure to decrease, while cooling causes it to increase.

Because of the influence of temperature on air pressure, cold surfaces, such as high latitude continental interiors during the winter, tend to develop shallow, thermally induced areas of high air pressure known as *thermal highs*. Similarly, the summertime heating of some land surfaces tends to foster the development of *thermal lows*. These thermal pressure centers greatly influence seasonal weather conditions in large parts of the world.

The second major cause of air pressure changes is variations in air density that result from the deflection of air into or out of certain latitudinal zones because of the earth's rotation. This factor is rather complex and its effects not yet entirely understood, but it is related to the Coriolis effect, discussed later in this chapter. For the present, it is sufficient to note that the earth's rotation causes air to accumulate in certain areas, forming high pressure systems, and deflects air away from other areas, producing low pressure systems.

Global Distribution of Air Pressure

The earth's rotation, coupled with the influence of warmer and colder surface temperatures, produces constantly changing patterns of air pressure in the lower troposphere. The resulting high and low pressure systems, often referred to simply as "highs" and "lows," are each associated with characteristic weather conditions. The high pressure in a "high" is produced by descending air that is compressed as it approaches the earth's surface (see Figure 4.1). The compression of the air makes it warmer and relatively drier. Highs are therefore normally associated with fair weather, and a rising barometer is viewed as a sign of good weather to come. Lows operate in precisely the opposite manner (see Figure 4.2). They are areas of rising air, which expands and cools, thus reducing its ability to hold moisture (see Chapter 5). Lows are therefore normally associated with clouds and precipitation. A falling barometer, by signaling the approach of a low, often indicates that inclement weather is on its way. The alternating passages of highs and lows at intervals of a few days are largely responsible for the changing weather conditions of the middle and high latitudes.

The atmosphere at any one time contains a large number of high and low pressure systems that vary in size, shape, and intensity. On occasion, pressure systems remain nearly stationary over one area for an extended period. More commonly, they actively move, or "build," across the earth's surface at speeds ranging up to approximately 35 miles per hour (55 km/hr). The direction and speed of motion of the pressure systems are controlled largely by wind currents in the upper atmosphere. In general, the pressure systems of the middle and high latitudes are better defined and move more rapidly than those of the low latitudes.

With the passage of time, pressure systems form and disappear, strengthen and weaken, change size and shape, and vary in their speed and direction of motion.

The earth is thus covered by an ever-changing kaleidoscope of pressure patterns—one that contains an unlimited potential for variation and that never exactly repeats itself. The great potential for change as well as our incomplete understanding of the complex forces controlling not only pressure, but all weather systems, makes accurate weather forecasting highly challenging.

Although the precise distribution of pressure systems around the world varies constantly from day to day, the global pattern of air pressure is not entirely random. On a long-term basis, some parts of the world are dominated by high pressure; others are dominated by low pressure. In addition, some regions tend to experience marked seasonal differences in air pressures; others do not. The whole situation is somewhat analogous to the pattern of traffic on a busy highway. Although a limitless potential for variation exists in terms of the number, types, and placement of the vehicles, an underlying logic and order govern the pattern. Likewise, pressure systems, and weather systems in general, tend to follow certain principles that control their development and movement.

Global Surface Pressure Belts

Computing long-term averages of air pressure reveals a belted pattern of high and low pressure systems. These pressure belts encircle the earth in an east-west direc-

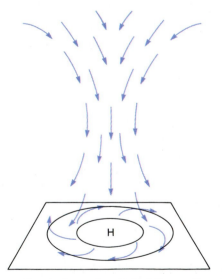

Figure 4.1 Three-dimensional wind-flow pattern in the northern hemisphere associated with a high pressure center.

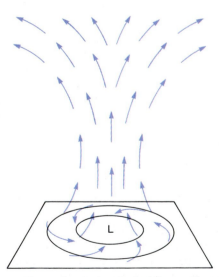

Figure 4.2 Three-dimensional wind-flow pattern in the northern hemisphere associated with a low pressure center.

tion. In a general sense, the arrangement of the world air pressure belts can be considered symmetrical with respect to the equator, with the northern hemisphere pattern being a mirror image of the southern hemisphere pattern. Each hemisphere contains three complete pressure belts, while a fourth straddles the equator and is shared by both hemispheres (see Figure 4.3). There are thus seven belts in all. The belts are in reality far from perfect in shape, especially in the northern hemisphere, and vary considerably from summer to winter in many areas. The reasons for this will be explained shortly.

The most equatorward of the global pressure belts dominates the region between about 5° N and 5° S. It is a broad but rather weak trough of low pressure termed the **Intertropical Convergence Zone (ITCZ).** This low pressure zone is thermally induced by the high sun angles and resulting year-round warm temperatures of the equatorial region and is associated with some of the world's rainiest climates.

Proceeding poleward in either direction, air pressures gradually rise until the centers of the subtropical highs are reached at latitudes ranging between 25° and 40°. Unlike the ITCZ, the two subtropical highs are produced largely by air at high altitudes that is deflected into the subtropics by the earth's rotation and subsequently sinks to the surface. They are therefore dynamically, or rotationally, induced. The subtropical highs vary regionally in the effectiveness with which they suppress precipitation, but in many places they cause arid or semiarid conditions to prevail. As a result, most of the world's deserts are located in the subtropics.

Air pressures typically decline in a poleward direction through the middle latitudes, until the subpolar lows are reached at latitudes of approximately 55° to 70°. These lower pressures are generated both by the earth's rotation, which deflects air away from this latitudinal zone, and by the lifting of warm air from the lower latitudes as it encounters denser, cold air from the polar regions.

The most poleward, and typically the most weakly developed of the four global pressure belts, are the polar highs. In actuality, they are not belts at all, but ideally form roughly circular caps over the polar regions. They are induced by the cold temperatures that prevail in this energy deficient region and are quite shallow. The South Polar high is the much better developed of the two, because Antarctica is considerably colder than the Arctic Ocean location of the North Polar high.

It should be stressed that the world pressure belts merely represent average pressure conditions. They

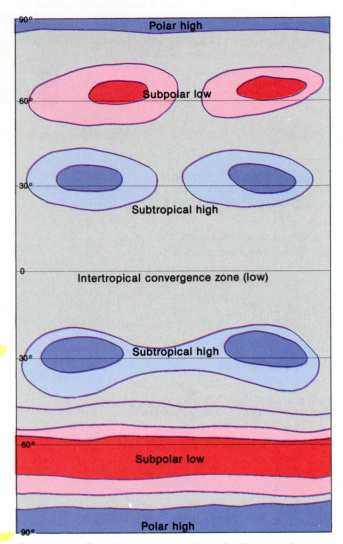

Figure 4.3 The idealized latitudinal distribution of global air pressure belts.

display considerable daily and seasonal variations in their location and development. In addition, numerous traveling highs and lows exist independently of the pressure belts. These pressure systems are especially common in the middle latitudes, where they largely control daily weather conditions. Their characteristics and influences on the weather will be explored more fully later in this chapter and in Chapter 6.

Seasonal Pressure Variations

The entire pattern of world pressure belts experiences a seasonal shift in latitude that mirrors, on a reduced

Air Pressure Measurement

The most frequently measured weather element, other than temperature, is air pressure; the instrument used to measure it is the *barometer*. Although several different types of barometers are in use, they may be grouped into two basic categories: mercurial barometers and aneroid barometers.

The mercurial barometer was invented by the Italian physicist Torricelli in 1643. Its operating principle is quite simple; changes in air pressure control the height of a mercury column in a graduated glass tube from which the air has been removed (Figure 4.4). A second type, the aneroid barometer, is currently much more popular and is found in many homes. It generally takes the form of a dial-type instrument that is compact, sturdy, adjustable, and accurate. The heart of an aneroid barometer is the sylphon cell: a collapsible, accordion-shaped box from which the air has been partially evacuated (see Figure 4.5). An increase in air pressure causes the sylphon cell to be compressed, while an air pressure decrease allows it to expand. It is connected through a series of chains, levers, and springs to a pointer on a dial.

The two standard scales used for barometric pressure readings in the United States are the inches of mercury and the millibar scales. The older inches of mercury scale simply measures the height to which the mercury column in a mercurial barometer will rise. Standard sea level air pressure of 14.7 lbs/sq in is equivalent to 29.92 inches of mercury. Many aneroid barometers as well as mercurial barometers use this scale. The second scale uses metric units of pressure termed millibars. Millibars are currently employed by meteorologists throughout the world. Standard sea level air pressure in millibars is 1013.2 mb.

On most maps (see Figures 4.6 and 4.7), air pressure patterns are depicted by the use of isobars. These are lines connecting points with the same air pressure; typically they are drawn at four-millibar intervals.

continued on next page

scale, the annual latitudinal shift of the sun's vertical rays. The extent that the belts shift either north or south from their mean positions totals only a few degrees over mid-ocean areas but can amount to 10 degrees or more over large continents.

The shift is thermally generated by the seasonal movement of surface temperature patterns with the sun. The uneven distribution of land and water bodies produces marked seasonal differences in temperatures over areas at similar latitudes. These temperature differences strongly influence air pressure patterns, resulting in the formation of thermal highs and lows. Because land areas heat and cool more readily than water bodies, land areas typically experience greater seasonal pressure fluctuations. Specifically, land areas are usually hotter than water bodies during the summer months and therefore tend to have lower surface pressures; during the winter months these areas are colder and have higher pressures. Seasonal heating and cooling variations are the dominant factor responsible for disrupting the orderly pattern of global pressure belts. Because the southern hemisphere has relatively little land, it heats and cools much more evenly than does the northern hemisphere. As a result, the air pressure belts of the southern hemisphere are better defined than those of the northern hemisphere. The January and July world

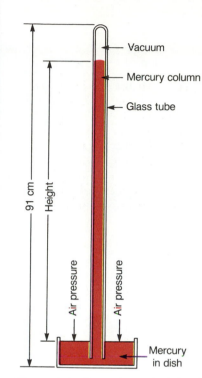

Vacuum

Mercury column

Glass tube

91 cm

Height

Air pressure

Air pressure

Mercury
in dish

Figure 4.4 Operating principle of the mercurial barometer. Air pressure exerted on the mercury in the bowl is transmitted to the mercury inside the glass tube. The mercury in the tube rises to the level at which its weight exactly counterbalances the air pressure.

Figure 4.5 The barograph pictured here is a recording aneroid barometer. The disk-shaped sylphon cell can be seen inside (*Courtesy of Qualimetrics, Inc.; Sacramento, CA*)

pressure maps (Figures 4.6 and 4.7) portray the global pressure patterns during the midsummer and midwinter months, when the temperature contrasts between land and water are most extreme.

WIND

Wind can be described simply as air in motion. This motion may be in any direction, but in most instances the horizontal component of wind flow far surpasses the vertical. The wind is tremendously important as a global distributor of both moisture and thermal energy. It carries water evaporated from the oceans onto the land, where much of it falls as precipitation. The wind also transports tremendous quantities of tropical heat to the higher latitudes, thereby moderating the temperatures of both latitudinal zones. The uneven solar heating of the earth system is largely responsible for generating the air pressure imbalances that cause the winds. The relationship between air temperature and wind is therefore complex and reciprocal: spatial differences in temperatures cause winds to blow, and the resulting winds transport heat and cold over the earth's surface, thereby influencing world temperatures.

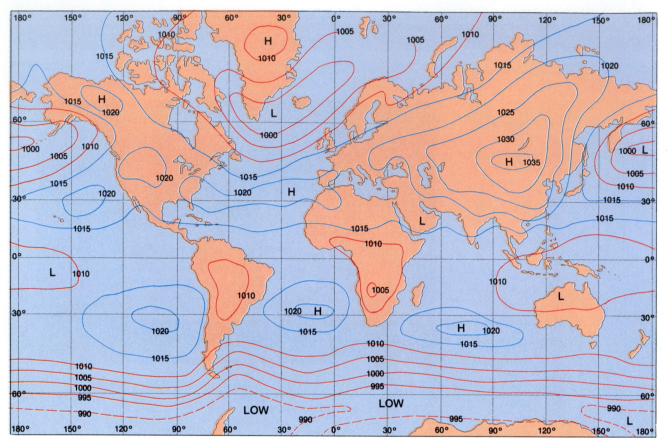

Figure 4.6 January global mean sea level air pressure values in millibars.

Factors Controlling Wind Speed and Direction

In a sense, the wind can be considered two weather elements in one because it has two distinct components: speed and direction. These components are not only measured and stated separately, but are also produced by differing combinations of controlling forces.

Wind Speed Controls

The wind blows because of differences in air pressure. It always blows from high to low pressure, and, if other factors remain constant, its speed is determined by the rate of air pressure change per unit of distance between the pressure systems. The situation is analogous to the flow of water down a hill. Just as water flows from higher to lower elevations at a speed that increases with the steepness of the slope, air flows from areas of higher to lower atmospheric pressure at a speed that increases with the steepness of the pressure slope or pressure gradient. On an air pressure map, the steepness of the pressure gradient is indicated by the spacing of *isobars,* or lines of equal air pressure. In areas where the isobars are closely spaced, the pressure gradient is steep, and strong winds can be expected. In areas of widely spaced isobars, the pressure gradient is gentle, and light winds typically exist (see Figure 4.8).

A second force that influences wind speed and that always partially offsets the pressure gradient force is friction. Some molecular friction occurs within an air stream, but friction between the atmosphere and the earth's surface is more significant. The amount of friction generated by the surface is a function of both windspeed and surface roughness. With increasing wind speeds, friction between air and the nonmoving surface increases rapidly. For this reason, a steepening of the pressure gradient will produce a less-than-proportional increase in the speed of the wind.

Air Pressure and Wind

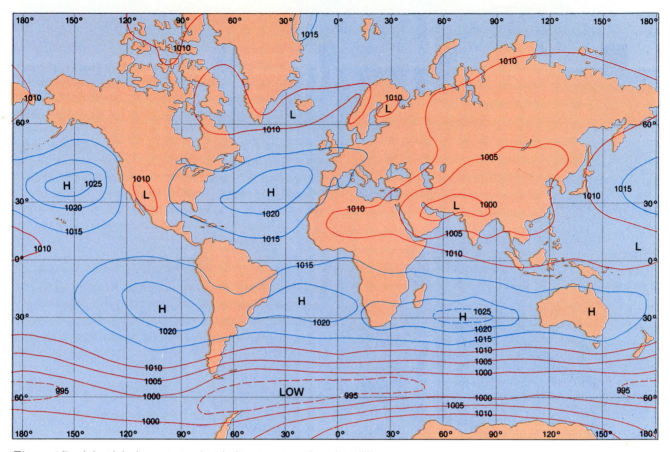

Figure 4.7   July global mean sea level air pressure values in millibars.

The frictional resistance provided by the surface is determined largely by its roughness. Land areas vary greatly in this respect. At one extreme, a smooth, ice-covered surface, such as the interior of Antarctica, offers very little frictional resistance to the passage of air; and winds there frequently attain very high velocities. Conversely, a forested, mountainous surface generates a high degree of friction, and wind speeds are correspondingly reduced. The generally smoother surfaces of water bodies cause average wind speeds to be greater over water than over land. This is often quite evident on the seashore or on the shore of a large lake. The highest average wind speeds occur in the upper atmosphere, above the frictional influence of the surface.

Wind Direction Controls

The direction of the wind is determined by three factors: the orientation of the pressure gradient, the Coriolis effect produced by the earth's rotation, and the frictional resistance to the flow of the wind.

The primary factor controlling wind direction is the orientation of the pressure gradient. Air not only flows from high to low pressure but tends to do so along the most direct route down the steepest pressure gradient. Again, an analogy with the flow of a stream can be drawn. Just as a stream tends to flow down the steepest slope from high to low elevation, wind tends to flow from high to low pressure down the steepest pressure gradient.

If the pressure gradient were the only factor controlling wind direction, the wind at any given location would blow perpendicularly across the isobars, just as a stream tends to flow perpendicularly across contour lines (see Figure 4.8). In the time it takes the wind to flow from a high pressure system to a low, however, both pressure systems are moving in a circular path because of the rotation of the earth. If the wind is to continue to flow down the steepest pressure gradient, it

59

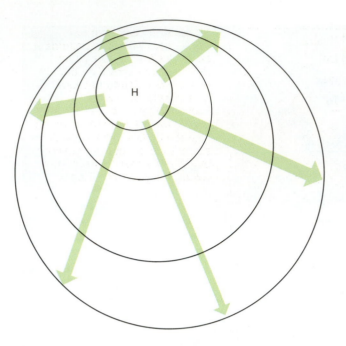

Figure 4.8 The more closely the isobars in an area are spaced, the steeper the pressure gradient and the greater the speed of the wind. In this diagram, wind speed corresponds to arrow thickness. Wind directions are shown as they would be with no Coriolis deflection.

The third influence on wind direction is friction. Friction interferes with the Coriolis effect, causing it to deflect the wind less than a full 90°. With increasing friction, the angle of deflection is progressively reduced. Friction reduces the deflection of the wind because, when wind speed is reduced, its momentum in that direction is also diminished, allowing it to turn more easily in the new direction of the pressure gradient orientation. A similar reduction in momentum, for example, makes it much easier for an automobile to navigate a sharp turn when it is traveling slowly than when it is speeding. Because most land surfaces provide more frictional resistance than water surfaces do, there is less deflection of the wind over land areas than over water. Deflection angles average about 60° from the orientation of the steepest pressure gradient over land areas in general, but can be reduced to as little as 45° over extremely rough terrain. Over water, deflection angles range between 70° and 75°. In the upper atmosphere, where friction is minimal, the wind is deflected by nearly 90° and flows nearly parallel to the isobars.

must gradually turn in a similar arc. Its angular momentum, though, keeps it from fully doing so. As a result, to an observer on the earth, it appears that the wind starts by blowing directly down the pressure gradient but gradually changes its direction to a path that increasingly parallels the pressure gradient (see Figure 4.9). In reality, however, the underlying earth, not the wind, is turning; the Coriolis effect, as this gradual deflection of the wind is called, results from the use of the earth's surface as a "stationary" reference for the determination of wind direction.

The counterclockwise rotation of the northern hemisphere, as observed from above the North Pole, deflects the wind to the right of its original path, regardless of its direction of flow. Conversely, the clockwise rotation of the southern hemisphere deflects all winds to the left. The ultimate tendency of the Coriolis effect is to deflect the wind by 90°, so that it flows parallel to the isobars (see Figure 4.10). The Coriolis effect at the equator is zero, so no wind deflection exists there. The Coriolis effect increases poleward, but even in the high latitudes, where it is strongest, the deflection is gradual. Local winds are therefore not significantly affected.

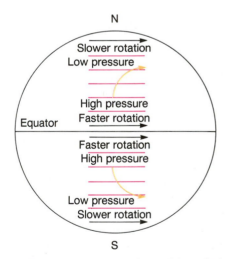

Figure 4.9 Coriolis effect deflection of the wind caused by the earth's rotation. Deflection occurs largely because the earth's rotational speed differs in differing latitudes. In the diagram, the wind currents in both hemispheres maintain some of their initial faster rotational momentum as they flow poleward to areas of slower rotation; they therefore end up "ahead" of their original targets. Conversely, winds flowing equatorward will end up "behind" their targets.

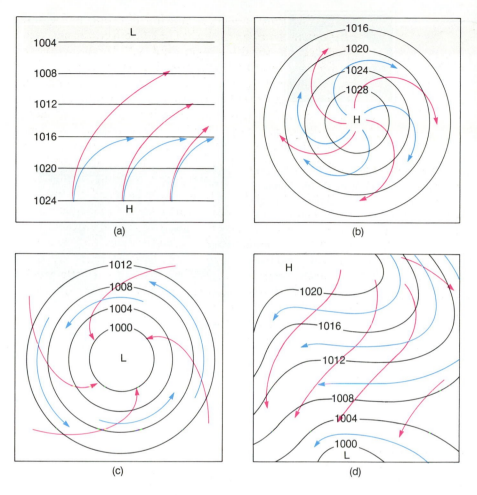

Figure 4.10 Northern hemisphere wind-flow patterns after a full 90° Coriolis effect deflection are shown by the blue arrows. Wind directions here parallel the isobars. Frictional resistance prevents a full 90° deflection from occurring near the earth's surface, so the actual wind directions are shown by the red arrows. Air pressure values are given in millibars.

The procedure for determining wind directions for both land and water surfaces is illustrated in Figure 4.14.

General Circulation of the Atmosphere

Because differences in air pressure cause the wind to blow, the global pattern of wind currents is closely related to the world air pressure pattern. Therefore, keeping the air pressure pattern (Figure 4.3) in mind will greatly facilitate the understanding of the major wind currents.

Two factors are primarily responsible for maintaining the pressure imbalances that produce the global wind systems: the uneven latitudinal heating of the surface and the Coriolis effect resulting from the earth's rotation. These two factors combine to produce a beltlike arrangement of global wind systems. In detail, however, the global wind pattern is made much more complex because of additional factors such as land/ water heating differences, the development and movement of weather fronts and storms, daily and seasonal variations in heating and cooling, and topographic influences. These factors, when coupled with sun angle and rotational influences, create a complex and ever-changing pattern of winds over the face of the earth.

The remaining portion of this chapter explores the basic characteristics of the existing wind-flow pattern of the troposphere. The discussion centers on four subjects: the global surface wind belts, winds associated with traveling anticyclones and cyclones, local and regional winds, and upper-level winds.

Global Surface Wind Belts

The global surface wind belts are the largest units of the earth's surface wind system. They are closely associated with the world pressure belts, with which they have

Wind Measurement

The wind varies in both direction and speed. Therefore, two different systems of measurement as well as two types of meteorological instruments must be employed to provide a full description of this important weather element. A crucial, and sometimes misunderstood, aspect of wind description is that its direction is always stated as the *direction from which the wind comes.* Hence, a west wind is traveling eastward, and a southeasterly wind is blowing toward a northwesterly direction. This practice sometimes confuses people unfamiliar with wind measurement because we tend, when traveling, to give directions in terms of where we are going, rather than in terms of where we have been. A very good reason, however, exists for emphasizing the wind's direction of origin. Air's temperature and moisture characteristics are determined by its previous location, not by its future destination. Thus a north wind tends to bring colder-than-normal temperatures to the middle latitudes of the northern hemisphere, while a southerly breeze is typically associated with warm weather.

The weather instrument commonly used to measure wind directions is the well-known *wind vane,* sometimes imprecisely termed a weather vane. This simple instrument consists of a flattened metal design that is mounted on a shaft about which it can rotate freely to point to the direction from which the wind is blowing (see Figure 4.11). The instrument most commonly used to measure wind

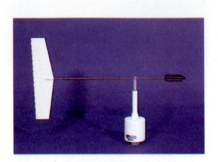

Figure 4.11 A wind vane is designed to point in the direction from which the wind is blowing. (*Courtesy of Qualimetrics, Inc.; Sacramento, CA*)

continued on next page

many features in common. Altogether, seven east-west-oriented wind belts are located in each hemisphere. They are zones of prevailing wind conditions, but wind patterns on any given day may differ considerably from the norm.

The two hemispheres, ideally speaking, contain a mirror-image reversal of the wind belt pattern, just as they do with the pressure belt pattern. As is true of the pressure belts, the wind belts of the southern hemisphere are better defined than those of the northern hemisphere. A seasonal shift in latitude of the wind belts also occurs along with, and because of, the thermally induced seasonal shift of the pressure belts.

Our examination of the wind belts begins at the equator and progresses poleward. Referral to Figure 4.15 will help clarify the positional relationships of the various belts.

Centered near the equator, and extending to about 5° N and S, is the *equatorial belt of variable winds and calms.* This is a weak wind belt associated with the low pressure center of the ITCZ. Winds here are normally light and variable in direction, with variations depending on small changes in pressure patterns from day to day. The calm central portion of this wind belt is sometimes referred to as the *doldrums.* Locally strong, gusty winds, however, are often associated with rain squalls and thunderstorms in this unsettled area.

Just poleward of the equatorial belt of variable winds and calms, and surrounding it to both the north and south, are the broad expanses of the **trade winds,**

speeds is the *anemometer*. It normally consists of three or four hemispherical cups mounted on a metal shaft so that they are free to rotate with the wind (see Figure 4.12).

Upper-level measurements are also of vital importance, not only to aircraft, but also to weather forecasters. Systematic observations are provided by the release of large hydrogen- or helium-filled balloons with a known rate of ascent. As they rise, these balloons are also carried downwind, so that the horizontal component of their motion indicates the wind directions and speeds for the various levels through which they ascend (see Figure 4.13). Instrument packages called *radiosondes* may be carried aloft by these balloons. They transmit temperature, humidity, and air pressure readings as they ascend; and their radio signals enable them to be tracked in order to determine wind direction and speed.

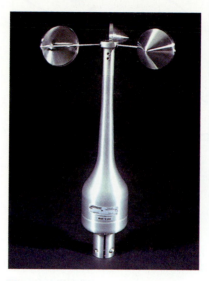

Figure 4.12 An anemometer. As the wind blows, air is trapped within the cups, causing them to spin at a speed proportional to the wind speed. (*Courtesy of Qualimetrics, Inc.; Sacramento, CA*)

Figure 4.13 The launching of a weather balloon carrying a radiosonde. As it ascends, the radiosonde automatically measures and transmits back information on temperature, air pressure, and humidity. (*Courtesy of NOAA*)

which dominate the zones between 5° and 25° N and S. The trade winds, so-named because they aided European sailing ships on trading voyages from Europe to the West Indies during the colonial period, are powered by the pressure gradient between the subtropical highs of each hemisphere and the low pressure of the ITCZ. The trade winds are noted for both their persistence and their steadiness of direction. In the northern hemisphere, the winds blow from a northeast or east-northeast direction, producing the *northeast trades*. In the southern hemisphere, the winds blow from the southeast or east-southeast to produce the *southeast trades*.

Within the centers of the subtropical highs, the winds again become light and variable due to the absence of a significant pressure gradient. The *subtropical belts of variable winds and calms*, as they are called, are from a meteorological standpoint, the most tranquil latitudinal zones on earth. They are characterized by clear skies, calm or light winds, dry air, and warm temperatures. They cover at most five degrees of latitude and are usually located between approximately 30° and 35° N and S. Their seasonal shift in latitude, however, typically carries them several degrees poleward in summer and equatorward in winter. Legend has it that, on occasion, the crews of ships bound for the New World with cargoes of horses exhausted their drinking water supplies and were forced to throw the horses overboard when the ships were becalmed for extended periods in the subtropical highs. As a result,

Figure 4.14 Illustration of the method for estimating surface wind directions from the isobar pattern on a weather map. The left map represents an ocean area in the northern hemisphere. In order to determine the approximate wind directions at Points A and B, first draw arrows (shown here as dashed lines) perpendicularly across the isobars from high to low pressure. Then rotate the arrows 70° in a *clockwise* direction to approximate the actual wind directions. On the right-hand map, representing a land area in the southern hemisphere, draw arrows directly across the isobars as before, then rotate them 60° *counterclockwise* to approximate the wind directions.

(a) Ocean area, Northern Hemisphere

(b) Land area, Southern Hemisphere

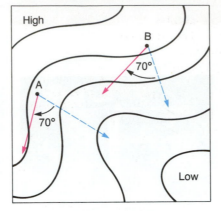

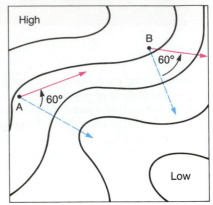

mariners have named this wind belt the "horse latitudes."

Occupying the zone between the subtropical highs and subpolar lows of each hemisphere are the ==wester-lies,== the major wind belts of the middle latitudes. They dominate the zones from 35° to 60° N and S and are second only to the trade winds in total areal extent. Although the westerlies are well-developed wind systems, they are neither as steady nor as persistent as the trade winds because the middle latitudes are much more prone to the development of traveling high and low pressure systems than are the tropics. Each such pressure system, while being carried eastward by the flow of the westerlies, maintains its wind circulation in much the same fashion that an eddy of water in a river maintains a circular flow pattern while being borne downstream. Although the constantly changing positions of these traveling pressure systems (discussed in the next section) can cause the wind to blow from any point of the compass, westerly winds are predominant.

Within the latitudinal zones dominated by the subpolar lows, centered at about 60° to 65° N and S, winds again become variable in direction. In this region, sometimes termed the *polar front zone,* prolonged periods of light winds are uncommon. In fact, the subpolar low pressure zone tends to contain well-developed traveling storm centers. The daily positions and intensities of these storms largely determine both the speed and direction of the winds. The storm centers and their associated winds are especially intense over ocean areas in the winter because the strongest latitudinal temperature contrasts exist at that time of year.

Proceeding farther poleward, a poorly developed wind belt termed the ==polar easterlies== is encountered. Dominating the latitudinal zones between 65° and 80° N and S, it exists because of the pressure gradient between the polar highs and subpolar lows. Winds of the polar easterlies theoretically come from a direction slightly poleward of due east, but the constantly changing positions and intensities of high and low pressure centers in these zones produce considerable short-term variation in wind direction and speed.

A final pair of wind systems exists within the centers of the two polar highs. These systems, which can be termed the *polar zones of variable winds and calms,* are not beltlike in shape but ideally form roughly circular caps over both polar regions in the vicinity of 80° to 90° N and S. In reality, they are poorly developed because the position and strength of the polar highs vary and also because the ice-covered, low-friction polar surfaces promote strong winds.

Traveling High and Low Pressure Systems

The wind-flow patterns of the middle and high latitudes of both hemispheres are modified by the development and general eastward movement of traveling high and

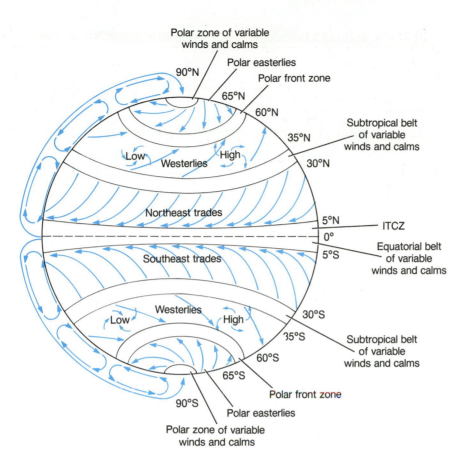

Figure 4.15 The generalized pattern of global wind belts. Air is exchanged between the surface and the upper atmosphere in the vicinity of the pressure belts.

low pressure systems that exist independently of the global pressure belts. Low pressure systems are also termed *cyclones* because they are characterized by a cyclonic pattern of converging and rising air (see Figure 4.2). High pressure systems, conversely, are termed *anticyclones* because of their reverse pattern of diverging and descending air (see Figure 4.1). The Coriolis effect deflection influences the wind fields associated with these pressure systems, giving clockwise or counterclockwise twists to their circulations, as shown in Figure 4.16. The development and characteristics of these weather systems, which play a vital role in our day-to-day weather conditions, are discussed further in Chapter 6.

Local and Regional Winds

Further contributing to the overall complexity of the earth's total wind pattern are various local and regional winds. Most of these winds are directly or indirectly produced by air pressure differences resulting from the differential heating or cooling of the earth's surface. Surface temperature differences, in turn, usually result from variations in surface materials, elevation, or slope directional orientation. This section briefly examines three of the most important of these winds.

Land and sea breezes are familiar to residents of coastal areas in many parts of the world. They are produced by daily differences in heating and cooling over adjacent land and water areas (see Figure 4.17). During the day, the land warms rapidly, heating the overlying air and causing it to expand and rise. The resulting reduction in surface air pressure causes somewhat cooler and denser air, called a *sea breeze*, to flow in from the adjacent water. (The same process also occurs along the shores of large lakes such as the Great Lakes, where the breeze is known as a "lake breeze.") During a sunny afternoon, a sea breeze may penetrate inland as far as 10 to 15 miles (16–24 km) before losing its identity. At night a reverse process occurs, with the more quickly cooling land fostering the formation of an

Figure 4.16 Wind-flow patterns for high and low pressure systems in the northern and southern hemispheres.

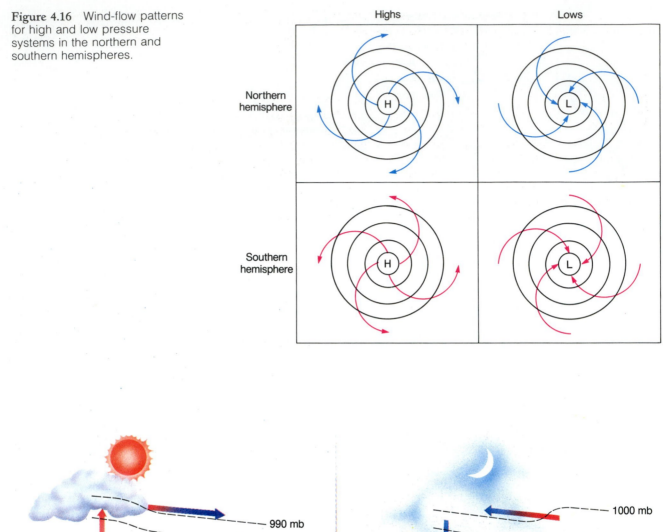

Highs
Lows

Northern hemisphere

Southern hemisphere

H

L

H

L

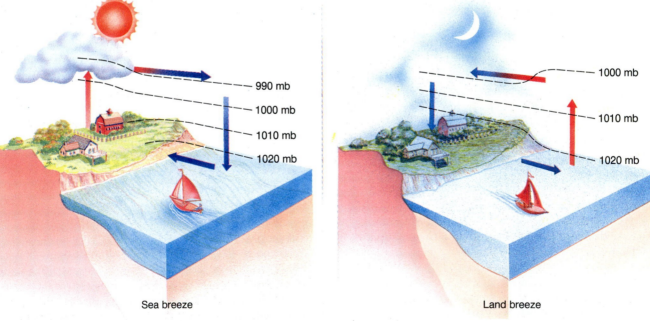

990 mb
1000 mb
1010 mb
1020 mb

1000 mb
1010 mb
1020 mb

Sea breeze

Land breeze

Figure 4.17 Land and sea breezes are produced by air pressure differences resulting from land/water temperature variations. During the day, air heated over the land expands and rises, producing an onshore sea breeze. At night, the cooler and denser air from the land spreads over the adjacent water, producing a land breeze.

overlying mass of relatively cool, dense air. By the late night hours, a *land breeze* may develop as the cooler and denser air from the land spreads over the coastal waters and displaces the lighter, warmer oceanic air. Land and sea breezes develop best when skies are clear, midday sun angles are high, and the regional pressure gradient is weak. Consequently, they occur most frequently in summer, especially in the lower latitudes, where they are a daily phenomenon in many coastal areas.

A group of mountain-related winds are known by such names as *Chinook* in the Rocky Mountains and *Foehn* in the Alps. They are not produced in the mountain areas they influence; rather, they result from the regional pressure gradient. However, these winds are modified by their passage over the mountains and subsequent descent into the adjacent lowlands. Typically, mild, moist air on the windward side of the mountains is carried over the mountain crests, where uplift causes precipitation and warming by the released heat of condensation (see Chapter 5). Upon descending the leeward flanks of the mountains, the air is further heated and dried by compression so that it enters the lowlands as a relatively warm, dry wind.

The most important world regional winds not associated with the semipermanent global wind belts are the monsoons. A *monsoon* is a seasonally reversing wind system produced by the differing thermal characteristics of continents and oceans. In terms of formation, it is the large-scale equivalent of a land and sea breeze system, where summer and winter are substituted for day and night (Figure 4.18). During the summer, land areas that experience a monsoon are strongly heated by the sun. This causes the overlying air to expand, producing a thermal low pressure area. The existence of higher pressure over the adjacent ocean induces an onshore flow of cooler, denser air that produces the summer monsoon. In winter a reverse process occurs as the land and air overlying it become cooler than the adjacent ocean. This results in the formation of a thermal high over the land, producing the seaward flow of dry air that characterizes the winter monsoon. The monsoon circulation tends to be most strongly developed over large land masses in the lower latitudes, which seasonally heat and cool to greater extremes than do small land masses. In addition, large land masses develop larger and stronger thermal lows and highs as a result of their seasonal heating and cooling. It is the size and strength of these thermal pressure centers that produce the pressure gradients that set the monsoonal winds in motion. Asia, by far the largest continent, has the

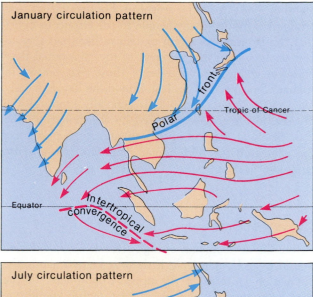

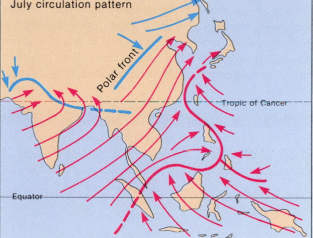

Figure 4.18 The summer and winter monsoon wind-flow pattern of Southeast Asia.

best-developed monsoon system. Monsoonal circulations are also well developed in northern Australia and in sub-Saharan North Africa. Most other large land masses, including North America, have weak monsoonal tendencies that alter somewhat their predominant wind directions between summer and winter.

The importance of the monsoon results primarily from its influence on the moisture content of the air. In areas where it is well developed, the summer monsoon causes the inland transport of vast quantities of moisture-laden oceanic air. The lifting of this air due to heating over land and the possible influence of mountain barriers generates clouds and often copious precipitation. The resulting association of the summer monsoon with the rainy season in Southeast Asia and other areas has led many people to the erroneous belief that the monsoon is a rain system rather than a wind system. In winter, the reversal of the monsoonal flow produces conditions opposite of those that prevail in summer. Skies tend to be clear, humidities low, and daily temperature ranges large. This generally produces the dry season for areas dominated by the monsoon. The Case Study at the end of this chapter further explores the characteristics and human impact of the Asiatic monsoon.

Upper-Level Winds

Before the development of high-altitude aircraft during World War II, relatively little was known about the circulation patterns of the upper troposphere. In the succeeding decades, though, a great deal of knowledge has been gained about the upper-level winds, and it has become apparent that they exert a controlling influence on the development and movement of surface weather systems.

GENERAL CIRCULATION   The air pressure and wind patterns of the upper troposphere are considerably simpler than those of the lower troposphere because of the absence of an adjacent surface. Because no land and water heating differences are present and because local variations in wind direction and speed caused by frictional and blocking effects are absent, wind in the upper troposphere normally flows as a smooth, steady current. In the absence of significant friction, moreover, the Coriolis effect is able to exert its full influence in deflecting the winds so that they blow at right angles to the pressure gradient. This causes the upper-level winds to be much more zonal than the near-surface winds. It should be stressed, however, that the upper-level winds are closely associated with the low-level wind systems and that they also are powered largely by the absorption of surface insolation. Air is constantly exchanged between the upper and lower levels by the rising air columns of low pressure systems and the descending air columns of highs.

THE JET STREAMS   The fastest-flowing upper level winds, and the systems with the greatest impact on surface weather conditions, are the **jet streams.** They consist of relatively narrow bands of very strong winds centered in the upper troposphere at altitudes of approximately 6 to 9 miles (9–13 km). When viewed three-dimensionally, a jet stream is tubular in shape, with wind speeds increasing toward the center from all directions (see Figure 4.19). Maximum wind speeds vary considerably along the path of a jet stream but usually range between 50 and 150 miles per hour

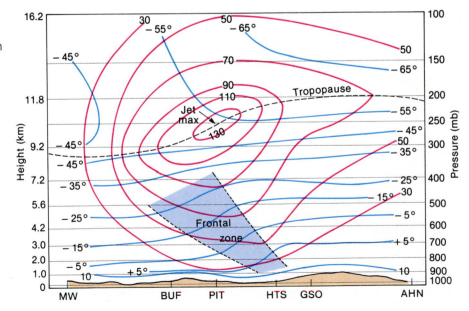

Figure 4.19   Vertical cross-section through the atmosphere along a transect from Ottawa, Ontario (MW) to Athens, Georgia (AHN) at 7 A.M. on October 16, 1973. The red lines are lines of equal wind speed, and the blue lines are lines of equal temperature (isotherms). Wind speeds are given in knots, and temperatures in degrees Celsius. The jet stream is centered over Pittsburgh at a height of about 10.5 kilometers.

(40–130 knots or 20–70 m/sec). The winds of the jet streams generally flow from west to east in both hemispheres but, like the upper-level westerlies, tend to follow a meandering track (see Figure 4.20). The entire undulating pattern of a jet stream typically shifts slowly eastward at speeds that average about 15 to 25 miles per hour (7–10 m/sec). As this shift occurs, existing waves change their amplitude, new waves may form, and old ones disappear. A jet stream thus exhibits a complex, dynamic flow pattern.

At most times, two separate jet streams are located in each hemisphere. These are termed the subpolar jet and the subtropical jet. The *subpolar jet* is thermally produced by the temperature contrasts between the high and low latitudes, and its position tends to coincide closely with the surface location of the polar front. It is strongest during the winter period because the temperature contrast between high and low latitude air masses is greater in winter than in summer. The jet stream also shifts equatorward during the winter because the frontal boundary between warm and cold air masses is located at a lower latitude. In North America, for example, the subpolar jet in midwinter typically extends over the southern United States; by midsummer it is usually positioned over southern or central Canada.

The *subtropical jets* are typically centered above the surface positions of the subtropical highs at a somewhat higher altitude than the subpolar jets. Their origin is less certain than that of the subpolar jets. Although they may be thermally generated to some extent, they are also apparently produced in part by the earth's rotation.

The practical significance of the jet streams is primarily associated with their influence on weather conditions at the earth's surface. They act as upper-level steering currents or "highways" that largely control both the speed and direction of movement of the migratory highs, lows, and fronts that influence our surface weather. These surface weather systems extend far enough into the atmosphere to be carried along by the jet streams. As a result, weather systems located beneath a jet stream tend to travel along its path, and their forward speed is determined by the jet stream's strength. The speed of movement of surface weather systems is normally between one-half and one-third the maximum speed of the jet stream because of the restraining influence of surface friction.

The subpolar jet is particularly important to the weather of the middle latitudes because it is located above the surface boundary between warm and cold air. An equatorward loop in this jet, for example, will draw a mass of air from the higher latitudes unusually far equatorward, producing an outbreak of cold weather. A poleward loop, conversely, will result in the poleward flow of a mass of warm air. In addition, the loops of the subpolar jet influence the locations of storm development. Because of the subpolar jet's critical importance

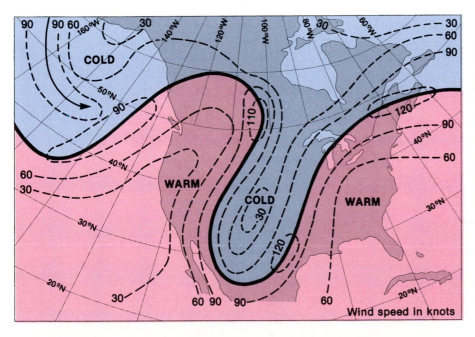

Figure 4.20 A deep trough in the subpolar jet stream associated with a cold air outbreak over the central United States appears on this map of February 22, 1975, showing wind-flow patterns at the 300 millibar pressure level. Dashed lines indicate wind speeds in knots. The jet stream maximum (thick line) separates warm surface areas (shaded pink) from cold surface areas (shaded blue).

to surface weather conditions, information on the jet streams is regularly gathered and mapped, and a great deal of research has been conducted in an effort to better understand the jet streams and predict their behavior.

Summary

This chapter has examined the causes and global distributions of two weather elements—air pressure and wind—that are perhaps most important because of their influence on other aspects of the weather. This is especially true of air pressure, which is the only weather element to undergo changes that can be neither seen nor felt. From a practical standpoint, however, air pressure patterns are of great significance because of their effects on wind direction and speed. The wind, in turn, is important both because of its direct effects on the surface and because it transports energy and moisture. Daily changes in temperature and the potential for precipitation are strongly influenced by short-term changes in wind patterns. On a global scale, regions of high and low precipitation are associated, respectively, with zones of rising and descending air.

Air pressure variations occur largely because of global heating differences and the earth's rotation. Heating causes the air to expand and the air pressure to fall, while cooling causes the air to contract and the pressure to rise. The earth's rotation produces the Coriolis effect. This causes the air to be deflected into some latitudinal zones, producing high pressure, and to be deflected away from other latitudinal zones, producing low pressure.

In general terms, a symmetrical pattern of global pressure belts exists. The Intertropical Convergence Zone (ITCZ), a zone of low pressure produced primarily by high temperatures, straddles the equator. Surface air pressures rise poleward until the centers of the subtropical highs are reached at about 30° to 35° N and S. Mean air pressure values decline poleward to about 60° to 65° N and S, where the centers of the subpolar lows are found. Finally, polar highs are located in the vicinity of the North and South Poles.

Wind is produced by differences in air pressure. Air tends to flow from high to low pressure at speeds proportional to the steepness of the pressure gradient between the two pressure systems. The wind speed is reduced, however, by frictional resistance.

Wind direction is controlled by three factors: the orientation of the pressure gradient, the rotation of the earth, and friction. The net result of these factors is that surface winds in the northern hemisphere are deflected clockwise in their flow from high to low pressure at angles varying from about 45° to 80° from the orientation of the steepest pressure gradient. In the southern hemisphere, a similar counterclockwise deflection occurs.

The pattern of global wind belts closely resembles that of the global pressure belts. Each pressure belt is associated with a central zone of variable winds. The variability results from a changeable, and often weak, pressure gradient. The broad zones of relatively steep pressure gradients between the pressure belts are occupied by belts of stronger and steadier winds. The trade winds occupy the latitudinal zone between the ITCZ and subtropical highs; the westerlies occupy the zone between the subtropical highs and the subpolar lows; and the polar easterlies occupy the zone between the subpolar lows and the polar highs.

The most important upper level winds are the jet streams. They are narrow, meandering belts of very strong westerly winds centered in the upper troposphere. The jet streams act as steering currents that influence the direction and speed of surface weather systems.

Review Questions

1. Explain why air pressure exists. How and why do temperature changes affect air pressure?
2. Describe the locations of the global pressure belts. In which hemisphere is the beltlike pattern best developed, and why?
3. What causes the wind to blow? What two factors must be considered when forecasting the strength of the wind? Explain the influence of each on wind speed.
4. What factors influence the direction of the wind? What is the Coriolis effect, and how does its influence on wind direction differ between the northern and southern hemispheres? Why is it more effective in the upper atmosphere than near the earth's surface?
5. Draw a diagram of the earth showing the idealized pattern of global surface air pressure belts. Label the wind belts that exist within each pressure belt and between each pair of pressure belts. Indicate with

arrows the windflow pattern for those wind belts that have a prevailing direction.

6. Explain the cause, seasonal flow pattern, and attendant climatic characteristics of the monsoon winds. In what parts of the earth are they best developed, and why?

7. What are the jet streams? Where and why do they form? Explain why a knowledge of the pattern of the jet streams is essential for accurate surface weather forecasting.

Key Terms

Air pressure
Pressure gradient
Intertropical
    Convergence Zone
    (ITCZ)
Subtropical highs
Subpolar lows
Polar highs

Coriolis effect
Trade winds
Westerlies
Polar easterlies
Summer and winter
    monsoons
Jet streams

CASE STUDY

The Asiatic Monsoon

The Asiatic monsoon is a thermally induced wind system. Climatically, however, it is most important as the mechanism controlling the annual precipitation pattern of southern and eastern Asia. The "monsoon lands" of Asia, where the effects of the monsoon are especially dominant, occupy a broad arc from Pakistan northeastward into extreme eastern Siberia. Approximately half the world's population lives within this area.

The Asiatic monsoon is caused by seasonal temperature differences between the air over the continent of Asia and the air over the adjacent Pacific and Indian Oceans. During the late fall and early winter, the Asian landmass cools rapidly, causing surface air pressures to increase. The coldest temperatures occur in eastern Siberia, producing the giant Siberian High. The resulting air pressure pattern produces cool, dry, north-to-northeasterly winds over most of monsoon Asia. The relative dryness of this continental air is increased by its descent from the land toward the sea; in many areas no rain may fall for several months.

As the sun climbs higher in the sky and the length of daylight increases during the late winter and early spring, a rapid warming occurs. This causes a reduction in air pressures over East Asia and weakens the northeast monsoon winds. By late spring, temperatures have become hot over southern Asia; and inland air pressures have become lower than those over the cooler adjacent oceans. This pressure gradient initiates an onshore flow of tropical maritime air over extreme southern and eastern Asia which gradually strengthens and expands northward and westward as the summer progresses.

In India, rains begin on the Bay of Bengal coast in early June and spread northwestward to the Pakistani border by mid-July. The onset of the rains associated with the summer monsoon is often very sudden and is sometimes referred to as the "burst" of the monsoon. In Bombay, for example, mean monthly rainfall totals increase from 0.7 inches (18 mm) in May to 19.1 inches (485 mm) in June (see Table 4.1). It is estimated that 85 percent of India's annual rainfall occurs during the three-month period from mid-June until mid-September. Rainfalls in most areas are not steady, but instead take

the form of frequent, often torrential, showers and thunderstorms. By September, the rains begin to retreat southeastward, ending where they began on the Bengali coast in November as high pressure again becomes dominant over Asia.

Several factors account for the intensity of the summer monsoon rains of Southeast Asia. One is the large water vapor content of the tropical oceanic air masses that are drawn into the continent by the monsoonal airflow. A second factor is that the ITCZ, with its associated wind convergence and uplift, is drawn northward far into Asia during the summer months. Third, and probably most important, is the forced lifting of the air as it passes from the ocean at sea level to the Asian landmass, much of which is at a high elevation. Rainfall totals can be extremely heavy where mountain ranges lie directly across the path of the monsoon winds. For example, the town of Cherrapunji, located on the southern slopes of the Khasi Hills in eastern India, has a mean annual rainfall of 428 inches (1087 cm). Cherrapunji holds the world's record for the most rainfall ever recorded in a single twelve-month span. From August of 1860 through July of 1861,

continued on next page

Table 4.1
Monthly Mean Precipitation Totals for Bombay, India (in Inches)

Jan.	Feb.	Mar.	Apr.	May	Jun	July	Aug.	Sept.	Oct.	Nov.	Dec.	Year
0.1	0.1	0.1	T	0.7	19.1	24.3	13.4	10.4	2.5	0.5	0.1	71

Cherrapunji received an amazing total of 1042 inches (2645 cm) of rain! This is about thirty times the world average annual total.

The monsoon precipitation cycle is especially critical to the inhabitants of Southeast Asia because of the dominance of subsistence agriculture in that part of the world. In India, for example, two-thirds of the population is engaged in agriculture; in Pakistan, three-fourths of the population is agricultural. Landholdings are small, and high population growth rates increasingly pressure the land to provide ever-larger crop yields.

The agricultural calendar of events is dictated by the rainy season. Little can be done in the hot, dry spring months when the land lies parched, cracked, and bare. The onset of the monsoon rains brings about a frenzy of planting and plowing, as the crops must have maximum exposure to the available moisture. Most har-vesting takes place in the fall, at the beginning of the winter monsoon dry season. As winter arrives, the land again lies dry and bare until the spring and the annual return of the rains.

Figure 4.21 Flooded fields near Ho Chi Minh City (Saigon), Vietnam, during the rainy season. *(Rodman Snead, © JLM Visuals)*

Chapter Five

Atmospheric Moisture

Outline

Focus Questions

1. Under what conditions do water phase changes occur, and what role do they play in the global distribution of energy?
2. What is the hydrologic cycle, and why does it occur?
3. How do clouds and precipitation form?
4. What is the geographic pattern of precipitation, and why does it exist?

Water covers most of the earth's surface and is also a vital component of the atmosphere. Atmospheric moisture is, however, highly variable in amount, comprising less than 0.1 percent by volume of very cold, dry air and as much as 4 percent of very warm, moist air. This variability exists because of the unique ability of water to enter or leave the atmosphere in large quantities and over comparatively short timespans. The continuous cycle of water between the earth's surface and atmosphere not only supplies life-giving moisture to the land, but also plays a crucial role in the processes of energy transfer throughout the earth system.

PHASE CHANGES OF WATER

The phase changes that water readily undergoes are the key to its ability to enter and leave the atmosphere. Matter can exist in three physical states, or phases; that is, a substance can be a solid, liquid, or gas (vapor). Although any substance, with sufficient gain or loss of energy, can theoretically change from any one of these states to any other, water is the only atmospheric component that undergoes phase changes within the existing range of atmospheric temperatures and pressures.

Energy is absorbed or released when water undergoes changes either in temperature or in phase. This energy is commonly measured in terms of calories. A *calorie* is the amount of heat needed to raise the temperature of one gram of liquid water by one Celsius degree (1.8 F°). Heat is absorbed by water when it is warmed and is released when water cools. Large quantities of energy are involved in the six phase changes in which water can participate. As with heating, energy is absorbed by water when it changes from a lower to a higher phase; energy is released to the environment when water changes from a higher to a lower phase. Figure 5.1 illustrates these changes. They consist of three pairs of opposite processes, which we shall now examine.

■ *Evaporation and condensation.* **Evaporation,** the change of liquid water to water vapor, occurs as individual molecules in a water body acquire enough kinetic energy through random molecular collisions to escape into the atmosphere. For every gram of water that evaporates at normal earth temperatures, approximately 590 calories of energy are transferred

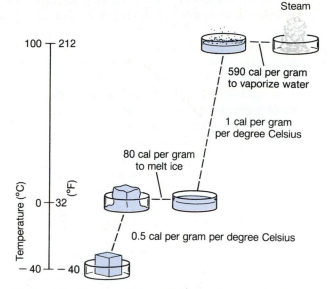

Figure 5.1 Temperature and phase change energy requirements for water. (*From* Introduction to Environmental Science *by Joseph M. Moran et al. Copyright © 1980 W.H. Freeman and Company. Reprinted with permission.*)

from the liquid water to the atmosphere.[1] Because most evaporating water is located on the earth's surface, this process produces a major transfer of energy from the surface to the atmosphere. The addition of this water vapor does not directly heat the atmosphere, because the absorbed energy is used to change the physical state of the water molecules rather than to raise their temperature.

■ **Condensation,** the change of water vapor back to liquid water, is the exact opposite of evaporation. When condensation occurs, the energy absorbed by the water molecules during evaporation is released to the atmosphere as the *heat of condensation*. Most condensation takes place well above the earth's surface during the process of cloud formation; it is therefore a major heating process for the troposphere.

■ *Melting and freezing.* A much smaller but still important quantity of energy is associated with the melting of ice and the freezing of water. This *heat of fusion* involves the transfer of 80 calories per gram of water. (The transfer of this energy to the ice makes

---

1. The lower the temperature of the evaporating water, the greater the quantity of energy required for evaporation. At the boiling point, 540 calories are required to evaporate a gram of water; at the freezing point, 600 calories are required.

your hand cold when you hold an ice cube as it melts.) When water freezes, this energy is released to the atmosphere.

■ *Sublimation.* Under certain conditions ice can change directly to water vapor, or vice versa, without passing through the liquid state. Both changes are known as sublimation. Sublimation involves the transfer of a greater amount of energy per unit of mass than does any other change of state: a total of 670 calories per gram. This is the sum of the heats of condensation and fusion. In the atmosphere, sublimation occurs primarily when ice crystal clouds form directly from water vapor at temperatures well below freezing. Although it liberates large amounts of energy per gram of water involved, sublimation is much less important as a source of atmospheric heating than is condensation because it occurs only at cold temperatures, which greatly reduce the capacity of the air to hold water vapor.

Taken together, the phase change processes perform three vital atmospheric functions. First, they transfer from the earth's surface to the atmosphere much of the energy that powers the atmospheric processes, especially storms. Second, they add water vapor to the atmosphere. Third, they make possible the conversion of water to the liquid and solid states, which results in precipitation.

THE HYDROLOGIC CYCLE

Earth's water supply is involved in an endless cycle of movement, called the **hydrologic cycle,** which is powered by solar energy and made possible by the ability of the water to change phase. As a result of this cycle, water is transported from the oceans to the continents; the land surface is sculptured by erosion; and the earth's supply of solar energy is spread more evenly over the face of our planet.

The hydrologic cycle is depicted in Figure 5.2. The cycle is actually more complex than it appears in the figure because a number of alternative routes, depending on local conditions, exist for the water. Basically,

Figure 5.2 The hydrologic cycle.

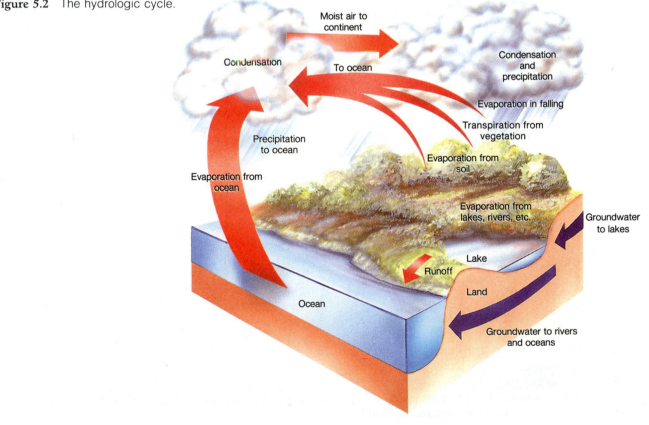

though, the cycle consists of five key steps that, when completed, will return the water to its point of origin and set the stage for a repetition of the cycle.

The hydrologic cycle involves the continuous recycling of the same water. It is therefore possible to begin the discussion with any of the cycle's five steps. Perhaps the best starting point, however, is with the ocean waters, which contain an estimated 96.7 percent of the earth's near-surface water supply. The evaporation of water from the oceans and its upward diffusion into the atmosphere represent, in our illustration, the first step of the hydologic cycle.[2] Rates of evaporation from oceans

in different parts of the world vary greatly, but a mean annual rate of 40 inches (100 cm) can be used as an approximation.

Most evaporation occurs from the warm oceans of the lower latitudes (see Figure 5.3). The leading source of the moisture received as precipitation for much of the world is the ocean surface in areas dominated by the subtropical highs. With their combination of clear skies, warm temperatures, and relatively dry air, the subtropics lose vast quantities of water to evaporation, and many subtropical areas receive little water back as precipitation. On a global basis, evaporation greatly exceeds precipitation at latitudes between 10° and 40° in both hemispheres, while precipitation exceeds evaporation in all other latitudinal zones, largely as a result of the transport of subtropical moisture into those zones, where it is released by storms. Because of temperature

---

2. Large amounts of water also evaporate from the land, but the total quantity is much less than that evaporated from the oceans. We shall temporarily ignore the land source in order to concentrate on the basic steps of the hydrologic cycle.

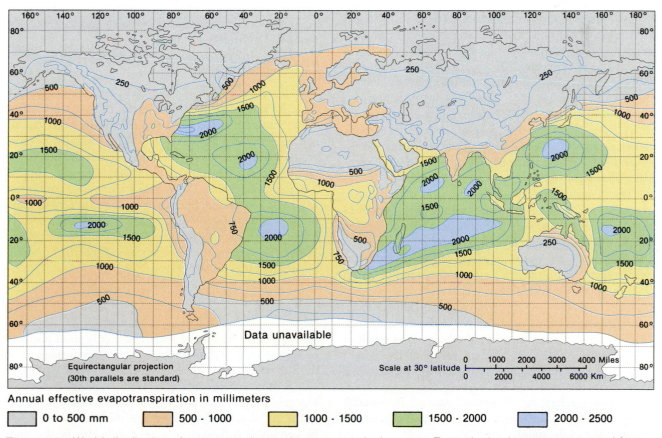

Annual effective evapotranspiration in millimeters

| 0 to 500 mm | 500 - 1000 | 1000 - 1500 | 1500 - 2000 | 2000 - 2500 |

Figure 5.3 World distribution of mean actual annual evapotranspiration rates. Transpiration is water evaporated from plants, and evapotranspiration is the combined surface water loss from evaporation and transpiration. Globally, it is estimated that 15 percent of the atmosphere's water vapor is derived from the continents (including inland water bodies), while 85 percent evaporates from the ocean.

and vegetative factors, evaporation rates outside the tropics are also generally higher in summer than in winter and higher during the day than at night.

The second step in the hydrologic cycle is the condensation (and/or sublimation) of atmospheric water vapor to form clouds. When condensation occurs, the water returns to its original liquid physical state, but is far from having its original appearance. Before, it was part of an extensive, interconnected body of water covering approximately 71 percent of the earth's surface; now, it is dispersed high in the atmosphere in myriad tiny water droplets centered around dust particles that serve as condensation nuclei.

Most clouds do not gain enough moisture through condensation or sublimation to produce precipitation, but simply evaporate back into the air. Clouds that do generate precipitation, though, return water to the surface. In contrast to its stay of perhaps thousands of years in the oceans, water remains in the atmosphere only a short time before precipitating out. Its estimated average residency time in the atmosphere is 10 days. If the clouds are still above the ocean, a three-step hydrologic cycle of evaporation, condensation, and precipitation results. Indeed, since 80 percent of the world's precipitation falls into the oceans, this is the most common pathway within the hydrologic cycle. We, however, are usually much more concerned with the 20 percent of the precipitation that falls on land. The third step of the hydrologic cycle therefore involves the transport of the moisture from sea to land by wind currents either before or after condensation has occurred.

The fourth step is the precipitation of the water onto the land surface. Land areas in different parts of the world receive greatly differing amounts of precipitation, but the average is approximately 30 inches (76 cm) of liquid water per year. Much of this moisture has been transported hundreds or even thousands of miles from the places where it was evaporated.

The final step of the hydrologic cycle is runoff, which eventually returns the water to the oceans. Once again, complications can arise; it is likely that much of the water will evaporate before reaching the ocean. Flow to the sea may be on or beneath the surface. The water can also be stored for varying periods in rivers, lakes, and glaciers, or as ground water. Eventually, though, the water completes its journey and is available to begin the cycle anew.

The hydrologic cycle, as described here, provides an outline for material covered in the remainder of this chapter and in several subsequent chapters of this book.

Each of the final three weather elements to be discussed—humidity, clouds, and precipitation—first appears at a particular point in the hydrologic cycle. Later material on surface and subsurface waters (Chapter 8), weathering and mass wasting (Chapter 13), fluvial processes and landforms (Chapter 14), glacial processes and landforms (Chapter 15), and coastal processes and landforms (Chapter 17) covers the surface stages of the hydrologic cycle as well as its direct effects on the earth's surface.

Humidity

When water evaporates, it contributes to the air's *humidity;* that is, the amount of water vapor contained by the air. The atmospheric water vapor content can be calculated in a variety of ways, so several different types of humidity measurements exist. The two most commonly used in the atmospheric sciences are specific humidity and relative humidity.

Specific Humidity

Specific humidity is defined as the ratio of the mass of water vapor in a parcel of air to the total mass of the air parcel, including the water vapor. It is usually given in grams of water per kilogram of air, so that the numerical value is in easily calculable parts per thousand. Specific humidity values give the actual water vapor content of the air. They are unaffected by changes in temperature or air pressure—factors that influence the values of most other humidity measurements, including relative humidity. The geographic distribution of specific humidity by latitude is shown in Figure 5.4.

Relative Humidity

Relative humidity is the most commonly used measure of atmospheric water vapor; it is also the measurement most familiar to the general public. Its popularity probably results from the fact that relative humidity readings correspond rather well to the sensible moisture content of the air; that is, to the apparent moisture content felt by people. It is, however, a poor indicator of the total water vapor content of the air.

By definition, the **relative humidity** is the percentage ratio of the amount of water vapor actually in the air

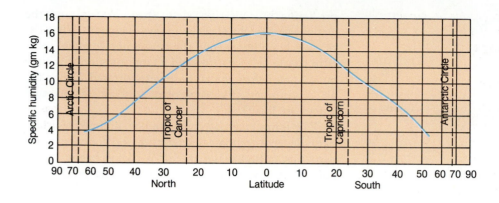

Figure 5.4 Mean specific humidity values by latitude. The dominant influence on this pattern is the effect of temperature on the water vapor holding capacity of the atmosphere.

to the maximum amount of water vapor the air could hold at that temperature and pressure. Put another way, the relative humidity indicates what percentage of its water vapor capacity the air is currently holding. The mean relative humidity of the air near the earth's surface is about 78 percent.

Unlike the specific humidity, the relative humidity is directly affected by changes in air temperature (see Figure 5.5). Because the air's capacity for holding water vapor increases with a rise in temperature, heating the air causes the relative humidity to fall. This occurs because the amount of water vapor contained by the air, though unchanged, now constitutes a smaller percentage of the air's increased capacity to hold this moisture. Similarly, if the air becomes cooler, its water vapor holding capacity is reduced; and the relative humidity rises, even though no new water vapor is added to the air. As a consequence, relative humidity readings usually fall during the day as the air is heated by insolation, and rise during the nighttime hours as the atmosphere cools (see Figure 5.6).

Figure 5.7 depicts the distribution of mean relative humidity values on a latitudinal basis. Whereas the global distribution of specific humidity values is closely correlated with temperature (warm regions tending to have higher readings than cold regions), the distribution of relative humidity values is strongly correlated with the global air pressure belts. The highest relative humidity values are thus associated with the ITCZ and subpolar lows, where cloudiness and precipitation are prevalent. The lowest values, conversely, are associated with the subsiding air of the subtropical highs. Relative humidity values in the southern hemisphere are higher on average than those in the northern hemisphere because of the presence of greater amounts of surface water in the southern hemisphere.

Condensation and Sublimation: Processes and Products

Water that evaporates from the earth's surface eventually returns as precipitation. In order for this to occur, water vapor, one of the atmosphere's lightest gases, must first be converted to the much denser liquid or solid states. The phase change processes involved in these conversions are condensation and sublimation.

Under what conditions do these processes occur? Obviously, condensation and sublimation do not continually remove water vapor from the atmosphere, because

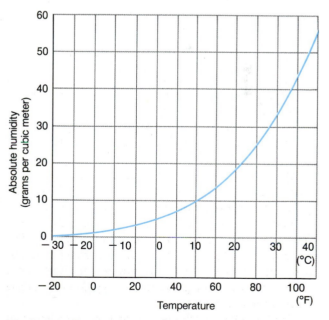

Figure 5.5 The maximum water vapor holding capacity of a cubic meter of air at sea level at differing temperatures.

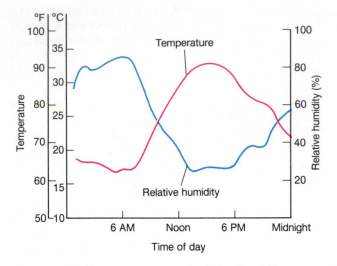

Figure 5.6 Air temperature and relative humidity readings at Tucson, Arizona, during September 18, 1982. Note the inverse relationship between the two graph patterns.

the weather is often fair and dry. At other times, though, precipitation or the formation of dew or frost shows that water vapor is indeed being removed from the air. The key to these processes is cooling. It was noted earlier that cooling the air reduces its water vapor holding capacity, causing an increase in the relative humidity. If air is cooled sufficiently, a temperature will be reached at which the relative humidity is 100 percent and the air is saturated with water vapor. Any further cooling will reduce the ability of the air to hold water vapor even more, producing an excess of moisture that must be removed by condensation or sublimation.

The temperature at which a given parcel of air will be saturated is its **dew point.** The dew point, by indicating how much the air must be cooled in order to initiate condensation or sublimation, is important in

predicting precipitation, dew, frost, or fog, and the amount and height of certain types of clouds. When the relative humidity is low, the dew point is much lower than the air temperature, and a large amount of cooling must occur before the air becomes saturated. When the air is saturated, the dew point and air temperature values are identical.

The forms taken by condensed or sublimed moisture vary, depending upon temperature and location. Condensation occurs when the air is cooled below a dew point that is near or above freezing. The products of condensation are dew, fog, water droplet clouds, drizzle, and rain. Sublimation normally occurs when the dew point is well below freezing. Its three major products are frost, ice crystal clouds, and snow. The remainder of this chapter examines the causes and global distributions of these forms of atmospheric moisture.

Adiabatic Temperature Changes

Atmospheric cooling that leads to the formation of clouds and precipitation nearly always results from the rising of air. As a parcel of air rises, it enters areas of reduced atmospheric pressure and consequently expands. The expansion of the rising air requires energy, which is taken from the kinetic (or thermal) energy of the air molecules, making the air parcel cooler. When this cooling has reduced the air temperature to the dew point, condensation and cloud formation begin. Clouds are thus the visible tops of rising air currents. The cooling of air resulting from its expansion is termed **adiabatic cooling.** It involves no actual loss of energy, because this energy has not left the air but has merely changed form. As long as the air keeps rising, it continues to cool, causing the condensation or sublimation of additional water vapor. This allows the cloud to

Figure 5.7 Mean relative humidity values by latitude. The dominant influence on this pattern is the distribution of the global air pressure belts.

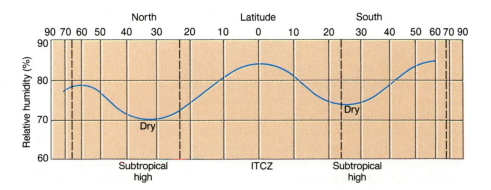

grow larger and its constituent water droplets and ice crystals to increase in size, so that eventually they may become large enough to fall as precipitation. Conversely, when the air descends, it undergoes **adiabatic heating** because the surrounding air forces the descending parcel to contract and it regains the energy it used in expansion.[3] As the descending air warms, its water vapor holding capacity increases, so that a cloud within a descending airflow will soon evaporate. This is why the descending air associated with high pressure systems produces fair, dry weather.

Dew and Frost

Dew and frost both form on the earth's surface. Because they do not fall from the atmosphere, they are not technically considered to be forms of precipitation. **Dew** consists of water droplets that have condensed on surface objects because the overlying air has cooled below the dew point. Dew nearly always forms at night and is most likely when the sky is clear, the wind light or calm, and the air moist. Under these conditions, terrestrial radiation causes the land surface to cool rapidly. Conduction with the cold surface then chills the overlying layer of air, forcing the excess atmospheric moisture to condense on available surfaces such as leaves and blades of grass. Dew normally evaporates in the morning as the air is warmed by the sun and its water vapor holding capacity increases. It is a relatively minor source of surface moisture, but typically provides land areas in the middle latitudes with 0.5 to 2.0 inches (12–50 mm) of water per year.

Frost forms under nearly identical conditions, but occurs when the dew point is below freezing. Frost is not frozen dew, but instead forms by sublimation directly from water vapor. All products of sublimation have a crystalline structure, and the feathery patterns formed by frost on windowpanes and other surfaces are well known.

Fog

Dew and frost form when a layer of air just above the surface is cooled below its dew point and the resulting moisture is deposited on objects at the surface. When a deeper layer of near-surface air is cooled, however, condensation or sublimation occurs on atmospheric dust particles. The result is fog. *Fog* is a visible accumulation of minute water droplets or ice crystals suspended in air immediately overlying the surface. It is actually a cloud at ground level.

The most common fog type over land is **radiation fog** (see Figure 5.8). It forms under the same clear, calm, moist nighttime conditions that result in the formation of dew and frost. Local topographic features often control the distribution of radiation fog. Over flat surfaces it can form an extensive blanket covering many square miles. In more rugged areas, though, the downhill drainage of air chilled by contact with the cold surface often causes radiation fog to accumulate in valley bottoms while the surrounding uplands remain clear.

Another important fog type, **advection fog,** forms when moist, mild air flows over a colder surface. Although it occurs less frequently worldwide, it typically develops much more extensively than radiation fog and occasionally covers hundreds or even thousands of square miles. The most favorable locations for advection fogs are ocean areas where warm and cold waters lie adjacent to one another. One site notorious for advection fog is the Grand Banks area southeast of Newfoundland, where air that has traveled over the warm Gulf Stream passes over the adjacent much colder water of the Labrador Current. Another well-known site is the California coast, where fog forms as mild Pacific air

Figure 5.8 A thin layer of radiation fog hugs the ground at dawn. (*Courtesy of NOAA*)

---

3. An analogy to a compressed spring might be made: the spring uses energy when it is released and expands, but regains this energy when it is again compressed.

is chilled by its passage over cold water that has upwelled near the coast. Extensive advection fogs can also be produced when mild maritime air is transported over cold, perhaps snow-covered land in winter or a glacier in summer. A southerly flow of air from the Gulf of Mexico often produces dense winter fogs over the central United States in this manner.

Clouds

Clouds, with their endless variety and constantly changing patterns of form and color, add substantially to the beauty of the natural environment. They also provide valuable information about the operation of often-complex atmospheric processes, and, most important, they produce virtually all the earth's precipitation.

Like fog, clouds are accumulations of minute water droplets or ice crystals, suspended in the air, that have condensed or sublimed around dust particles because of atmospheric cooling. However, clouds, unlike fog, are not in contact with the ground. The diameter of a typical cloud droplet is on the order of 1/2500 inch (1/100 mm), and its mass is about one-millionth of a gram. Particles of this size and mass can remain suspended in the air almost indefinitely.

Most clouds consist largely or entirely of water droplets because relatively warm air typically contains more water vapor for cloud formation and because tiny cloud droplets, with their high surface tension values, are able to remain in the liquid state at temperatures well below freezing. Because of the very cold temperatures needed for their formation, ice crystal clouds are primarily located in the upper troposphere.

Although an unlimited variety of clouds exists, ten basic types are commonly recognized on the basis of their altitude and their appearance to an observer at the surface. These cloud types are often separated into four categories on the basis of height, as indicated in Table 5.1. Each is illustrated and described briefly in Figure 5.11.

Clouds have a complex and constantly changing geographic distribution related to a number of influencing factors. These include surface features, pressure and wind systems, weather fronts, and daily and seasonal variations in temperature. Satellite observations have indicated that an average of about 52 percent of the earth is covered by clouds at any one time (see Figure 1.1). Distinctive cloud bands typically demarcate the locations of the ITCZ and the subpolar lows. Areas of cold ocean water in the subtropics and the tops of

Table 5.1
Major Cloud Groups, Types and Heights*

High Clouds	16,000–60,000 feet (4900–18,300 m)
Cirrus	
Cirrostratus	
Cirrocumulus	
Middle Clouds	6500–25,000 feet (2000–7600 m)
Altostratus	
Altocumulus	
Low clouds	0–6500 feet (0–2000 m)
Stratus	
Stratocumulus	
Nimbostratus	
Clouds of Vertical Extent	
Cumulus	2000–20,000 feet (600–6100 m)
Cumulonimbus	2000–60,000 feet (600–18,300 m)

*Height values are approximate and relate to the low and middle latitudes only. High latitude cloud heights are considerably lower.

high mountain ranges are also frequently cloud-covered. In addition, the middle and higher latitudes experience a constant progression of distinctive cloud patterns attending traveling cyclonic storms and their associated frontal systems. These weather systems are examined in more detail in the next chapter.

The tropopause forms a "lid" on vertical cloud development. Tropospheric conditions favor cloud formation because temperatures decline with altitude, thereby encouraging the uplift of relatively warm and buoyant near-surface air. Above the tropopause, however, the temperature trend is reversed (see Chapter 2). The characteristic flattened "anvil" shape associated with thunderheads (see Figure 5.11j) often develops as the rapidly rising tops of cumulonimbus clouds encounter the tropopause.

Precipitation

Precipitation is water, in either liquid or solid form, falling through the atmosphere toward the surface of the earth. If precipitation did not occur, life on land would likely never have evolved. Without the physical and chemical influences of water, earth's landforms would be strikingly different in appearance; and a true soil cover would not have formed. Even the oceans, without

Precipitation Measurement

The accurate measurement of precipitation is vital because of precipitation's influence on soil moisture levels and crop growth, water supplies, and flood potentials. As a result, precipitation amounts are measured at thousands of meteorological field stations around the world.

The instrument most commonly used for the collection and measurement of precipitation in the United States is the standard eight-inch *rain gage*. It is a simple apparatus consisting of a funnel-shaped collector with a diameter of 8 inches (20 cm) that feeds into a 20 inch (50 cm) long measuring tube (see Figure 5.9).

A much more sophisticated instrument now widely used to estimate the intensity of falling precipitation is *radar*. Radars send out pulses of microwave energy that are reflected from falling precipitation. The returning radar "echo" is plotted on a screen to depict the size, shape, and intensity of the precipitation area (see Figure 5.10). Observed over a period of time, changes in the radar echo patterns indicate the direction and speed of movement of storms.

Figure 5.9 A standard 8-inch rain gage consists of a narrow collecting tank with an 8-inch mouth. The barrel-shaped outer container reduces evaporation and provides basal support. (*Courtesy of NOAA*)

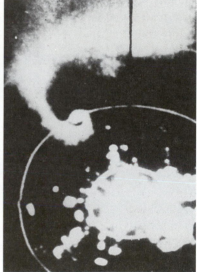

Figure 5.10 A radar screen showing the echo of a strong thunderstorm cell. The hook-shaped echo indicates that the thunderstorm may contain a funnel cloud or tornado. (*Courtesy of NOAA*)

the receipt of nutrients eroded from the land by runoff from precipitation, would be far less capable of supporting life. The condensation and sublimation from which precipitation is derived are also associated with the release of huge amounts of heat energy to the atmosphere. For example, a rainfall of one inch (2.5 cm) on one square mile (2.6 sq km) of surface is associated with the release of a quantity of heat equivalent to that produced by burning approximately 20,000 tons (9 million kg) of coal. Much of this energy escapes into space, but the rest is widely distributed to the earth's atmosphere and surface.

PRECIPITATION FORMATION AND TYPES Precipitation is usually produced in deep clouds that have been uplifted far enough for the air within them to have cooled adiabatically to well below the original dew point, forcing the condensation or sublimation of large quantities of water vapor. Because the upper portions of

83

Figure 5.11a Cirrus clouds are the highest of the ten major cloud types, and one of the most common. Like the other cirriform clouds, they are composed entirely of ice crystals. (*R. Walker/H. Armstrong Roberts*)

Figure 5.11b Cirrostratus clouds consist of a thin, smooth veil of ice crystals. They have the ability to refract light, often causing the sun or moon to be surrounded by a halo. (*Thomas M. Conrow, Educational Images*)

Figure 5.11c Cirrocumulus clouds consist of a large number of very small, rounded, and sometimes iridescent cloud masses arranged in groups or lines. (*Richard Jacobs, © JLM Visuals*)

Figure 5.11d Altocumulus clouds are rather similar to cirrocumulus in that they consist of cloud masses arranged in a linear pattern, but the individual masses appear considerably larger, darker in their centers, and have more distinct margins. (*John K. Nakata/ TERRAPHOTOGRAPHICS/BPS*)

Figure 5.11e Altostratus clouds typically form a smooth, textureless cloud layer over the entire sky. The sun may be visible as a diffuse bright spot, and a light, steady rain or snow may fall. (*Thomas M. Conrow, Educational Images*)

Figure 5.11f Stratus clouds, such as those shown here at the Delaware Water Gap, are low, dense, uniformly gray clouds that usually cover the entire sky and are capable of producing a light drizzle. (*John S. Shelton*)

Figure 5.11g Stratocumulus clouds consist of distinct cloud masses, usually in the form of rolls, between which patches of clear sky may show. They sometimes produce light rain or snow showers. (*B.F. Molnia*)

Figure 5.11h Nimbostratus clouds form a dark, generally textureless cloud layer often partially obscured by falling precipitation. They are the major cold season precipitation producers of the middle and high latitudes. (*Thomas M. Conrow, Educational Images*)

Figure 5.11i Cumulus clouds are the world's most common cloud type. They consist of distinct individual cloud masses with flat bases and uneven sides and tops, and are produced by masses of rising warm air. (*R. Krubner/H. Armstrong Roberts*)

Figure 5.11j Cumulonimbus clouds, the most impressive cloud type, are the chief precipitation producers of the tropics. They form in air that is unstable to considerable altitudes and are associated with heavy rain, lightning and thunder, and, occasionally, hail and tornadoes. (*Jerry Scott*)

clouds have risen and cooled the most, their water droplets and ice crystals are much larger than those of the lower portions. Most precipitation is therefore initiated near the tops of clouds. Because it originates in the cold upper portions of clouds, nearly all precipitation outside the tropics begins falling as snow and simply melts to become rain as it passes through warmer air on its way to the surface.

Precipitation can assume a variety of forms, depending on atmospheric conditions both within the clouds and between the clouds and the earth's surface. The two types of liquid precipitation are rain and drizzle. *Rain* is by far the most common and important form of precipitation; it comprises about 90 percent of the earth's precipitation total. By definition, it consists of liquid drops with diameters of at least 0.02 inches (0.5 mm). *Drizzle* consists of droplets with diameters less than 0.02 inches (0.5 mm). It falls only from stratus clouds developed from weak updrafts.

Four major forms of frozen precipitation are recognized. Rain falling onto a subfreezing surface will quickly freeze on contact with whatever object it strikes, forming a solid coating of ice. This makes *freezing rain* one of the most potentially dangerous forms of precipitation. Severe ice storms, although they may transform the landscape into a crystalline wonderland, are capable of causing widespread damage to power lines and trees because of the sheer weight of the accumulated ice (see Figure 5.12).

When a considerable depth of the lower atmosphere has subfreezing temperatures, raindrops will freeze while falling to form sleet. *Sleet* consists of small, hard pellets of ice that collect to form a crunchy, granular surface.[4] Both freezing rain and sleet indicate the

Figure 5.12 Heavily weighted tree branches following an ice storm. (*John S. Shelton*)

presence of a temperature inversion and form when above-freezing air overlies subfreezing air (see Figure 5.13).

The most common form of solid precipitation, comprising nearly a tenth of the earth's precipitation total, is snow. *Snow* consists of crystalline ice formed directly from water vapor by sublimation. Its occurrence indicates that subfreezing (or near-freezing) conditions exist from the upper cloud levels all the way to the surface.

The most spectacular of the major precipitation forms is hail. *Hail* consists of relatively large, rounded particles of ice that typically have a layered structure and are associated with severe thunderstorms. Hailstones form as the strong updrafts and downdrafts of a thunderstorm carry ice pellets alternatively above and below the freezing level in a cumulonimbus cloud. Water and ice crystals adhere to the outside of the hailstone and freeze onto it when it is carried above the freezing level. Hailstorms are most common in continental climate areas of the middle latitudes, especially the

---

4. In English-speaking countries outside of North America, sleet is commonly considered to be a mixture of rain and snow.

Figure 5.13 Vertical temperature profiles associated with various forms of precipitation.

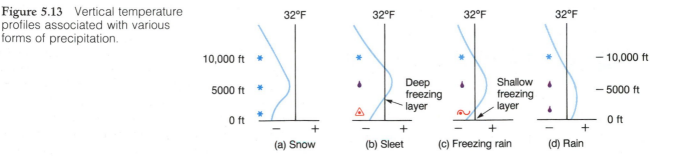

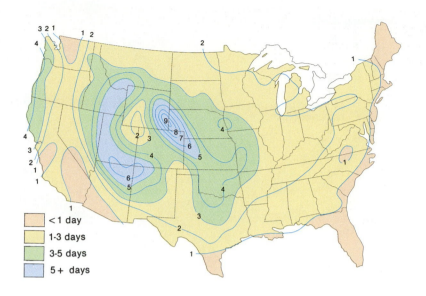

Figure 5.14 The average number of days each year on which hail occurs in the coterminous United States.

Legend:
- < 1 day
- 1-3 days
- 3-5 days
- 5+ days

High Plains of the west-central United States (see Figure 5.14). They occur most frequently during the spring and summer months, when the atmosphere is unstable. Most hailstones have diameters of well under an inch (2.5 cm), but some larger than baseballs have been recorded (see Figure 5.15). Severe hailstorms can do great damage to crops, livestock, and buildings.

CAUSES OF PRECIPITATION: ATMOSPHERIC LIFTING PROCESSES It has been noted that clouds and precipitation are produced by the rising and adiabatic

Figure 5.15 Hail of this size can cause severe damage to crops.

cooling of air, but the factors causing the air to rise have not yet been examined. Four basic atmospheric lifting mechanisms are capable of causing uplift sufficient to produce precipitation. They are termed convectional lifting, orographic lifting, frontal lifting, and cyclonic lifting.

Convectional lifting refers to the rising of air because it is warmer, and therefore lighter, than the surrounding air (see Figure 5.16). Air that rises because of its buoyancy is described as being unstable. The chief way by which the air is heated is by contact with a warm surface. Over land, this frequently occurs during the midday or afternoon hours because of surface heating by the sun. Over water, convectional lifting may occur at almost any hour of the day or night because of the constancy of surface temperatures; but it is largely restricted to areas of relatively warm water. Convectional lifting is associated with cumuliform clouds and localized, showery precipitation. The summertime afternoon showers and thunderstorms commonly experienced in the middle latitudes are mostly products of this atmospheric lifting process. The brevity of a convectional shower results largely from the fact that the shower cools and stabilizes the air, thereby reducing its ability to continue to rise and prolong the precipitation. The other three lifting processes are produced by the forced uplift of air; they are typically characterized by less intense and more-prolonged precipitation.

Because it is generally associated with warm, buoyant air, convectional lifting occurs most frequently in

Figure 5.16 Convectional lifting is initiated by surface heating, which causes the air to expand, become buoyant, and rise. Adiabatic cooling then causes the rising air to cool below the dew point, resulting in cloud development.

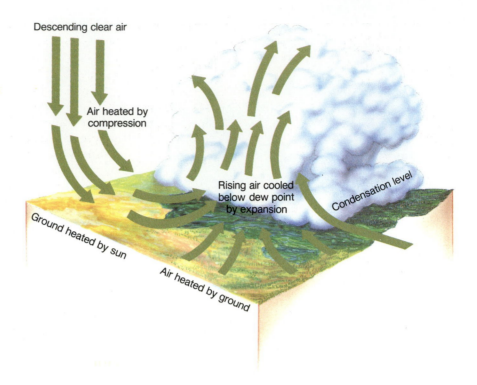

Descending clear air

Air heated by compression

Rising air cooled below dew point by expansion

Condensation level

Ground heated by sun

Air heated by ground

the tropics, where it is the leading cause of precipitation in most areas. In the middle and especially the higher latitudes, it is less common and is restricted largely to the warm season. Convectional precipitation, however, is not always associated with warm weather. This is illustrated by the heavy convectional snowfalls that may be triggered by the passage of cold arctic air over an unfrozen water body. (See the Case Study at the end of the chapter.)

Orographic lifting is the lifting of air over a mountain range or other elevated surface. In contrast to convectional lifting, it involves the forced ascent of air because the topography imposes a physical barrier to its movement. As an air mass approaches a mountain barrier (Figure 5.17), it begins to rise and to cool adiabatically. The amount of cloudiness and precipitation produced by this uplift depends on several factors. The most favorable conditions for heavy orographic precipitation include a high, continuous mountain range, the presence of warm, moist air, and strong winds blowing directly across the mountain range.

Most clouds and precipitation produced by the orographic effect are concentrated on the side of the mountain range facing the wind (the *windward* side) because the forced uplift occurs there. As the air passes over the crest and descends the downwind (or *leeward*) side of the mountains, it is adiabatically heated; and progressively drier conditions prevail. An area that has a dry climate because it is on the leeward side of an orographic barrier is said to experience a *rainshadow effect*. The orographic effect can therefore either increase or decrease precipitation amounts (see Figure 5.18).

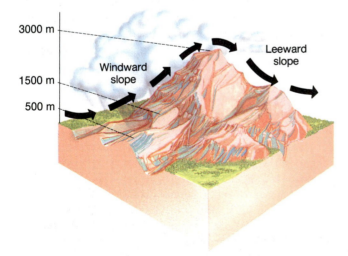

3000 m

1500 m

500 m

Windward slope

Leeward slope

Figure 5.17 The orographic effect. The windward sides of mountains are associated with rising air and precipitation; the leeward sides are associated with descending air and dry climates.

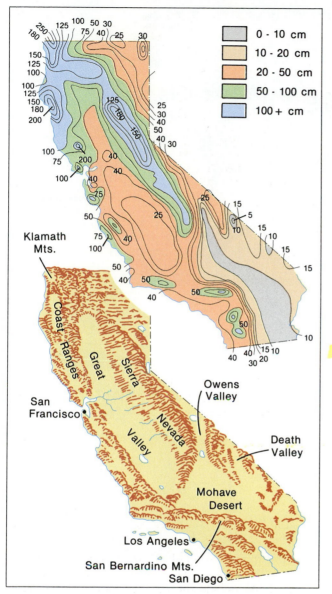

▨	0 - 10 cm
▨	10 - 20 cm
▨	20 - 50 cm
▨	50 - 100 cm
▨	100 + cm

Figure 5.18 The effect of surface topography on precipitation is clearly illustrated by a comparison of landform and mean annual precipitation maps of California. The highest precipitation totals are associated with the Coast Ranges and the Sierras. Precipitation amounts are in centimeters.

Although orographic lifting can be an extremely effective precipitation generator, it is the most geographically restricted of the four atmospheric lifting mechanisms. It is normally strongly developed only in the vicinity of relatively high mountain ranges or plateaus, which comprise at most 10 percent of the earth's surface. Because this lifting mechanism is tied to the earth's surface features, it is incapable of movement. The other three lifting mechanisms are readily transported by the wind currents, greatly increasing the total surface area they influence.

==Frontal lifting== is associated with the lifting of relatively light warm air by denser cold air along a weather front (see Figure 5.19). It is the primary cause of precipitation in most of the middle and higher latitudes. In the middle latitudes, frontal lifting is most prevalent in winter and early spring, when weather fronts are much stronger and more numerous than they are during the summer. In the high latitudes, fronts are fairly numerous throughout the year, but the lowered water vapor holding capacity of the cold air limits the amount of precipitation they produce. Fronts and frontal precipitation are generally unimportant in the tropics because cool air masses from the higher latitudes usually weaken and lose their identity before penetrating far into this area. The different types of fronts and their characteristics are examined in Chapter 6.

The final atmospheric lifting mechanism, ==cyclonic lifting,== is produced by the rising spiral of converging air associated with a low pressure system. In the middle and high latitudes, low pressure systems usually form on weather fronts, so that frontal and cyclonic lifting act together. Low pressure systems of the tropics, such as hurricanes, do not have fronts. Their cloudiness and precipitation are induced primarily by cyclonic uplift.

It should be noted that various combinations of lifting mechanisms may occur. Heavy precipitation in the Pacific Northwest, for example, may be generated when a weather front from the Pacific crosses mountain ranges that orographically intensify the front's rainfall or snowfall totals. Similarly, precipitation along a

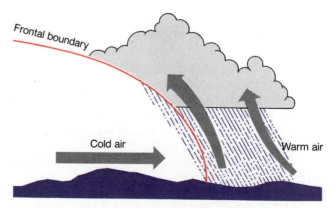

Figure 5.19 Frontal lifting occurs as a warm air mass is forced to rise over a denser cold air mass.

Figure 5.20 Mean annual world precipitation totals. Values are given in inches of liquid water.

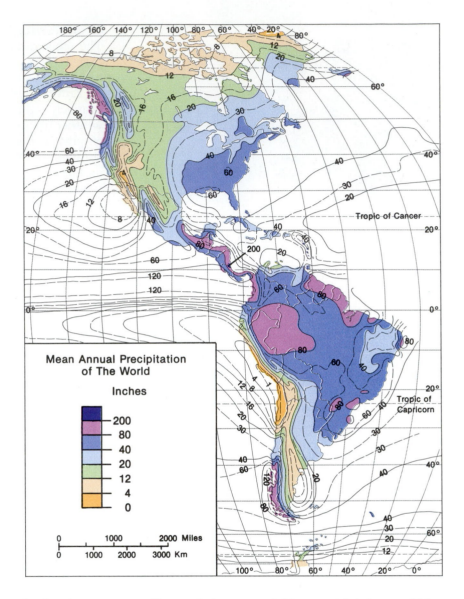

weather front is typically heavier during the day than at night because the daytime air is warmer and less stable.

WORLD PRECIPITATION DISTRIBUTION   The total amount of precipitation received at any given location depends upon the moisture content of the atmosphere and the frequency and intensity of atmospheric lifting. Figure 5.20 depicts the mean annual precipitation totals for the land areas of the world. The estimated average annual precipitation for the earth as a whole is 35 inches (90 cm). This value provides a useful point of comparison for assessing the annual total at any given site.

In the next several paragraphs, the basic characteristics and causes of the global pattern of precipitation

are discussed. As you read this material, it is a good idea to refer frequently to Figure 5.20. More-detailed information on regional precipitation patterns is provided in Chapter 7, which covers the subject of world climates.

The equatorial zone, located between approximately 15° N and 15° S, is the world's rainiest zone of latitude. Most equatorial locations receive between 50 and 120 inches (125 to 300 cm) of rainfall annually. The air in the tropics is very warm, moist, and unstable, giving it a high rainfall potential. Rainfalls are further augmented by the convergence of air from the trade wind zones of both hemispheres into the ITCZ. The heaviest rainfall averages occur in areas where high mountains produce substantial orographic uplift.

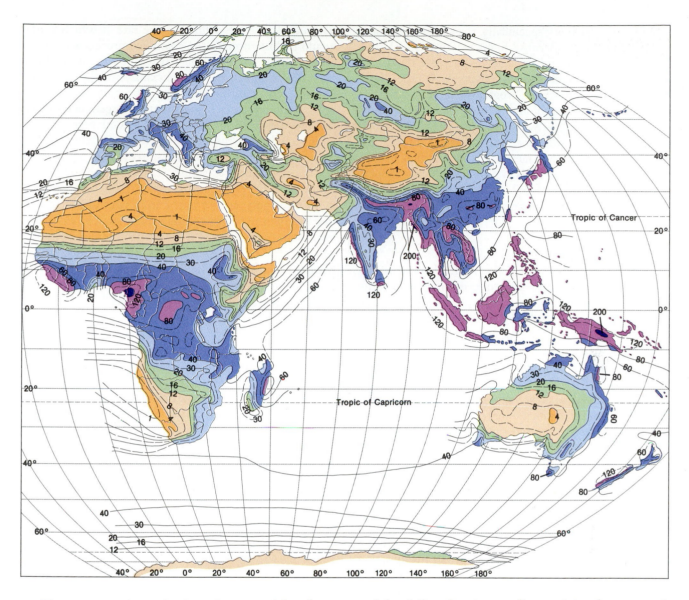

The outer tropics and subtropics, comprising those areas between 15° and 35° N and S, are substantially drier as a whole than the equatorial zone, although large regional variations in precipitation means exist. The predominant precipitation influences here are the subtropical belts of high pressure, with their subsiding air. In places where the highs are especially well developed, precipitation totals are so greatly reduced that arid or semiarid conditions prevail. Along continental west coasts paralleled by cold ocean currents, atmospheric lifting is further discouraged by the chilling of the air at the surface, and the world's driest deserts exist here. In contrast, the warm ocean currents that flow along the east coasts of continents in the subtropics serve to warm and destabilize the air as well as to inject large quantities of water vapor into it. Precipitation in these areas is generally substantial, especially if orographic or monsoonal influences are present.

The middle latitudes, which include areas between 35° and 65° N and S, are characterized by moderate precipitation amounts. Many areas receive annual precipitation totals near the global mean of 35 inches (90 cm) per year. This is due largely to two factors. First, temperatures in the middle latitudes are mild because of the intermediate quantities of insolation received. This causes the air's water vapor holding ability, and therefore its potential for precipitation, to be moderate. In addition, the middle latitudes are located between high and low pressure belts; specifically, the subtropical highs and subpolar lows. Because highs act to decrease precipitation amounts and lows act to increase them, the tendency for moderate totals exists. As is true of all latitudinal zones, however, orographic factors and proximity to water can profoundly influence precipitation receipts.

In the high latitudes, extending from 65° N and S to the poles, precipitation means are lower than in any other latitudinal zone. This is primarily due to temperature. The cold air overlying these areas simply does not contain enough water vapor to generate large precipitation totals.

OTHER IMPORTANT PRECIPITATION VARIABLES
Up to this point, the discussion of precipitation patterns has centered on the amounts received in different parts of the world. While quantity is undoubtedly the most important of the geographic precipitation variables, several others are also noteworthy.

A factor with a profound effect on water supplies and agricultural activities is the *seasonal distribution* of precipitation. Many areas have precipitation relatively evenly distributed throughout the year, but other areas have distinctive wetter and drier seasons. In southern California, for example, nearly all precipitation occurs during the colder half of the year. Where precipitation is highly seasonal in nature, the maintenance of a continuous water supply is a problem, and cycles of human activities, especially those relating to agriculture, are strongly affected.

The *annual variability* of precipitation refers to the average percentage deviation of an area's precipitation from the mean. For example, if a location has a mean annual precipitation variability of 30 percent, a 50 percent chance exists that the total precipitation it receives in any given year will be at least 30 percent higher or lower than its long-term annual mean. A low annual variability of precipitation means that the area can more confidently expect to receive a quantity of precipitation close to its mean value. A high variability, on the other hand, means that the amount of precipitation to be received is much less certain and that floods or droughts are more likely to occur. Figure 5.21 depicts the pattern of the mean annual precipitation variability for the world.

A third factor that is not actually a precipitation factor, but that nonetheless has a tremendous effect on surface water supplies, is the *potential evapotranspira-*tion rate. It may be defined as the rate at which water would be lost through the combined processes of evaporation and *transpiration* (water evaporating from plants) if the surface were constantly supplied with an unrestricted quantity of water. The balance between precipitation and evapotranspiration largely determines the "wetness" of an area. Of the total precipitation falling on land areas of the world, approximately 75 percent is evaporated or transpired; only 25 percent reaches the oceans as runoff. Therefore, although precipitation is the chief means by which the land gains its moisture supply, evapotranspiration is the most important mechanism by which this moisture is lost. Potential evapotranspiration rates are influenced primarily by the air temperature and, to a lesser extent, by the relative humidity and total insolation receipt. In general, these rates are highest in the tropics and subtropics, especially in dry areas, and become progressively lower through the middle and high latitudes.

The necessity of examining both precipitation amounts and evapotranspiration rates in order to assess the moisture balance of an area is well illustrated in the case of Antarctica. The interior of this continent annually receives the liquid equivalent of only a few inches of precipitation, but is currently buried under approximately 2 miles (3 km) of ice. The Antarctic ice has formed from the accumulation of precipitation over many thousands of years, coupled with the virtual absence of losses through evaporation or sublimation. Conversely, areas with high annual or seasonal temperatures, such as the southeastern United States, require substantial precipitation to support a continuous vegetation cover.

Summary

This chapter has examined the role of water in the atmosphere. Water (H_2O) is a unique atmospheric component because of its ability to undergo phase changes between solid, liquid, and gaseous states. In so doing, it produces a variety of atmospheric phenomena, helps distribute heat energy derived from insolation throughout the earth-atmosphere system and, most importantly, makes possible the transfer of great quantities of water from the oceans to the land.

The continuous closed cycle of water transport between the surface and atmosphere is termed the hydrologic cycle. Although a number of alternate routes can be employed, the following five steps comprise the

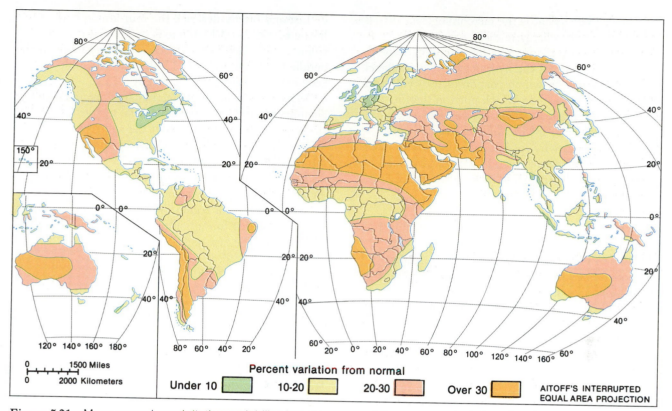

Figure 5.21 Mean annual precipitation variability. Values represent the average deviation of a given year's precipitation total from the long-term mean.

basic elements of this cycle as it involves land areas: (1) evaporation of ocean water; (2) condensation of this water vapor to form clouds; (3) transport of the clouds to the land by wind currents; (4) precipitation of the moisture onto the land; and (5) surface and subsurface runoff of the moisture back to the ocean.

Clouds form when air is cooled below the temperature at which it can hold all its moisture as water vapor. When this happens, the excess moisture condenses or sublimes onto dust particles. Most cooling which results in cloud formation occurs adiabatically as the air rises and expands. The air rises because of one or more of the following factors: atmospheric heating and resulting instability (convectional lifting); lifting along a weather front (frontal lifting); lifting caused by passage over a topographic barrier (orographic lifting); or lifting within the rising air spiral of a low pressure system (cyclonic lifting). Precipitation forms as cloud water droplets or ice crystals become large enough for gravity to cause them to fall toward the earth's surface.

The global precipitation pattern is complex because it is affected by many factors. In general, the largest precipitation totals occur in the tropics, where the air is very moist and low pressure prevails. Most portions of the subtropics experience varying degrees of aridity because of the influence of the subtropical highs. The middle latitudes typically receive moderate precipitation amounts caused largely by traveling frontal cyclones and summertime convectional showers. The high latitudes are the driest latitudinal zone because of the limited water vapor holding capacity of cold air. In general, precipitation totals are increased in areas near warm ocean currents and on the windward sides of mountain ranges.

Review Questions

1. List and define each of the six phase change processes involving water. How many calories of energy are transferred by each process per gram of water involved? What is the meteorological significance of the energy released when water undergoes a phase change to a lower energy state?

2. Draw a diagram of the hydrologic cycle and label the various steps. In which portion of the cycle does water remain longest? In which portion does it remain for the briefest period, on the average?

3. Define the terms *specific humidity* and *relative humidity*. Explain why specific humidity varies more by latitude than relative humidity, and why relative humidity varies more during the day than specific humidity.

4. What must normally happen to the air temperature in order for condensation or sublimation to occur? (Use the term *dew point* in your explanation.) List and briefly describe the major products formed by the condensation and sublimation of atmospheric water vapor.

5. What is fog, and why does it form more frequently at night than during the day? What areas of the United States are particularly prone to frequent fog, and why? In what respect is the method of formation of clouds similar to that of fog? In what respect is it different?

6. Describe the six basic forms of precipitation and briefly explain the atmospheric and/or surface conditions under which each occurs.

7. List and briefly describe the four atmospheric lifting processes. Which one is predominant in the tropics, and why? Which ones are predominant in the middle latitudes, and why? Which one is most geographically restricted, and why?

8. What factors account for the large precipitation amounts received in most equatorial regions? Why are the high latitudes and much of the subtropics dry? Why do large areas of the middle latitudes receive intermediate precipitation amounts?

9. What precipitation factors other than the annual mean totals received have important implications for the inhabitants of a region? Why is more precipitation generally needed for agricultural purposes in the low latitudes than in the high latitudes?

Key Terms

Evaporation	**Dew**
Condensation	**Frost**
Sublimation	**Radiation fog**
Hydrologic cycle	**Advection fog**
Specific humidity	**Convectional lifting**
Relative humidity	**Unstable air**
Dew point	**Orographic lifting**
Adiabatic temperature	**Frontal lifting**
changes	**Cyclonic lifting**

CASE STUDY

Snow Belts of the Great Lakes

The Great Lakes moderate the climate of the areas that surround them. During the spring and summer, the lake waters are colder than the land; and the shorelines receive cooling lake breezes. During the fall and winter, the lakes are warmer than the surrounding land and serve to warm the areas to their east. This warming has the beneficial effect of delaying autumn frosts and has permitted the planting of extensive orchards of cold-sensitive fruit trees. The same warming influence, however, is responsible for the formation of the heavy lake-effect snowstorms that plague the lee shores of the Great Lakes each winter. Each of the five lakes has a distinctive snow belt along its southern and eastern shores.

The lake-effect snowstorms are highly localized storms of a predominantly convectional nature. They form following the passage of strong cold fronts, when frigid arctic air is carried on gusty west or northwest winds over the unfrozen waters of the Great Lakes. The air picks up large quantities of water vapor during its passage over the lakes, providing the moisture for the heavy snow squalls.

The key factor responsible for lifting this moistened air is the air's great temperature contrast with the waters of the Great Lakes. Arctic air, as it reaches the lakes, may have temperatures well below 0° F ($-18°$ C). As it passes over the water, which usually has Fahrenheit temperatures in the low or mid-30s (0° to 4° C), it suddenly encounters a surface 30 to 60 F° (17 to 33 C°) warmer than it is. Large quantities of this heat are conducted to the overlying air, making it much lighter and more buoyant than the unmodified air farther aloft. This produces conditions of extreme instability. Convectional uplift of the air results, producing cumuliform clouds and heavy snow squalls. Latent heat energy, released as the water vapor condenses and sublimes, enhances conditions for uplift.

One of the most remarkable aspects of the lake-effect snowstorms is their persistence in any given area, despite their localized nature and the rapid movement of the clouds that produce them. In most parts of the world, a convectional storm will influence a given location for only a short period of time, such as an hour or less. This occurs largely because these storms are produced by the surface heating of the air, and the heat is soon dissipated by the cooling effect of the storm. The waters of the Great Lakes, however, generate a continuous supply of heat; and as long as the wind direction is constant, new snowclouds form along the same band to replace those that are streaming inland and dissipating. Under extreme conditions, this can produce continuous heavy snowfall for a period of several days, resulting in amazingly large snow depths. Table 5.2 lists the dates and maximum snow depths of some of the more noteworthy lake-effect snowstorms experienced in upstate New York over the past sixty years.

The winter of 1976–77 was a record-setting year for lake-effect snowstorms in many portions of the eastern Great Lakes region because of a jet stream pattern that produced a nearly continuous flow of arctic air over the Great Lakes for most of the winter. The season was capped by the great snowstorm of January 28 through February 1, which produced snowfall totals exceeding 100 inches

continued on next page

Table 5.2
Major Lake-Effect Snowstorms of the Last 60 Years in Upstate New York

Dates	Maximum Snow Depths in Inches
Oct. 18–19, 1930	48" in suburbs of Buffalo
Jan. 18–23, 1940	96" along east shore of Lake Ontario
Dec. 5–11, 1958	72" just south of Buffalo
Jan. 27–31, 1966	102" at Oswego
Jan. 9, 1976	68" at Adams
Jan. 28–Feb. 1, 1977	100+" near Watertown

(250 cm) in some locations southeast of Watertown, New York. This storm, which paralyzed the city of Buffalo, produced snowdrifts over 25 feet (7.6 m) high and resulted in 29 deaths. During the entire 1976–77 winter season, Buffalo received a record 199.4 inches (506.5 cm) of snowfall; but Hooker, New York—a small town east of Lake Ontario—was buried under 466.9 inches (1185.9 cm) of snow! Figure 5.22 depicts the 1976–77 snowfall totals for the snow belt areas of Lakes Erie and Ontario. The much higher totals to the east of Lake Ontario resulted from the fact that relatively shallow Lake Erie froze over in midwinter, cutting off its moisture supply to the atmosphere, while Lake Ontario, which is much deeper, remained unfrozen.

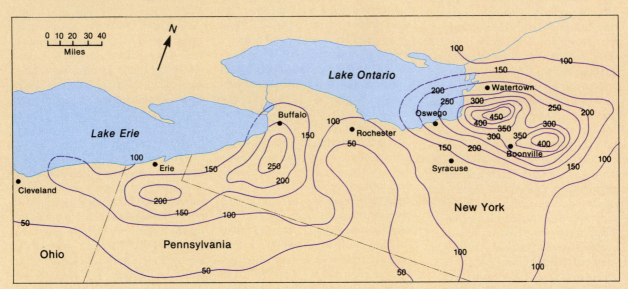

Figure 5.22 Snowfall totals in inches for the 1976–77 winter season in the snow belt region of Lakes Erie and Ontario. (*Weatherwise, "Lake-effect Snowstorms . . ." by Kenneth F. Dewey, Dec. 1977. Reprinted with permission of the Helen Dwight Reid Educational Foundation. Published by Heldref Publications, 4000 Albemarle St., N.W., Washington, DC © 1977.*)

Chapter Six

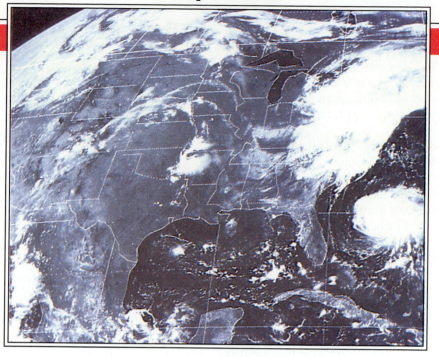

Weather Systems

Focus Questions

1. What are air masses? How and why do they form, and what influence do they have on our weather?

2. What are the most important weather systems of the middle and higher latitudes, and what types of weather conditions are associated with them?

3. How do thunderstorms and tornadoes form, and what areas are most affected by them?

4. How and why do weather conditions in the tropics differ from those of the higher latitudes?

5. What are hurricanes? How and where do they form, and what land areas do they influence?

The preceding three chapters have discussed the characteristics, causes, and world distributions of the six basic elements of weather and climate. The weather elements, however, do not occur separately. At any location and at any given moment, the air has a temperature, pressure, and relative humidity. Additionally, in most cases, wind is blowing from a given direction and at a certain speed; clouds of one or more types are present; and precipitation may be falling. Over large areas, these combinations of weather elements are organized into patterns that form weather systems such as air masses, fronts, and storms. Now that the components of the weather have been discussed, it is time to put them together to discover how and why weather systems develop, and what areas of the world they influence.

Some weather systems, notably air masses, occur in all parts of the world. Because they influence weather conditions everywhere, they will be discussed first. Most other weather systems are restricted primarily to certain general zones of latitude. For example, weather fronts, frontal cyclonic storms, and tornadoes are predominantly nontropical phenomena; easterly waves and hurricanes originate within the tropics. Following the discussion of air masses, the weather systems of the middle and higher latitudes will be treated; this will be followed by a discussion of the systems of the tropics.

AIR MASSES

Television weather forecasters make it obvious that air masses are important controllers of present and future weather conditions. Statements such as "A frigid Canadian air mass should reach our area by tonight . . ." or "A continued onshore flow of cool, damp Pacific air can be expected . . ." are the forecasters' stock in trade. But what exactly are air masses, and why do they form?

An air mass is simply a large body of air with relatively uniform temperature and moisture characteristics. A typical air mass has a diameter between 500 and 1500 miles (800–2400 km). Its properties develop gradually over several days as the air lies relatively stationary over a large land or water body. During this period, the temperature and moisture characteristics of this surface area, termed the *source region* of the air mass, are gradually imparted to the overlying air.

Air Mass Formation and Types

Air masses form with much greater frequency over some areas than over others. Like the air masses that form over it, a source region must be large in size and homogeneous in its temperature and moisture characteristics. The principal source regions for the world's air masses are located in the low and high latitudes, not in the middle latitudes (see Figure 6.1). The middle latitudes lack some of the basic criteria for air mass formation. For example, the large differences in insolation experienced over relatively short north-south distances cause the middle latitudes to lack the constancy of temperatures needed for air mass formation. In addition, the frequent presence of strong upper-level winds, especially the subpolar jet stream, does not often permit air to remain over a single location long enough to acquire the characteristics of the surface.

The virtual elimination of the middle latitude source regions means that air masses must form over large land or water bodies in the high and low latitudes. This has led to the development of a simple four-type classification system for air masses based on the nature of their source regions; the classifications also implicitly describe the temperature and moisture characteristics of the air masses themselves. Each classification consists of two words; the first indicates the surface nature of the source region. An air mass that forms over land is called *continental* and can be expected to contain dry air. A *maritime* air mass forms over water and can be expected to have a high water vapor content. The second word indicates the general latitudinal zone in which the air mass forms. A *polar* air mass forms over a polar or subpolar source region and is relatively cold. A *tropical* air mass forms over a tropical or subtropical source region and is characterized by warm temperatures.

Combining these terms, we have an air mass classification that consists of four types. On weather maps, these air masses are often abbreviated by using lowercase *c* and *m* for "continental" and "maritime," and uppercase *P* and *T* for "polar" and "tropical" as follows: continental polar (cP), continental tropical (cT), maritime polar (mP), and maritime tropical (mT).[1] The source regions

---

1. Meteorologists recognize a fifth type of air mass, which can be considered a variety of continental polar air: continental arctic (cA) air, which is extremely cold and dry. It forms only in the winter season over northern Canada, Alaska, and Siberia and is responsible for the occasional winter cold waves that affect most portions of the United States.

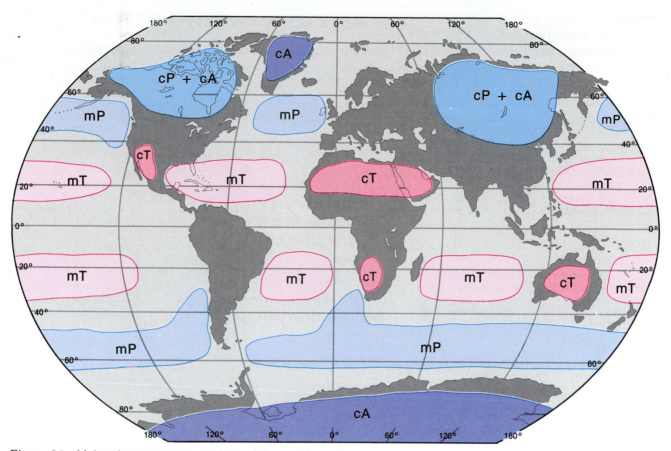

Figure 6.1 Major air mass source regions of the world.

Air Mass Movement and Modification

and the summer and winter characteristics of each of these air masses are summarized in Table 6.1.

Air masses are frequently transported out of their source regions by the upper-level winds. As air masses move, they bring their weather characteristics to new areas; a basic task of weather forecasters is to predict accurately the direction and speed of these air mass movements. Although air masses rarely form in the middle latitudes, those latitudes are constantly invaded by air masses from both the high and low latitudes. Indeed, the conflict between dissimilar air masses makes the weather of the middle latitudes more variable on a day-to-day basis than that of either the high or low latitudes. The boundaries between air masses are indicated on maps by weather fronts, and the differences in temperature and humidity conditions attending a frontal passage result largely from the change in the controlling air mass.

An air mass moving out of its source region is gradually modified as it passes over surfaces of differing temperature and moisture characteristics. Topographic barriers may also remove moisture through the orographic lifting process and may generate vertical wind currents that mix air in the lower levels of the air mass. Just as the air mass took several days to gain its characteristics, however, it normally takes several days to lose them. As a result, when it arrives over a new location, an air mass has temperature and moisture values that reflect both the conditions of its source region and those of the areas over which it has passed. The degree of modification that takes place depends on the speed at which the air mass moves, the distance it travels, and the extent of the dissimilarity between the

Table 6.1
Air Mass Characteristics

	Continental Polar (cP)	Maritime Polar (mP)	Maritime Tropical (mT)	Continental Tropical (cT)
Source Regions	Continental interiors in high latitudes	Oceans in subpolar regions	Oceans in tropics and subtropics	Continental interiors in dry portions of tropics and subtropics
Basic Attributes	Cold and dry	Cool and moist	Warm and moist	Warm to hot and very dry
Associated Summer Weather in Middle Latitudes	Pleasantly cool, low humidities, excellent visibilities, scattered cumulus	Cool to mild, moderately high humidities, mostly sunny in lowlands, showers likely in mountains	Uncomfortably warm and humid, hazy conditions inland, considerable cumulus cloudiness, scattered showers and thunderstorms	Very hot and dry, good visibilities, scattered daytime cumulus, very rare showers or thunderstorms
Associated Winter Weather in Middle Latitudes	Cold, dry, and often windy; excellent visibilities; clear to scattered or broken cumuliform cloudiness, snow showers likely in mountains or lee shores of warm water bodies	Chilly and damp, frequently windy, considerable cloudiness, rain or snow showers especially in mountains	Mild and moist, fog or low stratiform cloudiness common especially night and morning hours; substantial precipitation along frontal boundaries	Does not reach middle latitudes in winter

areas over which it passes and its source region. As an illustration of the modification of a southward moving cP air mass, note in Figure 6.8 on page 106 the progressively warmer temperatures that exist from Canada southward to Texas. (The temperature, in degrees Fahrenheit, for each reporting station is to the upper left of the circle that indicates the station.)

Eventually, an air mass is so extensively modified that it becomes, in effect, a different air mass. The formation, movement, modification, and dissolution of air masses is a continuous process that produces a constantly changing and never exactly repeated pattern of weather conditions over the earth's surface.

SECONDARY CIRCULATION SYSTEMS OF THE MIDDLE AND HIGH LATITUDES

The major air pressure and wind belts, which play a crucial role in the pattern of global climatic characteristics, are sometimes referred to as primary circulation systems of the atmosphere. While the primary systems exert their control over the climates of the world, a traveling and constantly changing array of smaller secondary circulation systems is responsible for the variations in daily weather conditions at a given site.

The changing patterns of secondary circulation systems are especially important in controlling the weather of the middle and high latitudes, which tends to be much more variable than that of the tropics. Three major types of systems can be recognized: anticyclones (high pressure systems), fronts, and cyclones (low pressure systems). Outside the tropics, all three are transported in a generally eastward direction by the upper-level steering currents; and each is associated with its own sequence of weather conditions. The causes, characteristics, and distributions of these weather systems will now be examined.

Traveling Anticyclones

An anticyclone, or high pressure system, can be described as a mound of air under relatively higher pressure than the air of surrounding areas. Highs develop where atmospheric conditions favor the upper-level convergence and subsidence of air. Because highs are generally located in the centers of air masses, they normally contain air that is relatively homogeneous in temperature and water vapor content. The wind-flow pattern associated with anticyclones was described in Chapter 4.

Anticyclones affecting the United States generally come from one of three source regions. During the colder portion of the year, when weather systems are typically in active movement across the country, high pressure systems containing Pacific mP air masses frequently enter the western United States. As they cross the western mountains they dry out and are reclassified as cP air masses, but their air remains relatively mild for the time of year. Southward loops in the jet stream frequently cause cold cP or cA anticyclones from central or northern Canada to invade the United States during the winter season (see Figure 6.2). They may bring subfreezing conditions deep into the South and East.

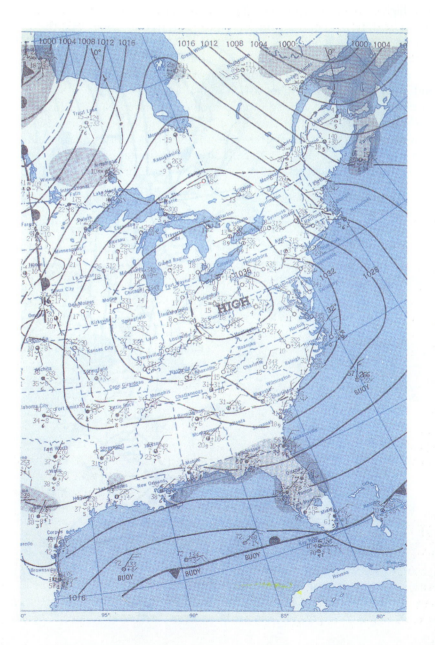

Figure 6.2 A portion of the surface weather map for March 28, 1982, showing a large Canadian anticyclone consisting of cP air centered over the central Appalachians. (*Courtesy of NOAA*, Daily Weather Maps, *March 28, 1982*)

During the summer, the southeastern and south-central portions of the country normally come under the domination of a quasi-stationary poleward extension of the northern subtropical high. This Bermuda High, as it is called, contains warm, humid mT air derived largely from the Gulf of Mexico (see Figure 6.3).

An orderly sequence of weather conditions is typically associated with the approach and passage of a cold Canadian anticyclone. As the cold front that precedes the high pressure system passes through the area, the weather turns cold, with blustery northwest winds and considerable cumuliform cloudiness. As the high pressure center approaches and the pressure gradient relaxes, the winds decrease in speed, the sky clears, temperatures remain cold, and visibilities are excellent. If the center of the high passes nearby, the winds become light and variable for a time, and then turn gradually to the south or southwest as the center moves to the east of the area. The southerly windflow, coupled with the gradual modification of the cP or cA air within the system, causes a warming trend. As the high continues moving eastward, an increase in cloudiness is likely as the next low pressure system, with its associated fronts, approaches.

Weather Fronts

A *weather front* is a boundary between adjacent air masses. It is a relatively narrow zone where air masses of different temperature and moisture characteristics come into contact. Because it is situated between the high pressure centers of two air masses, a front is a trough of low pressure. This low pressure causes air to flow from both air masses into the frontal zone. As the air converges, the warmer and lighter air mass is forced to rise over the cooler and denser air mass; this typically leads to the formation of clouds and frontal precipitation (see Figure 6.4).

The surface position of a front appears as a line on a weather map, but a front is actually a three-dimensional surface with height and width as well as linear extent. From its surface location, the frontal boundary extends aloft at a very gentle angle from the horizontal, so that

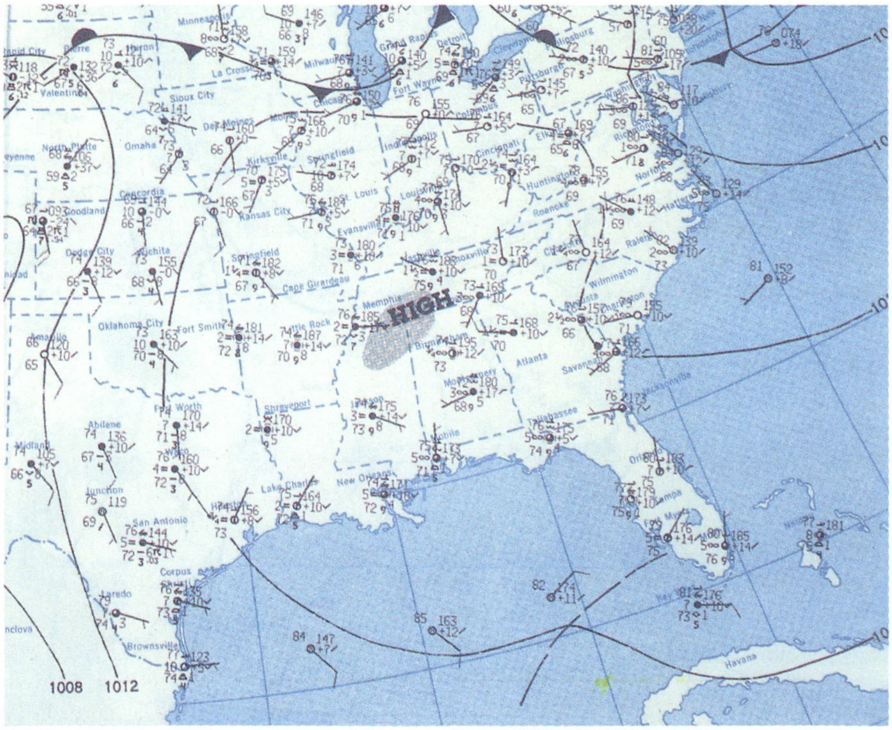

Figure 6.3 The surface weather map for July 17, 1983, showing a typical summertime "Bermuda High," composed of mT air, blanketing the Southeast. (*Courtesy of NOAA,* Daily Weather Maps, *July 17, 1983*)

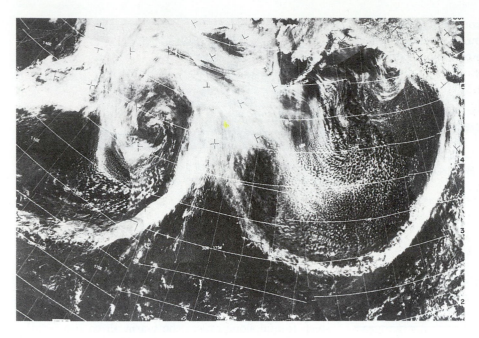

Figure 6.4 This satellite photo taken over the Pacific shows two low pressure circulation systems and their associated looping frontal cloud bands. (Clouds appear white.) (*Courtesy of NOAA*)

it resembles an inclined ramp. As a result, a few miles above the surface the front may be located several hundred horizontal miles from its surface position. This frontal inclination is caused by the differing densities of the two air masses, with the warmer air, which is lighter, always lying above the colder air. Fronts also have width because some mixing of the adjacent air masses takes place, and several hours may be required for the full transition of air masses to occur at a specific site following a frontal passage.

Fronts are especially numerous and important in the middle latitudes because this portion of the world is located within reach of both the cold air masses of the polar and subpolar regions and the warm air masses of the subtropics. The rapid weather changes for which the middle latitudes are noted are usually associated with frontal passages. These weather changes are most dramatic from late autumn through early spring, when the latitudinal variation in insolation, and therefore in temperature, is especially large. The stronger upper-level wind currents of the winter season also transport fronts at faster speeds than in the summer, allowing less time for modification and mixing of the opposing air masses.

The primary frontal boundary separating air masses of subpolar and subtropical origin (usually cP and mT air) is called the *polar front*. Although subject to large daily and seasonal fluctuations in position, it is generally located within the middle latitudes. This front is not straight, but develops large-scale waves that travel eastward along it, causing it to shift alternately northward and southward.

The passage of a front is usually attended by a number of changes in weather conditions. The extent and speed of the changes are subject to considerable variation, depending on such factors as the temperature contrast on opposite sides of the front and its speed of motion. These weather changes can be grouped into the following categories:

1. temperature
2. atmospheric water vapor content
3. wind direction and speed
4. air pressure tendency
5. cloud cover and precipitation

By observing changes in these weather elements, a knowledgeable weather observer, even without sophisticated meteorological instruments, can usually tell when frontal passages have occurred (see Table 6.2).

Types of Fronts

Weather fronts, which separate warmer and colder air masses, are named according to which air mass, if either, is advancing. Four types of fronts are recognized.

The **cold front** is the most common and typically the most sharply defined of the four. By definition, it is a surface along which advancing cold air is displacing

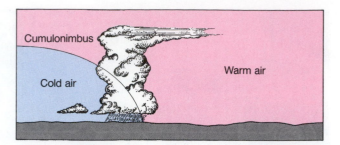

Figure 6.5 Profile (side) view of a cold front, showing the relative positions of the warm air, cold air, clouds, and precipitation.

Table 6.2
Idealized Frontal Weather Conditions*

Cold Front

Temperature: Warm before passage; progressively colder after passage.

Moisture: High humidity and dew point before passage; progressive lowering after passage.

Wind direction: Southwest winds before passage; northwest winds after passage.

Air pressure: Slow fall before passage; rapid rise after passage.

Clouds and precipitation: Brief period of heavy showers or a thunderstorm; rapid clearing after passage.

Warm Front

Temperature: Cold before passage; warmer after passage.

Moisture: High relative humidity but low dew point before passage; high relative humidity and high dew point after passage.

Wind direction: Easterly winds before passage; southwest winds after passage.

Air pressure: Rapid fall before passage; steady readings after passage.

Clouds and precipitation: Gradual increase and lowering of cloudiness followed by extended period of steady precipitation; slow clearing after passage.

Stationary Front

Weather: Extended period of cloudiness and light precipitation; temperature, moisture, wind, and air pressure remaining relatively steady.

Occluded Front

Temperature: Cold before and after passage.

Moisture: High relative humidity but moderate dew point before passage; lowering relative humidity and low dew point after passage.

Wind direction: Southerly before passage; west to northwest after passage.

Air pressure: Falling before passage; rising after passage.

Clouds and precipitation: Gradual increase and lowering of clouds, followed by extended period of precipitation; short period of heavy precipitation at end, followed by clearing.

*Conditions described are typical, but do not occur in all cases. For example, weak fronts or fronts separating dry air masses may pass with no precipitation and little cloudiness. Wind directions are those typical of the northern hemisphere; the north-south directional component should be reversed for southern hemisphere fronts. Terms such as warm and cold are *relative* to those that normally occur.

warm air. The denser cold air wedges beneath the warm air, lifting it well above the surface (Figure 6.5). If the warm air has sufficient water vapor, this uplift will produce frontal clouds and precipitation. Surface friction tends to slow the advance of the cold air near the surface, steepening the slope of the front. This relatively steep frontal slope causes the warm air to rise rapidly, typically producing a period of brief but heavy precipitation followed by rapid clearing. Fair, cooler, and drier weather may be expected over the next few days.

Along a **warm front,** advancing warm air is displacing cold air. Warm fronts tend to move considerably more slowly than cold fronts because in this case the advancing air is lighter than the air it is displacing (see Figure 6.6). The warm air rides up over the retreating wedge of cold air, causing the front aloft to be tilted in the direction of motion. (Note that a cold front aloft is tilted in the opposite direction.) As the warm air rises, it is cooled adiabatically and, with sufficient moisture, clouds and precipitation again develop. By retarding the movement of the lower air, surface friction stretches and flattens the front, rather than steepening it. Because the warm air gradually slides up this gentle incline, warm frontal precipitation usually falls from nimbostratus clouds and tends to be lighter in intensity and longer in duration than cold frontal precipitation. If, however, the rising warm air is unstable, heavier showers and possible thunderstorms may be interspersed within the area of precipitation.

The mean forward speed of cold fronts is about 25 miles per hour (40 km/hr). That of warm fronts is about 15 miles per hour (25 km/hr). Speeds average somewhat faster in winter than in summer because of the greater strength of the upper-level steering currents.

A **stationary front,** as the name implies, shows little or no present movement. Although the front itself may be stationary, warm air continues to drift up the

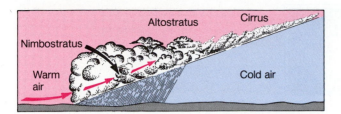

Figure 6.6 Profile view of a warm front. Note that a warm front, unlike a cold front, is tilted forward from its surface position.

frontal surface, producing cloudiness and precipitation. Fronts that remain stationary for several days tend to weaken and eventually lose their identity as the opposing air masses gradually mix and moderate. On the other hand, a front that has been stationary will frequently begin to move as an imbalance between the opposing forces develops; as it moves, it becomes either a cold front or a warm front.

It should be noted that any of these three types of fronts can, by a change in its motion, become any of the other types. In fact, different portions of the same front can be classified, because of their motions, as different types of fronts. This situation is analogous to that of a battlefront between opposing armies. On one day, the battlefront may move in one direction; on the next day, a counterattack may move it in the opposite direction. In addition, an army may simultaneously advance the front in one location and lose ground in another. Weather "fronts" received their name because these air mass boundaries, like the battlefronts of World War I, are linear zones of conflict between opposing forces.

The fourth type of front is different from the others in that it is formed by the meeting of three, rather than two, air masses. It was noted earlier that cold fronts tend to move somewhat faster than warm fronts. On occasion, a cold front will overtake a warm front, lifting the intervening mass of warm air entirely above the surface (see Figure 6.7). The joining of the warm and cold fronts forms an occluded front. Although a newly formed occluded front may produce substantial precipitation, the warm air that supplies most or all of this moisture has been cut off from its surface supply. As a consequence, precipitation amounts soon diminish as the remaining moisture is exhausted. Occluded fronts are weak fronts, since they separate two cold air masses at the surface. Weak fronts tend to become weaker as the opposing air masses mix and modify; therefore, these fronts usually lose their identity within a few days.

Frontal Cyclones

The importance of weather fronts is not solely the result of their associated precipitation and the changes in air mass that accompany their passage. They also produce conditions favoring the development of **frontal cyclones.** These storms are among the most important secondary circulation features of the atmosphere and are the dominant weather disturbance in both size and significance for areas outside the tropics. Frontal cyclones produce most of the large storms of rain and snow experienced in the middle and high latitudes. Their low air pressures cause diverse air masses to converge, often setting the stage for severe weather. In addition, they transport thermal energy from the surface to the upper atmosphere and help to carry oceanic moisture to the continents.

Formation and Developmental Cycle

A weather front provides a favorable setting for the development of cyclonic disturbances because it is a trough of low pressure and therefore a zone of air convergence. Low pressure centers can form only in areas where convergence occurs. In addition, through the process of frontal lifting, adjacent air masses of differing densities on opposite sides of a front provide a mechanism for releasing the heat of condensation of atmospheric water vapor.

The cycle of formation and dissolution of a frontal cyclone can be compared to that of an ocean wave, because both tend to build, break, and dissipate. In Figure 6.9 panels a through f, various stages of the cycle are diagrammed as they would appear on a weather map. Since the entire process normally takes about four or five days to complete, the time interval between each diagram is approximately 20 to 24 hours. It should be

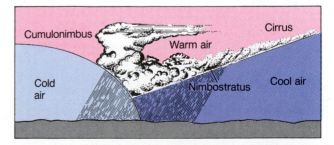

Figure 6.7 Profile view of an occluded front. These fronts are formed by the meeting of three, rather than two, air masses.

The Weather Map as a Forecasting Tool

The modern science of forecasting meteorology is highly complex, and meteorologists use a variety of sophisticated instruments to gather and analyze the tremendous quantity of weather data needed for accurate weather forecasting. Most weather analysis and prediction, in fact, is currently preformed by high-speed computers that can simultaneously digest and synthesize data from a large number of sources.

Despite this fact, a relatively old forecasting tool—the surface weather map—is still widely used by forecasters. Although it appears complex, a map such as that shown in Figure 6.8 merely displays the current weather at a large number of reporting stations in the United States and Canada. A person who can translate the weather information shown at each station and who has a basic knowledge of weather

continued on next page

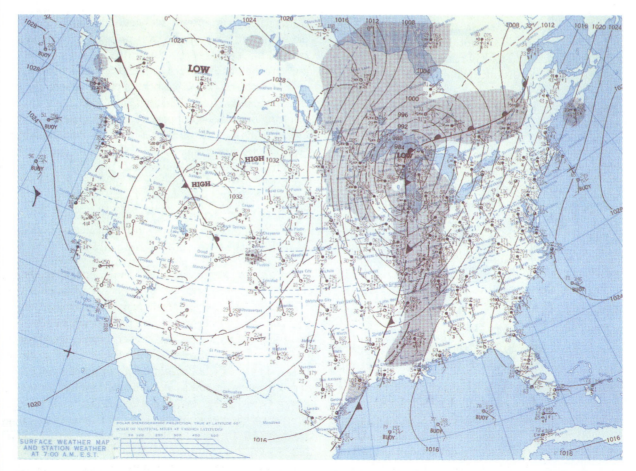

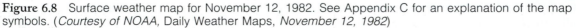

Figure 6.8 Surface weather map for November 12, 1982. See Appendix C for an explanation of the map symbols. (*Courtesy of NOAA,* Daily Weather Maps, *November 12, 1982*)

systems such as highs, lows, fronts, and air masses can often make reasonably accurate short-term forecasts using only the surface map. (Appendix C provides more detailed weather map information.)

Before we try a couple of sample forecasts using Figure 6.8, note the following points:

1. As shown in the station model on page 399 of Appendix C, each reporting station is indicated by a circle. The amount of cloudiness is indicated by the proportion of the circle that is blackened in. The temperature in degrees Fahrenheit is to the upper left of the station. The direction the wind is *coming from* is indicated by the "shaft" protruding from the station circle. The wind speed is indicated by the number of "feathers" at the end of the shaft, with each full feather representing a wind speed of an additional 10 knots (11.5 mph).

2. Areas of current precipitation on the map are shaded, and precipitation at any station is also indicated to the left of each station circle. Dots indicate rain, and stars indicate snow.

3. Weather fronts are shown as thick lines. Triangles on the lines represent cold fronts, half circles indicate warm fronts, and an alternation of these symbols indicates stationary fronts.

4. Air pressures are given in millibars and are shown by the use of isobars, the thin lines extending throughout the map.

5. Weather systems in the middle latitudes typically travel eastward at an average speed of 20 to 30 miles per hour, or roughly 500 to 600 miles per day.

Let's now try to forecast the weather for the next 24 hours at a couple of representative cities.

Oklahoma City. We see that Oklahoma City lies to the west of the cold front extending through the center of the country. The sky is currently mostly cloudy, the temperature is 46° F, and the wind is from the northwest at about 15 knots. As the cold front continues to move away and the large high pressure area consisting of cold cP air approaches, we would forecast that the skies will gradually clear and that temperatures will turn colder.

Pittsburgh. Pittsburgh lies in the warm mT air mass to the east of the cold front. Currently, the sky is mostly cloudy; the temperature is 58° F; and the wind is light and southerly. We would expect the front to pass through the Pittsburgh area within the next 24 hours, placing the city in the cP air mass. The forecast, then, should go something like this: "Cloudy and mild with increasing southerly winds over the next few hours. A period of rain or showers will begin shortly and last for several hours, followed by winds shifting to the northwest and becoming strong. The rain will end, but skies will remain mostly cloudy; and it will turn colder."

You may wish to try your own forecasts for Atlanta, New York City, and your own city.

stressed that both the diagrams and the description of the cycle in the following paragraphs pertain to "model" cyclones, and that no two frontal cyclones are exactly alike. Nevertheless, the differences are more in degree than in kind, and most frontal cyclones go through the basic stages described here.

In panel a, a portion of the polar front extends in an east-west direction through an area in the middle latitudes of the northern hemisphere. This front delineates the boundary between cold cP air to the north and warm mT air to the south. The wind direction in the opposing air masses parallels the front, with only a slight tendency toward convergence. This slight convergence is responsible for a zone of cloudiness and light precipitation along the front. A balance of forces exists between the two air masses, so the front is stationary.

In panel b, a bend, or *wave,* has formed on the front, perhaps in response to the approach of a trough in the jet stream. The position of the wave is indicated by both the actual bend, or wave, on the front and by the formation of a center of slightly lower barometric pressure. This low pressure begins to cause the air on opposite sides of the front to flow inward, initiating a counterclockwise spiral. Because warm air east of the center of the wave flows northward into the front, it begins to move northward as a warm front. At the same time, cold air west of the center is being directed southward, causing that portion of the front to advance as a cold front. The area of cloudiness and precipitation associated with the front undergoes a marked expansion as the wind-flow patterns of the two opposing air masses converge more directly on the front, thus increasing

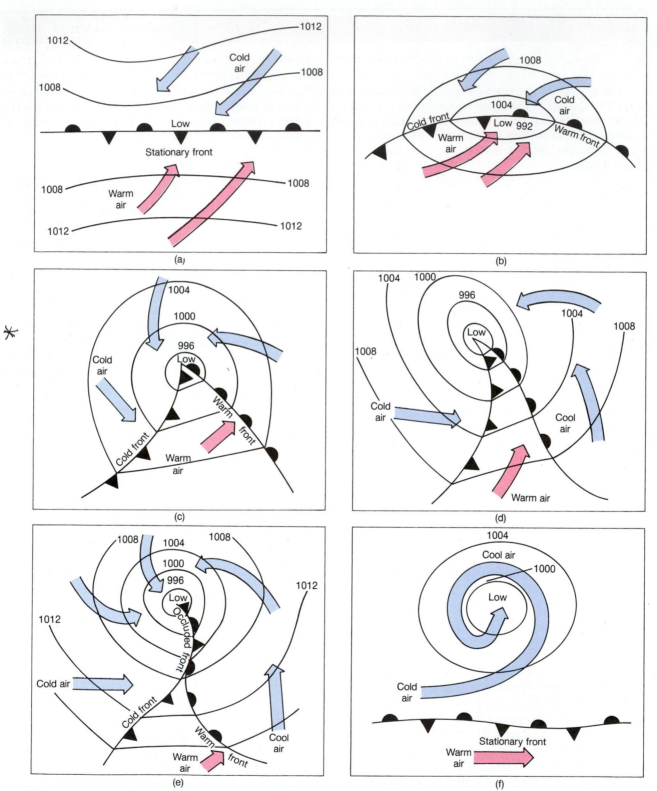

Figure 6.9 Stages in the "life cycle" of a frontal cyclone.

frontal lifting of the warm air mass. This, in turn, increases the release of heat of condensation available to power the low.

Panels c and d illustrate stages in the development of a mature open-wave cyclone (one that has not yet occluded). The developing wave is carried northeastward by the upper-level winds and continues to intensify as more and more heat energy is released by condensation and sublimation. This causes air pressures within the cyclone to fall and wind speeds to increase. The warm front, extending southeastward from the center of the low, advances northwestward, preceded by a broad band of precipitation. The trailing cold front, which extends southeastward from the low's center, advances southeastward at a somewhat more rapid rate, producing a narrow band of heavy showers and thunderstorms. Because the cold front advances more rapidly than the warm front, the mT air that supplies moisture and energy to the storm is gradually reduced to a narrowing wedge, and the storm center is increasingly surrounded by cold air.

In panel e, the cold front has overtaken the warm front in the vicinity of the low pressure center to form an occluded front; and the process of occlusion is continuing at the point of intersection between the warm and cold fronts. As this occurs, the warm air is lifted from the surface, and its supply of water vapor is cut off. The low pressure center gradually weakens as its energy source is depleted.

Panel f depicts the dissipating stage of the cyclone. The occluded front has weakened and lost its identity, and the low pressure center, now entirely surrounded by cold air, continues to fill and weaken. The polar front has once again stabilized, and, as the last remnants of the cyclonic circulation die away, the situation returns to that depicted in panel a. The stage is now set for the formation of another frontal cyclone that will repeat the pattern.

Frontal Cyclone Distribution and Movement

Frontal cyclones develop primarily within the middle latitudes. Their specific areas of formation vary during the course of the year because they are closely associated with the position of the polar front. The most vigorous storms develop during the winter and early spring, when the temperature contrasts between air masses of subpolar and subtropical origins are greatest.

Frontal cyclones affecting the United States in winter most commonly develop in the North Pacific, in the Gulf of Mexico, along the Atlantic coast, and in the eastern Rockies and adjacent sections of the Great Plains (see Figure 6.10). During the summer, the major storm track lies well to the north, near the border of the United States and Canada, and the storms are typically much weaker due to the reduced temperature contrast between the opposing air masses. The frontal cyclones

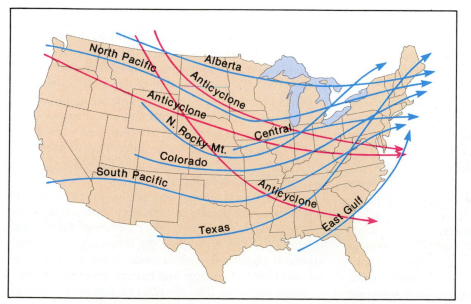

Figure 6.10 Typical winter tracks of cyclones (in blue) and of anticyclones (in red) across the United States. Note that while all systems track generally eastward, most cyclones have a poleward directional component, while anticyclones have an equatorward component of motion.

that develop in winter over the central and southern United States or along the Atlantic coast typically deepen and strengthen as they move northeastward. They eventually reach their peak development over the subpolar North Atlantic south of Greenland. In similar fashion, storms developing along the east coast of Asia move northeastward, often attaining massive proportions over the waters south of Alaska. In the southern hemisphere, the great temperature contrast between the frigid continent of Antarctica and the relatively mild adjacent ocean water often causes that landmass to be surrounded by a ring of well-developed frontal cyclones.

Localized Storm Types of the Middle and High Latitudes

Frontal cyclones have a profound impact on the middle and high latitudes because of their large size, widespread geographical distribution, and frequency of occurrence. In terms of their effect on people and property, though, two much smaller and generally more intense storm types, the thunderstorm and the tornado, are especially noteworthy. In reality, all three storms are interrelated, because thunderstorms outside the tropics are often associated with frontal cyclones, and tornadoes are produced only by severe thunderstorms.

Thunderstorms

A **thunderstorm** is a convectional storm accompanied by lightning and thunder and associated with cumulo-nimbus clouds and unstable air. Most thunderstorms produce heavy precipitation, usually in the form of rain, but are quite localized, with diameters of only a few miles. Severe thunderstorms are also capable of producing hail and tornadoes.

The thunderstorm is the world's most common storm type, with a nearly global distribution. It is estimated that, at any given moment, some 1800 thunderstorms are in progress over the earth's surface. They have undoubtedly impressed more people with the power of the forces of nature and have fostered more general interest in the weather than any other single meteorological phenomenon (see Figure 6.11).

Two categories of thunderstorms are commonly recognized by meteorologists. *Air mass thunderstorms* are strictly convectional in origin and may form anywhere when the atmosphere is sufficiently warm, moist, and unstable. Over land areas, they are most numerous

Figure 6.11 Time-lapse photograph of a thunderstorm at night, showing multiple ground-to-cloud lightning strokes. (*E. Simonsen/H. Armstrong Roberts*)

in the afternoon and early evening hours, when insolation has increased the temperature of the lower troposphere to its daily maximum value. Air mass thunderstorms are usually distributed randomly within an area; currently it is impossible to predict accurately the precise localities they will affect on any given day. *Frontal thunderstorms* are triggered by the frontal lifting of unstable air. Unlike the air mass type, these storms tend to form in a line, which may be located either along the front itself or at some distance ahead of it as a pre-frontal *squall line*. Some of the world's most violent and destructive thunderstorms develop along squall lines that form about 150 miles (240 km) in advance of strong cold fronts in the middle latitudes.

Both air mass and frontal thunderstorms are similar in structure. Each is composed of one or more *cells* that originally formed from a roughly circular updraft of air a few miles in diameter. Larger thunderstorms or those that have formed into a line are composed of multiple cells located adjacent to one another. The life cycle of a thunderstorm cell, which typically lasts less than an hour, is illustrated in Figure 6.12.

Thunderstorms occur in nearly all parts of the world, but vary greatly in frequency in different areas. Figure 6.13 illustrates two major aspects of their global distribution. The first is that the frequency of thunderstorms is correlated with latitude. They are most numerous near the equator and become progressively less common poleward. This distribution is related to that of

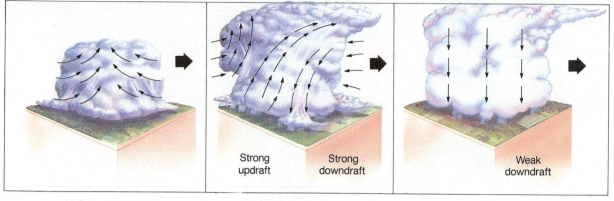

(a) Cumulus stage   (b) Mature stage   (c) Dissipating stage

Figure 6.12 The three stages of a single-cell thunderstorm "life cycle." During the *cumulus stage*, the developing cell consists of a single strong updraft of air that has been heated by both the ground and by the released heat of condensation. No rain has yet reached the surface. During the *mature stage*, the frictional drag of heavy rainfall columns produces strong downdrafts that gradually choke off the initial updraft feeding the cloud. In the *dissipating stage*, the downdraft has spread throughout the base of the cloud. This causes the lower portion of the cloud to evaporate as its upper ice-crystal portion gradually spreads out. Rainfall tapers off and eventually ends.

both global mean temperatures and specific humidity values—an indication of the great importance of heat and moisture to thunderstorm development. The second distributional aspect indicated by this figure is that thunderstorms are far more common over land than over water. This is primarily a thermal factor. Land areas attain maximum temperatures considerably higher than those experienced over water, making the air over land correspondingly lighter and more unstable. This same factor causes thunderstorms over land to form most frequently during the afternoon hours. Orographic lifting can also trigger thunderstorms when the air is unstable. This lifting mechanism is, of course, restricted to land areas.

The global pattern of thunderstorms is shown in more detail in Figure 6.14. As can be seen, the continent of Africa leads the world in the frequency of thunderstorms, followed closely by the world's wettest continent, South America. Thunderstorms are almost daily occurrences in many parts of the tropics and provide much of the rainfall received in that part of the world. In the middle and higher latitudes, thunderstorms occur primarily during the late spring and summer months, when insolation is sufficiently strong to make the air unstable. Most of the middle latitudes have an intermediate number of thunderstorms, while the high latitudes, some desert regions, and areas dominated by cold ocean currents have the lowest numbers.

In the United States, thunderstorms are most common in the Gulf Coast area, especially in central Florida, where strong solar heating in summer induces the convergence of unstable mT air from both the Atlantic and the Gulf of Mexico (see Figure 6.15). These storms are least frequent along the Pacific coast, where the cold California Current keeps the air cool and stable throughout most of the year.

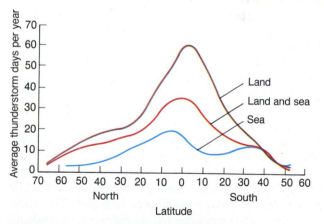

Figure 6.13 Latitudinal distribution of mean annual thunderstorm days for land areas, water areas, and the earth's total surface. Note that thunderstorms are most common near the equator and are also far more abundant over land than water.

111

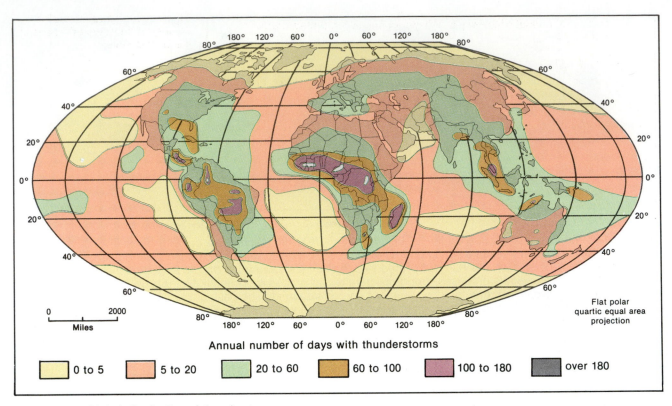

Annual number of days with thunderstorms

0 to 5	5 to 20	20 to 60	60 to 100	100 to 180	over 180

Figure 6.14 Global distribution of thunderstorms.

Tornadoes

The **tornado** is by far the smallest of the world's major storm types, but it is also the most violent. It takes the form of a tubular vortex of whirling air surrounding a central core of extremely low pressure, which extends downward from the base of a cumulonimbus cloud (see Figure 6.16). In a sense, the tornado should not be considered a separate storm type. Along with lightning, thunder, and hail, it is one of the phenomena uniquely associated with thunderstorms. Because the specific causes, characteristics, and world distributions of tornadoes differ from those of thunderstorms, however, they are treated here as a distinctive storm type.

Most tornadoes form in thunderstorms associated with squall lines that have developed in advance of strong cold fronts in the middle latitudes. The atmosphere must be very unstable, giving the air a tendency to rise rapidly. For this reason, tornadoes are most frequent during the spring, when a large temperature contrast exists between opposing mT and cP air masses. In addition, the air near the surface in spring heats much more rapidly than it does in the upper troposphere,

producing greater instability than is typical during the summer months. The tornado season begins in March in the Gulf Coast area and advances slowly northward with the retreat of the polar front, reaching the vicinity of the Canadian border in June or July.

A tornado is apparently initiated in the middle levels of a mature cumulonimbus cloud when a vigorous updraft begins to rotate. This occurs as updrafts feeding into the cloud are given a cyclonic spiral by wind shear conditions (opposing directions of wind flow) at differing altitudes. The area of rotation gradually extends downward through the cloud and narrows, increasing its rotational speed. Eventually, the vortex protrudes from the base of the cloud as a *funnel cloud,* which becomes a tornado if it reaches the ground.

A tornado funnel generally tapers from cloud to ground somewhat like an elephant's trunk. It moves along the surface at speeds usually between 20 and 45 miles per hour (9 to 20 m/sec). Typical diameters of tornado funnels at the surface are between 300 and 2000 feet (100–600 m), although diameters of up to a mile (1600 m) have been observed (see Figure 6.17). Wind speeds associated with tornadoes are difficult to

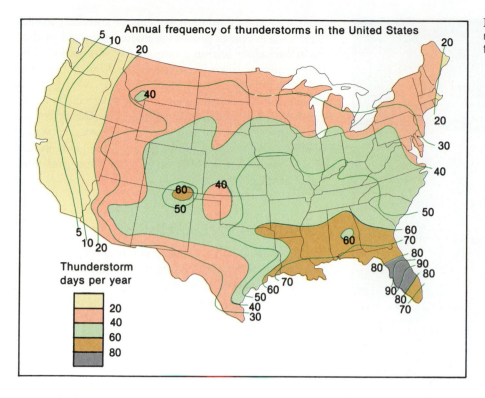

Annual frequency of thunderstorms in the United States

Thunderstorm
days per year

	20
	40
	60
	80

Figure 6.15 Mean annual number of thunderstorm days in the United States.

determine with any accuracy, but are currently believed to range from about 50 miles per hour to a maximum of 300 miles per hour (22 to 134 m/sec).

The paths of tornadoes are frequently highly erratic and vary greatly in length. On the average, tornadoes are on the ground for a distance of only about 4 miles (7 km) over a time span of perhaps 10 minutes. The record is held by a tornado that, on May 26, 1917, carved a path 292 miles (470 km) long through Illinois and Indiana. It was on the ground for over seven hours.

The great destructiveness of tornadoes is attributable to several factors. One is the winds of the tornado funnel, which are the most powerful of any storm type on earth. These winds are capable of blowing down buildings, leveling swaths of trees, and even lifting railroad locomotives from their tracks. Another tornado attribute that can have a deadly effect on large structures is the wind shear within the funnel, which can literally rip buildings apart. A third factor, which is largely responsible for the substantial annual death toll associated with these storms, is their unpredictability. The extremely localized nature of tornadoes, coupled with their erratic movement, rapid formation, and generally short existence, make accurate tornado forecasts extremely difficult at this time. Although the ability to reliably pinpoint specific areas that will expe-

rience tornadoes lies well in the future, the recent development of Doppler radar holds considerable promise. Doppler radar, which can remotely measure and display the speed of winds moving toward or away from

Figure 6.16 Tornado-like storms at sea are termed waterspouts. This photograph shows two waterspouts in contact with the sea surface, while two more funnel clouds are evident. (*Richard Jacobs © JLM Visuals*)

Figure 6.17 The narrow but destructive path of a tornado that sliced through Edmond, Oklahoma, on May 8, 1986. (*Chris Johns © 1987 National Geographic Society*)

the radar site, can detect larger-scale (mesocyclonic) rotary circulations developing within severe thunderstorms up to 20 minutes before they form tornadoes.

Tornadoes are largely restricted to land areas and are experienced mostly within the middle latitudes, where air masses of subtropical and subpolar origin are most likely to collide. The United States, with an annual average of over 700 tornadoes, has the dubious distinction of experiencing more of these storms than the rest of the world combined. The concentration of tornadoes within the United States results largely from its geographic position between major source regions of both cP and mT air masses, coupled with an absence of east-west-oriented mountain barriers. These two contrasting air masses are often drawn into contact and

conflict by the circulation systems of traveling frontal cyclones, sometimes producing major outbreaks of dozens of tornadoes within a one- or two-day period. The distribution of tornadoes within the United States is shown in Figure 6.18. Tornadoes occur with greatest frequency in the southern Great Plains, especially in the states of Texas, Oklahoma, and Kansas. Tornadoes are also quite common throughout the southeastern states. Within a given region, tornadoes are more likely in flat, treeless areas than in hilly or forested areas that offer higher surface friction.

TROPICAL WEATHER

The subtropical belts of high pressure serve as a sort of meteorological barrier dividing the earth into two zones that display marked differences in weather and climate. On the equatorward side of the subtropical highs, the weather is much less susceptible to dramatic change, especially with respect to temperatures. The tropics also experience their own storm types, which are triggered much more by the instability of the intensely heated atmosphere than by the conflict between opposing air masses.

General Weather Characteristics of the Tropics

When weather and climate between tropical and nontropical regions are examined over an extended period, several important differences become apparent. Nearly all the differences in the following list result from the year-round receipt by the tropics of large quantities of insolation.

1. Yearly temperatures within the tropics are warmer on average than those in the higher latitudes. No cold season occurs, so annual temperature ranges tend to be small and are often exceeded by daily temperature ranges.
2. Day-to-day temperature fluctuations are small and are usually related to variations in cloud cover. The constancy of temperatures exists largely because cold fronts do not normally penetrate into the low latitudes. In fact, genuine weather fronts of any type are rare within the tropics; and mT or cT air masses permanently control most areas.
3. While temperature is the leading seasonal climatic variable in the higher latitudes, precipitation is the

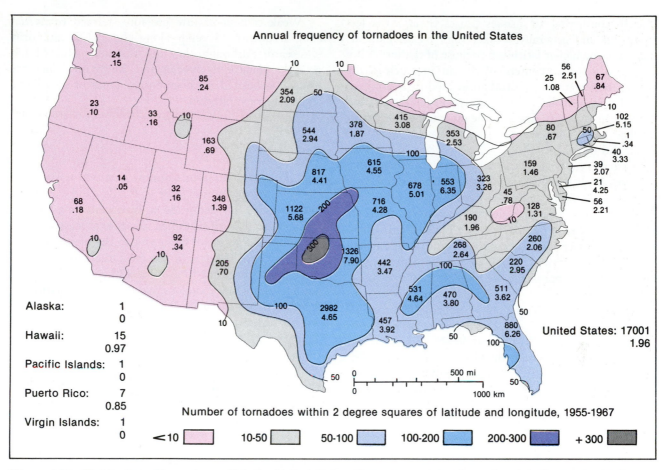

Figure 6.18 Distribution of tornadoes within the United States. The upper number in each pair indicates the total number of tornadoes within 2-degree squares of latitude and longitude between 1955 and 1967. The lower number represents the average annual number of tornadoes per 10,000 square miles.

leading seasonal climatic variable of most places in the tropics. In most tropical locations, summer is the wettest season; winter is the driest.

4. Whereas most precipitation in the higher latitudes is generated by frontal lifting, most tropical precipitation is convectional. As a result, it tends to be showery and often intense.

5. Frozen moisture forms, such as frost, snow, and sleet, do not occur in most tropical areas because of a lack of subfreezing temperatures. The only natural form of ice ever seen by many residents of the tropics is hail.

6. Although the prevailing direction of movement of wind and weather systems in most nontropical areas is from west to east, the prevailing direction is from east to west within the tropics. The weather systems of the tropics tend to move relatively slowly as well, because the upper tropospheric steering currents are

weaker, on the average, than those of the higher latitudes.

All in all, the weather of the tropics is much more constant than that of the higher latitudes. Most tropical areas experience a well-defined, repetitious daily cycle of morning heating, afternoon clouds and scattered showers, and nighttime clearing and cooling.

Weak Tropical Weather Disturbances

The predominant tropical weather disturbances are localized convectional showers and thunderstorms. Convectional activity in general and thunderstorms in particular are more common in the tropics than anywhere else on earth. Large-scale weather disturbances also develop within the lower latitudes, but the relative lack of conflicting air masses causes them to be less numerous and generally weaker than weather disturbances of the higher latitudes. Because of the instability and high moisture content of the air in much of the tropics, however, these disturbances produce substantial rainfall and contribute significantly to the precipitation totals of many areas. Three of the more important larger tropical disturbances are easterly waves, equatorial lows, and hurricanes.

Easterly waves are weak weather disturbances that are most likely to develop during the summer and early autumn over ocean areas. They consist of north-south-oriented troughs that drift slowly westward in the trade wind belts of each hemisphere (see Figure 6.19). Their western sides contain diverging air and generally fair weather. To the east of the trough line, the inflowing air converges as it approaches the axis of the wave, reducing the air's stability and producing numerous showers and thunderstorms. Important producers of precipitation for the islands and east coasts of continents that lie between approximately 10° and 30° N and S, easterly waves are watched closely during the summer and early autumn because of their occasional tendency to develop into hurricanes.

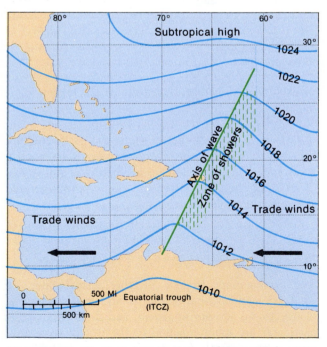

Figure 6.19 Weather map depiction of an easterly wave in the western Atlantic.

Weak centers of low pressure, termed *equatorial lows,* frequently develop along the ITCZ and drift westward. Although they are normally situated too near the equator to develop into hurricanes, they are usually associated with extensive areas of cloudiness and heavy precipitation.

Hurricanes

The hurricane, the world's most destructive storm, is the only strong storm of large size to develop within the low latitudes. By definition, a **hurricane** is a tropical cyclonic storm with sustained winds of at least 75 miles per hour (33.5 m/sec or 65 knots). In the western North Pacific, where hurricanes are most numerous and intense, they are termed *typhoons.* Other names for this storm type include "cyclone" in the Indian Ocean and "chubasco" off the west coast of Mexico.

Hurricanes gain energy and moisture through a combination of convectional and cyclonic lifting. The tremendous quantities of warm, moisture-laden air that spiral into these storms because of their very low barometric pressures are lifted into the upper troposphere and are cooled adiabatically to release torrential rainfalls.[2] The heat of condensation released during this process powers the storms and helps to maintain the global heat balance by transferring excess thermal energy from the surface to the upper atmosphere and, as the storms progress poleward, to the higher latitudes.

Development and Characteristics

Hurricanes form only over tropical or subtropical oceans with surface water temperatures of 79° F (26° C) or more. Unlike the frontal cyclones of the middle and higher latitudes, a hurricane does not develop on the frontal boundary between opposing air masses, but instead forms within a single mass of maritime tropical air. In its earliest stages of development, it is little more than a large, unorganized mass of convectional showers and thunderstorms. The rising of the air within the showers reduces surface pressures and initiates a weak

---

2. World records for both low barometric pressures and high rainfall amounts have been set in hurricanes. The world's lowest sea level air pressure of 25.69 inches of mercury (870 mb) was recorded during a 1979 typhoon in the Philippine Sea. A record 24-hour rainfall total of 73.62 inches (187 cm) occurred on the island of Réunion, in the Indian Ocean, during a 1952 storm.

Figure 6.20 Satellite photo taken on July 11, 1978, showing four tropical systems in differing stages of development in the eastern Pacific. (*Courtesy of NOAA*)

Tropical storm
Emilia

Hurricane

Tropical
depression

Tropical disturbance

inflow of air, which is given a cyclonic rotation by the Coriolis effect. The development of a cyclonic wind flow raises the disturbance to the category of a *tropical depression*. When its sustained winds reach 40 miles per hour (18 m/sec), the developing storm is given a name and is classified as a *tropical storm*.[3] Upon reaching sustained winds of 75 miles per hour (34 m/sec), the storm officially becomes a hurricane (see Figures 6.20 and 6.21). If conditions remain favorable for intensification, the hurricane may eventually attain wind speeds exceeding 150 miles per hour (67 m/sec); the most powerful storms on record had sustained winds in the vicinity of 200 miles per hour (90 m/sec).[4]

Hurricanes develop over warm tropical oceans at latitudes between 5° and 30° N and S (see Figure 6.22). They occur most frequently on the west sides of oceans, both because of the prevalence of warmer water and

because of the general westward movement of weather systems in the tropics; they are most frequent of all in the vast reaches of the western tropical and subtropical North Pacific. Cool water temperatures inhibit hurricane formation poleward of 30° latitude, although already-developed storms are often able to move well poleward of this limit before losing intensity. Oddly, the very center of the tropics—the area within five degrees of latitude of the equator—is immune from hurricanes. This immunity results because the Coriolis effect is not strong enough to impart the necessary rotary motion to weather disturbances this close to the equator.

The paths followed by most hurricanes are rather similar and are controlled by the upper-level windflow patterns. Most hurricanes form within the trade wind belt. Here, the upper-level winds that steer the surface weather systems blow toward a direction slightly poleward of due west. As a result, most hurricanes take an initial course toward the west-northwest in the northern hemisphere and toward the west-southwest in the southern hemisphere at speeds of perhaps 10 to 20 miles per hour (4–9 m/sec). As they move poleward, they approach the subtropical highs. The steering currents in the subtropics are weak and often erratic, but most hurricanes gradually turn more poleward and reduce their forward speed to 5 or 10 miles per hour (2–4 m/sec). Eventually, the storms enter the westerly wind belt and recurve toward the northeast in the northern

---

3. The practice of naming tropical storms and hurricanes according to alphabetical lists was begun in 1953 by the U.S. National Weather Service. Separate lists are maintained for the western Atlantic area, the eastern Pacific, and the western Pacific.

4. Only one storm this powerful is known to have occurred during the twentieth century in the Atlantic Ocean. This was the Labor Day Storm of 1935 that hit the Florida Keys. However, three relatively recent hurricanes—Camille in 1969, Allen in 1980, and Gilbert in 1988—had peak winds of at least 185 miles per hour (160 knots). Hurricane Camille is spotlighted in the Case Study at the end of the chapter.

Figure 6.21 Cross-sectional view through the center of a hurricane. Note the bands of cumulonimbus clouds alternating with clear air.

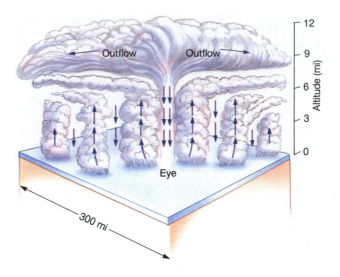

hemisphere or toward the southeast in the southern hemisphere. Their forward speed gradually accelerates to perhaps 30 to 40 miles per hour (13–18 m/sec), carrying them rapidly poleward. Eventually they pass over colder water, which causes them to lose strength and eventually to dissipate because of the loss of the water vapor that supplies their energy.

Although the majority of hurricanes reach colder water and dissipate without ever encountering land, others strike coastal areas, often with disastrous consequences. The great destruction and loss of life associated with the landfall of these storms results from three factors. The best known but often least destructive of these are hurricane-force winds, which are capable of

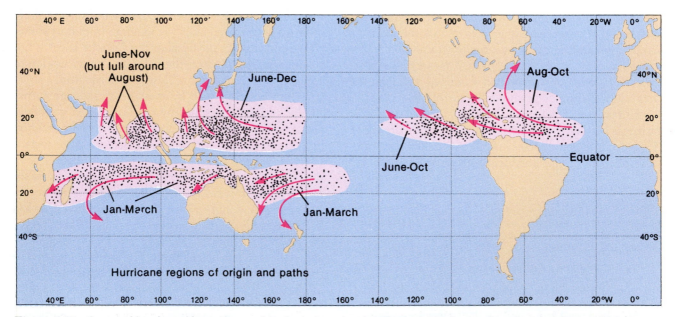

Figure 6.22 Areas of hurricane formation and typical storm tracks. Each dot indicates the site where a tropical storm initially reached hurricane intensity.

uprooting trees, collapsing weaker buildings, and turning loose objects such as boards and strips of sheet metal into lethal missiles. A second factor is the extremely heavy rainfall associated with hurricanes. Rainfall totals, which may exceed 10 inches (25 cm) with slow-moving storms, frequently cause extensive flooding, landslides, and loss of life. By far the most deadly feature of hurricanes, however, is the *storm surge* that accompanies their landfall. The very low air pressure within a hurricane, coupled with strong air convergence, causes the ocean surface to bulge upward as much as 30 feet (10 m) above normal water levels. This water may then be blown onshore by winds well in excess of 100 miles per hour (45m/sec), inundating the coast. Flooding is most extensive on low-lying coastlines and on those occasions when a hurricane crosses the coast during the time of normal high tide. The greatest loss of life caused by a hurricane in the United States occurred in Galveston, Texas, in 1900, when 6000 people were drowned by a storm surge. The largest numbers of human casualties on record are associated with two hurricanes that struck the Bay of Bengal coast in what is now Bangladesh in 1737 and in November 1970. The number of lives lost in each of these storms was estimated at a staggering 300,000! Nearly all deaths in both cases were produced by the storm surge, which in the 1737 storm was said to have raised water levels by 37 feet (12 m), causing the ocean to advance inland for many miles.

Summary

This chapter has examined the earth's major weather systems. These systems are in a constant state of flux—forming; changing size, shape, and intensity; and dissipating—as they drift across the face of the earth, largely in response to the upper level steering currents. As they pass over any given location, their attendant weather conditions influence that location for a period of a few hours or days. Each weather system is an organized assemblage of the weather elements discussed in the previous chapters, and each exists because of the differential receipt of insolation in various portions of the earth system. The seven weather systems discussed in greatest detail in this chapter are air masses, anticyclones, weather fronts, frontal cyclones, thunderstorms, tornadoes, and hurricanes.

Air masses are large bodies of air that are relatively homogeneous in their temperature and water vapor content. They form over large land or water bodies that gradually impart their surface temperature and moisture conditions to the overlying air. After they have formed, air masses may be transported to other parts of the world by the upper-level winds. Most air masses form in either the high or low latitudes. The middle latitudes experience frequent incursions of both polar and tropical air masses.

The major large-scale weather systems of the middle and higher latitudes are anticyclones (high pressure systems), weather fronts, and frontal cyclones. Anticyclones are generally associated with the centers of air masses. They are areas of descending and diverging air and generally produce fair weather. Weather fronts are linear boundaries where air masses from different source regions meet. They are typically associated with cloudiness and precipitation as air from the warmer, and therefore more buoyant, air mass flows over the adjacent colder, denser air mass. Frontal cyclones are low pressure systems that form on weather fronts. Their convergent wind-flow patterns typically draw in warm air on their equatorward and eastern sides and cold air on their poleward and western sides, forming warm and cold fronts, with their associated precipitation.

Thunderstorms and tornadoes are relatively small but intense storms associated with unstable air. They are most likely to form when the air has been strongly heated at the surface, especially during the afternoon hours. This produces vigorously rising convectional air currents which, in the case of tornadoes, are apparently given a cyclonic spiral by wind shear conditions (opposing directions of wind flow) at differing altitudes. Thunderstorms are most numerous over land areas in humid tropical regions and generally decrease in frequency poleward. Tornadoes, however, are most frequent in the middle latitudes, especially in the central United States, where they typically form in advance of strong cold fronts.

Tropical weather conditions differ from those of the higher latitudes largely because the tropics have large insolation receipts throughout the year. This makes the air warmer and more buoyant and increases its water vapor holding capacity. As a consequence, tropical temperatures are warm throughout the year, weather fronts are generally absent, and precipitation is often heavy, coming largely in the form of afternoon convectional showers.

Hurricanes are the most powerful storm type of the tropics. They are cyclonic storms that form over ocean areas at latitudes between approximately 5° and 30°.

They are initiated when a region of moist, unstable air containing numerous showers and thunderstorms develops a cyclonic wind-flow circulation that gradually intensifies. Like most tropical weather systems, hurricanes tend to move in a westerly direction; so the east coasts of land masses are most frequently affected. If they reach the middle latitudes, however, they usually recurve to a poleward and easterly direction. They typically dissipate after striking land or passing over colder water.

Review Questions

1. What is an air mass? Where and how does it attain its characteristics?

2. Name the four basic types of air masses and briefly describe the temperature and moisture characteristics of each. Which type of air mass dominates in your area in summer? In winter? What are the source regions of these air masses?

3. Describe the wind-flow patterns associated with anticyclones and cyclones in each hemisphere. Why are the patterns different in the northern and southern hemispheres?

4. What is a weather front? Explain how weather fronts are classified and briefly describe each of the four types of fronts.

5. During what time of year are thunderstorms most frequent in your area? Explain why. What essential characteristics must the atmosphere exhibit in order for thunderstorms to develop?

6. What is a tornado? Under what meteorological conditions are tornadoes likely to form? Why do they occur more frequently in the United States than in other parts of the world?

7. Discuss the basic ways in which weather conditions would differ from those typical of your own area if you lived near the equator.

8. What is a hurricane? Where do these storms form, and why? Why do you think they are frequently experienced on the East Coast of the United States, but not on the West Coast?

Key Terms

Air mass	Occluded front
Anticyclone	Frontal cyclone
Cold front	Thunderstorm
Warm front	Tornado
Stationary front	Hurricane

CASE STUDY

Hurricane Camille of 1969

The entire Atlantic and Gulf Coast of the United States, from New England to the Mexican border, is vulnerable to the devastation produced by hurricanes. Despite continuing advances in hurricane-forecasting technology, the unprecedented pace of human development of this coastal region continues to increase the risk to lives and property. For the past few decades, hurricanes have been especially numerous and intense within the Gulf of Mexico, while the Atlantic coast has received relatively few major storms. One of the most powerful and destructive Gulf hurricanes on record was Hurricane Camille, which struck the Mississippi coast in August of 1969.

Hurricane Camille began as a weak tropical disturbance in the eastern Atlantic off the West African coast (see Figure 6.23). First detected on August 5, 1969, it drifted west-northwestward across the tropical North Atlantic and into the Caribbean Sea while still weak and poorly organized. On August 14, the disturbance was centered near the Cayman Islands in the northwestern Caribbean. At this point, it assumed a more northwesterly course and began to intensify rapidly, soon reaching tropical storm status. Camille, as the storm was named, attained hurricane strength early on August 15, and later that day crossed the western tip of Cuba with 115 mile-per-hour (50

m/sec) winds. Emerging into the warm waters of the Gulf of Mexico, the storm resumed its rapid intensification, with wind speeds reaching a phenomenal 185 to 200 miles per hour (83–89 m/sec) by the evening of August 16. Winds of this magnitude are very rare because of the tremendous amount of friction generated between the atmosphere and the ocean surface; nearly ideal conditions must therefore have existed for Camille's development at this time. The intensity of the storm is indicated by the extremely low barometric pressure of 26.73 inches of mercury (905 mb) measured in Camille's eye by a Hurricane Hunter plane on the evening of August 16.

Hurricane warnings were posted for the central Gulf Coast, and on August 16 and 17 a massive evacuation took place. A few residents, however, chose to "ride out" the storm, a

continued on next page

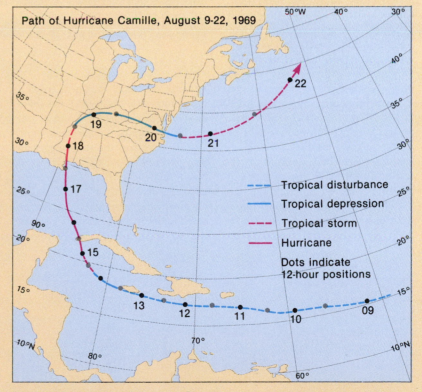

Figure 6.23 Path of Hurricane Camille.

fatal decision for some of them. Camille moved ashore in the vicinity of Gulfport, Mississippi, at about 11:00 P.M. on the seventeenth, accompanied by winds of at least 185 miles per hour (83 m/sec) and a 23-foot (7 m) storm surge. The combination of winds and the storm surge resulted in some 171 deaths in Mississippi and Louisiana, mostly by drowning; but timely warnings are credited with having saved as many as 100,000 lives.

As Camille moved northward through Mississippi on the eighteenth, its winds rapidly diminished. On the nineteenth, the weakening storm, now a tropical depression, curved eastward through Tennessee and Kentucky, producing heavy, but not excessive, rains. Camille still had one more surprise left, however. During the night of August 19–20, while the storm's remnants were crossing the mountains of West Virginia and Virginia, its rainfalls briefly intensified to unprecedented levels. Precipitation totals exceeded 30 inches (76 m) in a small area of west-central Virginia within a six-hour period, producing massive flooding and landslides (see the Case Study at the end of Chapter 13) and causing 158 deaths.

The storm passed off the Virginia coast on the afternoon of August 20. As it moved over the waters of the

Figure 6.24 A shopping center in Biloxi, Mississippi, destroyed by Hurricane Camille. (*NOAA, JLM Visuals*)

Atlantic, it soon regained tropical storm intensity, but, under the influence of strong upper-level winds, was carried rapidly northeastward. Within two days the storm began to pass over cold waters off the coast of Newfoundland, where it weakened and dissipated.

Hurricane Camille will be remembered as one of the great hurricanes of the twentieth century. Total property damage attributed to the storm was estimated at nearly $1.5 billion, of which $950 million occurred in Mississippi. In all, 326 human lives were lost. In retrospect, this figure is surprisingly small and is a tribute to both the accuracy of the warnings and the efficiency of the evacuation which preceded the landfall of this great storm.

Chapter Seven

Climates of the World

Outline

Focus Questions

1. What are the chief advantages and shortcomings of climate classification systems?
2. What are the major climatic characteristics of each of the global climate regions, and why do these characteristics exist?

In this chapter, the global pattern of climates is examined on a regional basis. A region's climate consists of its characteristic long-term weather conditions. Separating weather and climate is often difficult, however, because only the time factor differentiates the two subjects. Much of the material presented in the previous chapters, especially the discussions of the global distributions of the various weather elements, was actually climatic information. In contrast to the approach of the previous four chapters, however, where weather and climate phenomena were discussed in a topical fashion, we now examine the long-term condition of the atmosphere in a spatial context by dividing the earth into a number of climatic regions.

An understanding of the global pattern of climate is vital to the geographer because climate is the leading independent environmental variable. It exerts a powerful and sometimes controlling influence on the other components of the environment. For example, climate is the primary factor controlling the world distribution of natural vegetation. Together, climate, vegetation, and rock type largely determine the type of soil that will develop within a given area. Climate also exerts a major influence on landform characteristics.

The distribution of earth's human population is also strongly influenced by climate. Especially important in this respect are climatic characteristics, such as extremes of temperature and precipitation, that restrict the growth of the cultivated plants that form the basis of our food supply. Continuing rapid advances in science and technology are reducing the degree to which climate influences the distributional pattern of the human population, but its continuing importance cannot be denied.

CLIMATE CLASSIFICATION

Examined in detail, the climate of any locality is unique. The key to understanding regional climate patterns and to developing climatic classification systems lies in generalization. Climate types are of necessity based not on specific values, but on ranges of values. All areas having conditions falling within the stated ranges are considered, perhaps somewhat arbitrarily, to have the same type of climate. Climatic systems are devised to simplify the study of climate by reducing the number of climate types to a manageable total.

The division of the earth into climatic zones allows regional associations to emerge. It becomes apparent that the global distribution of climates is not random; rather, climates are spatially organized, with similar climatic conditions occurring in widely separated parts of the earth. This, in turn, points to the existence of unifying controls that operate similarly in areas that may be distant from one another.

The use of classification systems to regionalize climates also has limitations that need to be understood by those employing them. One of the most important shortcomings is loss of detail. The more that climatic characteristics are generalized, the larger the individual climate regions tend to become, and the more detail is sacrificed. A second shortcoming is the danger of taking climatic regions too literally. The regions and boundaries shown are correct only for the system being used, and other, equally valid systems may divide the area under consideration into quite different regions. It must also be realized that climates do not change suddenly as boundary lines are crossed, but tend to vary gradually over large distances.

The data employed to develop climate classification systems also have limitations. Inadequate climatic data are available for large portions of the earth. This includes the 71 percent of our planet covered by oceans as well as extensive areas in the polar regions, subtropical deserts, and tropical forests. As data from these areas become increasingly available, climatic boundaries are occasionally revised.

Because of the complexity of world climate patterns and the diversity of human interests, no single classification system satisfactory to everyone can be devised. A variety of systems has been developed through the years, and the choice of the system depends largely on its intended purpose. Our need in this chapter is for a climate classification system that is relatively simple, stresses geographic patterns of climates, and corresponds to the world patterns of natural vegetation and soils that are examined in subsequent chapters. The system used in this chapter divides the world into eleven climatic regions that are defined largely on the basis of their temperature and moisture characteristics. For each region, stress is placed on the following factors:

1. geographic distribution
2. basic atmospheric temperature and moisture characteristics
3. the reasons for these characteristics

Following the precedent established in the discussion of air pressure and wind belts, our examination of world

Climagraphs

A common technique for providing a clearer understanding of climate characteristics is to display climatic data for representative reporting stations in graphic form on **climagraphs,** such as the one displayed in Figure 7.1. Most climagraphs depict monthly mean values of temperature and precipitation. For clarity, the temperature data are usually indicated by points that are connected to form a line graph, while the precipitation values for each month are indicated by the respective heights of twelve bars. Additional information that may be provided on climagraphs includes station latitude and longitude, elevation, a location map, annual temperature means and ranges, annual precipitation totals, and climate type. Climagraphs providing this information are used throughout the chapter to illustrate representative conditions in each of the climatic regions discussed.

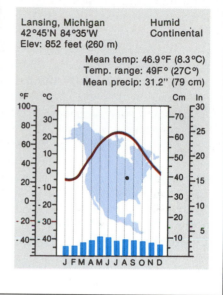

Figure 7.1   Climagraph for Lansing, Michigan. Mean temperatures throughout the year are depicted by a line graph, and mean precipitation amounts for each month are indicated by bar graphs.

climatic regions begins at the equator and progresses poleward. The locations of the various climate types are mapped in Figure 7.2.

however, climatic conditions are controlled by the seasonal migration of these two pressure belts. This produces a pattern of alternating rainy and dry seasons.

THE LOW LATITUDE CLIMATES

Approximately half the earth's surface is located within the low latitudes. The basic weather characteristics of this part of the world were discussed at the end of the preceding chapter; in essence they consist of continuously warm temperatures, along with showery precipitation that is predominantly convectional in nature. The location of the subtropical highs and the Intertropical Convergence Zone (ITCZ) is largely responsible for the differences between the three major climates of the low latitudes. Near the equator, the ITCZ produces year-round cloudy and humid conditions, with generally abundant rainfall. Conversely, within the subtropical highs, clear skies, low relative humidities, and scanty rainfalls generally prevail. In much of the low latitudes,

Tropical Wet Climate

The Tropical Wet Climate is the most equatorward of the earth's climatic types. It occurs in extensive areas between 10° N and S, and in some coastal areas it extends to the vicinity of the Tropics of Cancer and Capricorn (Figure 7.2). Three large land areas are contained within the Tropical Wet Climate: the Amazon Basin in South America; the northern Congo Basin in western Africa; and the majority of the islands of the western Pacific "East Indies," including most of Malaysia and Indonesia, and the Philippines.

The Tropical Wet Climate may be characterized as warm and rainy. Most locations near sea level have monthly mean temperatures near 80° F (27° C) throughout the year. Constantly high midday sun angles and

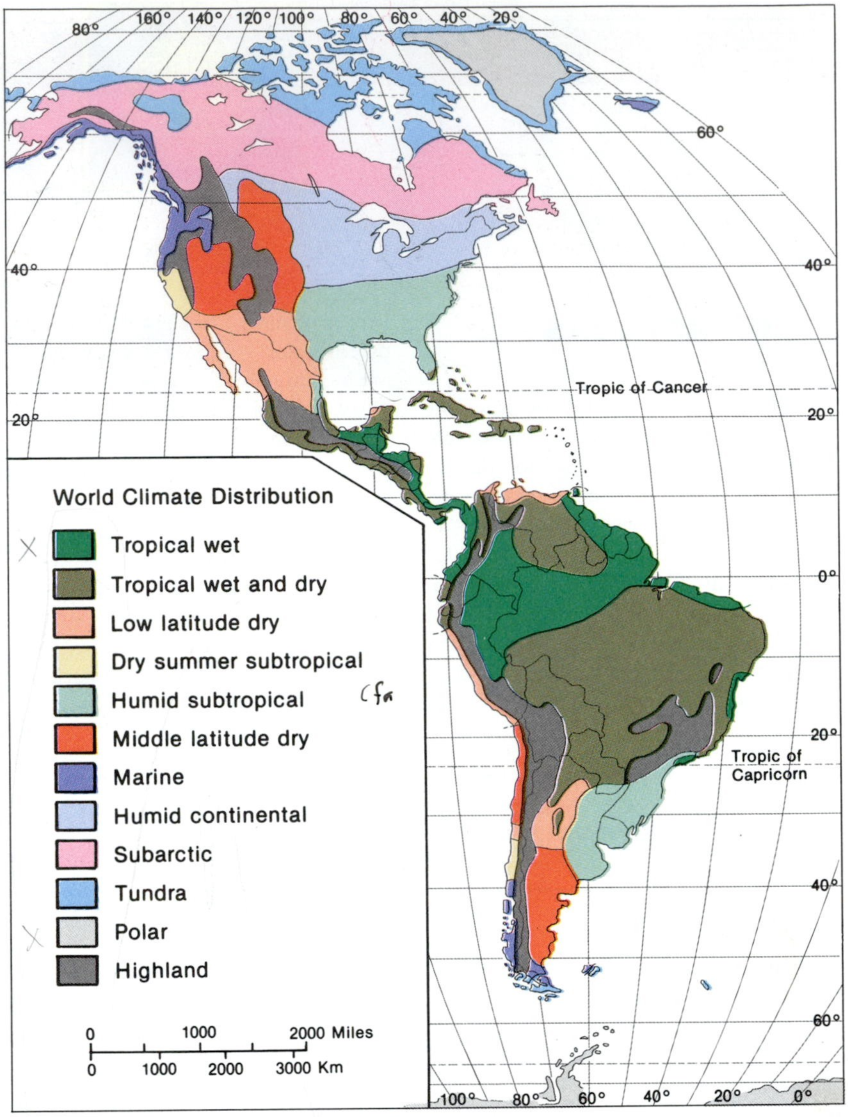
World Climate Distribution
Tropical wet
Tropical wet and dry
Low latitude dry
Dry summer subtropical
Humid subtropical
Middle latitude dry
Marine
Humid continental
Subarctic
Tundra
Polar
Highland
0
1000
2000 Miles
0
1000
2000
3000 Km
Tropic of Cancer
Tropic of Capricorn
60°
40°
20°
0°
20°
40°
60°

Figure 7.2 World climate types.

daylight periods of nearly equal length throughout the year are responsible for the warm temperatures as well as the very small annual temperature ranges, which are 5 F° (3 C°) or less for most locations. Even areas situated near the poleward margins of the climate almost never experience annual ranges of over 10 F° (6 C°).

Not only are seasonal differences in temperature very small, but day-to-day variations are minor and result primarily from differences in cloud cover. Air masses of nontropical origin rarely, if ever, penetrate into the inner tropics to create frontal temperature contrasts; and maritime tropical air dominates throughout the year. The result is that one day is very much like another with respect to temperature. Only the constant rhythm of daily heating and nightly cooling provides thermal variety, leading some writers to describe day and night as the "seasons" of the tropics. In all portions of the Tropical Wet Climate, daily temperature ranges substantially exceed annual ranges.

Yearly rainfall totals in the Tropical Wet Climate usually average between 50 and 200 inches (125–500 cm), making it the world's rainiest climate type. The heavy precipitation totals result from the high moisture content of the air, intense surface heating, and the proximity of the ITCZ. Most rainfalls occur as heavy showers and thunderstorms of brief duration

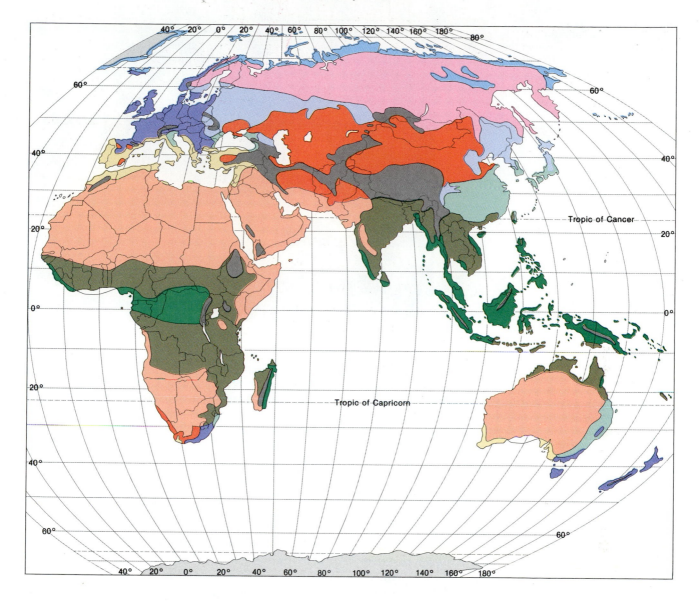

interspersed with periods of sunshine. Along coastal areas, showers may occur at any hour of the day or night. Inland areas, on the other hand, generally have distinct afternoon precipitation maxima caused by surface heating, which results in the convective uplift of the unstable air. Especially heavy rainfalls occur where mountains provide an additional lifting mechanism.

Most locations within the Tropical Wet Climate experience distinct periods of increased and decreased rainfall during the course of the year. The chief cause of the seasonal variability in precipitation is the proximity of the ITCZ. This zone of converging air is associated with heavy rainfalls, and its seasonal shift in latitude causes the area of heavy rains to oscillate alternately northward and southward, affecting different portions of the Tropical Wet Climate. Some locations near the equator, particularly in Africa, experience two periods of increased rainfalls, one in the spring and one in the fall. Such a double pattern occurs in Kisangani, Zaire (see Figure 7.4).

Except for local gusts in heavier showers, winds in many areas within the Tropical Wet Climate are light and variable. Daily sea breezes develop in most coastal locations, providing welcomed relief from the heat. Locations in Southeast Asia experience the strongest winds because of their monsoonal wind-flow pattern.

Figure 7.3 A village on the banks of the Amazon, within the Tropical Wet Climate. (*Breck Kent, © JLM Visuals*)

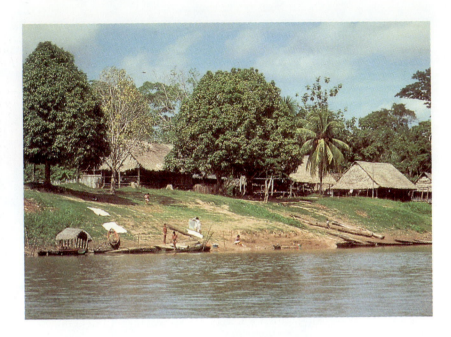

Tropical Wet and Dry Climate

Proceeding poleward in either direction from the Tropical Wet Climate, one enters regions dominated by alternating wet and dry seasons. This climate type, commonly known as the Tropical Wet and Dry Climate, exists in most areas between 5° and 20° N and S. It is the chief climate of the trade wind belt and is located between the ITCZ and the subtropical highs of each hemisphere.

From both a geographic and a human standpoint, the Tropical Wet and Dry Climate is one of the world's most important climate types, as it occupies an extensive and heavily populated segment of the earth. Its largest areas are in Latin America and Africa. In the Americas, it dominates a vast area of central and southern Brazil as well as most of Bolivia. It also occupies much of northern South America and coastal Central America and most of the islands of the Caribbean area, as well as the southern tip of Florida. It covers nearly 40 percent of the African continent, forming a broad arc that nearly surrounds the relatively small central core of Tropical Wet Climate. Also within the Tropical Wet and Dry Climate are densely populated areas of South and East Asia, including most of India and extending eastward to Vietnam and extreme southern China. Finally, northern Australia and the southernmost islands of the East Indies are located within this climate type.

The annual precipitation pattern of the Tropical Wet and Dry Climate is its dominant characteristic and is also largely responsible for the annual temperature distribution. The well-defined alternation of wet and dry seasons is caused by the thermally induced seasonal shift in latitude of the ITCZ and the subtropical belts of high pressure; in many areas this alternation is further enhanced by a monsoonal tendency. The wet season nearly always occurs during the summer half of the year and usually centers on the midsummer months, when the ITCZ makes its deepest penetration into the hemisphere. Conversely, the driest season in most locations occurs during midwinter, when the subtropical high has shifted to its most equatorward position. The resulting pattern can be readily seen on the climagraph for Bombay, India (see Figure 7.5).

The Tropical Wet and Dry Climate is transitional between the constant wetness of the equatorial regions and the aridity of the subtropics. Annual precipitation means are mostly between 35 and 70 inches (90–180 cm). Precipitation amounts decline poleward, as the subtropical high is approached; the climate ends at the ill-defined line where semiarid conditions and an annual water deficit begin. During the rainy season, conditions in the Tropical Wet and Dry climate are much like those of the Tropical Wet Climate. Skies are mostly cloudy, the air is very warm and humid, and convectional showers and thunderstorms are numerous. Especially

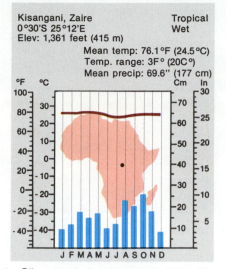

Figure 7.4 Climagraph for Kisangani, Zaire. The double precipitation maximum is caused by the semiannual passage of the ITCZ through this area as it follows the sun.

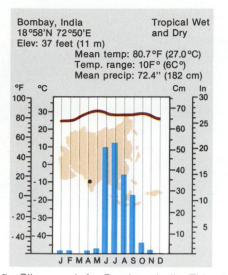

Figure 7.5 Climagraph for Bombay, India. This city's strong summertime precipitation maximum is caused by a combination of the shift of the ITCZ and the summer monsoon.

numerous showers, and sometimes even steady rains, are associated with the occasional passage of weak equatorial lows or easterly waves; and in some areas tropical storms or hurricanes may bring torrential rainfalls in the late summer and autumn. During the dry season, conversely, sunshine prevails, humidities are lower, and no rain may fall for periods of weeks or even months (see Figure 7.6).

The annual temperature pattern is controlled by a combination of seasonal sun angle variations and the alternation of wet and dry seasons. Winter is usually the coolest season, especially in the more poleward locations where sun angles are lowest. Temperatures rise rapidly during spring as the sun moves higher in the sky, the length of daylight increases, and the weather stays dry and often virtually cloudless. The hottest temperatures of the year commonly occur during middle or late spring, just before the onset of the summer rains. Note, for example, that Bombay's warmest month is May. Temperatures usually fall somewhat during the summer rainy season because of the reduction of sunshine. Despite the lower temperatures, the combination of continued warmth and very high humidity produces sticky, uncomfortable conditions. The disruption of the normal summer temperature maximum by the rainy season is the most notable thermal characteristic of the Tropical Wet and Dry Climate.

Low Latitude Dry Climate

In the dry climates, potential evapotranspiration rates exceed precipitation amounts, producing long-term conditions of water shortage. In assessing the dryness of an area, net radiation and resulting temperatures are crucial because they are the primary factors affecting water loss from vegetation and soil. The warmer a region, the more precipitation it can receive and still be "dry." For this reason, the world's dry climates are found in the low and middle latitudes—not in the high latitudes.

The dryness of an area influences such factors as the extent of soil development, vegetation type, and potential for many types of human use. It is common practice to divide the dry lands into two categories. Areas that receive significant amounts of precipitation and are only moderately dry are termed **semiarid**. They commonly have well-developed and often fertile soils, support a natural grass cover, and offer considerable potential for certain types of agricultural development. The very dry areas are termed **arid** and can generally be described as deserts. Arid regions typically have poorly developed soils, contain only a scattering of drought-resistant natural vegetation (if any at all), and are normally capable of supporting only limited human populations.

The Low Latitude Dry Climate is centered between 20° and 30° N and S, but in places extends equatorward to 15° and poleward to 35° (see Figure 7.7). It is the

Figure 7.6 Wet and dry season contrasts within Mikumi National Park, Tanzania. This area has a Tropical Wet and Dry Climate. (*Wayne McKim*)

chief climate of the subtropics, although it should be noted that large portions of the subtropics have humid climates. The climate is especially extensive on the west sides of the continents and adjacent east sides of the oceans, where the subtropical highs are most strongly developed and cold ocean currents help stabilize the atmosphere.

In North America, the areas with a Low Latitude Dry Climate include most of the southwestern United States and northern Mexico. In South America, this climate is primarily restricted to the narrow Atacama Desert of coastal Peru and northern Chile. By far the most extensive land areas with this climate are to be found in Africa, Asia, and Australia. In Africa, it includes the vast expanse of the Sahara and its borderlands, the Kalahari region of Southwest Africa, and the easternmost African "horn." The low latitude dry lands of Asia are in the southwestern part of the continent and include the Arabian Peninsula, most of Iraq, Iran, Pakistan, and western India. In Australia, sometimes called the "desert continent" because a greater proportion of this continent is desert than any other, all of the interior and much of the western and southern margins are dry. Finally, vast ocean areas within the tropics and subtropics experience a dry climate.

The relative lack of precipitation in Low Latitude Dry Climate areas results primarily from the year-round dominance of the subtropical highs. The dryness is further accentuated in most areas by high temperatures, high potential evapotranspiration rates, continentality, and, in some cases, by mountain barrier effects. Arid areas, which typically have annual precipitation means

averaging less than 10 inches (25 cm), form the central or core areas within the regions having this climate. These core areas are surrounded by broad peripheral rims of semiarid conditions that generally have annual precipitation means between 10 and 30 inches (25–75 cm).

Within the arid core of this climate type, rainfalls are very scanty and no distinctive seasonal precipitation pattern may be evident. The passage of time commonly produces an endless succession of days with deep blue skies, perhaps with scattered cirrus or afternoon cumulus clouds, and excellent visibilities. On occasion, however, visibilities may be reduced by airborne dust particles or even full-fledged dust storms. During the summer, intense heating may produce a few scattered convectional showers and thunderstorms, especially over upland regions. Often the air between the clouds and ground is so dry that the rainfall evaporates as it descends. By contrast, any rain falling in winter is generally produced by the unusually deep equatorward penetration of frontal cyclonic storms originating in the middle latitudes. These rainfalls are lighter and more widespread than those produced by summer season convectional storms. The climagraph for Yuma, Arizona (Figure 7.8), is typical of a low latitude desert station.

The Low Latitude Dry Climate is as well known for its heat as for its aridity. The subtropical deserts are the source regions for continental tropical air masses, and their summer temperatures are the hottest in the world. Many Low Latitude Dry Climate stations have means in the 90s (32°–38° C) for several consecutive summer months, and a few reach a mean of 100° F (38° C) in their hottest month. Daytime highs of 115° to 120° F

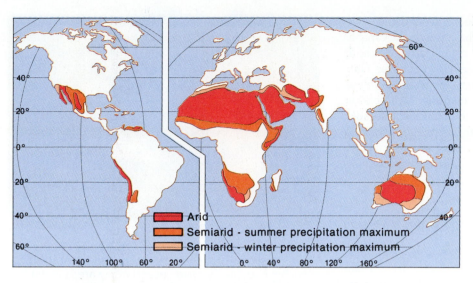

Figure 7.7 Distribution of arid and semiarid subtypes of the Low Latitude Dry Climate.

Arid
Semiarid - summer precipitation maximum
Semiarid - winter precipitation maximum

(46°–50° C) are not uncommon in some inland desert areas at low elevations, and these daytime maximum readings are often remarkably consistent for weeks at a time.

Associated with the extreme summer heat are very low relative humidity values. Low humidities are due not so much to a lack of atmospheric water vapor as to the very high water vapor holding capacity of the air at such hot temperatures. Daytime relative humidity readings in the desert often fall below 10 percent in the afternoon, and a low of 2 percent has been recorded. The relatively dry air may make the heat somewhat more bearable, but it results in extremely high evaporation rates from moist surfaces such as reservoirs and irrigated fields and necessitates the human consumption of large quantities of liquids. Semiarid areas, with their somewhat larger amounts of evaporable surface moisture, experience slightly less extreme heat.

In contrast to the excessive summer heat, winter temperatures within the Low Latitude Dry Climate tend to be surprisingly cool. The combination of clear skies and dry air makes the atmosphere relatively transparent to outgoing terrestrial radiation, while the lower sun angles and shorter periods of daylight considerably reduce insolation receipts. As a result, midwinter temperatures average in the 50s or low 60s (10°–18° C) in all but the equatorward portions of this climate. Winter days are generally pleasantly warm, especially in the sun, but nighttime lows can be quite cold; frost and freezing temperatures occur occasionally in many of the more poleward locations. Daily temperature ranges

within the deserts of the subtropics during the winter half of the year are among the highest in the world. These ranges are usually between 30 F° and 60 F° (18°–35 C°), and temperatures have been known to rise from subfreezing readings to highs exceeding 100° F (38° C) in a single day. Little wonder, then, that stories of the desert sometimes describe it as "broiling during the day and freezing at night"!

Strong surface heating during the summer half of the year makes the air highly unstable at the lower levels,

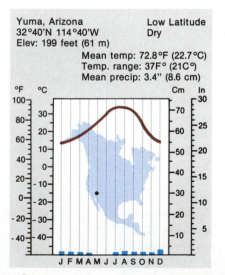

Yuma, Arizona
32°40'N 114°40'W
Elev: 199 feet (61 m)

Low Latitude
Dry

Mean temp: 72.8°F (22.7°C)
Temp. range: 37F° (21C°)
Mean precip: 3.4" (8.6 cm)

Figure 7.8 Climagraph for Yuma, Arizona, a station dominated by the northern hemisphere subtropical high.

131

resulting in strong, gusty daytime winds and the occasional formation of dust devils. These winds are capable of picking up clouds of dust and fine, stinging sand, adding one more discomfort for the desert traveler. Hot, dust-laden winds from the Sahara, for example, blow periodically all along that desert's margins. Atmospheric dust problems have been compounded in recent decades by the large-scale human removal of protective vegetation from low latitude semiarid areas for agricultural purposes and for fuel (see the Case Study following Chapter 16).

THE MIDDLE LATITUDE CLIMATES

Between the zones dominated by the subtropical highs and the subpolar lows lie the broad expanses of the middle latitudes. This area is often subject to fast-changing weather conditions as air masses and cyclonic storms with their associated fronts pass by in endless succession in the westerly wind flow. Climatic conditions are determined largely by continentality and latitude and in some places are strongly modified by the topography. In general, more poleward locations are farther from the subtropical highs and experience increased cyclonic and frontal activity. The decline of temperatures in a poleward direction, however, eventually significantly limits the atmosphere's water vapor content and therefore the potential for precipitation. The net result of the lowered temperatures and increased storminess is that annual precipitation totals exceed potential evapotranspiration rates within most of the middle latitudes, bringing a return to humid or subhumid climatic conditions.

Dry Summer Subtropical Climate

The Dry Summer Subtropical Climate (also called the Mediterranean Climate) is found in the western portions of continents in the lower middle latitudes, generally between 30° and 45° N and S. This climate is transitional in location between the subtropical deserts and the humid middle latitudes; in effect, the climatic characteristics of each region dominate for a portion of the year. The mountainous nature of most land areas in which this climate occurs typically restricts it to rather narrow coastal zones. As a result, the Dry Summer Subtropical Climate occupies the smallest land area of any climatic type. Its human significance, however, is great since most areas in which it occurs are densely populated and economically prosperous (see Figure 7.9).

Figure 7.9 View of the Parthenon, in Athens, Greece. The combination of hot, dry summers and a long history of human occupancy have caused many areas in the Dry Summer Subtropical Climate to have thin, stony soils and meager vegetation. (*Rodman Snead, © JLM Visuals*)

Portions of five continents have a Dry Summer Subtropical Climate. Although separated by vast distances, these areas have similar relative geographical positions and remarkably similar climatic conditions. In North America, the climate extends from just south of Los Angeles northward to near the Oregon border. In general, it is restricted to the coastal zone west of the Sierra Nevada. In South America, it is located in central Chile west of the high Andes. It occurs throughout the Mediterranean Basin region of southern Europe and northwestern Africa, eastward into parts of the Middle East. This is the only land area where there are both enough land and a sufficient lack of restricting mountain barriers to allow the climate to cover a large areal extent. Finally, the Dry Summer Subtropical Climate occupies the coastal portions of southern and southwestern Australia and brushes the southwestern tip of Africa in the vicinity of Cape Town.

Temperatures within the Dry Summer Subtropics are generally moderated by the presence of nearby water bodies. The maritime influence lowers both daily and annual ranges. During the summer, large temperature differences frequently exist between locations near the coast and those farther inland. Although daytime highs in the summer commonly rise into the 90s and sometimes exceed 100° F (38° C) in interior locations, nighttime low temperatures in most areas are pleasantly cool, with readings in the 50s or lower 60s (10°–18° C). Coastal areas, in contrast, are often kept quite cool all day by the adjacent water. Temperature differences are especially striking along coasts paralleled by cold ocean currents and backed by hills or low mountains that block the inland penetration of cool ocean breezes. Such is the case in California and, to a lesser extent, in Chile and Portugal. An indication of the resulting temperature contrasts may be gained by comparing the climagraphs for San Francisco and Red Bluff (Figures 7.10 and 7.11).

The Mediterranean coast has much warmer summer means than do locations, such as San Francisco, situated on an oceanic coast. This is because the Mediterranean Sea is a restricted interior body of warm water that experiences intense summer insolation and receives only a very limited inflow of cooler Atlantic water. Split, a city on the Adriatic coast of Yugoslavia, provides a typical example (see Figure 7.12).

Winter temperatures within the Dry Summer Subtropics average mostly in the 40s and lower 50s (5°–12° C). Temperature variations between coastal and inland locations are much smaller at this time of year than they are in summer, and the interiors are now generally cooler. Maritime polar air masses dominate all areas during the winter, but on the west coast of North America and in the northern Mediterranean Basin area, occasional invasions of continental polar air from the northeast are capable of bringing killing frosts even to the coasts.

The most distinguishing characteristic of the Dry Summer Subtropical Climate is its annual precipitation pattern of dry summers and humid winters. During the

summer season, the subtropical high in each hemisphere shifts poleward with the sun to blanket the Dry Summer Subtropics. The summer months are almost rain-free along cold water coastal regions such as California and Chile, where cool temperatures further stabilize the atmosphere. The climagraph for San Francisco illustrates this situation. Summer advection fogs are prevalent along these coasts because of the cold water. In more poleward locations, the influence of the subtropical highs is not as strong; occasional rainfalls from convectional showers or weak frontal passages may occur. Monthly precipitation means, however, are generally less than an inch (2.5 cm).

During the winter, the subtropical highs retreat equatorward, and a moderate onshore flow of less stable maritime polar air develops. Fronts and frontal cyclones also influence the subtropics at this time of year, providing the lifting necessary to trigger periods of precipitation. Storms typically occur every three or four days, producing moderate to occasionally heavy rainfall. The periods between storms usually consist of sunny, mild weather. Most areas in the Dry Summer Subtropics receive at least 75 percent of their total precipitation during the winter half of the year.

Annual precipitation totals in most cases average between 15 and 35 inches (40–90 cm), with the lower totals most common in the more equatorward locations. Yearly variations in the position of the subtropical high and the storm tracks result in a large annual variability

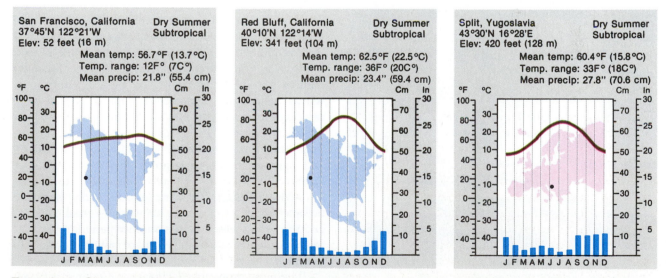

Figure 7.10 Climagraph for San Francisco, California. San Francisco's exposed coastal location causes it to have unusually cool summers and mild winters for its latitude. **Figure 7.11** Climagraph for Red Bluff, California. Red Bluff's interior location causes it to have three times the annual temperature range of San Francisco.
Figure 7.12 Climagraph for Split, Yugoslavia.

in precipitation amounts received in all areas. As a result, droughts and floods are not uncommon.

Humid Subtropical Climate

The Humid Subtropical Climate is the east coast counterpart of the Dry Summer Subtropics. It occupies the eastern sides of continents in the lower middle latitudes between approximately 25° and 45° N and S. The leeward position of these areas results in a moderate continental influence, but this is tempered by large adjacent bodies of warm water (usually oceans with warm currents) that supply sufficient water vapor to produce humid conditions.

The most extensive land area within the Humid Subtropics is the southeastern one-third of the United States extending from extreme southeastern New England southward through most of Florida and westward to Texas and eastern Kansas. Another large area occupies southeastern South America from southern Brazil

Figure 7.13 Key Biscayne, Florida, is a popular beach resort in the Humid Subtropics. (*F. Sieb/H. Armstrong Roberts*)

through Uruguay and northeastern Argentina. In Asia; much of eastern China, southern Japan, and extreme southern Korea experience a Humid Subtropical Climate. A fourth relatively extensive area is found in eastern Australia.

Temperatures within this climate are relatively similar to those experienced in inland portions of the Dry Summer Subtropics, with warm-to-hot summers and cool-to-mild winters, depending upon latitude and elevation. Higher humidities, however, make the Humid Subtropics less comfortable, especially in summer. Poleward areas also generally have colder winters and larger annual temperature ranges than do comparable west coastal locations.

The long period of summer heat and humidity is one of the most notable characteristics of the Humid Subtropics. Oceanic centers of the subtropical highs far to the east pump in a nearly continuous flow of maritime tropical air. Adding to the discomfort produced by humidity are hot temperatures resulting from high sun angles and long daylight periods. Although weak incursions of continental polar air bring occasional interludes of pleasant weather to the more poleward areas, the heat and humidity are almost unbroken equatorward of approximately 35° N and S for a two- or three-month period. Daily highs usually average from the mid-80s to lower 90s (29°–34° C), but in some areas the hottest summer days have readings in excess of 100° F (38° C). Nights are warm and sultry, with overnight lows frequently remaining in the 70s. The climagraphs for Baltimore, Maryland, and Jacksonville, Florida (Figures 7.14 and 7.15), illustrate the influence of latitude on the length and intensity of the warm season.

Autumn is a delightful time of year in the Humid Subtropics, with extended periods of fair, pleasantly mild weather. Weather fronts become active during the winter half of the year, resulting in substantial daily temperature variations. Northern hemisphere areas are especially prone to sharp temperature fluctuations because they are within reach of arctic (cA) air masses, and most northern hemisphere locations experience winter days in which temperature readings remain near or below freezing. Temperatures normally begin to become milder a few weeks before the vernal equinox, and the spring season brings another period of pleasant temperatures before the return of summer heat and humidity.

The most pronounced climatic difference between the Dry Summer Subtropics and the Humid Subtropics is in their annual precipitation patterns. Whereas the

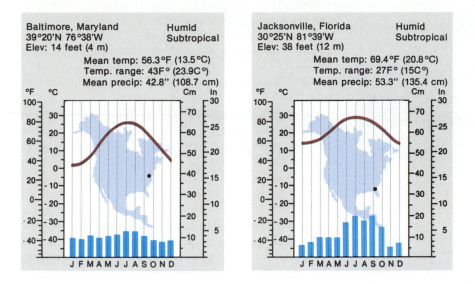

Baltimore, Maryland   Humid
39°20'N 76°38'W   Subtropical
Elev: 14 feet (4 m)
              Mean temp: 56.3°F (13.5°C)
              Temp. range: 43F° (23.9C°)
              Mean precip: 42.8" (108.7 cm)

Jacksonville, Florida   Humid
30°25'N 81°39'W   Subtropical
Elev: 38 feet (12 m)
              Mean temp: 69.4°F (20.8°C)
              Temp. range: 27F° (15C°)
              Mean precip: 53.3" (135.4 cm)

Figure 7.14 Climagraph for Baltimore, Maryland.
Figure 7.15 Climagraph for Jacksonville, Florida. The differences between Baltimore's and Jacksonville's temperature and precipitation means result largely from their differing latitudes.

former climate experiences a well-defined dry season in summer, the latter receives substantial amounts of precipitation throughout the year. Annual precipitation means within the Humid Subtropics are about twice those of the Dry Summer Subtropics, with most areas receiving between 30 and 65 inches (75–165 cm).

During the summer, the subtropical highs dominate; but their centers are situated well to the east, producing a weak onshore flow of unstable mT air. The resulting weather is warm, humid, and mostly sunny; but daytime heating frequently produces scattered afternoon and evening convectional showers and thunderstorms. Rainfalls are highly variable in amount because the storms are localized in nature and are slow moving due to the weak upper-level winds that prevail at this time of year. During the winter, most precipitation is frontal and cyclonic in nature, causing it to be lighter, steadier, and much more widespread than the convectional rains of summer. Most winter precipitation falls as rain, but occasional snowstorms occur, especially in the more poleward areas.

Along the coasts of the United States, Asia, and Australia, summer and early autumn bring the threat of hurricanes. These tropical storms occasionally produce great destruction and loss of life within the Humid Subtropics, as illustrated by the Case Study at the end of Chapter 6.

Mid Latitude Dry Climate

Nearly three-fourths of the world's dry climates—those areas where potential evapotranspiration rates exceed precipitation amounts on a yearly basis—are located within the low latitudes. The remaining one-fourth of the dry climates occurs in the middle latitudes, mostly within continental interiors in the northern hemisphere. Their dryness is not associated with the subtropical highs, but instead results from orographic barriers on their windward sides. A secondary factor in many cases is their remoteness from oceanic sources of maritime air.

Three widely separated areas are included within the Mid Latitude Dry Climate. The largest is the central portion of Eurasia extending from just north of the Black Sea eastward some 5000 miles (800 km) to north-central China. This vast region is cut off from moisture-bearing winds by an imposing rim of mountains extending along its southern and southwestern margins. Central Asia, in the heart of this area, is more remote from oceans than any other place in the world. In North America, the Great Plains and Great Basin are largely blocked from Pacific moisture by the mountains to their west. The third, and smallest, area having this climate type is the Patagonian region of southern and western Argentina. It is geographically distinct from the other two areas in that it is located on a relatively narrow body of land. Its dryness is primarily associated with the rainshadow effect of the Andes.

Most portions of the Mid Latitude Dry Climate experience highly continental conditions because maritime air masses normally do not enter these areas. Annual mean temperatures vary considerably from place to place, as illustrated by the climagraphs for Saratov, USSR; Reno, Nevada; and Mendoza, Argentina

Figure 7.16 Black Angus cattle graze in the rolling wheat land of central Montana, located within the Mid Latitude Dry Climate. (*R. Lamb/H. Armstrong Roberts*)

(Figures 7.17–7.19). Readings are largely dependent upon latitude, degree of continentality, and elevation.

Summer temperatures in the Mid Latitude Dry Climate are generally warm, with occasional hot spells. Clear skies and dry air allow daytime readings to soar well above 100° F (38° C) on occasion in the hottest areas. Nighttime lows, however, generally fall to comfortable readings, especially in locations at higher latitudes and elevations. Even in the warmest locations, the summer heat is not constant since these areas, unlike the dry climates of the low latitudes, are within reach of occasional air masses of subpolar origin.

Winter temperatures are cold in most places, but the degree of coldness is highly dependent upon location. The mildest of the three regions is Argentina, where virtually all areas have temperature means above freezing. In contrast, most stations in North America and virtually all of those in Eurasia have subfreezing temperatures during the winter months, with January means below 0° F (−18° C) in portions of eastern Asia. Both North America and Eurasia are within reach of wintertime arctic air masses and experience cold waves when this frigid air moves southward from its source regions. Coupled with the bone-chilling cold are winds that frequently reach high velocities over these flat, treeless surfaces, producing potentially deadly wind-chill values for humans and unprotected livestock. Interspersed with the cold periods, however, are intervals of milder weather as air masses from the south or west penetrate into the regions. Sudden onsets of Chinook winds in the northern Great Plains, for exam-

ple, are capable of dramatically raising temperatures as they displace arctic air masses. An extreme example of a Chinook occurred in Spearfish, South Dakota, on January 22, 1943. At 7:30 A.M., the temperature was −4° F (−20° C). Two minutes later, at 7:32 A.M., it had risen to 45° F (7.2° C)!

Precipitation amounts within different portions of the Mid Latitude Dry Climate depend largely on the strength of the rainshadow effect. Both arid and semiarid conditions occur. As an approximation, arid areas receive less than 8 inches (20 cm) of precipitation annually; semiarid areas receive between 8 and 20 inches (20–50 cm). As in the subtropics, the geographic pattern of the Mid Latitude Dry Climate is arid core areas surrounded by extensive semiarid rims. Each of the three continents with this climate has one or more arid cores (see Figure 7.20).

As is typical of the middle and higher latitudes, most precipitation in the Mid Latitude Dry Climate is associated with traveling frontal cyclones. These storms are usually capable of generating only light amounts of rain or snow because they do not have access to major sources of moisture. Winter snowstorms, however, occasionally produce substantial accumulations, with accompanying strong winds creating blizzard conditions.[1] Spring sometimes brings turbulent weather because of the strong temperature contrasts that develop at this time of year between opposing warm and cold air masses. In the U.S. Great Plains, especially, this conflict can produce severe hailstorms and tornadoes. During the summer, weather fronts are weaker and less numerous, but the air may be sufficiently moist and unstable to trigger convectional showers and thunderstorms. Most locations receive a warm season precipitation maximum largely because the air contains more water vapor at this time of year.

Marine Climate

Moving poleward into the upper middle latitudes, we enter the heart of the westerly wind belt. Here, frontal cyclonic storms are both numerous and well developed, and weather systems tend to move rapidly eastward under the influence of strong upper-level winds. Situated in this zone on the west, or windward, sides of

---

1. A *blizzard* is a heavy snowstorm accompanied by winds of over 35 miles per hour (16 m/sec) and visibilities reduced to below 500 feet (150 m) by falling and blowing snow.

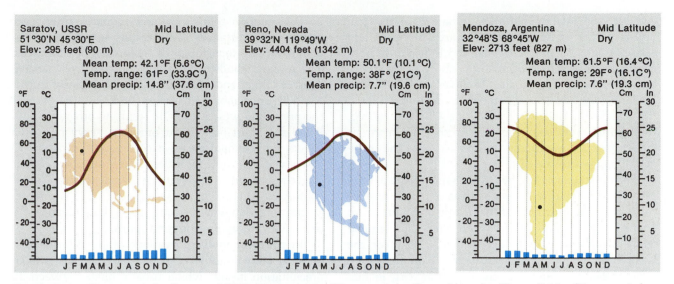

Figure 7.17 Climagraph for Saratov, USSR. **Figure 7.18** Climagraph for Reno, Nevada. **Figure 7.19** Climagraph for Mendoza, Argentina.

continents generally located between 40° and 60° N and S, the aptly named Marine Climate has the strongest maritime influence of any of the climate types. An almost constant onshore flow of maritime polar (mP) air provides adequate moisture to allow precipitation totals to exceed potential evapotranspiration rates throughout most or all of the year, so that no dry season occurs. In addition, the onshore wind flow produces cool summers but unusually mild winters.

Geographically, the Marine Climate is generally located just poleward of the Dry Summer Subtropics.

Like that climate, the Marine Climate is restricted in size, except in Europe, by the blocking influence of coastal mountain ranges and, in the southern hemisphere, by a lack of landmasses at the proper latitude. In North America, the Marine Climate occupies a narrow coastal strip beginning in extreme northern California and extending northward along the west coast of Canada and the south coast of Alaska to the Aleutian Islands. In South America, it occurs in a similarly restricted zone in the portion of Chile situated west of the Andes and south of 40° S. The Marine Climate

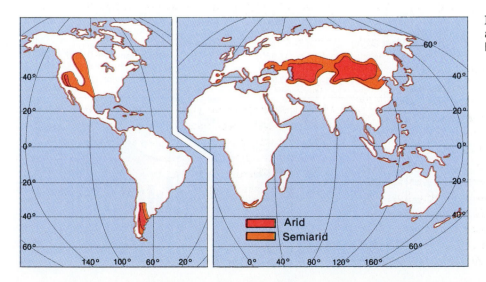

Figure 7.20 Distribution of arid and semiarid regions within the Mid Latitude Dry Climate.

extends well inland in Europe, where it exists from southern Iceland southward to northern Spain and eastward into Poland, Romania, and Bulgaria. In both Africa and Australia it occurs in portions of the southern coastal regions; it also occupies most of New Zealand. It should also be noted that vast ocean areas in the middle latitudes have a Marine Climate. Indeed, these areas are the source regions for the mild, moist air masses that are eventually transported onto the continental margins to the east.

Perhaps the most noteworthy characteristic of temperatures in Marine Climate areas is their relative constancy. The resistance of the nearby ocean waters to temperature changes causes both daily and annual temperature ranges to be exceptionally small for the latitude. As might be expected, temperature ranges are lowest along the immediate coasts and increase inland. As a result, coastal regions experience cooler summers and milder winters than interior locations do.

The Marine Climate is characterized by a large amount of cloudy and rainy or drizzly weather. In fact, the mean annual percentage of cloudiness and total duration of precipitation exceed those of any other climate. Most precipitation is frontal and cyclonic in nature, but orographic lifting also plays an important role in many areas. The storms that affect the west coasts are usually large and often deeply occluded; they tend to produce prolonged periods of leaden skies, light rain or drizzle, and strong winds. Convectional precipitation is minimized by the cool ocean temperatures, which stabilize the air.

Precipitation totals vary tremendously, often over rather short distances. This results largely from the highly varying topography of these west coastal areas. The heaviest totals occur along the coasts of Canada, Alaska, Chile, and New Zealand, where lofty north-south-oriented mountain ranges lie across the path of the prevailing westerlies. The resulting orographic uplift of the moist oceanic air produces heavy rains and snows, with annual means exceeding 200 inches (500 cm) of water in a few localities. These means are greater than those of any other locations outside the tropics and subtropics. Ketchikan, Alaska (Figure 7.22) is a representative station with a strong orographic influence and excessively heavy precipitation totals.

In contrast, lowland portions of the Marine Climate, notably in Europe, have relatively low annual precipitation means, which usually range between 20 and 30 inches (50–75 cm). These modest amounts, however, are generally more than sufficient to maintain adequate surface moisture supplies, because evapotranspiration rates are restricted by the combination of high humidities, cool temperatures, and abundant cloudiness (see Figures 7.21 and 7.23).

Precipitation is usually rather evenly distributed throughout the year. Coastal regions, however, tend to receive a winter maximum because of the frequency of coastal storms triggered by temperature contrasts between the cold land and warmer ocean. Most coastal storms, even in winter, bring only rain and drizzle. Some of the heaviest snowfalls in the world occur in the high mountains backing these coasts, but their elevation

Figure 7.21 Climagraph for Brest, France.
Figure 7.22 Climagraph for Ketchikan, Alaska. Ketchikan's heavy precipitation is primarily a result of orographic lifting.

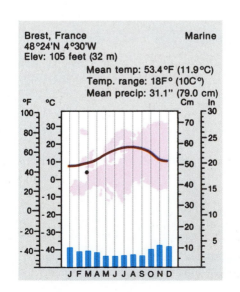

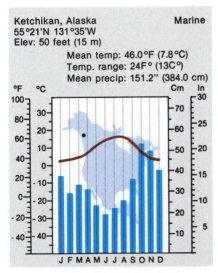

Figure 7.23 Glenborrodale, on the coast of Scotland, has a well-developed Marine Climate. (*K. Scholz/ H. Armstrong Roberts*)

causes these areas to have winter temperatures too low to be classified as Marine Climates. In the inland portions of Europe, the larger annual temperature ranges and reduced frequency and intensity of winter storms (compared with the coasts) typically result in a weak summer maximum of precipitation.

Two additional noteworthy weather phenomena within the Marine Climate are strong winds and fog. High winds are commonly associated with the deep storms that affect the coastal regions in winter. The combination of rugged, rocky coasts and frequent gales has led to many shipwrecks over the centuries. Perhaps the most famous was the destruction of the Spanish Armada off the coast of Scotland in 1588.

The combination of constant dampness and frequent land/water temperature contrasts also favors the formation of dense advection and radiation fogs. Human urban and industrial activities frequently enhance conditions for fog formation through their large-scale production of hygroscopic (water-attracting) condensation nuclei. The infamous smogs of London and the Rhine Valley are largely of human origin.

Humid Continental Climate

The interiors and east coasts of continents in the middle latitudes experience some of the most changeable weather on earth. These regions lie in the latitudinal zone most frequented by the polar front and are in the collision zone between air masses of subtropical and subpolar origins. Because these air masses have not passed over moderating bodies of water, as in the Marine Climate, they largely retain the thermal characteristics of their source regions. As the polar front shifts northward and southward, these differing air masses alternately occupy a given locality, bringing about sharp changes in weather conditions.

Geographically, the Humid Continental Climate occupies the interiors and eastern (or leeward) coasts of large continents in the middle and upper portions of the middle latitudes. It occurs only in the northern hemisphere, in the continents of North America and Eurasia, because the higher latitudes of the southern hemisphere lack sufficient land to produce continental climate conditions. Specifically, the climate is located in three widely separated areas at latitudes generally ranging from 35° to 55° N. In North America, it extends from the Upper Midwest of the United States northward into south-central Canada and eastward through the Great Lakes region, New England, and the Canadian Maritime Provinces. A second area begins in eastern Europe, extending from southern Scandinavia southward into Bulgaria and eastward into the central Asiatic Soviet Union. The third area, located in eastern Asia, occupies portions of northeastern China, northern Korea, and northern Japan.

The great variability of temperatures on both a seasonal and short-term basis is an important thermal

Figure 7.24 A checkerboard pattern of productive farmland occupies a vast area of the midwestern United States within the Humid Continental Climate. The scene here is near Fredonia, Wisconsin. (*Richard Jacobs, © JLM Visuals*)

characteristic of this climate type. Large annual temperature ranges result from major seasonal variations in insolation received by continental landmasses that readily heat and cool. Ranges are further increased by the shift of the polar front, which is generally situated to the north of the Humid Continental areas in summer and to their south in winter. Most locations have annual ranges between 40 and 75 F° (22–42 C°), with the greater values occurring at higher latitudes, especially in more interior locations. The climagraphs for Des Moines and Harbin in Figures 7.25 and 7.26 illustrate the large annual temperature ranges typical of this climate type.

During the summer, the weather is almost continuously warm and humid near the southern margin of the Humid Continental Climate, where conditions are similar to those of the Humid Subtropics. Northward, however, temperatures fall to pleasantly cool values near the climate's poleward margins. Even the most poleward areas have hot spells, with daytime temperatures reaching the 90s (32°–38° C). Unlike the southern margins, however, these areas are also within reach of refreshingly cool summertime cP air masses that frequently produce highs in the 60s or lower 70s (16°–23° C).

Winters are cold, with the lowest means occurring in poleward and inland locations. Temperatures average well below freezing everywhere, and the ground is frozen and snow-covered for one to five months. Fronts are very active during the winter, so day-to-day temperatures are subject to large variation, depending on the dominating air mass. The dominant wintertime air mass is cP, which is generally associated with clear, moderately cold weather. On occasion, though, outbreaks of frigid arctic (cA) air bring subzero temperatures to all areas; and nighttime lows in the coldest regions can plummet to −40° F (−40° C) or even lower. At the opposite extreme, midwinter thaws sometimes occur when traveling cyclones draw mT air northward. Variations in the flow of the jet stream may cause extended periods, or even the entire winter season, to be unusually cold or mild.

Because available moisture is limited, precipitation totals within Humid Continental Climate areas are generally modest. The long, cold winters reduce evapotranspiration rates sufficiently to produce an annual moisture surplus, but summer water deficits occur in all areas except near-coastal regions. Precipitation means are greatest along southern and coastal regions such as Japan and New England, and gradually decline poleward and inland.

Seasonal precipitation distributions are most strongly influenced by continentality. Along east coasts and in the eastern Great Lakes area of the United States and Canada, winter storms are frequent; and substantial cold season precipitation occurs. Progressing inland, the tendency for a summer precipitation maximum increases because the much higher temperatures of summer greatly increase the water vapor holding capacity of the air and reduce its stability. The tendency for a summer

Figure 7.25 Climagraph for Des Moines, Iowa.
Figure 7.26 Climagraph for Harbin, China.

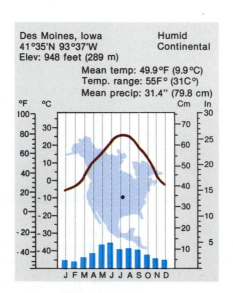

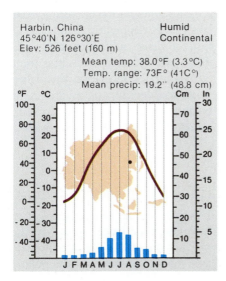

maximum is strongest in continental East Asia, as in Harbin (Figure 7.26), where the monsoonal tendency produces a winter dry season.

Precipitation during the fall, winter, and spring is almost entirely cyclonic and frontal and normally occurs at intervals of two to four days. Most winter precipitation is in the form of snow. Winter snowstorms are sometimes heavy and, if accompanied by strong winds, can produce blizzard conditions capable of bringing outdoor activities to a standstill. In summer, cyclonic and frontal activity weakens and shifts northward; and most areas come under the dominance of mT air. The weather is much clearer at this time of year than during the cold season, but strong surface heating frequently triggers convectional showers and thunderstorms. In contrast to the general rains and snows of winter, the localized nature of these storms may produce greatly varying precipitation amounts and leave localized pockets of drought. More general droughts occur in years when shifts in the pressure and wind patterns draw in cT, rather than mT, air masses.

THE HIGH LATITUDE CLIMATES

The high latitude climates are dominated by cold weather during most or all of the year. Low temperatures produce difficult living conditions and, along with perennially frozen subsoil conditions, severely restrict agricultural activities. As a consequence, most areas in the high latitudes are lightly populated or uninhabited. Low temperatures limit the water vapor content of the air, so precipitation totals are generally low, despite the extended periods of rain or snow that frequently accompany high latitude storms. Evapotranspiration rates are also greatly reduced by the cold air, so that an annual moisture surplus exists nearly everywhere.

Subarctic Climate

The Subarctic Climate exists only in the northern interiors of North America and Eurasia, where it is found in a broad zone situated mostly between 50° and 70° N. As is true of the Humid Continental Climate, no Subarctic Climate regions exist within the southern hemisphere because of the absence of sufficient land in the proper latitudes. West winds deliver enough oceanic warmth to extend the climate poleward on the western sides of the continents, and latitudes gradually decline eastward. In North America, the climate extends from

Figure 7.27 A springtime view of the Canadian Shield wilderness in the Northwest Territories, within the Subarctic Climate. The larger lakes are still covered by winter ice. (*David Butler*)

central Alaska eastward through most of interior Canada to Newfoundland. In Eurasia, it exists from northern and eastern Scandinavia eastward in a widening swath through the northern interior of the Soviet Union to the Kamchatka Peninsula. Climagraphs of three subarctic stations (Figures 7.28 to 7.30) illustrate the diversity of ranges which can occur in a subarctic climate.

Winter, the dominant season in the Subarctic Climate, is long, dark, and extremely cold. The cold winter temperatures, lower than those found anywhere else in the world except for the ice plateaus of Antarctica and Greenland, result from a combination of several factors. One is the lack of insolation. Most areas with a Subarctic Climate are located near or above the Arctic Circle and receive at most only a few hours of low-intensity heating from a sun barely above the horizon. A second factor is the continentality produced by locations, for most subarctic areas, in the interiors of large, snow-covered landmasses. Finally, the thermal compaction of the cold air forms high pressure centers over Canada and Siberia. These high pressure systems are generally associated with clear skies and light winds that favor surface radiation.

Winter readings in most locations average below 0° F (−18° C), and subfreezing mean temperatures last anywhere from five to eight months. The lowest temperatures of all occur in eastern Siberia; in the Soviet city of Verkhoyansk, for example, the January mean temperature of −58° F

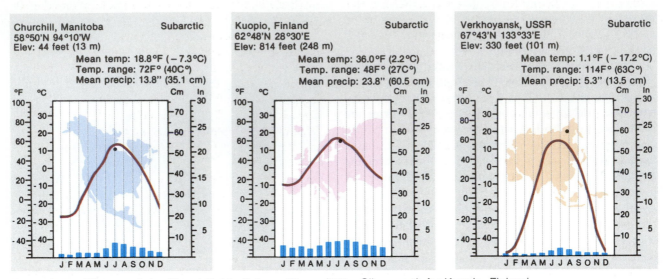

Figure 7.28 Climagraph for Churchill, Manitoba. **Figure 7.29** Climagraph for Kuopio, Finland.
Figure 7.30 Climagraph for Verkhoyansk, USSR.

(−50° C) makes it one of the coldest permanently inhabited locations in the world (Figure 7.30). In western Canada and Alaska, northern Europe, and the east coasts of Canada and Siberia, the proximity of water raises winter means (see the climagraph for Kuopio in Figure 7.29), but the reduction in cold is partially offset by an increased prevalence of clouds, winds, and heavy snows.

Summers are brief but very mild in contrast to the frigidity of the winters. At this time of year the midday sun is fairly high in the sky, and daylight is almost continuous. Temperature means are in the 50s and 60s (10°–18° C), with the warmest readings in the southern interiors. The great contrast between summer and winter temperatures gives this climate type the world's largest annual temperature ranges. Because fronts are active throughout the year, day-to-day temperature variations are also substantial.

The combination of cold temperatures and continentality results in generally light precipitation amounts. Heaviest totals occur along the coasts, where annual mean temperatures are warmest, water sources are present, and winter storms are frequent. Annual means of up to 40 inches (100 cm) occur in eastern Newfoundland and northern Norway. Precipitation declines rapidly inland and, in the northern interiors, generally amounts to less than 10 inches (25 cm).

Most precipitation is frontal and cyclonic. Despite the low totals, storms with limited moisture supplies frequently pass through the subpolar regions, producing a large number of days with cloudy skies and light rain or snow. Except in coastal areas, a summer maximum of precipitation occurs almost everywhere. Winter snows are generally light, but they are frequently associated with deep low pressure systems that produce strong winds and blizzard conditions. Coastal areas sometimes receive heavy snowfalls from storms with access to oceanic moisture. The coasts in general are cloudier, foggier, and damper year-round than the interiors.

Tundra Climate

The last two climates in our world survey are differentiated only by degrees of coldness. Located in the subpolar and polar regions, they receive smaller quantities of insolation than any other climates on earth. As noted earlier, the high latitudes have a net radiation deficit. A substantial proportion of their thermal energy is supplied by the poleward advection of heat from the lower latitudes by winds and ocean currents.

Within the Tundra Climate, the mean temperature of the warmest month is between 32° and 50° F (0°–10°C). As in the previous two climates, the great majority of land areas experiencing a Tundra Climate are located within the northern hemisphere. Most form a coastal rim around the Arctic Ocean. Included are northern Canada and Alaska, coastal Greenland, northern Iceland, and the north coast of Eurasia from Norway eastward to the Bering Strait. Land areas within the southern hemisphere are limited to the southern tip of South America, the Antarctic Peninsula, and a number of small islands. Most high latitude ocean areas also have a Tundra Climate. Finally, this climate occurs in high mountains and plateaus in various parts of the world, notably on the Tibetan Plateau. Its characteristics within these areas are discussed more fully in the Highland Climates section of this chapter.

The Tundra Climate can be described as a high latitude maritime climate. Few areas are over 100 miles (160 km) from the ocean. The proximity to water reduces annual temperature ranges from the extremes encountered in the Subarctic Climate, making summers colder, but winters less frigid than those in the former climate, despite the higher latitude. The major seasonal temperature control is the duration of sunlight. Most areas are poleward of the Arctic Circle; as a result, sunlight varies from 24 hours per day near the time of the summer solstice to 0 hours per day near the winter solstice.

During the summer, most temperatures average in the 40s (4°–10° C). The sun, though rather low in the sky, is out all, or nearly all day. As a result, insolation receipts in late spring and early summer can reach daily values equivalent to those of the tropics. A south wind in mainland North America or Eurasia will bring in warm continental air and can occasionally raise temperatures to the 70s and 80s (22°–31° C). The ice-covered Arctic Ocean is always nearby, though, and a shift to

northerly winds can quickly lower readings into the 30s (0°–4° C), with the likelihood of frost.

Winter temperatures are regionally much more variable than those of summer. For most of the tundra lands of North America, Eurasia, and northern Greenland, winter means are far below zero (−18° C). The climagraph for Chesterfield Inlet, on the northwest shore of Hudson Bay, serves as an example (see Figure 7.31). In these areas, subfreezing mean temperatures are experienced for eight or nine months of the year; and readings can fall as low as −50° to −60° F (−46° to −50° C). Some areas, however, are near water bodies that remain partially or entirely unfrozen throughout the winter; temperatures here are considerably less cold. One large area that experiences such conditions is located where the North Atlantic merges with the Arctic Ocean. Here, the last remnants of the Gulf Stream cause water temperatures to remain above freezing and air temperatures to stay near or above zero (−18° C) on the adjacent land. The climagraph for Svalbard, an island in the Arctic Ocean north of Norway, provides an example (see Figure 7.32).

Cold temperatures help reduce annual precipitation means to below 15 inches (38 cm) in most places. The exceptions are mountainous windward coasts such as southern Greenland, northern Norway, parts of western Alaska, and the southern hemisphere islands. These same coastal areas generally have a winter maximum caused by storms associated with the subpolar low and receive heavy winter snowfalls. Most areas, though,

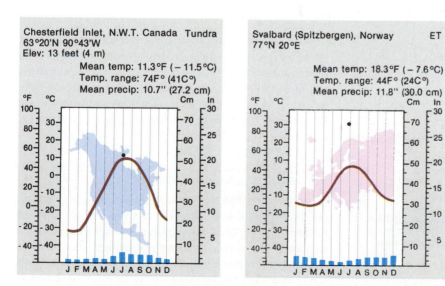

Figure 7.31 Climagraph for Chesterfield Inlet, Northwest Territories, Canada.
Figure 7.32 Climagraph for Svalbard, Norway.

have a strong summer maximum (as in the case of Chesterfield Inlet), because the winter cold causes the Arctic Ocean to freeze over, removing the primary moisture supply. Summer precipitation is predominantly rain, but some falls as wet snow. Virtually all precipitation is frontal and cyclonic. During the fall, winter, and spring, cyclonic storms can be very strongly developed, generating blizzard conditions across the flat, treeless tundra surface, which offers little frictional resistance.

Polar Climate

The Polar Climate is the most poleward and the coldest of the world's climate types. It is also the only climate with no permanent human inhabitants. Monthly mean temperatures are below freezing all year, and the landscape is covered almost everywhere by a deep layer of ice. Two widely separated areas—Antarctica and the interior of Greenland—have a polar climate; together they comprise about 10 percent of the world's land surface.

The extreme coldness of this climate is only partially caused by low sun angles. It is also produced by the high reflectivity of the snow cover and by the high elevation of the surface. Most areas are situated between one and two miles (1.6–3.2 km) above sea level.[2]

Although the unrelenting cold is by far the dominant characteristic of this climate, the extreme variation in the duration of daylight is its major seasonal attribute. Except for southern Greenland, all areas lie poleward of the Arctic or Antarctic Circles and experience a full 24-hour range in the duration of sunlight between the times of the solstices.

Summer temperatures average below freezing on a monthly basis, but in some parts of coastal Antarctica and the margins of the Greenland ice sheet, mild spells occasionally bring above-freezing conditions lasting for hours or even several days. The high-elevation interiors of both areas remain well below freezing, as the climagraph of Eismitte, in central Greenland, illustrates (see Figure 7.33). In interior Antarctica, temperatures re-

main below 0° F (– 18° C) throughout the year.

Winter temperatures, especially in Antarctica, are the lowest in the world. The coldest readings occur not at the South Pole but at the so-called "Pole of Cold" located in the center of the continent near 78° S 107° E. In the 1950s, the Soviets established a research station named Vostok at this location, which has recorded the lowest temperatures in the world. The current record is – 128.6° F (– 89.2° C), set on July 21, 1983.

The Polar Climate is not only the coldest, but probably also the driest of the world climate types in terms of precipitation receipts. Evapotranspiration rates are so low, however, that a major surface moisture surplus exists. The answer to the poetic question "Where are the snows of yesteryear?" is obvious in Greenland and Antarctica (see Figure 7.35). They bury the surface to a depth of up to 2 1/2 miles (4 km)! These areas illustrate the inadequacy of measuring aridity solely in terms of total precipitation, without taking temperatures and evapotranspiration rates into account.

Polar precipitation is cyclonic and frontal in nature and, of course, virtually all is frozen. The combination of cold, continentality, and thermal high pressure blocks most storms from entering the interior areas, especially in Antarctica. As a result, precipitation is heaviest near the coasts and decreases rapidly inland. Amounts received are uncertain due to a lack of reporting stations and the difficulty involved in accurately measuring windblown snowfalls. It is believed that precipitation totals in the interiors are mostly between 2 and 5 inches (5–13 cm) per year in terms of water equivalent.

Winter storms frequently produce gale-force and even hurricane-force winds, largely because the snow-and-ice-covered surface offers a lower frictional resistance than any other surface on earth, including water. Coastal storms are also numerous and often very strongly developed; some coastal stations in Antarctica report gale-force winds 9 days out of 10. Under such conditions, the air becomes filled with fine, horizontally driven snow and ice crystals, which sometimes produce a "whiteout" that reduces visibilities to zero.

HIGHLAND CLIMATES

It is widely known that mountain regions have climatic conditions distinctly different than those of adjacent lowlands. Mountains are popular summer vacation destinations for people seeking to escape the heat of lowland areas, and a rapidly growing number also visit

---

2. Because of their elevation, they could well be included within the Highland Climates classification, discussed in the next section, as are the glaciated areas of the lower latitudes. By convention, however, because of their location and large size, they are treated as a distinct climatic type.

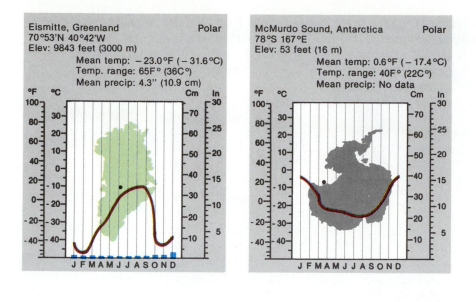

Eismitte, Greenland | Polar
70°53'N 40°42'W
Elev: 9843 feet (3000 m)
Mean temp: −23.0°F (−31.6°C)
Temp. range: 65F° (36C°)
Mean precip: 4.3″ (10.9 cm)

McMurdo Sound, Antarctica | Polar
78°S 167°E
Elev: 53 feet (16 m)
Mean temp: 0.6°F (−17.4°C)
Temp. range: 40F° (22C°)
Mean precip: No data

Figure 7.33 Climagraph for Eismitte, Greenland.
Figure 7.34 Climagraph for McMurdo Sound, Antarctica.

mountainous areas in winter to take advantage of winter sports opportunities. The highland climates as a group differ in one basic respect from the world climate types already discussed. Instead of consisting of a single large, relatively homogeneous region, they form a complex mosaic of diverse climate types located in close proximity to one another. In fact, it is their diversity, rather than their homogeneity, that causes them to be grouped together here. Climatic conditions in the world's highland areas vary so greatly and change so rapidly with distance that their total pattern is too complicated to depict on a world map or to analyze fully in a section of a chapter. It should be stressed that the highland areas do not contain any new climate types that have not been examined. We have already covered the entire range of world types, from those that are hot and wet to those that are cold and dry.

In North America, the Rockies, Sierra Nevada, Cascades, and the mountains and interior plateaus of Mexico are climatically distinct from the surrounding lowlands. In South America, the Andes produce a continuous wide band of highland climates almost 5000 miles (8000 km) long. The world's greatest concentration of highland climates occurs in southern Eurasia, extending from western China to northern Spain, and includes the Himalayas and its associated ranges, the Tibetan Plateau, the Zagros and Elburz Mountains of Iran, the Caucasus of the Soviet Union, and the European Carpathians, Alps, and Pyrenees. In Africa, important areas include the Atlas Mountains and the Ethiopian Highlands.

Lower mountain and upland areas that modify their climates to a lesser degree are also extensive. Some of the most notable are the Appalachians, the highlands of southeast Brazil and southern Africa, the Soviet Urals, the Italian Apennines, the mountains of Scandinavia, and the Great Dividing Range of eastern Australia. Smaller highland areas, often located on mountainous islands, are too numerous to list fully, but include portions of Japan, Madagascar, the East and West Indies, and New Zealand. It is evident that the areas of

Figure 7.35 The constancy of subfreezing temperatures causes most of the continent of Antarctica to be perennially covered by ice and snow.

Figure 7.36 Heavily glaciated Mt. Rainier, located in Washington State, towers above the surrounding evergreen forest. Major variations in climate often exist over short distances in mountainous regions. (*H. Armstrong Roberts*)

the world with topographically modified climates are extensive and widely distributed.

The climate of any specific highland location is controlled by the combined influences of elevation, latitude, orographic effect, and local topography. The most important of these is elevation. Its significance stems from the fact that, in the troposphere, the temperature declines at an average rate of 3.5 F°/1000 feet (6.4 C°/1000 m). As a result, temperature means in highland areas are substantially lower than those of adjacent lowlands. An example is provided by the climagraph for Bogotá, Colombia (Figure 7.37). Most sea level stations located at Bogotá's latitude of 4 1/2° N experience annual means near 80° F (27° C). Bogotá's elevation of over 8000 feet (2500 m), however, causes its mean temperatures throughout the year to remain in the cool mid- to high 50s (13°–14° C). In the highest mountains, temperatures can be reduced sufficiently to produce polar climatic conditions, along with permanent ice and snow, even at the equator.

Latitude, a second important climatic control, affects highland areas in much the same way as it does lowlands. As latitude increases, decreased insolation produces lower temperatures. As a result, the higher the latitude of a place, the lower the elevation at which a given climate may be expected to occur (see Figure

7.39). An increase in latitude is also generally associated with larger annual temperature ranges. For example, Bogotá, despite its cool temperatures, has the very small annual temperature range that one would expect of a city at its low latitude. Säntis, in the Swiss Alps (Figure 7.38), is at nearly the same elevation as Bogotá, but its latitudinal position causes it to have a much lower annual mean and a much larger annual range of temperature.

The orographic effect is the dominant control on precipitation amounts and distributions within a highland area. In general, the windward sides of mountains tend to be wet, and the leeward sides dry; large differences in amounts typically exist over small distances. The combination of orographic uplift and reduced evapotranspiration rates resulting from cooler temperatures generally causes highlands to be much more humid than nearby lowlands. Most of the world's major river systems, in fact, have their headwaters in highland regions.

The last of the major climatic controls is the local topography. Topographic variation is obviously much greater in mountainous areas than in flat areas, and the variety and complexity of localized climates are correspondingly increased. Two of the most important topographic factors were discussed in Chapter 3. These are

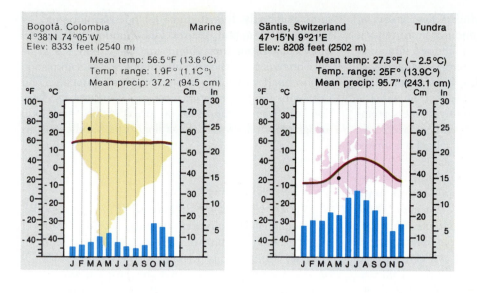

Figure 7.37 Climagraph for Bogotá, Colombia.
Figure 7.38 Climagraph for Säntis, Switzerland.

(1) slope aspect, which causes slopes facing the sun to have warmer temperatures and generally drier surface conditions than do more shaded slopes, and (2) the nature of the surface, which can vary greatly in material constituency and vegetative cover. For example, a bare, rocky slope will likely be warmer, drier, and windier during the day than a forested slope. A third topographically related factor is a site's extent of exposure. This especially influences wind speed and direction. A sheltered valley nestled beneath high mountains, for instance, is protected from strong winds; but an isolated peak is exposed to the full force of winds unreduced by surface friction. A good example of the effects of exposure has been provided by a weather station located atop New Hampshire's Mt. Washington, New England's highest peak, which recorded the world's highest wind speed of 231 miles per hour (90 m/sec).

Summary

This chapter has explored the global pattern of climates. World climatic characteristics and resulting climate regions are distributed in a systematic geographic pattern. This pattern, in turn, results largely from the controlling influences of spatial variations in insolation,

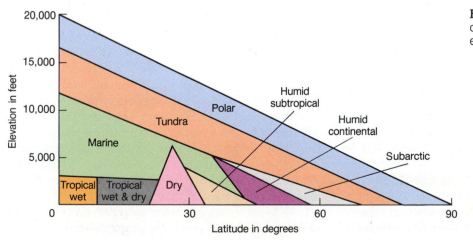

Figure 7.39 Relationship of climate type to latitude and elevation.

continentality, air pressure and wind patterns, and, in some areas, orographic influences and ocean currents.

Climate classification systems are devised largely as an organizational tool for the systematic analysis of world climates. Their basic purpose is to simplify the study of climate by reducing the number of climatic types to a manageable total. By so doing, these systems reveal geographical patterns of climatic characteristics; and a greater understanding of global climate controls can be gained.

In this chapter, world climates were grouped into low latitude, middle latitude, and high latitude types, with highland climates treated as a separate category. Three low latitude climate types exist. The Tropical Wet Climate, under the domination of the ITCZ, is characterized by continuously warm temperatures and by abundant precipitation of a predominantly convectional nature. The Tropical Wet and Dry Climate also has warm temperatures throughout the year, but displays a distinctive pattern of rainy summers and dry winters. The Low Latitude Dry Climate, which is dominated by the subtropical highs, is characterized by mild winters, very hot summers, and year-round dryness.

The middle latitudes contain five climate types. The Dry Summer Subtropical Climate has a pattern of dry, warm summers dominated by the subtropical high and mild, humid winters under the influence of the westerlies, with their traveling cyclones and anticyclones. The Humid Subtropical Climate has hot, humid summers with scattered convectional precipitation and cool-to-mild winters with alternating periods of fair weather and frontal and cyclonic precipitation. The Mid Latitude Dry Climate is cut off from major moisture sources by orographic barriers and is therefore dominated by dry weather with large annual temperature ranges. The Marine Climate, in contrast, is characterized by adequate-to-abundant precipitation from cyclonic storms and by small annual variations in temperature. The most poleward of the middle latitude climates is the Humid Continental Climate, which has cold winters, mild-to-warm summers, and modest but generally adequate precipitation of predominantly frontal and cyclonic origin.

Three high latitude climates were discussed. All are characterized by varying degrees of coldness and by rather low precipitation amounts from cyclonic storms and their associated fronts. The Subarctic Climate has severely cold winters and cool-to-mild summers of short duration. The Tundra Climate experiences above-freezing mean temperatures for only a few months in summer. Finally, the Polar Climate is characterized by subfreezing temperatures and a surface cover of ice and snow throughout the year.

The climates of the high mountain regions of the world display great variability, often over small distances. This variability exists largely because of differences in elevation, orographic influences on precipitation, and local topographic conditions such as slope aspect and exposure.

Review Questions

1. What are the chief advantages of using a regional climate classification system to study world climate characteristics? What are the shortcomings of such systems?

2. Why could it be said that the Tropical Wet Climate is the world's most monotonous climate? What are the controlling factors for this climate's temperature and precipitation characteristics?

3. Explain how the geographical location of the Tropical Wet and Dry Climate controls its annual precipitation distribution pattern.

4. Why are the world's hottest temperatures recorded in the Low Latitude Dry Climate, rather than nearer the equator? Are winter temperatures in this climate also hot? Why or why not?

5. Describe the annual precipitation distribution pattern of the Dry Summer Subtropical Climate. What causes this pattern?

6. How and why does the annual precipitation distribution pattern of the Humid Subtropics differ from that of the Dry Summer Subtropics?

7. How do the causes of the dryness of the Low Latitude Dry Climate and the Mid Latitude Dry Climate differ? In which climate are the annual temperature ranges larger? Why?

8. Why is the only extensive land area with a Marine Climate located in Europe? What changes in climatic conditions occur as one travels eastward from the west coast of France to the western border of the Soviet Union? Why do these changes occur?

9. Why do the Humid Continental and Subarctic Climates exist only in the northern hemisphere?

10. Why do the world's largest annual temperature ranges occur within the Subarctic Climate? How do these large ranges of temperature influence the seasonal distribution of precipitation?

11. Why do Tundra Climate areas remain cold during the summer, despite receiving large amounts of

insolation? Why do the Polar Climates remain so cold in summer?

12. List several reasons why climatic characteristics frequently vary significantly over small distances in mountainous areas.

Key Terms

climagraph

semiarid

arid

CASE STUDY

Urban Climates

The Case Study at the end of Chapter 3 dealt with possible global temperature changes resulting from human activities. While it now appears that temperatures on a global scale may, indeed, be rising in response to human actions, the current impact of such changes is minor, and the chief concern lies with the potentially much greater changes to come. Human beings, however, have already produced important climatic changes in urban areas, where a large and rapidly growing segment of the world's population resides.

The most significant impact that an urban area has on its climate is to increase local temperatures by producing what has been termed an *urban heat island*. A combination of factors is responsible for this. First, and probably most important, large quantities of energy are produced by human urban activities and eventually released to the atmosphere. Most of these activities, such as heating homes and businesses, running factories and power plants, and operating vehicles, involve the burning of fossil fuels. Heat-producing operations have an especially large impact in the winter, when heating needs are greater and proportionally less solar energy is present. Studies have shown that the production of artificial heat in cities in northern Europe and North America during the midwinter months may rival or exceed that provided by insolation.

A second factor aiding in the formation of a heat island is the abundance of urban construction materials such as brick, concrete, and asphalt. These materials readily absorb heat and reradiate it for long periods. A third factor is the buildup of heat-absorbing pollutants over a city. The major product of fossil fuel combustion is carbon dioxide, the gas chiefly responsible for the greenhouse effect. A final important factor is the surface dryness of a city. The materials from which a modern city is constructed are relatively impervious to moisture, and most rainwater quickly flows into storm sewers. As a result, the surface soon dries; subsequent insolation cannot be used to evaporate water, and more insolation is converted to sensible heat. In contrast, the evapotranspiration of water from soil and vegetation in rural areas utilizes a large proportion of the sun's energy—often more than is converted directly to heat.

The urban heat island is typically best developed at night, largely because energy absorbed by city materials during the day continues to be released at this time. The difference in nocturnal temperatures between nearby urban and rural areas is sometimes large enough that forecasters give separate predicted lows for each. Cool temperatures, clear skies, and light winds also aid in heat island development. Cool temperatures result in a greater artificial production of urban heat, clear skies allow the day's heat in rural areas to radiate

quickly into space, and light winds keep the warmer city air from being blown away.

The heat island patterns of a number of cities have been mapped (Figure 7.40). Most patterns are roughly concentric. The highest readings occur in the center of the city, where the heat-producing factors are most concentrated; progressively lower values are found toward the suburbs. In detail, the patterns are complicated by the presence of especially warm temperatures over industrial or other areas with high concentrations of activity and by cool pockets, often associated with open areas like parks and cemeteries.

Urban climates also differ from those of rural areas in several aspects involving atmospheric moisture. Relative humidities are somewhat reduced by warmer city temperatures, which increase the air's water vapor holding capacity. On the other hand, urban specific humidity values are at least as high as those of surrounding rural areas, because urban combustion processes release large amounts of water vapor. Mean amounts of cloudiness, fog, and precipitation are all higher in urban areas than in rural ones. One factor encouraging the formation of these condensation products is the large quantity of dust released by human activities. Dust particles within large cities are sometimes present in sufficient quantities to restrict visibilities greatly and to form what has been termed an "urban dust dome." The dust particles reduce the inten-

continued on next page

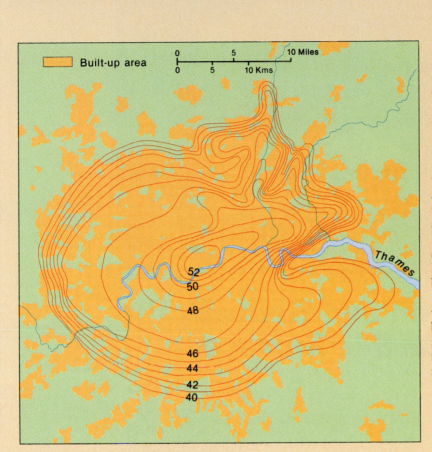

Figure 7.40 Map of the London urban heat island, in degrees Fahrenheit, as it developed on a clear night.

percent increase in mean annual precipitation totals in, and immediately downwind, of large North American cities.

Wind patterns in a city also differ considerably from those of the surrounding countryside. A large city has an extremely irregular surface profile, and the large amount of surface friction reduces average windspeeds by 20 or 30 percent. On the other hand, the wind is often redirected down streets, which are literally "avenues of least resistance" to the wind. Windspeeds may therefore actually be higher on average in areas frequented by pedestrians than in rural areas.

The combination of explosive world population growth, major urban immigration, and a continued rise in per capita energy consumption will further increase the extent and significance of urban climatic modification.

sity of sunlight, and many are also hygroscopic (water-attracting), so that they serve as effective condensation nuclei. City heat also encourages atmospheric lifting, especially in summer, resulting in the development of cumuliform clouds and occasional convectional precipitation. Studies have determined that urban heat production is responsible for a 5 to 10

Chapter Eight

Water on and Beneath the Earth's Surface

Focus Questions

1. What factors control the movement and storage of subsurface water?
2. How do lakes form and disappear?
3. What factors determine geographic and temporal variations in stream discharge?
4. What are the chief physical and chemical characteristics of the ocean water?
5. What are the causes of the three oceanic motions of currents, waves, and tides?

Water in the atmosphere has been discussed in detail in the last several chapters. At any given time, however, only 0.001 percent of our planet's near-surface water is located in the atmosphere; the overwhelming majority is to be found on and beneath the earth's surface (see Table 8.1). In this chapter, we begin the study of the terrestrial portion of the hydrologic cycle by examining the location, characteristics, and movement of the earth's surface and subsurface water supplies. Succeeding chapters examine the crucial roles this water plays in influencing the planetary distributions of vegetation, soils, and landforms.

Most precipitation falling on land does not flow immediately into rivers; instead, it enters the soil and sometimes the underlying rock layers, where it may stay for an extended period before eventually returning to the surface. The first portion of this chapter is therefore devoted to the earth's subsurface waters. We next examine lakes and rivers, which derive their water from a combination of surface and subsurface sources. The final topic is the ocean, which is the ultimate destination of runoff from the land and which dominates the surface geography of our planet.

SUBSURFACE WATER

Although large quantities of fresh surface water exist in lakes and rivers, the great majority of the earth's liquid fresh water is located beneath the surface. The total quantity of this water has not been accurately determined, but its volume is estimated at some 2,400,000 cubic miles (about 10^7 km^3). This amount is approximately 77 times greater than the total in all the earth's surface rivers and lakes and is about one-third the quantity locked up in the Antarctic and Greenland ice caps. It is enough water, if evenly spread over the earth's surface, to cover continents and oceans alike to a depth of about 35 feet (10.5 m). Nearly half of this water is situated within a half mile (0.8 km) of the surface and is considered to be accessible for extraction.

It is convenient to separate the earth's subsurface water into two categories based on depth, movement, and origin. Nearest the surface is **soil water,** which occupies the pore spaces between individual soil particles. Soil water occurs in the *zone of aeration,* where both water and air occupy the pore spaces between the individual soil particles in constantly varying proportions. Below this is the *zone of saturation.* Here, all pore spaces or other openings in the soil or rock are completely filled with **ground water** (Figure 8.1). The zone of saturation, which extends to about two miles beneath the surface, contains most of the earth's fresh subsurface water. Soil water and ground water are both derived primarily from precipitation and are in constant slow movement. Most of this water eventually returns to the surface to continue its participation in the hydrologic cycle.

Soil Water

Although nearly all water in the soil is derived from precipitation, much precipitation on land does not fall directly onto the soil. Instead, it is **intercepted** by the surface cover of vegetation and vegetative litter. These overlying materials must be thoroughly wet before an excess drips or flows down to the soil level. This water may collect on the surface and eventually evaporate, it may flow directly into rivers or lakes, or it may be absorbed by the soil.

Water that had infiltrated the soil is drawn slowly downward by gravity. The water moves in very circuitous routes around the soil particles, and some moisture adheres to the surfaces of the soil grains. With depth, soils become increasingly compacted by the weight of the overlying material, resulting in a progressive decline in their ability to absorb and transmit soil water.

During a prolonged rainstorm, the continued downward percolation of soil water may produce temporary saturation in the lower soil. Because this water can no longer move downward rapidly enough to offset the addition of water from above, it is diverted laterally so

Table 8.1
Distribution of Global Near-Surface Water Supplies

Location	Volume	Proportion
Oceans	286,230,000 cu mi	96.7%
Fresh water	9,769,600 cu mi	3.3
Distribution of Fresh Water		
Ice caps and glaciers	7,327,500 cu mi	75.00%
Ground water below 2500 ft	1,327,000 cu mi	13.60
Ground water above 2500 ft	1,074,700 cu mi	11.00
Rivers and lakes	32,000 cu mi	0.32
Atmospheric moisture	3,400 cu mi	0.03
Soil moisture	5,000 cu mi	0.05
Total	9,769,600 cu mi	100.00

Figure 8.1 Water may be intercepted or may be stored on the surface before infiltrating the soil. During rainy periods, considerable subsurface water travels laterally near the surface as throughflow, rather than descending to the water table to become ground water. Most throughflow soon returns to the surface.

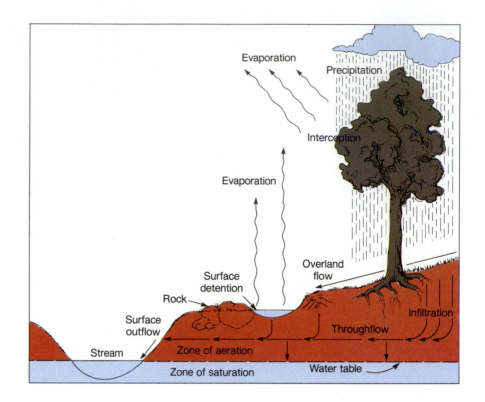

that it flows parallel to the surface as **throughflow.** Throughflow is probably the major type of hillside flow in humid climates. This flow may intersect the surface along a stream bank or other steep slope and seep out to continue downhill as surface flow (Figure 8.1).

Soil water that does not seep out at the surface or adhere to the soil particles continues to sink slowly downward until it reaches the **water table,** which is the upper boundary of the zone of permanently saturated soil or rock. In most cases, water reaches the water table within 48 hours of entering the soil. At that point the soil contains only water that adheres to the soil particles and that can therefore be held indefinitely against the pull of gravity. This "held" water is considered to be an integral constituent of the soil. Although the amount of soil water varies because of evaporative losses and extraction by plant roots, it is always present to some extent, even in the driest areas. Its characteristics and behavior are further considered in Chapter 10.

Ground Water

Ground water saturates the soil and rock layers below the water table. In general, it is most abundant in areas underlain by extensive layers of porous rock such as sandstone. The quality and quantity of water present as well as the location and extent of the zone of saturation, however, vary greatly in different localities. In some places, the upper surface of the zone of saturation intersects the earth's surface, producing an outflow of ground water. Elsewhere, it exists thousands of feet beneath the surface. Any hole, such as a well, that is dug into the zone of saturation will fill with water up to the level of the water table.

In arid areas little or no water is currently being added to existing ground water supplies. Ground water in such areas is generally a legacy of more humid climatic conditions in the past and is sometimes referred to as "fossil water." Like oil, it is a nonrenewable resource because it cannot be replaced under the existing climate. The extraction of this water results in a permanent lowering of the water table and a reduction of ground water supplies.

Even in semiarid and subhumid regions, such as the southern Great Plains of the United States, the large-scale extraction of water for irrigation is producing rapidly declining water tables and the threat of eventual ground water exhaustion. Shallow wells in such regions

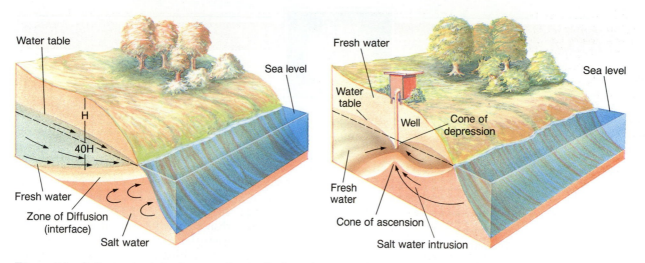

Figure 8.2 Saline water intrusion usually results from the excessive withdrawal of fresh ground water from a near-coastal site.

generate surrounding "cones of depression" in the water table, as the water is withdrawn more rapidly than it can be replaced (see Figure 8.2). If wells are located too close to one another, adjacent cones of depression can coalesce to produce a regional lowering of the water table.

In humid areas, the water table continually fluctuates in level. During wet periods, the addition of water from precipitation raises the level until it reaches the surface in valleys or other low-lying areas, sometimes producing an outflow of ground water known as a **spring.** If the outflow occurs in a valley that continues downhill, the spring will form or add water to a river or stream (see Figure 8.3). If it occurs in an enclosed depression, a lake or swamp will be formed. In this fashion, ground water returns to the surface to continue its movement through the hydrologic cycle.

Most stream surfaces in humid areas are at the water table level and have a substantial portion of their total volumes of flow provided by the outflow of ground water. This provides a dependable supply of water, causing larger streams to be *perennial streams,* which continue to flow even during extended droughts. The much higher flow volumes during wet periods are derived primarily from the temporary additions of surface runoff and throughflow. In contrast, the water table is frequently far below the level of the river valleys in arid regions, causing such regions to contain intermittent or *ephemeral streams,* which flow only during wet periods when they are supplied by surface or soil water sources. For the world's land areas as a whole, it is estimated that ground water outflow supplies 30 percent of total stream flow.

As already noted, different earth materials vary tremendously in their ability to store and transmit ground water. A rock layer that contains large quantities of ground water and permits its ready movement is called an **aquifer.** In many parts of the world, large aquifers, often consisting of deep layers of sandstone or limestone, act as subsurface reservoirs that hold tremendous volumes of fresh water. Wells penetrating these strata may be expected to yield large and dependable water supplies.

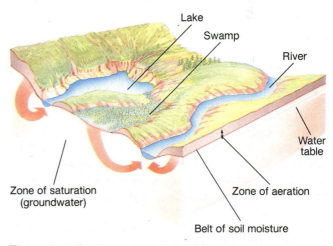

Figure 8.3 Relationships between the surface topography, the level of the water table, and the movement of ground water.

155

Human Impact on Ground Water

Ground water has been extracted by humans since the beginning of civilization, primarily for the irrigation of crops. Until less than a century ago, only near-surface water supplies were tapped through the digging of shallow wells. In recent decades, however, water-extraction technology has advanced at a rapid pace. The development of deep drilling methods and powerful pumps has made large quantities of ground water available, but has significantly lowered the water table in virtually all areas where ground water is extracted on a large scale.

Ground water supplies are often most extensively developed in arid and semiarid areas where surface supplies are inadequate or nonexistent. This creates a problem in maintaining reserves, since the ground water in these areas is generally fossil water. The overpumping of water in numerous drier areas of the world is producing rapidly declining water tables, increased pumping costs, and decreasing water yields. Three areas within the United States where these problems are especially acute are the southern Great Plains, southern Arizona, and the Central Valley of California.

Other problems resulting from the overuse of ground water are saline water intrusion and land subsidence. *Saline water intrusion* occurs in coastal regions where so much fresh water has been removed that the normal seaward flow of fresh water is reversed, and saline water from the ocean is instead drawn in (Figure 8.2).

When ground water saturating certain fine-grained sediments is withdrawn, an irreversible dewatering and compaction of the sediments occurs, resulting in the sinking, or *subsidence,* of the surface. Surface subsidence has produced serious problems in such places as Mexico City, Bangkok, Ven-ice, Southeast Texas, and parts of the Central Valley of California. In parts of Mexico City, for example, subsidence has lowered the surface by more than 25 feet (8 m); and the process continues at a rate of about an inch a year. Subsidence in Venice, Italy, is especially critical, because much of the city is situated at or near sea level; as a result, the famed city of canals is gradually being inundated by the Adriatic Sea.

An additional ground water problem resulting from human activities is pollution. An increasing variety of liquid contaminants is finding its way into ground water supplies, especially in urban and industrialized areas with humid climates such as the eastern United States and western Europe. The chief sources of pollutants are waste disposal sites such as landfills, lagoons, and disposal pits. Nonpoint sources such as urban runoff and water from agricultural land treated with fertilizers and pesticides also produce pollution problems.

The world's largest known aquifer, the Nubian Sandstone, underlies much of the western Sahara. It contains an estimated 144,000 cubic miles (600,000 km^3) of water. Although its present recharge rate is almost zero, this vast underground reservoir is still virtually untapped and constitutes a natural resource with enormous potential for the countries of northwestern Africa.

The Ogallala aquifer, which underlies much of the southern and central Great Plains, is the most economically important aquifer within the United States. In contrast to the Nubian Sandstone, this aquifer has in recent decades been greatly lowered in many areas by large-scale withdrawals of water for irrigation agriculture.

LAKES

A *lake* is a landlocked body of water with a horizontal surface water level. Freshwater lakes cover an estimated 320,000 square miles (825,000 sq km), or about 1.5 percent of the earth's land surface. They contain over 30,000 cubic miles (125,000 km^3) of water, which represents more than 95 percent of the earth's surface supply of fresh liquid water. Although probably more than a million lakes exist, roughly four-fifths of the earth's total volume of fresh surface water is found in the 40 largest lakes.

The global distribution of lakes, like that of ground water and the ocean, but unlike that of rivers, is related more closely to the presence of basins suitable for water accumulation than it is to climate. Most of the world's freshwater lakes occur on three continents: North America, Africa, and Asia (see Table 8.2). Although freshwater lakes are much more numerous and important for human activities than those containing salt water, some of the world's largest lakes are saline. Saline lakes cover an estimated total area of 270,000 square miles (700,000 sq km) and contain approximately 25,000 cubic miles (105,000 km³) of water. Both the combined areal extent and volume of saline lakes are therefore about 85 percent of the respective totals for freshwater lakes. Saline lakes are located mostly in Asia and are dominated by the Caspian Sea, which, despite its name, is considered the world's largest lake. By itself, the Caspian Sea contains 76 percent of the total volume of all the world's saline lakes.

Lakes with outlets nearly always contain fresh water, while lakes with no outlets are usually saline. In cases such as the Caspian and Aral Seas, lakes are salty because they are cutoff arms of the ocean. In most cases, however, the salt has been supplied gradually by inflowing rivers and has accumulated because the evaporation of water has left the salt behind. Some lakes are only slightly saline, whereas others, like the Dead Sea, are saturated with salt. The Dead Sea is the world's most saline large lake. It contains about 24 percent dissolved solids, making it six times as salty as the ocean.

Origins of Lakes

Only two criteria are necessary for the formation of a lake. One is the presence of a relatively nonporous surface depression or basin; the other is the availability of sufficient water from either surface or subsurface sources to fill the depression at least partially. Surface depressions capable of trapping water are formed in a variety of ways, but the two most important by far are movements of the earth's crust and glacial action.

Crustal movements, the origins and characteristics of which are discussed in Chapters 11 and 12, can form lakes of large size and sometimes of great depth. In some instances, the gradual broad uplift of a portion of the

Table 8.2
The World's Largest Lakes

Name	Basin Origin	Area (sq mi)	Maximum Depth (ft)	Volume (mi³)
Caspian Sea*	Uplift (cut off from sea)	168,450	3,104	19,040
Superior	Glacial erosion	32,150	1,007	2,880
Victoria	Uplift of margins	26,560	259	650
Aral Sea*	Uplift (cut off from sea)	23,900	223	230
Huron	Glacial erosion	22,970	732	1,100
Michigan	Glacial erosion	22,330	869	1,380
Tanganyika	Faulting (graben)	13,120	4,823	4,550
Baikal	Faulting (graben)	12,160	5,712	5,500
Nyasa	Faulting (graben)	11,890	2,316	2,020
Great Slave	Glacial erosion	11,600	2,014	—
Great Bear	Glacial erosion	11,400	—	—
Erie	Glacial erosion	9,970	210	130
Winnipeg	Glacial erosion	9,470	62	750
Ontario	Glacial erosion	7,240	738	410
Ladoga	Glacial erosion and uplift	7,230	820	220

*Saline lakes.

Source: G. E. Hutchinson, *A Treatise on Limnology,* vol. 1 (New York: John Wiley & Sons, 1957), pp. 168–9.

earth's surface cuts off an arm of the sea to form a lake. The Caspian and Aral Seas formed in this fashion when crustal uplift gradually severed their connection with the Mediterranean. Tectonic activity can also result in a narrow wedge of land (a *graben*) dropping sharply downward between a pair of faults. Lakes resulting from this type of movement tend to be long, narrow, and exceptionally deep. Lake Baikal in the Soviet Union was formed this way, but the world's largest group of fault-produced lakes is located along the Great Rift Valley, which extends through Asia and Africa from Syria to Mozambique. Major lakes within this valley include the Dead Sea, Lake Albert, Lake Tanganyika, and Lake Nyasa.

Lakes resulting from glacial erosion and deposition are more numerous than those of all other origins combined (see Figure 8.4). They vary greatly in size, and most are located in northern portions of North America and Eurasia, which were covered by continental glaciers between 10,000 and 20,000 years ago (see Chapter 15). Most glacial lake basins were produced by the scooping action of ice either in areas of soft bedrock or in loose ice-deposited sediments. Others resulted from the disruption of surface drainage by glacial deposits. Literally hundreds of thousands of glacial lakes, large and small, are found within North America north of the Ohio and Missouri Rivers. Included are some of the world's largest lakes, such as the five Great Lakes, Great Slave Lake, Great Bear Lake, and Lake Winnipeg. Large numbers of glacial lakes also exist in northern Europe, especially in Scandinavia.

Lakes can also form in depressions produced by numerous other natural processes. Foremost among these are solution depressions in regions underlain by highly soluble rock types such as limestone and dolomite. The characteristics and distributions of these rock types are discussed in Chapter 13. Other lakes result from volcanic eruptions, landslides, or other mass movements that may block drainage outlets; changes in river flow patterns; wind erosion or deposition; and meteorite impacts.

It should also be noted that man-made lakes have become important landscape features in many areas. Some are produced inadvertently as a result of earth-moving activities such as mining or quarrying. Others, often of large size, are purposefully created. Large reservoirs such as Lake Powell and Lake Mead in the western United States serve a variety of purposes, including flood control, water supply, hydroelectricity generation, transportation, and recreation.

Figure 8.4   This aerial view of the southern Canadian Shield in Ontario shows dense coniferous forest and numerous lakes formed by glacial erosion. (*Richard Jacobs, © JLM Visuals*)

Lakes are relatively short-lived landscape features because the relative lack of motion of lake water makes the basin a sediment trap. Sediments carried by inflowing streams are deposited on the floor of the lake, gradually filling it. This process has shortened to only a century or two the expected useful lifespans of even large artificial lakes in various parts of the world. The process of infilling is often hastened by the growth of aquatic plants and the consequent deposition of organic matter. Most shallow lakes gradually become *swamps* when their water depths decline sufficiently to allow water-tolerant vegetation to cover the entire lake surface; with continued infilling they eventually become dry land areas.

RIVERS

Rivers and *streams* are channeled bodies of flowing water. They engage in the gravitational transport of excess surface moisture derived from precipitation. The water is transported either to interior basins on land or, more commonly, to the sea. Although the two terms are often used interchangeably, a "stream" has no particular size connotation, whereas a "river" is generally considered to be large, often the master stream in a drainage system.

In addition to carrying water, streams also transport sediments derived from the land. In doing so, they serve

The Great Lakes Become Greater

Following an extended period of cool, wet years, the Great Lakes in the mid-1980s rose to their highest levels in 125 years of record keeping. In late 1986, their surface water levels averaged six feet higher than low points reached in 1964. In October 1986, the surface of Lake Michigan was at an average elevation of 581.6 feet, over a foot higher than a year earlier. Drier conditions in the last few years have lowered lake levels somewhat from the record values of 1986, but this decline may only be temporary.

The recent rise in water levels has occurred because the Great Lakes, like many lakes, respond in a highly sensitive fashion to the balance between water gains, resulting from precipitation and water inflow, and losses, resulting from evaporation and outflow. Evaporation is responsible for an estimated 40 percent of water losses from the Great Lakes, and outflow is responsible for the remaining 60 percent. For 15 of the 18 years from 1969 through 1986, precipitation in the Great Lakes area was above normal, while cooler-than-normal temperatures reduced evaporational losses.

The rising water levels pose serious problems for the millions of people who live along the shores of the lakes in both the United States and Canada. Lakefront property is heavily built up right to the water's edge in many urban areas such as Chicago, Milwaukee, and Cleveland, so the effect of any change

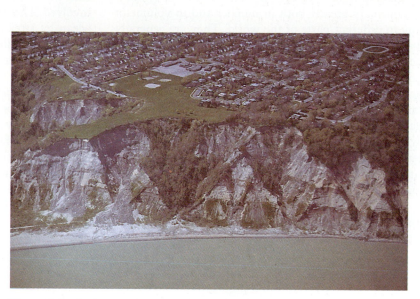

Figure 8.5 Active erosion along the north shore of Lake Ontario near Toronto keeps these bluffs steep and unstable. (*B.F. Molnia*)

in water level is felt immediately. During the winter storms that are common in this area, giant waves destroy seawalls and toss sheets of water onto streets and buildings, breaking windows and flooding basements. Chicago's famed Lake Shore Drive has been closed on occasion because of high water. Along less-developed stretches, storm waves have caused major property losses by undermining lakeshore bluffs. The problems are worst when strong winds blow onshore, piling up the water even higher than normal.

It may be wondered why these water levels are able to rise substantially, since the Great Lakes drain into the Atlantic by way of the St. Lawrence River and into the Mississippi River system through the Chicago River.

These outlets, however, are simply too small to carry off the water at the rate at which it has been accumulating. One expert likened their potential for draining the lakes to "pinholes in a bathtub." For example, a recent proposal to triple the outflow of Lake Michigan water through the Chicago River would not only produce flooding along that river, but would also remove only enough water to lower Lake Michigan's water level by 2 1/2 inches in a fifteen-year period.

Scientists agree that no quick solution to the problem is in sight. Lake levels should remain high for the foreseeable future and may even rise several feet, resulting in extensive flooding along the south shores of all five lakes.

as the major agent responsible for the gradual reduction of the earth's land surface by erosion. Stream-produced landscapes are so important that an entire chapter (Chapter 14) is devoted to them. In this section we will concentrate on the hydrologic aspects of streams.

Geographic Distribution and Significance

Streams vary greatly both in number and in volume of flow in different parts of the world. They are generally absent only from the 10 percent of the earth's land surface that is perennially covered by ice. Even those areas, located mostly within Antarctica and interior Greenland, have excess precipitation removed by rivers; the only difference is that they are rivers of frozen water, or glaciers. Elsewhere, as might be expected, the extent of river development is determined largely by the balance between mean precipitation amounts and evapotranspiration rates. Even the driest deserts, however, have stream channels that occasionally carry water.

The volume of water contained in the world's streams at any one time is very much smaller than the amount held in lakes and as ground water. It amounts to only about 0.025 percent, or 1/4000th, of the earth's total near-surface fresh water supply. The practical significance of this water, though, is much greater than its volume implies, both because of its ready accessibility and because of its speed and power of flow. While water in lakes and beneath the surface can be considered to be in at least temporary storage, the water in streams is in active movement; in time, large quantities of water are involved. The total volume of water discharged into the ocean by streams during a single year is estimated at 8,000 cubic miles (33,000 km^3). This is approximately one-third of the total volume of the earth's freshwater lakes.

Although virtually all stream flow is ultimately derived from precipitation, the water in a direct sense comes from several different sources. The most obvious of these is surface runoff from storms, which in some areas supplies the majority of the water entering stream systems. It is currently believed, however, that in most humid tropical and temperate regions, the largest volume of water in streams is supplied by the outflow of ground water and by throughflow. Some streams are also fed by glacial meltwater, and a small quantity of

Table 8.3
The World's Major Rivers

Name	Drainage Area, (thousands of sq mi)	Percentage of World's Land Area Drained	Length (mi)	Mean Discharge, (thousands ft^3/sec)	Percentage of World's Total Discharge
Amazon	2,722	4.8	4,000	6,350	19.2
Rio de la Plata/Paraná	1,600	2.8	2,485	777	2.3
Congo	1,314	2.3	2,914	1,458	4.4
Nile	1,293	2.3	4,132	110	0.3
Mississippi/Missouri	1,244	2.2	3,741	650	2.0
Ob/Irtysh	1,149	2.0	3,362	558	1.7
Yenisey	996	1.7	3,332	671	2.0
Lena	961	1.7	2,734	575	1.7
Yangtze	756	1.3	3,434	1,200	3.6
Niger	730	1.3	2,600	215	0.7
Amur	716	1.3	1,755	438	1.3
Mackenzie	711	1.2	2,635	400	1.2
Ganges/Brahmaputra	626	1.1	1,800	1,360	4.1
St. Lawrence/Great Lakes	565	1.0	2,500	360	1.1
Volga	525	0.9	2,293	282	0.9

Source: Encyclopaedia Britannica.

additional water is added directly by precipitation falling into streams.

A stream's **discharge,** or volume of flow past a given point in a specified time interval, normally increases downstream because of the increase in the total area drained. Losses occur through seepage and especially through evaporation, however, and streams flowing through dry climate regions actually decrease in discharge in a downstream direction, sometimes even drying up completely. Two well-known major rivers that decrease naturally in flow because their courses traverse arid regions are the Nile and the Colorado. (In recent times, the volumes of flow of both of these rivers have further diminished because of water withdrawals for human use. At present, in fact, no water from the Colorado River reaches its mouth at the head of the Gulf of California except during times of exceptionally high flow!)

The total global discharge of streams into the ocean is believed to be slightly less than 30 percent of all the precipitation that falls on the land areas of the earth. Most of the remainder is evaporated; this makes evaporation the leading process by which moisture leaves the land. A relatively small amount of water also enters the ocean as subsurface flow.

Although many thousands of streams of all sizes flow into the ocean, total global discharge is dominated by a small number of large rivers (see Table 8.3). The earth's greatest river system in all respects except length is the Amazon (see Figure 8.6). By itself, it drains 4.8 percent of the earth's land area and contributes nearly one-fifth of the total global discharge, a figure over four times that of the second largest river, the Congo. The Mississippi, North America's dominant river, ranks only seventh in total discharge among the rivers of the world. The Nile, with a length of 4132 miles (6650 km), has the distinction of being the world's longest river; it also has the fourth largest **drainage basin,** or area drained. Because of the arid nature of much of its course, however, it is not among the top twenty rivers in discharge.

The areal distribution of stream discharge volumes is strongly correlated with climate, especially with precipitation. As might be expected, regions with high precipitation totals generally have streams with large discharge volumes. Temperature and humidity levels are also important criteria, since they largely determine evapotranspiration rates. Geographically these two factors increase discharge at any given level of precipitation in areas at high latitudes and elevations.

Temporal Variations in Stream Flow

Although the size of a stream can be described in a general way by citing its mean volume of flow, the discharge of most streams varies considerably on both a seasonal and on a short-term basis. The causes of these variations are mostly climatic, because the great majority of the earth's land surface has distinctive fluctuations in precipitation, temperature, or both. Variations in the quantities of water transpired by plants are also significant in areas of seasonally active vegetation.

Near the equator, large rainfall totals produce high runoff rates throughout the year. Somewhat farther poleward, in the seasonally wet and dry tropics, streams tend to have a single well-developed summer discharge maximum. Many smaller streams here are intermittent, with no flow at all near the end of the dry season. Within the middle and higher latitudes, flow regimes vary considerably, depending on precipitation patterns; but streams most commonly exhibit winter and spring discharge maxima and late summer and autumn minima. High discharges in spring are associated with the combination of spring rains, snowmelt, the possible presence of impermeable frozen subsoils, and relatively low evapotranspiration rates. With increasing latitude, peak flows tend to occur progressively later in spring, as warm moist air and melting conditions advance poleward with the sun's rays.

In dry climates, especially within the subtropics, only larger rivers flow year-round. These rivers are *exotic;* their water is derived from runoff from more humid areas at much higher elevations or in differing latitudes. (The Nile and Colorado Rivers again serve as examples). Smaller local streams in these areas are usually situated well above the water table and are ephemeral, flowing only for short periods during and immediately following storms. In most cases their water evaporates, is absorbed by the ground, or flows into saline lakes occupying basins of interior drainage (see Chapter 16).

Short-term variability in flow is characteristic of virtually all streams and is especially significant for those fed principally by surface runoff. Changing stream-flow volumes result from variations in the quantity of water supplied to the stream systems primarily by storms and sometimes by snowmelt. The irregularity in the timing and quantity of these water influxes makes it much more difficult to predict short-term stream flow than seasonal variations in flow.

The surface runoff from an intense storm is relatively rapid and usually does not travel far before it

Figure 8.6 The drainage basin of the Amazon River, shown here in green, is the world's largest.

enters a tributary of the local drainage system. Small streams respond quickly and generally *crest,* or reach peak flow, shortly after the period of most intense rainfall. Crests are normally reached after progressively longer time intervals in streams draining larger basins, since most water must travel greater distances, and the rise in water levels is more gradual. Following precipitation, stream levels within drainage basins of all sizes tend to fall more gradually than they rose. The gradual decline occurs because the much slower movement of throughflow and ground water outflow normally causes water inputs from these sources to peak after the surface flow has begun to decline. For perennial streams draining humid areas, the flow eventually subsides to a stable base flow fed almost entirely by ground water.

One of the most practically important reasons that stream discharge rates are carefully monitored is the danger of flooding. A **flood** occurs when a stream's volume of flow exceeds the carrying capacity of its channel, and excess water spills over the banks to cover adjacent areas. Occasional flooding is characteristic of nearly all rivers and for some is an annual event.

Figure 8.7 A flood in Wisconsin devastates a small town. (*WI Dept. of Natural Resources, JLM Visuals*)

Major river floods rank with earthquakes and hurricanes as one of the world's foremost natural hazards (see Figure 8.7). In the 20-year period between 1947 and 1967, for example, floods claimed an estimated 173,000 human lives. In the United States, floods are responsible for average annual property losses of well over $100 million.

The flooding potential of streams is generally increased by human activities associated with agriculture and especially with urbanization. The removal of vegetation and the compaction of the surface by the passage of humans, livestock, and vehicles reduce the soil's infiltration capacity and cause increased surface runoff. In suburban and urban areas, extensive areas are surfaced with impervious materials such as asphalt and concrete. This results in greatly increased surface runoff, which then enters storm drains designed to conduct the water as quickly as possible to stream systems. As a result, many urban streams that in the past rarely overtopped their banks now do so on a frequent basis, adversely affecting large numbers of people.

THE OCEAN

The *ocean* is the interconnected body of salt water that dominates the surface of our planet. The earth is much more a world of water than of land, since the ocean covers nearly 71 percent of the surface and an additional 2 percent is covered by rivers and lakes. Even the quarter of the earth's surface that consists of dry land

owes some of its most basic characteristics to the ocean. Life on earth, which currently abounds in water and on land alike, is believed to have begun in the ocean. In addition, as we have seen, the ocean waters exert a tremendous influence on global climates.

The ocean is commonly considered to be both the beginning and the end of the hydrologic cycle. It provides most of the water that evaporates to enter the atmospheric portion of the cycle, and runoff from the land returns to it to complete the cycle. The ocean is sometimes referred to as the "hydrologic sink" because some 96.7 percent of all water at or near the surface of the earth is located in the ocean at any one time and because millennia may pass between the time a water molecule enters the ocean as precipitation or runoff and the time it evaporates to begin another journey through the hydrologic cycle.

Origin, Distribution, and Extent

The ocean water originated in the same way as the surface and subsurface water of the land. It was released from its original solution in rock deep within the earth as the rock moved surfaceward as magma; subsequently, it entered the atmosphere in the form of water vapor during volcanic eruptions. Cooling and condensation of the water vapor caused it to precipitate to the surface, where it accumulated in enclosed basins. The geographic distribution of the ocean waters is determined by differences in the thickness and density of the rocks comprising the earth's crust (see Chapter 11). These differences cause the earth's lithospheric surface to lie at varying elevations in different areas, and the ocean has simply covered the lowest 71 percent of the earth's surface. This includes approximately 60.6 percent of the northern hemisphere and 81.0 percent of the southern hemisphere (see Figure 8.8).

Several major subdivisions of the ocean exist. They are also termed oceans, and their areas and mean depths are listed in Table 8.4. As the table indicates, the mean depths of the water in the various oceans, with the exception of the comparatively shallow Arctic Ocean, are rather consistent. Depths range from 0 feet along the coasts to a maximum of 36,198 feet (11,033 m) in the Mariana Trench in the western Pacific. The estimated mean oceanic depth of 12,680 feet (3865 m) is almost five times the average elevation of the world's land areas, which is estimated at 2760 feet (841 m). The vertical dimensions of the ocean are therefore even more

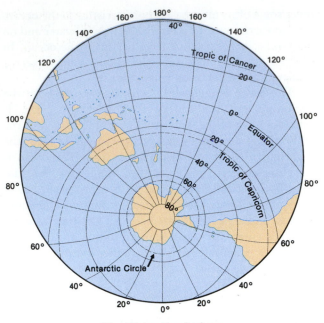

The Water Hemisphere

Figure 8.8 The "ocean hemisphere." This orientation of the earth stresses the areal predominance of the oceans over the continents.

predominant over those of the continents than are its horizontal dimensions (see Figure 8.9).

Physical and Chemical Characteristics

Ocean water displays significant spatial variation in a number of physical and chemical characteristics. Among the most important are chemical composition, temperature, and the presence or absence of ice.

Seawater contains an average of about 96.5 percent water and 3.5 percent dissolved solids by weight. Nearly every known substance is soluble to at least some extent in water, and about two-thirds of the 92 naturally occurring chemical elements have been found in seawater. The great majority of these solids are ions of salts.[1] Chlorine and sodium are the two dominant elements by far and, upon evaporation of sufficient water, will combine to form common table salt, or sodium chloride (NaC1).

---

1. An *ion* is an electrically charged atom or group of atoms. The solution of minerals in seawater causes them to dissociate into their constituent atoms, which will recombine if the water is evaporated. A *salt* is formed when the hydrogen atoms of an acid are chemically replaced by atoms of a metal.

The origin of the salinity of the ocean is similar to that of saline lakes. Over hundreds of millions of years, runoff from the land has dissolved and carried to the sea tremendous quantities of dissolved salts. Because there is no natural mechanism for removing these salts until saturation occurs (and the oceans are still well below the salt saturation level), salinity continues to increase gradually. The quantity of salt in the ocean is now so great that, if removed, it would exceed in volume the portion of the continent of Africa that is above sea level.

The temperature of the ocean surface varies considerably in different parts of the world (see Figure 8.10). As might be expected, the temperature distribution is generally similar to that of the atmosphere. Extensive areas of tropical ocean water have surface temperatures between 80° and 85° F (26.5° and 29.5° C), and values generally fall with increasing latitude to readings near or slightly below 32° F (0° C) in the subpolar and polar regions. The mean temperature of the ocean surface water as a whole is about 70° F (21° C).

Despite the transparency of the ocean, most solar energy is absorbed in the top 5 feet (1.5 m) of water. As a consequence, outside the polar regions, water temperatures decline everywhere with depth. The warm surface water, lighter and more buoyant than the underlying water, remains at the surface. Conversely, cold water produced in the high latitudes becomes dense enough to sink and spread outward at depth. As a result, all ocean water at depths greater than 10,000 feet (3000 m) is derived from the polar regions and has a constant temperature of approximately 35° F (2° C).

Table 8.4
Areal Extent and Mean Depths of the Ocean Basins

Ocean	Area (sq mi)	Mean Depth (ft)
Pacific	63,855,000	13,350
Atlantic	31,744,000	11,950
Indian	28,371,000	12,770
Arctic	5,427,000	3,950
"Antarctic"*	5,731,000	12,240
Other**	3,991,000	—
Total***	139,119,000	12,680

*This refers to the oceand water surrounding Antarctica.

**Includes restricted bodies of ocean water such as the Mediterranean Sea, Black Sea, and Red Sea. All figures are approximate because boundaries between adjacent ocean basins are arbitrary.

***Estimated total volume of water is 286,230,000 cubic miles (1,192,625,000 km³

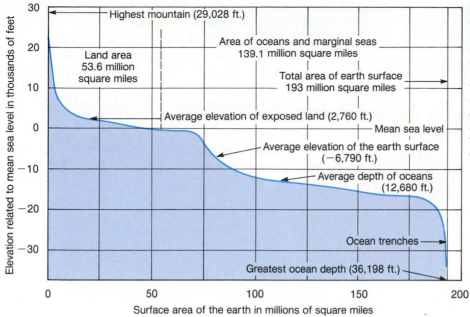

Figure 8.9 Hypsographic curve of the earth's surface. The proportions of the total global surface at differing elevations and ocean depths are indicated by the slope of the curve. The two relatively level portions of the curve—near sea level and at ocean depths between 12,000 and 16,000 feet—indicate that these surface levels are especially widespread.

Although most of the world's ice is located on the continents, the oceans of the higher latitudes contain both sea ice produced by the freezing of seawater and glacial ice derived from the land. **Sea ice** is extensively distributed in the oceans of the high latitudes and covers most of the Arctic Ocean, parts of the extreme northwestern Atlantic, and much of the water around Antarctica. The extent and thickness of sea ice varies considerably between summer and winter, but most is less than 11 feet (3.5 m) thick.

Icebergs are masses of freshwater ice produced by the breakup of glaciers as they enter the sea. Unlike sea ice, glacial ice is derived from the compaction of snowfall on land (see Chapter 15). Icebergs differ considerably in appearance from sea ice because they do not form continuous thin sheets but commonly form as individual rugged, islandlike masses (see Figure 8.11). Sizes vary greatly, but large icebergs may be over 300 feet (100 m) high and have above-water lengths exceeding 1 mile (1.6 km). Between 80 and 85 percent of an iceberg (depending on the amount of entrapped air) is hidden below the water line, so the visible portion is indeed "just the tip of the iceberg."

The chief problem posed by icebergs is their danger to shipping. In the northern hemisphere, most icebergs form from glaciers entering Baffin Bay from the west coast of Greenland. They drift southward in the Labrador Current to the Grand Banks area southeast of Labrador, where they are picked up by the Gulf Stream, carried eastward, and eventually melt. In this area, they are within the North Atlantic shipping lanes. One of these icebergs was responsible for the sinking of the ocean liner *Titanic* in 1912, with a loss of 1517 lives.

Motions of the Ocean Waters

The ocean is never completely still. A variety of forces originating on, above, and below the surface keep it in constant motion. These water movements are commonly divided into the three categories of currents, waves, and tides.

Currents

Ocean **currents** are continuous movements of water streams of similar temperature and density. As noted in Chapters 3 and 5, these currents help transport excess energy from the tropics to the higher latitudes. The presence of ocean currents of differing temperatures profoundly influences global patterns of both temperature and precipitation.

The most important ocean currents are those existing at and near the surface because they interact with the atmosphere and coast and affect shipping. The energy that sets these currents in motion is the frictional drag of the wind on the water surface. As a result, both the speed and direction of the surface currents are determined largely by the global wind belts; and there is

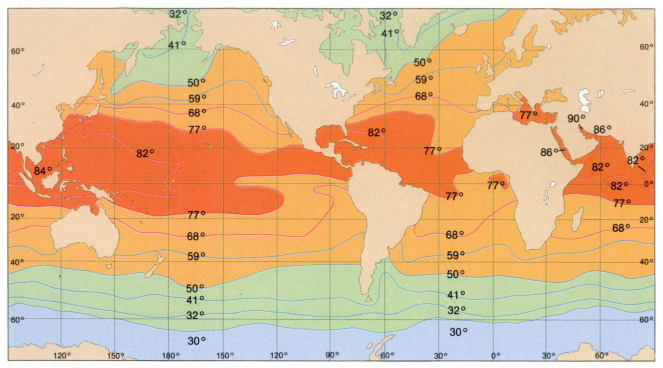

Figure 8.10 Mean ocean surface water temperatures in degrees Fahrenheit.

considerable similarity in the planetary patterns of winds and ocean currents. The Coriolis effect, produced by the earth's rotation, also influences the pattern of ocean surface currents. It deflects currents to the right of the prevailing wind direction in the northern hemisphere and to the left in the southern hemisphere. The average angle of deflection from that of the prevailing wind direction is 45°.

Figure 8.11 A large iceberg that has grounded at Godhavn, on the West Greenland coast, dwarfs the houses in front of it. (*Peter J. Bryant, Biological Photo Service*)

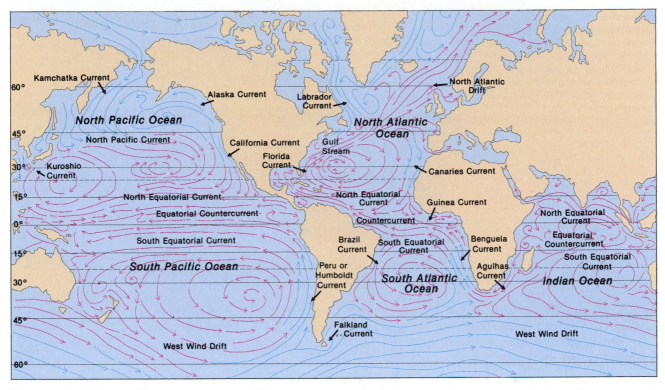

Figure 8.12 Global surface ocean currents. Warm currents are shown in red; cold currents are in blue. The centers of the major current circulation patterns, or gyres, roughly correspond to the centers of the subtropical highs.

The vast inertia of the ocean water causes the ocean currents to be more consistent in their position, speed, and direction of flow than are the wind currents. Satellite imagery and surface observations have revealed, however, that the ocean currents develop constantly changing patterns of swirls and eddies along their routes and that extensive eddy fields generated by the flow of the currents involve much of the ocean surface waters in slow swirling motions. The ocean currents flow at a much slower average speed than the wind; flow rates average only 1 or 2 miles per hour (1.5–3.0 km/hr) and do not exceed 5.5 miles per hour (9 km/hr).

If the three largest oceans are divided by the equator, a total of five extensive ocean basins are produced: the North and South Pacific, the North and South Atlantic, and the South Indian Oceans. (The North Indian Ocean is too small to be considered an additional member.) The basic pattern of currents for each basin is essentially similar; it consists of a giant elliptical loop, called a *gyre,* elongated in an east-west direction (see Figure 8.12). The oceanic gyres are centered at roughly 30° N and S somewhat to the west of the midpoints of their respective basins. They correspond closely in position to the oceanic

centers of the subtropical highs and are, in fact, produced largely by the wind circulations around these highs. Like the wind, the currents flow clockwise around their centers in the northern hemisphere and counterclockwise in the southern hemisphere.

Currents largely unrelated to those at the surface exist at depth in the oceans (see Figure 8.13). Unlike surface currents, the oceanic *deep currents* are produced by density differences resulting from variations in temperature and salinity. Deep currents have significant vertical as well as horizontal components of motion. A slow circulation occurs between the surface water and the deep water. Surface water descends to great depths in zones of convergence, mostly located where cold water from the high latitudes encounters and plunges beneath warmer water from the middle latitudes. Extensive convergence occurs in the waters around Antarctica and in the North Atlantic south of Greenland. These zones of convergence are the sources of the cold deep water that exists throughout the ocean. Deep water eventually returns to the surface in zones of *upwelling.* Most of these zones are situated near the west coasts of continents in the subtropics, where equatorward-flowing water is redirected westward by

Figure 8.13 A profile view of the Atlantic, showing the major deep ocean currents. Most of the oceanic deep water is derived from the subpolar regions, because only here is the water sufficiently cold to become dense enough to sink.

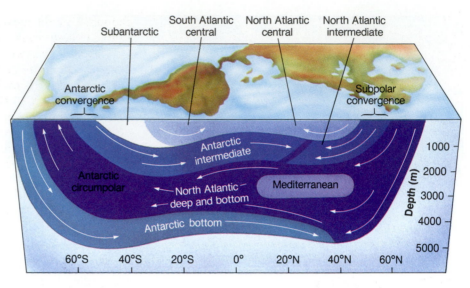

the trade winds. The seaward push of this water causes cold water from great depths to well up to replace it.

A period of several centuries may be required for one complete cycle of convergence, descent, and upwelling. Slow as it is, this vast vertical circulation system is crucial to sea life. The descent of surface water in areas of convergence carries into the depths dissolved carbon dioxide and oxygen needed by marine plant and animal life. Conversely, the rise of bottom water in zones of upwelling transports nutrients from the sea floor to the surface. As a result, upwelling zones are the sites of some of the world's most productive fishing grounds (see the Case Study at the conclusion of this chapter). Water in the vast mid-oceanic basins is largely cut off from this circulation. As a consequence, it is low in dissolved gases and nutrients and nearly devoid of life.

Waves

Waves are the most familiar of the ocean movements and have the greatest influence on the characteristics of the coastline (see Chapter 17). They are generated by the frictional drag of the wind blowing over the water surface. The largest wind-generated waves ever reliably reported have been 55 feet (17 m) high. Large storm waves are capable of damaging or sinking even large ships and of producing tremendous erosional and structural damage along coasts (see Figure 8.14).

The waves that reach our coastlines are at any given time generated only in certain restricted areas, usually located within storms. They are able to travel hundreds or even thousands of miles before they eventually reach distant shores. Large waves along the California coast, for example, are commonly produced by storms near Alaska. On occasion, waves reaching Hawaii are generated in storms off the coast of Antarctica.

Waves lose little energy as they travel through deep water, but this situation begins to change as they approach the shore and move into shallow water. When the water depth decreases to less than half the wave length, the wave starts to "feel bottom" as its forward progress is slowed by friction (see Figure 8.15). The retarding influence and upward push of the bottom causes the wave to rise gradually and tilt forward. When the water depth has been reduced to about 1.3 times the wave height, the upper portion of the wave pitches forward and breaks. (The process is much like the motion of a person who falls forward because he or she has been tripped.) The water now becomes foam-filled surf, which for the first time experiences significant net horizontal motion as its forward momentum causes it to rush up the beach as *swash* and then, its energy spent, to return as *backwash*. During this process, the energy imparted to the wave days earlier in a distant storm is finally expended on the coastline.

Tides

The ocean **tides** are the rhythmic rise and fall of the ocean produced by the gravitational influence of the moon and sun. Unlike waves and surface currents, the tides affect the ocean throughout its entire vertical

Figure 8.14 Large waves such as these are generated in storms that may be centered thousands of miles from shore. (*H. Armstrong Roberts*)

extent. The ebb and flow of the tides have been observed, pondered, and put to practical use by seafarers for thousands of years. Only in recent times, however, have sufficiently detailed records been compiled in most areas to allow for precise tidal predictions.

Although the sun is much more massive than the moon and has a much greater surface gravity, the tidal influence of the moon is over twice that of the sun because it is so much closer to the earth. The ocean water on the portion of the earth directly beneath the moon is pulled upward, producing a high tide when the moon is overhead. A second high tide occurs on exactly the opposite side of the earth because of the rotational inertia of the earth-moon system and because that area is farthest from the moon, making its pull weakest there. Midway between the two areas of high tides are zones of low tides (see Figure 8.16).

The combination of the earth's axial rotation and the moon's revolution around the earth causes the moon to take 24 hours and 50 minutes to return to the same longitudinal position over the earth. The pattern of high and low tides rotates with the moon's changing position, so most places experience two high tides and two low tides during this period. Therefore, a timespan of 6 hours and 12 minutes normally elapses between subsequent high and low tides.

Although the basic pattern of semidaily high and low tides is relatively simple, several additional influences combine to make the world pattern of oceanic tides exceedingly complex and difficult to predict. One of the most important is the gravitational influence of

the sun. Like the moon, the sun produces a pair of high tides in those areas facing directly toward and away from it, with an intervening pair of low tides. The tides experienced at any one place therefore result from the combined influence of the moon and sun. When the two are aligned either on the same side of the earth (at the time of the new moon), or on opposite sides (at the time of the full moon), their gravitational effects work in concert; and tidal ranges are largest. This produces the so-called **spring tides,** whose occurrences are actually unrelated to the season of the year. When the sun and moon are situated at right angles with respect to the earth (at the times of the half moon), their gravitational influences are opposed to each other; and tidal ranges are smallest. These are called **neap tides.** Tidal ranges are also affected by the varying distances of the sun and moon from the earth.

Probably the most important and certainly the most complex of the tidal influences is the geographic pattern of the ocean basins and coastal margins. Their irregular shapes redirect or reflect the tides in countless ways and often augment, diminish, or even cancel them altogether in some places. Highly uneven coastlines frequently experience tides that vary greatly both in range and in time of occurrence over small distances. In broad water bodies, including the major ocean basins, the tides display a circular pattern, with high and low tides rotating around one or more nodal points or lines that experience no tides at all (see Figure 8.17).

Tidal ranges of between 3 and 6 feet (1–2 m) are most common for the world as a whole. The highest

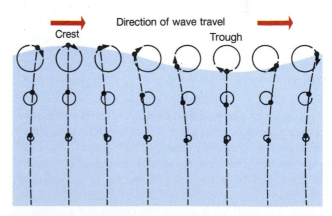

Figure 8.15 The water in waves moves in circular paths that gradually decrease in magnitude with depth. Waves begin to rise, tilt forward, and eventually break as they enter areas where the bottom is too shallow to permit this orbital motion to continue unimpeded.

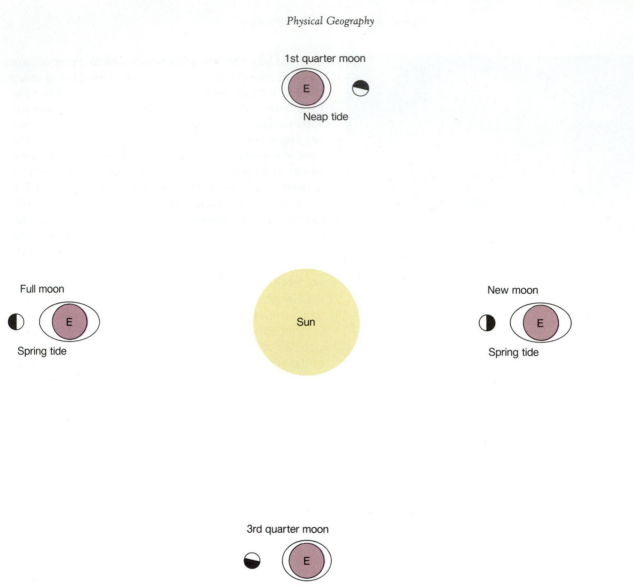

Figure 8.16 The ocean tides are caused by the gravitational pull of the moon and sun. The diagram shows the relative positions of these bodies with respect to the earth at the times of the spring and neap tides. The more pronounced bulges around the earth at the time of the spring tides indicate that tidal ranges are highest then because the gravitational attractions of the moon and sun are working in conjunction.

tides occur in funnel-shaped bays or estuaries, where the tidal influence is focused by the converging shorelines. Nova Scotia's Bay of Fundy has the highest tides of all, with one site experiencing a mean spring tidal range of 47.5 feet (14.5 m) and an extreme range of 53.5 feet (16.5 m).

Summary

This chapter has examined the characteristics and geographical distribution of earth's surface and near-surface waters. This water occurs in quantity in several different sites. It exists in a dispersed state in virtually all soils, and substantial volumes of deeper ground water occur within porous rock and sediment layers in many regions. Surface water sources include lakes, rivers, and the ocean. With the exception of some of the deeper ground water supplies, all of these sites contain water that participates in the hydrologic cycle, and all continually experience both additions and losses of water.

Subsurface water enters the ground primarily through the soil pore spaces. Large quantities of water

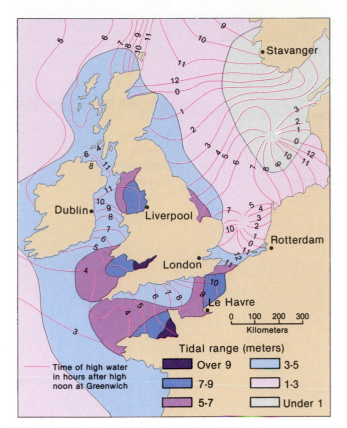

Figure 8.17 Map of tides in the North Sea. The tidal range is given in meters, and the time of the high tide is in hours after Greenwich noon. Note the rotation of the tides around the two nodal points.

temperatures. Nearly all streams display large short-term variations in flow as a result of precipitation events, and many streams display major seasonal variations in discharge. These variations are correlated with seasonal fluctuations in precipitation, temperature, and vegetative cover.

The ocean is the interconnected body of saline water that covers the lowest 71 percent of the earth's surface. It consists of approximately 96.5 percent water and 3.5 percent dissolved solids, mostly chlorine and sodium. Ocean surface water temperatures are warm in the tropics and decline with latitude, but the deep ocean water everywhere has a temperature near 35° F (2° C). The Arctic Ocean and much of the water around Antarctica is covered with sea ice or contains ice shelves or icebergs derived from coastal glaciers.

The three major oceanic motions are currents, waves, and tides. Currents are continuous movements of water streams of similar temperature and density. Surface currents are set in motion by the wind and display a flow pattern generally similar to that of the global wind currents. Deep currents are produced largely by water density variations resulting from differences in temperature and salinity. Waves are oscillations of the surface water generated by strong winds associated with storms. The tides are produced by the gravitational pull of the moon and, to a lesser extent, the sun. This pull results in a rhythmic rise and fall of the ocean surface.

adhere to the soil particles, while the remainder makes its way to the water table to become ground water. Ground water normally moves slowly downhill under the influence of gravity and hydrostatic pressure. Large volumes of ground water are confined more or less permanently within aquifers.

Lakes occupy enclosed basins that are supplied with enough water to fill them at least partially. These basins form in a number of ways, but the two most important are by glacial erosion and deposition and by crustal movements, especially faulting. Lakes have relatively brief spans of existence because their basins are eventually filled by sediments and organic matter.

Streams transport excess precipitation to enclosed interior basins or, more commonly, to the ocean. The discharge of a stream is determined primarily by climate and by the size of its drainage basin. Streams with drainage basins of similar size tend to have greater discharges in areas with heavy precipitation and cool

Review Questions

1. What are the two categories of subsurface water? Describe the location and characteristics of each.

2. How and why does ground water return to the surface? To what places is it most likely to go?

3. What determines whether a lake will contain fresh water or salt water?

4. What are the two chief methods by which lake basins are formed? Give some examples of large lakes produced by each method. How else may lake basins be formed?

5. What basic factor controls the geographic distribution of rivers? Is this same factor the major distributional control for lakes, ground water, and the ocean? Explain why or why not for each.

6. What climatic factors determine the proportion of a region's total precipitation that will be discharged by streams? How much of the world's land

precipitation is transported to the ocean by rivers? How is the remainder removed from the surface?

7. In what ways are ocean temperatures, both vertically and horizontally, similar to those of the troposphere? In what ways are they different? Why are the deep oceanic waters cold everywhere?

8. Describe, in general terms, the pattern of surface ocean currents within the five major ocean basins. What factors are responsible for this pattern?

9. How are waves generated? In what parts of the ocean are large waves produced?

10. Explain the cause of the tides. Why do they vary in height in most places at regular intervals

throughout the month? What factors complicate the global pattern of tides?

Key Terms

Soil water	Drainage basin
Ground water	Flood
Interception	Sea ice
Throughflow	Iceberg
Water table	Currents
Spring	Waves
Aquifer	Tides
Stream discharge	

CASE STUDY

El Niño—An Ocean Current with Worldwide Environmental Implications

Off the west coast of South America lies a vast sweep of northward-drifting cold water known as the Humboldt Current. The best-developed and most extensive of the world's cold ocean currents, it provides the coast from southern Ecuador to northern Chile with a cool, misty, virtually rain-free climate (Figure 8.18). Water temperatures off the coast of Peru range from as low as 50° F (10° C) in the south to about 71° F (22° C) in the north. These readings are at least 10 F° (5.5 C°) colder than the means for their latitudes.

In normal years the flow of the Humboldt Current is strongly maintained by the counterclockwise wind-flow around the South Pacific center of the subtropical high, located some 2000 miles (3200 km) west of South America. These winds blow northward along the coast, inducing an equatorward flow of cold water originating off the coast of Antarctica. As they reach the latitude of the Peruvian coast, the winds are gradually deflected more westward as they are incorporated into the southeast trade winds. This change in wind direction, coupled with the influence of the earth's rotation, slowly swings the northward-flowing water toward the west. The westward flow of the surface water, in turn, causes large volumes of cold water to well up from great depths. As this deep water slowly rises, it carries to the surface a great quantity of nutrients, mostly phosphates and nitrates, derived from the sea floor. Microscopic plants and animals abound in these rich waters and provide sustenance for larger forms of sea life, making this area, until recently, one of the most productive fishing grounds in the world. In the early 1970s, over 20 percent of the world's total fish catch was taken from these waters; and Peru was the leading fishing nation.

continued on next page

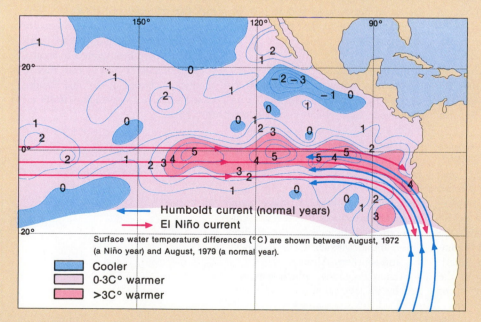

Figure 8.18 The effects of the 1972 El Niño on the surface water temperatures off the South American coast.

Humboldt current (normal years)
El Niño current

Surface water temperature differences (°C) are shown between August, 1972 (a Niño year) and August, 1979 (a normal year).

Cooler
0-3C° warmer
>3C° warmer

As the trade winds off the South American coast blow seaward, their westward transport of cool water slightly lowers the ocean level. This factor induces the upwelling of bottom water. At the same time, the trade winds produce a pileup of tropical water in the western Pacific as the water is pushed toward the land. The water here is too warm to sink, and mean sea levels are raised about 16 inches (40 cm) above normal.

Once every five years or so, for reasons still unknown, air pressure patterns over the South Pacific undergo a major alteration, and the southeast trade winds weaken and die, often to be replaced by light winds from the west. The warm water that has accumulated in the western Pacific, no longer pressed against Asia and the East Indies, surges eastward. It crosses the Pacific in about 2 months, overrunning the denser, cold water of the Humboldt Current to produce a southward flow of tropical water down the South American coast. Sea levels may rise by up to one foot (30 cm), water temperatures rise as much as 14.5 F° (8 C°) above normal, and the warm water may advance down the coast as far as 600 miles (960 km) south of the equator. This phenomenon often begins during the Christmas season and has been named *El Niño,* the Spanish name for the infant Jesus.

The more recent major occurrences of El Niño have been in 1957, 1965, 1972, 1976–77, 1982–83, and 1987–88. Although they have differed in detail, the consequences of each occurrence have been widespread and locally disastrous. The effects are generally felt most strongly in the immediate vicinity of the South American coast. The overspreading of the ocean surface by tropical water keeps upwelling nutrients from reaching the surface and reduces the water's dissolved oxygen supply. This results in the death or departure of commercial fish and spells economic disaster for the Peruvian fishing industry.

The warm water also moistens and destabilizes the air, which, when it penetrates inland and is lifted by the Andes, often unleashes torrential rains and severe windstorms. The resultant flooding is especially devastating because the affected desert surface lacks stabilizing vegetation and a well-developed system of river channels to transport the runoff.

The El Niño of 1982–83 was the most strongly developed and disastrous on record. At its height, the South Equatorial Current reversed direction across the entire Pacific to produce an eastward flow of tropical water some 8000 miles (12,800 km) long. Associated weather calamities affected every continent except Antarctica and Europe, causing 1500 human deaths and inflicting damage estimated between $2 and $8 billion. Among the areas hardest hit were the following:

■ Peru and Ecuador, where the fishing industry that was just beginning to recover from the 1972 El Niño was destroyed, and where tremendous floods triggered by El Niño left many dead and homeless.

■ Australia, where a record drought produced over $2 billion worth of destruction to farmland, crops, and livestock, and where 75 people were killed by brushfires. Severe drought also affected Indonesia, the Philippines, India, Mexico, Central America, and southern Africa.

■ French Polynesia, which was visited by six tropical storms and hurricanes; this followed a period of 75 years in which the territory had experienced none. The Hawaiian Islands also experienced a rare hurricane.

■ The U.S. West Coast, which received a spate of severe winter storms resulting in much coastal flooding and erosion, numerous landslides, and a record mountain snowpack. The Gulf coastal states also experienced torrential rainfalls and severe flooding.

The most recent advent of El Niño took place in 1987 and early 1988. It was neither as strong nor as destructive as the 1982–83 occurrence, but was linked to severe droughts in India and the U.S. Pacific Northwest.

El Niño is considered by some experts to be the single most important natural disruptive influence on world climate patterns and has recently become a major topic for research by scientists in several disciplines. An improved understanding of El Niño would help us predict future occurrences and better prepare for future climatic disasters. The nearly global impact of El Niño graphically illustrates the fact that all earth phenomena are interrelated to form a single complex system, and that a significant change in any one component can have far-reaching effects on the others.

Natural Vegetation

Focus Questions

1. What environmental factors chiefly control the global distribution of natural vegetation?
2. What are the major plant associations and subassociations, and what are their basic botanical characteristics?

Vegetation covers approximately 80 percent of the earth's land surface and is remarkable in its diversity. Plants range in size from microscopic algae, little changed from the earliest lifeforms to evolve on our planet, to the giant sequoia trees that are the largest living organisms on earth. Plants grow from the margins of glaciers to the heat of the equator, from areas submerged in water to almost rain-free deserts. They not only comprise the vast bulk of life on earth; they have also made possible the evolution of animals, including human beings.

The global cover of natural vegetation, however, has been profoundly modified by human activities. The extent of these changes is indicated by the fact that most of the earth no longer has an undisturbed natural vegetation cover. As a result, many of the plant assemblages discussed in this chapter no longer dominate extensive portions of the earth's surface, as they did before their alteration or removal by human beings. Much of this modification has unfortunately had adverse environmental consequences.

The chapter is divided into three sections. First discussed are the major characteristics of natural vegetation and vegetation assemblages, including their alteration through time as a result of evolution and changing site conditions. Second, the important environmental factors controlling the geographic distributions of vegetation assemblages are identified. Finally, the characteristics and distributions of each of these assemblages are examined.

GLOBAL VEGETATION CHARACTERISTICS

The first plants evolved very early in the earth's history—probably at least 3.5 billion years ago. Between that time and the present, continuing evolution has produced all the plants currently occupying the earth's surface. The two keys to the evolutionary process are (1) the biological changes that take place through mutations and (2) the ability of species that are better adapted to their environment to survive while those not so well adapted perish. The *mutations* that have produced the world's plant (and animal) species involve molecular rearrangements in their genetic material that produce changes in their physical characteristics. During the long biological history of the earth, literally millions of successful mutations have produced new species of gradually increasing variety and complexity,

culminating in the plants and animals that currently inhabit the earth.

Most plants produce the bulk of their solid matter from water and carbon dioxide in the presence of light. This process, termed **photosynthesis,** is accomplished by chlorophyll, a green-colored organic chemical. Photosynthesis not only enables plants to survive, it has made it possible for animals, including human beings, to evolve. The photosynthetic process releases free oxygen (O_2) to the atmosphere. This gas, essential to all animal life, did not exist in quantity in the earth's early atmosphere. Plants must therefore have evolved well before animals, which appeared only after plants had had sufficient time to produce a breathable supply of atmospheric oxygen. In addition, animals use plants directly or indirectly for food.

It is estimated that nearly 600,000 species of plants currently inhabit the earth. The systematic study of plants forms the subject matter of *botany,* a major subdivision of biology. The geographer is concerned chiefly with the spatial aspects of vegetation, and the branch of physical geography that is devoted to this subject is known as **biogeography.** Biogeographers strive to develop a better understanding of the global distributions of associated groups of vegetation and to better comprehend the interaction of plant groups with their environment.

The global distribution of natural vegetation is not random. Each plant species has specific environmental requirements, so different vegetative groupings or assemblages exist in differing environmental settings. A plant assemblage whose members occupy a particular environmental setting of restricted size is referred to as a *plant community.* The members of such a community share their environment, and the various habitats within it are divided so as to use the available space and resources as completely and efficiently as possible.

Carrying the idea of the biological organization of the local environment further, we have the concept of the ecosystem. An ecosystem is a functioning entity consisting of all the organisms in a biological community, as well as the environment in which and with which they interact. A balanced ecosystem is largely self-sustaining in its resources and is relatively stable in its assemblage of lifeforms. Two of the essential characteristics of a healthy ecosystem are the maintenance of flow cycles of energy and of nutrients (see Figure 9.1). Energy is derived from insolation, which is fixed or "captured" by plants during the photosynthesis process. It may then pass to herbivores, to carnivores, and, upon

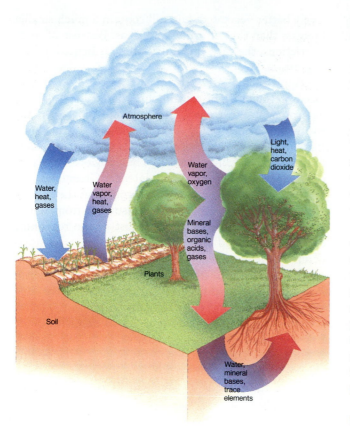

Figure 9.1 Representation of the flow of energy and materials between the atmosphere, soil, and vegetation.

their deaths, to bacteria and the soil. Mineral nutrients originally derived from the weathering of bedrock (see Chapter 13) are also cycled through vegetation, animals, the soil, and back again to vegetation. This often occurs with little loss of nutrients, and what losses do occur are replaced by minerals released by subsequent weathering or animal input.

Ideally a plant community is relatively stable. If environmental conditions are significantly altered, how-

ever, the community may undergo drastic changes. Disturbances such as high winds, floods, droughts, fires, insect invasion, or human interference can remove some plant species and encourage the growth of others, thereby greatly modifying the makeup of the community. If the disturbance is minor, and original environmental conditions return, the plant community will tend to revert gradually to its original composition. Reversion occurs because the original mix of vegetation formed the most stable and efficient community under the existing environmental conditions, and a return of those same conditions should produce the same vegetative response.

The gradual change in the composition of a plant community in response to changing environmental conditions following a disturbance is referred to as **vegetative succession** (see Figure 9.2). Greatly differing successional sequences occur under dissimilar environmental conditions. Eventually, succession produces the plant community best suited to survive under existing environmental conditions. This **climax community,** as it is called, should theoretically occupy the area until the next disturbance or environmental alteration takes place.

A climax vegetation community has several important attributes. It usually contains the largest *biomass,* or mass of living organisms, that the area is capable of supporting. A high degree of species diversity typically exists, so all available habitats are effectively utilized. The composition and relative distribution of species tend to be stable, and birth and death rates are balanced. Finally, most energy and nutrient needs are supplied and recycled by the organisms within the community, so losses of these resources are minimized.

Many areas of the world do not presently have climax vegetation communities. This is primarily the result of widespread human disturbance and, to a lesser extent, of geologically recent worldwide climatic changes (see Chapter 15). Nevertheless, it is possible to

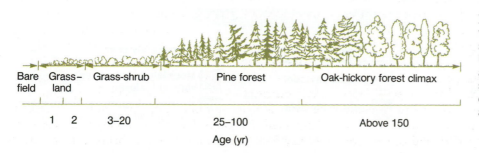

Figure 9.2 An idealized vegetative successional sequence in the southeastern United States following the abandonment of farmland. Each successional stage alters the environment in such a way as to allow the next stage to develop.

177

divide natural vegetation groupings into several major global-scale communities, or **associations,** that exist in relative equilibrium with their environment. These associations will be identified and described shortly. Before doing so, however, it is necessary to examine the basic environmental controls that are responsible for their geographic distribution.

NATURAL VEGETATION DISTRIBUTION CONTROLS

Despite the successful adaptations of vegetation to diverse and often adverse environments, large portions of our planet's surface are unvegetated. For the earth as a whole, the leading factor by far in restricting the areal distribution of vegetation is the presence of surface water. Oceanic plants are quite primitive and are more limited in variety and abundance than those on land. Although many ocean areas contain a sparse mixture of floating plant types such as phytoplankton and algae (which include seaweed), vast expanses are virtually devoid of life. Bottom-dwelling plants are restricted by their need for light to shallow, generally near-coastal waters; and waves are destructive of surface vegetation. Inland water bodies such as lakes and rivers are rela-

tively better vegetated, but still contain a much smaller biomass than that of many land areas. Because of space restrictions, the subsequent discussion is limited to land vegetation.

Approximately 80 percent of the world's land surface is vegetated. The remaining 20 percent is subject to environmental extremes that prohibit not only the survival of plants, but that of most animals as well (see Figure 9.3). About half of this area is too cold for plants to carry on their life processes. These regions, primarily consisting of Antarctica and interior Greenland, are mostly covered by snow and ice throughout the year, so they lack soil or even a bedrock surface from which plants could extract nutrients. Most of the remaining surface area devoid of vegetation is too dry and consists of almost lifeless desert. Most notable are portions of the central Sahara, the Arabian Peninsula, and interior Asia. Perhaps an additional 1 or 2 percent of the land surface lacks natural vegetation because it is too steep, contains toxic surface materials such as excessive quantities of salts, or has been subject to recent natural disturbances that have temporarily removed the vegetation cover.

From the preceding paragraph, it can be deduced that moisture, temperature, slope, and soil are among

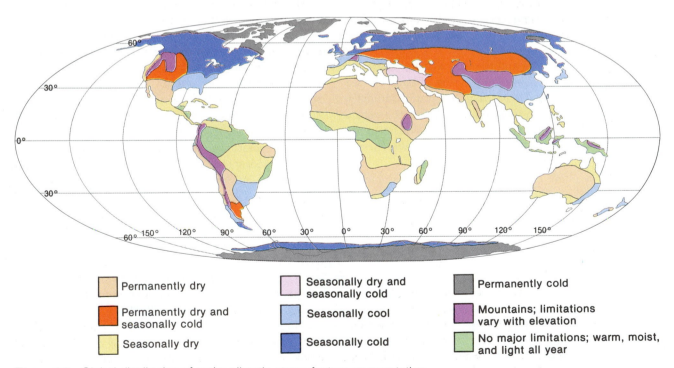

Figure 9.3   Global distribution of major climatic stress factors on vegetation.

(a)

(b)

Figure 9.4 Contrasting vegetation assemblages growing in response to differing moisture conditions: (a) Temperate rainforest in coastal Washington State. (b) Desert scene in Monument Valley, Arizona. (*a: © JLM Visuals; b: E.D. McKee, USGS*)

the factors that strongly influence the world distribution of natural vegetation. In all, it is possible to identify four major categories of vegetative controls: climate, topography, soil, and biological organisms. We now examine, in turn, the influence of each on the distributions and types of natural vegetation.

Climatic Influences

Climatic characteristics, especially moisture and temperature, are the dominant environmental controls on the global pattern of natural vegetation. The dependence of natural vegetation upon climate is so strong that considerable similarity exists between the geographic patterns of climate and vegetation. Natural vegetation is typically most abundant and diverse in areas where there are no serious climatic stresses. These stresses, the most crucial of which are the occurrence of cold or dry seasons, greatly reduce the variety of species of both plants and animals. As a result, the variety and quantity of lifeforms generally decline as one proceeds from wet to dry and from warm to cold regions of the earth.

Moisture

The availability of moisture is the single most important criterion controlling large-scale patterns of natural vegetation. Moisture availability is considered a climatic factor here because it is largely determined by the balance between precipitation amounts and evapotranspiration rates, but it is also significantly influenced by soil and topographic factors and will be discussed again under those headings. The total biomass within an area normally increases with available moisture until a point of excess is reached, after which the quantity of vegetation again diminishes. Perhaps the most striking characteristic of a desert is its relative lack of vegetation, while the density and lushness of a well-watered forest are equally obvious and impressive (see Figure 9.4).

Plants are sometimes categorized on the basis of their adaptations for coping with available moisture. **Xerophytes** (from the Greek *xeros* "dry" and *phytos* "plants") are plants especially adapted to regions deficient in water. These regions are mostly arid deserts, but also include such areas as beaches or steep slopes, where the nature of the soil, rather than the climate, limits moisture availability. Some of the water-saving characteristics commonly displayed by xerophytes include extensive root systems, water-storage ability, waxy leaves, thickened epidermal layers designed to reduce transpiration, small leaves or none at all (as in the case of cacti), and the ability to minimize insolation receipts by turning leaves edgewise to the sun. In addition, xerophytes tend to be small in size and to have slow growth rates.

Hygrophytes (*hygros* "wet") are plants especially adapted to excessively moist environments. Because the environmental stress factor is the opposite of that for xerophytes, many of the physical characteristics of

The Environmental Consequences of Acid Rain

The phenomenon of acid rain has become a major environmental concern in recent years. A complex problem that is far from completely understood, it seems to be caused largely by the combustion of fossil fuels. Natural factors, including soil and bedrock characteristics and wind and precipitation patterns, also influence the severity of this problem in different regions.

Although natural rainwater is mildly acidic, the atmospheric presence of industrially derived acids can increase the acidity of precipitation to the point that it becomes a corrosive substance potentially toxic to plant and animal life. The release of sulfur dioxide gas from coal-fired electric power plants is believed by most experts to be the major single cultural factor contributing to the formation of acid rain. Within the atmosphere, this gas is converted to liquid sulfuric acid droplets that are carried to the surface by precipitation. In many areas, the effects of acid precipitation on the soil are not a concern. This is because the surface is naturally buffered by the presence of lime or other alkaline substances in the rock or soil. However, in other areas such as the northeastern United States, southeastern Canada, and much of Europe, the presence of numerous coal-burning power plants, coupled with naturally acidic rock and soil types, have resulted in a serious acid rain problem.

Several major adverse environmental consequences of acid rain have been widely reported. In large portions of Scandinavia and Germany, needleleaf forests have been slowly dying. Needles near the tops of the trees turn yellow and fall off, and eventually the entire tree may be killed. In the Federal Republic of Germany, this phenomenon was first noticed in the Black Forest in the early 1970s; by the late 1980s over half the nation's trees were affected. In Eastern Europe, entire forests are reported to have been destroyed. Forest dieback in portions of the southern Appalachian Mountains in the United States has also been recently reported. Although the exact mechanism by which acid precipitation kills trees remains uncertain, it is believed that the acidification of the soil kills organisms that participate in the soil-plant nutrient cycle and also mobilizes potentially toxic aluminum and heavy metal ions that are then absorbed by the trees.

Acid rain has also long been blamed for the destruction of fish and other aquatic life in New England, upstate New York, eastern Canada, and Scandinavia. Many lakes within these areas, which previously supported large fish populations, have been virtually sterilized.

A cultural concern resulting from acid rain has been the damage suffered by numerous priceless works of art in Italy, Greece, and several other countries. Although acid rain can corrode metals, fade paints, and even etch glass, marble statues are especially affected

continued on next page

hygrophytes are also directly opposed to those exhibited by xerophytes. For example, hygrophytes tend to have shallow root systems, large leaves with numerous stomata (pore spaces), thin epidermal layers, and the ability to turn leaves directly toward the sun. In addition, they usually grow rapidly and may attain a large size when mature. Large size helps these plants compete for sunlight—an important consideration because vegetation is usually dense in areas where hygrophytes grow.

A third category of plants, the mesophytes (*meso* "middle"), have physical characteristics between those of xerophytes and hygrophytes. They typically occupy well-drained habitats in portions of the low and middle latitudes with ample year-round precipitation and without excessively low temperatures.

A fourth category, the tropophytes (*tropos* "to turn"), are defined somewhat more broadly than the other three. Tropophytes are adapted to alternating favorable and unfavorable climatic conditions for growth. These conditions generally consist of seasonally varying warm and cold periods, such as those of the middle and high latitudes, or wet and dry periods, such as those experienced, for example, in the Tropical Wet and Dry and the Dry Summer Subtropical climates.

Figure 9.5 Forest destruction caused by acid fog and rain on Clingmans Dome in the Great Smoky Mountains of North Carolina. (*Breck, P. Kent, © JLM Visuals*)

because the acid chemically converts the marble to calcium sulfate, which is soft and soluble. This process, when combined with the effects of freezing water, causes the outer portions of the marble to be shed in layers, allowing progressively further penetration of acidic water.

The release of sulfur dioxide by power plants in the American Midwest has been much criticized by residents of the Northeast and adjacent southeastern Canada, who have experienced severe acid rain problems in the last few decades. The Midwest contains a large concentration of coal-burning power plants. Midwestern coal is plentiful and easy to extract, but it unfortunately has a high sulfur content. In an effort to meet Environmental Protection Agency regulations, these plants have been installing expensive scrubbing systems that reduce the sulfur content of the coal before it is burned and are also increasingly importing low-sulfur coal from the western United States. The complete abandonment of the productive, high-sulfur coal deposits, however, would have a devastating impact on the economy of many regions; and further reductions in its use are being vigorously resisted as long as any doubt exists about the degree to which the burning of this coal affects acid rain contamination farther to the east.

The acid rain controversy illustrates an important environmental dilemma that societies face in the pursuit of nearly all major economic activities. It involves the "acceptable" degree to which we should allow the natural environment to be degraded as a consequence of our actions. Some degradation is inevitable, no matter how careful we are. However, the minimization of the environmental consequences of human activities, although frequently viewed as highly desirable, can be tremendously expensive both monetarily and in terms of human livelihoods.

Because most areas experience at least one of these two types of seasonal alterations, the tropophytes are more extensively distributed than the other three plant groups.

Most tropophytes are large leafy plants commonly categorized as trees or shrubs. Many tropophytic bushes and trees, especially those with broad leaves, are **deciduous;** that is, they lose all their foliage at the beginning of the unfavorable season. Conversely, most needleleaf bushes and trees are **evergreen;** they retain their foliage throughout the year. Tropophytes, whether deciduous or evergreen, are more versatile than xerophytes, hygro-phytes, and mesophytes in that they are able to grow vigorously during favorable seasons and to protect themselves by becoming dormant during unfavorable seasons.

Temperature

The second dominant climatic control on the distribution of natural vegetation is temperature. Plants vary greatly in thermal tolerance, so only the coldest areas do not support vegetation. Some plant species are adapted to a broad range of thermal environments, while others are restricted to a narrow range. Each plant species, though, has a maximum and minimum thermal limit beyond which it cannot survive, as well as an optimum temperature range within which it grows most vigorously.

Many plants outside the tropics require seasonally varying temperatures to initiate different crucial biological processes such as germination, fertilization, flowering, fruiting, and foliage growth. Because most plant growth stops when temperatures fall below 42° F (6° C), little vegetation exists in areas where temperatures never rise above this level. As temperatures increase (assuming other vegetative controls present no hindrance), growth rates and biomass concentrations increase to a maximum at about 86° F (30° C).

In regions that experience the climatic stress of a cold season with subfreezing temperatures, it is necessary for plants to adjust their life cycles in order to survive. One group of plants, the **annuals,** do not have mechanisms that enable individual plants to survive the rigors of the cold season. Instead, they have a life cycle that is sufficiently short to enable them to mature in one *growing season* and to produce the seeds that ensure the growth of the next generation of plants the following year. Most of our food crops, including grains and vegetables, but excluding tree crops, are annuals; they must be replanted each year.

A second group of plants, the **perennials,** have multiple-year life spans. In areas with a climatic stress, they have developed protective mechanisms that permit their survival as individuals. Most broadleaf trees and shrubs, for example, lose their leaves and become dormant during the autumn season in the middle and higher latitudes, resuming their growth and producing new foliage when warm weather returns the following spring. Dormancy and a deciduous habit therefore protect plants from the stresses of either a dry season or a cold season. Perennials have a tremendous advantage over annuals in terms of size, and therefore in terms of biomass and spatial dominance, because they can add to their growth over an extended period. As a result, most of the natural vegetation associations and subassociations are dominated by perennials; all the world's larger plant species are of this type.

Other Climatic Influences

Light is a third climatic necessity for vegetation. Most plants use light directly as the energy source for the process of photosynthesis. The competition for light provides the stimulus for vertical growth in plants. In densely vegetated areas, the heights of the various plant species are generally closely associated with shade tolerance. Plant species are frequently excluded from a given habitat solely because of their inability to secure the required amount of sunlight. Like temperature, seasonal differences in the duration of daylight trigger biological activities such as budding, flowering, and leaf growth or shedding.

Wind is another climatic element with both favorable and unfavorable effects on vegetation. On the positive side, it increases the carbon dioxide intake of plants and is essential for pollen and seed dispersal for many plant species. Conversely, the persistent high winds experienced along some exposed high latitude coasts or in mountainous areas can lead to excessive transpirational losses. Under these conditions, trees may not grow or may be stunted or deformed (see Figure 9.6).

Topographic Influences

Vegetation is influenced by a variety of topographically related factors. These include soil characteristics, insolation receipts, surface temperatures, and several factors relating to water availability, such as precipitation amounts, evapotranspiration, and runoff rates. Large-scale topographic influences are mostly climatic and are associated with the cooler and moister conditions that normally exist in mountain regions as compared with

Figure 9.6 Stunted and wind-deformed trees at the tree line in the Colorado Rockies. Branches grow only on the downwind side of the trees. (*Ralph Scott*)

adjacent lowlands. On a much smaller scale, significant differences in vegetation are frequently encountered on different portions of a hill or between an area of sloping land and a nearby flat stream valley. The primary effect of the topography is to produce a variety of differing microhabitats for vegetation. As a result, vegetative patterns tend to be much more diverse in rugged areas than in areas of low topographic relief. Slope steepness affects vegetation primarily through its influence on water drainage and soil erosion. As slope angles increase, gravity becomes increasingly effective in causing excess water to drain rapidly downhill. As a result, steep slopes may experience such rapid runoff that little water remains to enter the soil and become available to plants. Rapid runoff also produces a high erosion potential for the slope, so that an inverse correlation exists between slope steepness and soil depth. The net effect of steeply sloping terrain on vegetation is a reduction in the size and diversity of plant species present, as well as a significant lowering of the total biomass. Steep mountain slopes with little soil are likely to be devoid of vegetation or to contain a sparse cover of xerophytic species.

The directional orientation, or *aspect,* of a slope influences its vegetative cover by controlling the duration and intensity of sunlight. Slopes inclined toward the sun are substantially warmer during the day than are slopes inclined away from the sun. Sunward slopes also have drier surface conditions because evapotranspiration rates are greatly increased by the warmer temperatures and strong insolation. These contrasting microclimatic conditions may produce major vegetative differences on different sides of the same hill (see Figure 9.7). The significance of slope aspect increases with latitude as the angle of the sun's rays become progressively lower and more concentrated on the equatorward-facing slopes. In the northern hemisphere, south-facing slopes have the warmer and drier conditions, while this situation is reversed in the southern hemisphere.

Soil Influences

Plants obtain required water and nutrients from the soil. The nutrients are absorbed through the plant roots after being dissolved in soil water. It is therefore essential to vegetation that soil not only contains the needed minerals, but that it contains adequate, but not excessive, quantities of water. Minerals required by

Figure 9.7   Vegetation differences resulting from variations in slope aspect in southwestern Colorado. The north-facing slopes on the left are the most densely vegetated. (*J.K. Nakata: Sight & Sound Productions © 1990*)

plants are derived from the two solid soil-forming constituents: bedrock and organic matter. Plant species differ in their mineral requirements, so soils formed in areas with different types of bedrock may encourage the growth of different vegetative assemblages. The *fertility* of the soil, or its productivity with respect to vegetation, is a somewhat relative term because a given soil may contain adequate nutrients for one plant species but not for another. Some types of rock, however, are noted for forming soils of generally higher fertility than are others.

Assuming that the important plant nutrients are present in the soil, the soil's ability to support vegetation is still critically dependent upon the presence of sufficient water to dissolve the minerals. Soils in arid environments are often fertile, but they receive insufficient moisture to dissolve the available nutrients and hence are highly limited in their ability to support vegetation. Conversely, soils in excessively wet climates are frequently infertile because most soluble minerals have been removed by the downward percolation of soil water. A much more comprehensive discussion of soil characteristics relating to vegetation is provided in the next chapter.

Influences of Organisms

The joint occupation of most areas by a diversity of plant and animal species modifies the environment in a variety of ways. The presence and activities of other lifeforms serve to increase the favorability of the habitat for certain plant species. For example, the shade provided by a forest helps conserve soil moisture and affords protection from sunlight for many shade-loving plants of the forest floor. In addition, various plants utilize the mobility of animals to perform such tasks as pollination and seed dispersal. Organic matter produced by the decomposition of vegetation also releases to the soil nutrients for use by new generations of plants.

On the other hand, the presence of numerous plant and animal species within a region of limited size and resources necessarily results in competition for survival. Even if the climate, topography, and soil type are all favorable for a given plant species, it cannot grow within a given area unless it can successfully occupy a habitat despite the likely presence of competitors. Of all the plant species with access to a given habitat within a climax vegetative community, only those best adapted to the environment will be successful in occupying it. If a species is out-competed by other plants (or animals) within an area, it will be eliminated from that area. The removal of organisms less able to compete makes room for species that are better suited to survive within a given environment; such elimination is the means by which biological diversity and ultimate advancement occurs.

Table 9.1
Natural Vegetation Associations and Subassociations*

I. Forest
  A. Tropical Rainforest
  B. Tropical Semideciduous Forest
  C. Tropical Scrub Woodland
  D. Subtropical Broadleaf Evergreen Forest
  E. Mediterranean Woodland and Scrub
  F. Mid Latitude Deciduous and Mixed Forest
  G. Mid Latitude Evergreen Forest
    1. West Coast Coniferous Forest
    2. Southern Pine Forest
  H. Northern Coniferous Forest
II. Grassland
  A. Savanna
  B. Steppe
    1. Low Latitude Steppe
    2. Mid Latitude Steppe
  C. Prairie
III. Desert
IV. Tundra

*Roman numerals precede natural vegetation associations; letters precede subassociations. Plant associations are also sometimes referred to as *biomes.* Because of its diverse nature, mountain vegetation does not fit into this classification.

WORLD DISTRIBUTION OF NATURAL VEGETATION

Natural vegetation may be divided into large-scale groupings called associations. Four associations are commonly recognized: forest, grassland, desert, and tundra. At the scale of the association, major internal vegetative differences exist. Each association is therefore separated into several subassociations that exhibit a greater degree of structural similarity in their vegetative characteristics (see Table 9.1). The various subassociations of the forest and grassland associations are sufficiently large and important to merit individual treatment in this section.

The global pattern of natural vegetation is complex and varied when examined in detail (see Figure 9.8). Such complexity makes any limited classification system such as the one used in this book merely a convenient abstraction of reality. In actuality, there is no set number of vegetation groupings, and the character of the vegetation normally does not suddenly change when vegetative "boundaries" are crossed. The formulation of

natural vegetation groupings, however inadequately they may reflect the botanical complexity that actually exists, does serve useful purposes. It indicates that vegetation differs greatly on a regional basis and facilitates a generalized discussion of these differences. In addition, a geographic analysis of vegetative patterns emphasizes the crucial fact that the distribution of natural vegetation is not random, but is closely related to other environmental parameters, especially global patterns of temperature and available moisture.

In the following discussion of the various vegetation associations, an attempt is made to provide geographical continuity by proceeding in a general poleward direction and by stressing the factors responsible for the basic changes in vegetative characteristics from one grouping to the next. We begin with the forest association, which not only dominates the equatorial regions, but is the most areally extensive and important association from both a biological and a human occupancy standpoint (see Table 9.2).

Table 9.2
Comparative Areal Extent of Vegetative Regions

Vegetative Region	Area (millions of sq mi)
Tropical forest	9.5
Temperate forest	4.6
Taiga (subpolar forest)	4.6
Woodland and shrubland	3.9
Tropical grassland	5.8
Temperate grassland	4.0
Desert and semidesert	16.2
Tundra and mountain	3.1
Cultivated land	5.4
Swamp and marsh	0.8

Source: Robert H. Whittaker, *Communities and Ecosystems,* 2d ed. (New York: Macmillan, 1975).

Forest Association

A *forest* is an assemblage of trees that grow sufficiently close together for their foliage to overlap and largely shade the ground for at least a portion of the year. Trees are not the only type of vegetation present in a forest, but they are the dominant form in terms of areal extent.

In their total biomass, trees far surpass all other terrestrial lifeforms combined. The key to their success in the competition with other forms of vegetation is their size, which has enabled them to secure the sunlight they need and to shade out rivals. In essence, forests dominate in all parts of the world where the characteristics of the soil, the topography, and especially the climate allow their growth. Forests, however, are subject to two important climatic limitations which exclude them from large areas. The first is that trees need a warm growing season of reasonable length. Even the most cold-tolerant forests cannot grow in areas where the warmest month has a mean temperature below 50° F (10° C). The second and more geographically important limitation is the large water requirement of forests. Because they are large perennials, trees need substantial amounts of moisture during the growing season and at least some water even when they are dormant. This restriction excludes forests from extensive areas with perennial or seasonal soil moisture deficiencies.

Trees are commonly classified on the basis of leaf characteristics. **Broadleaf** trees, as the term implies, have leaves of considerable width, while **needleleaf** trees have leaves that are very narrow and sometimes pointed. Many species of needleleaf trees bear seeds in cones and are therefore described as *conifers*. Familiar middle latitude examples of broadleaf trees include oak, maple, and hickory; and well-known needleleaf trees include pine, spruce, and fir. The broadleaf trees, as a group, are more complex and biologically more advanced than the needleleaf trees and, under favorable environmental conditions, hold the competitive edge because of their higher shade tolerance. As a result, they are more widely distributed than needleleaf trees. On the other hand, the needleleaf trees are more tolerant of cold climates, drought, and infertile soils; they therefore gain a competitive advantage when any of these conditions prevail. Trees and shrubs are also classified as either deciduous or evergreen, depending on whether or not they undergo a seasonal loss of foliage. These two pairs of opposing characteristics permit trees and shrubs to be separated into four groups in a simple classification system. These groups are broadleaf evergreen, broadleaf deciduous, needleleaf evergreen, and needleleaf deciduous.

In order to examine more closely the characteristics and global distribution of forests, they are divided in this book into eight subassociations based on their structure and leaf characteristics. Each subassociation occurs in several widely separated areas of the world that have similar climatic conditions. This indicates that climate is the major natural vegetation distribution control. The following vegetative descriptions are of climax or near-climax plant assemblages in areas undisturbed by human activities.

Tropical Rainforest

The Tropical Rainforest is the most equatorward and the most extensive forest subassociation. It covers 6 percent of the earth's surface and contains up to four million plant and animal species, which collectively comprise 80 percent of the world's total biomass. The geographic distribution of the Tropical Rainforest coincides closely with that of the Tropical Wet Climate (compare Figures 7.2 and 9.8). The three largest areas with this forest type are the Amazon Basin of northern South America, the Congo Basin of central Africa, and most of the East Indies.

A climate highly favorable for vegetative growth is responsible for the lushness of the rainforest (see Figure

Figure 9.8 Global distribution of natural vegetation groupings.

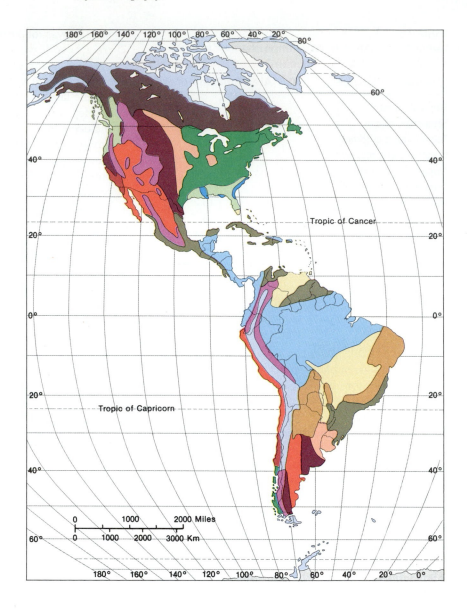

9.9). The absence of both a cold season and a lengthy period of drought means that no major climatic stresses are present. As a result, a 12-month growing season exists; vegetation is dense; and the structure of the forest is determined primarily by the competition for light.

The Tropical Rainforest is characterized by an amazing variety of broadleaf evergreen trees. Whereas an acre of land within a typical middle latitude forest may contain a dozen or so tree species, an area of similar size within the rainforest may contain a hundred or more species. Some of the better-known varieties include mahogany, rosewood, and ebony, all prized sources of

wood for furniture; ironwood, which has wood so dense that it cannot float; balsa, the lightest wood in commercial use and familiar to makers of model planes; and cinchona, the source of quinine used by victims of malaria. The numerous tree species are well integrated, so "stands" of a single type do not occur. This characteristic has made human exploitation of the forest for specific tree species difficult.

Despite the great diversity of types of wood produced by the trees of the rainforest, the trees are surprisingly similar in structure. The forest is tall and very dense. Trees are straight and relatively slender, branching only near their crowns and typically produc-

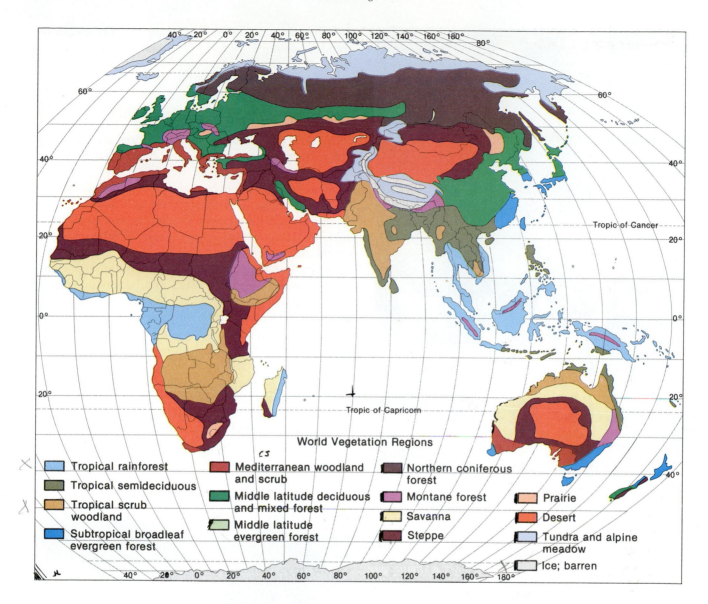

World Vegetation Regions

- Tropical rainforest
- Tropical semideciduous
- Tropical scrub woodland
- Subtropical broadleaf evergreen forest
- Mediterranean woodland and scrub
- Middle latitude deciduous and mixed forest
- Middle latitude evergreen forest
- Northern coniferous forest
- Montane forest
- Savanna
- Steppe
- Prairie
- Desert
- Tundra and alpine meadow
- Ice; barren

ing an interlacing canopy of leaves and branches sufficiently dense to shade the ground completely. Larger trees have often developed buttress roots for support in the soft, damp soil. Because of the density of the vegetation, success in the competition for sufficient sunlight to carry on photosynthesis becomes the key to survival for the plants of the rainforest. The forest typically exhibits a layered appearance caused by the tendency of large numbers of trees to attain similar heights. This produces canopies of leaves and branches at two, three, and sometimes four distinct levels.

The floor of the Tropical Rainforest is surprisingly open, allowing for reasonably easy passage on foot. This openness results from the density of the vegetative canopy, which normally allows less than 1 percent of the available light to penetrate to the surface. The forest floor is a still, damp, and dim world, permeated with the moldy odor of decomposing vegetation and containing a scattering of only the most shade-tolerant plants. Light, life, and growth are concentrated in the treetops. Here reside a tremendous assortment of plants and animals that greatly exceed, in the number of species represented, the biota associated with any other vegetative assemblage. Many thousands of the plant and animal denizens of the Tropical Rainforest have not as yet even been biologically classified and cataloged. Among the

Figure 9.9 Tropical Rainforest in French Guiana, northeastern South America. (*Pamela Easy, © JLM Visuals*)

countless millions, far outnumbering all other species of animals combined.

Tropical Semideciduous Forest

Progressing poleward from the equatorial regions dominated by rainforest vegetation, a climatic stress begins to appear in the form of a winter dry season of increasing length. Trees have sufficiently extensive roots to withstand relatively short periods of dry weather; as a consequence, rainforest vegetation extends into areas with up to two months without rainfall. Where the dry period increases to between two and five months, however, the vegetation begins to exhibit important differences in its characteristics and becomes Tropical Semideciduous Forest.

One of the most important differences between the Tropical Rainforest and the Tropical Semideciduous Forest is that the latter has a significantly reduced vegetative density and a much lower biomass. Trees are not crowded so closely, and, coupled with the smaller average size of the trees, the canopy of leaves and branches is much less dense and continuous (see Figure 9.10). Branches tend to start at relatively low levels, and tree crowns are much more rounded than those of the rainforest.

plants are giant woody vines called *lianas*, which festoon the trees and whose trunks can reach diameters of up to 8 inches (20 cm) at ground level. *Epiphytes*, or "air plants," are very common. These small plants typically grow in hollows or on the upper surfaces of horizontal branches and derive their moisture and nutrients from the air. Included in this category are mosses, bromeliads, ferns, and thousands of species of orchids whose colorful flowers relieve the monotony of the dark green rainforest vegetation.

Most animals also live in the treetops. Large mammals are relatively rare, but birds and monkeys fill the air with their raucous chatter, snakes slither silently along the branches, and a wide variety of other small mammals, reptiles, and amphibians are to be found. Most numerous of all, however, are the insects in their

Figure 9.10 The ruins of the Mayan city of Tikal, in Guatemala, have been invaded by a mixture of broadleaf evergreen and broadleaf deciduous trees typical of the Tropical Semideciduous Forest. (*J.K. Nakata: Sight & Sound Productions © 1990*)

As the dry season increases in length in a poleward direction, the forest undergoes a gradual transition from broadleaf evergreen to broadleaf deciduous, and progressively more trees lose their foliage and become dormant in order to conserve moisture. The term "semideciduous" indicates that this forest contains both evergreen and deciduous trees. A vertical zonation of tree heights is often still apparent, but usually only two foliage layers are present. The larger trees are most frequently deciduous in nature because of their greater transpirational demands, while the smaller trees are predominantly evergreen. The decreased density of the forest, coupled with the deciduous habit of many of the tree species, enables considerable sunlight to reach the forest floor, especially during the dry season. This in turn supports substantial amounts of undergrowth.

Tropical Semideciduous Forests are scattered in a number of widely separated regions of the tropics. In theory, this vegetation type should be found in continuous strips on both poleward margins of the Tropical Rainforests. The impact of dry season fires, both of natural and, increasingly, of human origin, however, has considerably reduced the areal extent of this vegetation type. Where fires have been numerous, the trees have been largely destroyed and the growth of savanna grass encouraged. Currently, large expanses of Tropical Semideciduous Forest are located in Southeast Asia, India, the southern East Indies, portions of northern South America and Central America, and southern Brazil.

other instances, it consists of a variable mixture of small trees, shrubs, and grasses. In all cases, the vegetation consists predominantly of tropophytes that grow actively during the rainy season and become dormant during the long, often hot, dry season.

The geographic distribution of the Tropical Scrub Woodland subassociation has also been complicated and generally reduced by the destructive effects of fire, so only several large but detached areas currently exist. One of the largest is in portions of South and Southeast Asia, where bamboo thickets are often encountered. In northern Australia, species of eucalyptus trees dominate the landscape (see Figure 9.11). A large section of southern and eastern Africa is dominated by species of acacia trees. In portions of eastern Brazil, Paraguay, and northern Argentina, a mixture of thorny bushes, cacti, and trees such as the quibracho makes some areas almost impassable.

Subtropical Broadleaf Evergreen Forest

As the subtropical high pressure belt is approached, a dry season of increasing length is responsible for the sequential transition of Tropical Rainforest to Tropical Semideciduous Forest and then to Tropical Scrub Woodland. Progressing still farther poleward, the climate eventually becomes so dry that in most places trees cannot grow. A major climatically imposed gap thus

Tropical Scrub Woodland

Progressing into tropical environments with a still longer dry season, further modification of the natural vegetation occurs. Many areas with a dry season four to seven months long are covered with a mixture of small trees, bushes, grasses and other herbaceous (nonwoody) plants; these areas are termed Tropical Scrub Woodland. The term *woodland* indicates that although woody trees and shrubs are the dominant forms of vegetation, no continuous tree canopy exists. The term *scrub* refers to small, stunted trees and shrubs growing thickly together.

The vegetation within this subassociation varies greatly in composition and appearance in different areas. In some localities it consists of small individual trees surrounded by tall grasses and other low-growing plants. Elsewhere it takes the form of dense thickets of thorny shrubs separated by grassy openings. In still

Figure 9.11 A Tropical Scrub Woodland/Savanna border area in Australia. The trees are eucalyptus. (*R. Snead,* © *JLM Visuals*)

divides the world's forests into tropical and nontropical segments. The subtropical high, however, is not strong enough everywhere to produce this vegetative transition. Especially along the east coasts of continents paralleled by warm ocean currents, enough precipitation occurs that trees continue to flourish, resulting in a "connecting link" between the tropical and nontropical forest subassociations. This subtropical forest differs in several significant respects from those found at either higher or lower latitudes; it is sometimes referred to as the Subtropical Broadleaf Evergreen Forest.

The most extensive Subtropical Broadleaf Evergreen Forests grow in East Asia, particularly in eastern China, southern Korea, and southern Japan. Other areas include portions of the extreme southeastern United States, New Zealand, and southeastern Australia.

In contrast to the broadleaf evergreen forests nearer the equator, the Subtropical Broadleaf Evergreen Forest contains a limited number of tree species within any given locality; and large groups of a single species are common. The forest is not particularly dense, and average tree sizes are modest (see Figure 9.12). Leaves are smaller and more leathery than those of the rainforest. Some vertical zonation is normally apparent, with an upper story of larger trees that require substantial sunlight and a lower story of small, shade-tolerant trees,

shrubs, and occasional grasses. Although the exact vegetative composition varies considerably from place to place, the dominant tree species in many areas include live oak, magnolia, and laurel. The trees are often draped with lianas and epiphytes. One epiphyte familiar to residents of the southeastern United States is Spanish moss.

Mediterranean Woodland and Scrub

On the poleward sides of the subtropical highs, the climate again becomes humid enough to support the growth of trees, and forests return as the dominant vegetation type. Once again, the transition from dryland vegetation to humid forest is not abrupt, but proceeds through a series of intermediate stages. These stages are well illustrated within the Mediterranean Woodland and Scrub subassociation.

This vegetative subassociation occurs mostly on the west sides of continents and is closely associated with the Dry Summer Subtropical Climate. Areas included are southern and central California, the Mediterranean Basin region, central Chile, the southern tip of Africa, and much of southern Australia. These areas have dry, warm summers dominated by the poleward-shifting subtropical high and mild, humid winters influenced by traveling cyclonic storms within the westerly wind belt. The dry summer season, usually lasting from four to seven months, is the major climatic stress factor. It is largely responsible for a vegetative subassociation dominated by shrubs and small trees that somewhat resembles the Tropical Scrub Woodland.

The composition of the constituent plant species varies greatly between the widely separated regions of the Mediterranean Woodland and Scrub subassociation. Despite their biological differences, however, the species exhibit a remarkable physical similarity in vegetative forms and community organizations in response to similar environmental conditions. Most trees and shrubs are classified as *sclerophylls* (from the Greek *sclero* "hard," and *phyllos* "leaf"). They have small, hard, thick, leathery leaves that minimize transpirational losses. Other common xerophytic adaptations include deep root systems, viscous fluids, waxy leaf coatings, and thick bark. Many plants also have thorns as a protection from grazing animals.

Better-vegetated regions typically display one of two vegetation assemblages. *Mediterranean Woodland* consists of a park-like combination of scattered small-to-

Figure 9.12 Live oak trees draped with Spanish moss are a common sight in the Subtropical Broadleaf Evergreen Forest of the southeastern United States. (*W. Metzen/H. Armstrong Roberts*)

medium-sized oak trees surrounded by grasses and flowering herbaceous plants. An assemblage known as *chaparral* in California and *maquis* in the Mediterranean region is somewhat more common (see Figure 9.13). It consists of an often dense cover of woody shrubs and small trees ranging up to 20 feet (6 m) in height. Species of oak most frequently predominate, although in Australia eucalyptus and acacia are most frequent.

Climax vegetation is extremely rare within this subassociation because of a long history of human occupancy and exploitation of a fragile ecosystem. The removal of trees for lumber and firewood and overgrazing by sheep and goats, especially in extensive areas within the Mediterranean Basin, have seriously decreased both the quality and quantity of the vegetation. Fire, too, has played a major destructive role. Brushfires in the tinder-dry vegetation are an annual late summer occurrence in many regions, but are probably most notorious in California and Australia. Today, those areas most adversely affected by overgrazing and fire are covered by a dense growth of thorny shrubs commonly less than 3 feet (1 m) in height.

Mid Latitude Deciduous and Mixed Forest

Continuing poleward from the subtropics into the middle latitudes, the dryness produced by the subtrop-

Figure 9.14 Autumn begins to color the trees in a rural portion of the Mid Latitude Deciduous and Mixed Forest within the eastern United States. (*Richard Jacobs,* © *JLM Visuals*)

ical highs is left behind, and conditions become more favorable for forest growth. At the same time, however, a new climatic stress appears in the form of a cold season of increasing length and severity. The result is a period of vegetative dormancy, during which the cold-sensitive broadleaf trees that dominate the middle latitude forests must reduce their moisture content and shed their leaves in order to survive.

The Mid Latitude Deciduous and Mixed Forest, as this subassociation is often termed, is the most extensive forest subassociation in the middle latitudes. Three large regions of this forest type within the northern hemisphere are the eastern United States, much of Europe and a portion of the Asiatic Soviet Union, and east-central mainland Asia. In the southern hemisphere, the subassociation's only significant area of occurrence is in the southern Andes of Chile and Argentina.

During the summer season, the closed canopy of broadleaf trees in the Mid Latitude Deciduous and Mixed Forest bears some similarity to that of the Tropical Rainforest (see Figure 9.14). Important differences exist, though, between the two forest types. One difference is the much lower vegetative biomass of the middle latitude forests. Mature trees reach only intermediate heights of perhaps 100 to 150 feet (30–45 m). No clear-cut vertical zonation of heights is normally observable, although there is frequently an understory of shade-tolerant trees and shrubs. The variety of tree

Figure 9.13 A chaparral-covered hillside in southern California east of San Diego. (*John Shelton*)

species is rather limited in any one area, and often only three or four species are locally dominant. Likewise, fewer vines and epiphytes use the trees for support, although certain climbing plants like the honeysuckle, poison ivy, and Virginia creeper of the eastern United States may be locally abundant. Although the forest is dense enough to produce a closed canopy, even in summer patches of sunlight penetrate to the forest floor, which is not dark and gloomy like that of the rainforest. The intensity of shade does, however, inhibit the growth of near-surface vegetation.

Probably the major difference between the Tropical Rainforest and the Mid Latitude Deciduous and Mixed Forest is that the former is evergreen while the latter experiences a seasonal loss of most foliage. While the rainforest is constant and rather timeless in appearance, the deciduous forest undergoes a seasonal transformation, in visual terms, between life and death. Because the trees spend much time in a dormant state, growth rates are much slower than those of the tropics.

Diversity in the mix of tree species gives a distinctive character to the forests of different areas. In the southeastern United States, oak, hickory, and poplar are among the dominant tree types. Farther north, maple, birch, and beech are dominant. Many of these trees are noted for their brilliant autumn colors and are an important tourist attraction in New England and the Great Lakes states. In Eurasia, species of oak, beech, and ash are most common; in southern Chile, beech trees are dominant.

Throughout many portions of the Mid Latitude Deciduous and Mixed Forest, a scattering of coniferous needleleaf evergreen trees exists. These trees are generally out-competed by the broadleafs but are more tolerant of adverse conditions such as cold temperatures and rocky, sandy, or highly acidic soils. They therefore tend to be more numerous and even locally dominant in mountainous areas within the higher Appalachians and the Alps and in sandy areas such as the Pine Barrens of southern New Jersey.

The Mid Latitude Deciduous and Mixed Forest has been greatly altered and in many cases completely removed by human beings. Most regions containing this forest subassociation have dense human populations and now consist of farmland, roadways, and urban sprawl. The removal of the forests is most complete in East Asia, where only the most rugged and inaccessible regions contain substantial numbers of trees. Only in the United States is the majority of the surface still forested, and nearly all is in secondary growth.

Mid Latitude Evergreen Forest

Broadleaf trees predominate over needleleaf trees in most of the middle latitudes. However, two large areas, both in North America and primarily within the United States, contain needleleaf evergreen forests. The reasons for the existence of these forests are unclear, because they do not represent the climax assemblages that would be expected, given the climatic conditions, within their respective areas. It is believed that fire may contribute to their maintenance.

One of these two forests is the *West Coast Coniferous Forest* (see Figure 9.15). The only fully developed forest of this type extends along the west coast of North America from northern California to southern Alaska. The West Coast Coniferous Forest contains the world's most impressive assemblage of needleleaf evergreen trees. The forest has a high biomass, probably second only to the Tropical Rainforest. Trees are of very large size, with tall, straight trunks. Heights can reach well over 200 feet (60 m), with some coastal redwoods more than 300 feet (90 m) in height and 14 feet (4.5 m) in diameter. The size of these trees exceeds that of all other living organisms on earth.

The largest and densest stands occur on the moist western (windward) sides of the Cascades of the western United States and the Coast Mountains of British Columbia. In these areas, Douglas fir, western hemlock, and Sitka spruce are among the dominant species. In the drier lands to the east of the mountains are less dense forests of smaller Douglas fir, ponderosa pine, and white pine.

An extensive portion of the southeastern United States stretching from eastern Virginia southward to Florida and westward to Texas is covered by a forest composed mostly of species of pines (see Figure 9.15). Known as the *Southern Pine Forest,* it is closely associated with the sandy, drought-prone, and rather infertile soils of the Atlantic and Gulf coastal plain. Its existence can probably be attributed to the resistance of pine trees to droughty soil conditions, their low soil nutrient demands, and their ability to withstand ground fires.

The pine forests occupy well-drained sandy surfaces situated between stream valleys. The most common tree species include longleaf, shortleaf, pitch, and loblolly

Figure 9.15 (left) Coast redwood trees in Redwood National Park, California. (right) A stand of longleaf pine trees on the North Carolina coastal plain. Ground fires have blackened the lower trunks. (*left: Breck P. Kent, © JLM Visuals; right: Barbara J. Miller/Biological Photo Service*)

pines. These trees are valuable sources of lumber, pulp-wood, and naval stores such as turpentine and resin. Within the swampy stream valleys, hardwoods such as live oak, bald cypress, tupelo, and red gum predominate.

Northern Coniferous Forest

The most poleward forest subassociation is the *Northern Coniferous Forest,* also known by its Russian name *taiga.* It is second only to the Tropical Rainforest in total areal extent. The Northern Coniferous Forest occurs prima-rily within regions having a Subarctic Climate, with its short mild summers and long frigid winters, but the southernmost margins enter Humid Continental Cli-mate regions. It exists within two giant areas, both in the northern hemisphere. These are the northern inte-rior of North America, extending from Alaska to Labra-dor, and the even larger interior of northern Eurasia, reaching from Scandinavia to the Pacific. Logging activ-ities have removed or modified the forest over portions of its southern range. The central and northern portions of the forest, however, have been influenced relatively little by human activities, so this forest subassociation as a whole remains more nearly in a natural state than any other.

The Northern Coniferous Forest is dominated by cold-tolerant needleleaf evergreen trees (see Figure 9.16). Only a few tree species are able to cope with the severe climatic stresses imposed by the short growing season and extremely low winter temperatures of these areas. As a result, only one or two tree species typically dominate over large areas; and the forest is usually locally uniform in height and density. Spruce and fir trees are most common for the region as a whole. In the northern United States, these trees are liberally mixed with pine trees, especially white pine, red pine, and jack pine. In western Eurasia, Scotch pine and stone pine are locally abundant. Eastern Siberia, the coldest region of all, is unique in being dominated by two species of larch trees, which are needleleaf deciduous in nature. Appar-ently, winter temperatures here are too extreme even for most cold-tolerant needleleaf evergreens, and the larch becomes the most successful competitor.

The trees of the Northern Coniferous Forest are moderate in size near their southern margins but become progressively smaller poleward. Most attain heights of 50 feet (15 m) or less. Near their northern limits, they are dwarfed by the shortness of the growing season; and mature specimens as much as a century old may be only a few feet tall. The density of the forest also tends to diminish in a poleward direction. The forest floor is

Figure 9.16 Northern Coniferous Forest scene in Katmai National Park, Alaska. (*Breck P. Kent, © JLM Visuals*)

typically covered with a mat of lichens and mosses. Especially dominant in northern Canada is a pale-colored species of lichen referred to as "caribou moss."

Breaks in the continuity of the forest have resulted from the lasting effects of the continental glaciers that covered parts of the area as little as 5000 years ago (see Chapter 15). Their legacy is a landscape dotted by myriad lakes of all sizes. Lakes are relatively ephemeral landscape features; as they gradually fill with sediment, they are occupied by a succession of hygrophytic vegetation assemblages. Many small lakes of glacial origin within this region have already been converted to *muskegs,* which are swampy areas covered with mosses, swamp grasses, shrubs, and scattered water-tolerant trees such as larches. In Canada and Scandinavia, some severely glaciated areas still have patches of exposed bedrock on which little vegetation grows.

Grassland Association

The *grasses* are a family of plants with bladelike, opposing leaves connected to a jointed stem by a sheathlike attachment. A great number of grass species exist, including many that are commercially important. Both directly and indirectly, the members of the grass family are largely responsible for producing the basic foods on which humanity depends. For example, all the cereal grains such as rice, corn, wheat, oats, and barley are grasses. In addition, most animals raised for meat and dairy products feed primarily or exclusively on grasses.

Grasses grow on virtually all vegetated portions of the earth's surface and are the dominant form of natural vegetation on over a third of the world's land area. Their widespread geographical distribution results both from their hardiness and their diversity. Grasses are abundant from the equator to the tundra and from swamps to deserts. A key to their survival in environments with severe periodic climatic stresses is the fact that many grasses are annuals and can simply regrow from seeds during times of favorable conditions. Numerous other grass species, unlike trees, are capable of growing new shoots from their subsurface stem sections after the above-ground portions of the plants have been killed by adverse weather conditions or removed by grazing.

The chief competitive disadvantage of grasses is, of course, their size. Most grasses attain heights of only a few feet. Grasses are not sufficiently xerophytic to dominate the vegetation in most arid regions, although they are well represented in many semidesert areas. As a result, most of the world's grasslands are situated on the margins of forests, in those areas either too wet or too dry for trees. These areas generally include regions with subhumid, semiarid, and seasonally dry climates, as well as more restricted areas that experience poor drainage or periodic flooding.

Over the last several millennia, the world's grasslands have been gradually expanding in size as they have encroached upon their forest margins. This trend may be partially due to post-Pleistocene climatic changes but is believed to be primarily caused by human activities such as the gathering of firewood, overgrazing by domesticated animals, and especially the use of fire. Expansion has been occurring for such a long time that researchers are unsure about the original extent of the world's "natural" grasslands. It is a certainty that, especially in the tropics and subtropics, both grasslands and deserts will continue to expand at the expense of forests in the near future.

The Grassland vegetation association is divided into three subassociations: the savanna, steppe, and prairie. The savanna grasslands, occurring in the low latitudes, mostly within drier portions of the Tropical Wet and Dry Climate, will be discussed first. Next to be examined are the steppe grasslands, which are associated with semiarid climatic conditions in both the low and middle latitudes. The prairie, which is restricted largely to subhumid portions of the middle latitudes, is covered last.

Savanna

The *Savannas* are tropical grasslands that contain scattered trees or shrubs (see Figure 9.17). They generally occupy the drier portions of the Tropical Wet and Dry Climate, where the winter dry season lasts between six and eight months. The most extensive savannas are located on the continents of Africa, South America, and Australia. The African savannas are the largest and best known. Most occur in a broad arc bordering the Tropical Rainforest of the Congo Basin to the north, east, and south. These savannas, especially those of East Africa, are frequently seen in "big game" movies or on television documentaries because of their great herds of large animals. In South America, extensive savannas are

Figure 9.17 Typical savanna landscape in northern Ghana. (*Wayne McKim*)

located in Venezuela and Colombia as well as in southern Brazil and Paraguay. Large portions of northern, eastern, and western Australia are also often considered to be savanna, although the mix of plants here is somewhat different. Other smaller but noteworthy savannas occur in much of western India, western Madagascar, and the Everglades of southern Florida.

Savanna grasses survive and thrive because they become dormant during the dry season. The aboveground portions of the plants die, but the root sections send out new shoots with the onset of the summer rains to turn the landscape lush and green (see Figure 7.6). Great variations in grass heights occur in different areas; these heights range from 1 foot (0.3 m) to over 12 feet (4 m). Most savannas are dominated by coarse, short-bladed bunchgrasses that grow in dense tufts with intervening patches of bare earth. They have well-developed root systems that support rapid growth during the rainy season.

Trees may be scattered individually throughout the savanna to produce a park-like appearance, or they may exist in groups. The trees consist of stunted broadleaf deciduous and broadleaf evergreen species which, with their thick bark, lack of lower branches, and flattened or umbrella-shaped crowns, are very similar to those of the Tropical Scrub Woodlands. Trees do not dominate because the seasonal rainfall distribution is too uneven to readily support large annuals with year-round moisture requirements. In general, the number of trees declines toward drier areas, as does grass height.

Along the banks of streams, dense strips of trees commonly exist. Their presence indicates that when water is available, trees will assert dominance over grasses. Often, the trees along either bank are large enough so that their foliage intermeshes, enclosing the stream within a tunnel of vegetation. These strips of forest are therefore termed *galeria*, the Italian word for tunnel.

Steppe

The *Steppe* comprises the short grasslands of the low and middle latitudes. They are closely associated with semiarid climates. Many geographers use this term to refer only to the short grasslands of the middle latitudes, and, indeed, it is there that they are most distinct as a separate vegetation subassociation. Large areas of the tropics, however, are also covered primarily by short grasses because of similarly marginal amounts of rainfall; for this reason, it seems justifiable to include them within the steppe classification as well.

The low latitude steppes are located in regions with a dry, hot climate typically punctuated by a brief, often intense rainy season lasting from two to four months. Ideally, they occupy the semiarid margins of the subtropical deserts. Larger areas include the borders of the Sahara and Kalahari Deserts in Africa and portions of north-central Australia. The vegetation of these regions consists predominantly of a short grass cover that lies dormant for much of the year. The grasses are similar to those of the savanna in that they have a tufted habit, and are interspersed with patches of bare earth. Varying amounts of often-thorny broadleaf deciduous shrubs are interspersed with the grasses and may be locally dominant.

The steppes of the middle latitudes, like their low latitude counterparts, generally occupy the semiarid margins of deserts. On their humid sides, they grade into either tall-grass prairies or forests. Two principal areas exist: one occupies the Great Plains of the United States and south-central Canada; the other extends over 4500 miles (7200 km) from the north shore of the Black Sea eastward through the southern Soviet Union to northern China.

Most steppe lands of the middle latitudes consist of vast expanses of flat or gently rolling terrain covered almost exclusively by short grasses less than 2 feet (0.6 m) in height (see Figure 9.18). In some areas the grass is dense and continuous, and the combination of soil and matted grass roots forms a tough sod that protects the surface from erosion. In other areas, especially toward the desert margins, the vegetative cover consists predominantly of bunchgrasses. Most grasses are perennials that lie dormant during the winter and sprout vigorously in the spring. During the summer, high temperatures and evapotranspiration rates produce severe soil moisture deficits; and grass growth is greatly reduced. Along with the grasses are various flowering annuals as well as some shrubs. In most cases, trees are lacking in the middle latitude steppes except along the banks of streams, where such water-tolerant species as cottonwoods and willows are found.

Prairie

The *Prairie* is one of the smallest and most distinctive of the world's major vegetation groups. It has also been

Figure 9.18 Short steppe grassland in the foreground gives way to eroded sedimentary rock formations in Badlands National Park, South Dakota. (*B.F. Molnia*)

more nearly removed by human activities than any other plant subassociation. The underlying soils have proven so fertile that virtually all prairie areas have been plowed up and converted to cropland. Remaining small patches of prairie along roadsides, the edges of fields, and on steeper slopes can provide an idea of the original appearance of the vegetation; but they certainly cannot convey an adequate impression of the original immensities of shoulder-high grasses extending to the horizon (see Figure 9.19).

The prairie regions are located in the middle latitudes, mostly near the dry (subhumid) margins of the Humid Continental Climate. The largest area is centered in the midwestern United States, extending in a rough triangle from Indiana northwestward to southern Alberta, Canada, and southward to eastern Texas. Smaller patches within the United States include the "Black Belt" of western Alabama and the Palouse country of eastern Washington, which is transitional to steppe. A second extensive area of prairie is the pampas of South America, located within northeastern Argentina, Uruguay, and southern Paraguay and Brazil. A small but important prairie region also occupies the Danube plain in Hungary, and vegetation transitional to prairie is found along the humid northern margin of the steppe lands of central Eurasia.

The predominant vegetation of the prairie was a mixture of tall grasses that were sufficiently dense to form a continuous sod cover over large areas. The

Figure 9.19 Tucker Prairie, a protected tall grass prairie in central Missouri. (*Marbut Memorial Slide Collection, Am. Society of Agronomy*)

grasses were mostly 2 to 4 feet (0.6–1.2 m) high but were reputedly as tall as 10 feet (3 m) near the eastern margin of the prairie in Indiana and Illinois. They diminished in both height and density toward the drier steppe margins. A variety of annual and perennial species contributed to the grass cover, and in the American Midwest, species of bluestem were predominant. Interspersed with the grasses were a large number and variety of herbaceous flowering plants that produced a colorful landscape, especially in late spring.

On the original prairie, trees existed only in scattered patches near the forest margins and, somewhat in galeria fashion, in strips along streams. With settlement, large numbers of trees were planted for decoration and shade and as windbreaks; these trees have generally thrived and have often attained large dimensions. At present, the prairies consist largely of a checkerboard of fields of crops. It should not be surprising that cereal grains such as wheat and corn are currently most common, since these domesticated tall grasses are biologically quite similar to the wild tall grasses that once grew so profusely in the prairie.

Desert Association

Desert vegetation is closely associated with the world's arid climate regions. These regions, located mostly in the subtropics, also occur in portions of the middle latitudes. The largest desert area by far extends across the Sahara of North Africa through the Arabian Peninsula and well into southwestern Asia. Other extensive subtropical deserts include the Kalahari of southwestern Africa, the Mojave and Sonoran Deserts of the southwestern United States and adjacent Mexico, the Atacama of western South America, and the Australian Desert. Within the middle latitudes, deserts occur in drier portions of the American Great Basin, in southern and western Argentina, and in much of Central Asia.

Despite their aridity, most deserts contain a variety of vegetation (see Figure 9.20). The one-quarter, or less, of all deserts that are essentially devoid of plants exist mostly in areas that are extremely dry or that have shifting sands or steep, rocky slopes. Desert vegetation typically consists of a rather diverse mixture of small, predominantly xerophytic plants that are especially adapted to survive in an arid environment. Desert plants must also be tolerant of extremes of both heat and cold. Species vary greatly in different areas, but the similarity in environmental conditions leads to the development of plant groups that are similar in general appearance and characteristics.

Perhaps the most fundamental attribute of desert vegetation is its wide spacing, as most other plant assemblages produce a more-or-less continuous ground cover. The lack of surface moisture does not permit this in deserts, despite the xerophytic nature of the plants. In general, the spacing between plants increases with aridity; but even in better-watered areas, most of the surface is bare. Two other basic characteristics of desert plants are small size and slow growth rates. Most plants attain heights of only a few feet or less, and many take years to reach maturity.

Desert plants can be divided into several botanical groups. Most widespread and important are shrubs of both evergreen and deciduous varieties. They are very common in the southwestern United States, where familiar species include sagebrush and mesquite.

The best-known desert plants are the leafless evergreen herbs, which include most cacti. It is commonly believed that cacti are the dominant plants in deserts. Actually, they are native only to the Americas and are abundant in relatively few localities. Their striking appearance and often rather large size, however, cause them to stand out from the more mundane desert plants.

A third group, the *halophytes* (from the Greek *halos* "salt"), are salt-tolerant plants that normally occupy lowland basins of interior drainage. Because these plants

Figure 9.20   Similarities and contrasts in desert environments: (left) Sonoran Desert, Arizona. (right) Central Nevada. (*left: Breck P. Kent, © JLM Visuals; right: J.K. Nakata: Sight & Sound Productions © 1990*)

have developed an ability to cope with highly saline surface conditions, they are able to grow where no other plants can survive and are therefore free from competitors.

A fourth group of desert plants are the *ephemerals*. They are small, short-lived plants that go through their life cycles quickly following a substantial rainfall. The ephemerals take advantage of those brief, rare periods when the desert is not dry; they therefore cannot be classified as xerophytes. They include a wide variety of plants; most are grasses or flowering herbs whose seeds or bulbs have lain dormant for months or even years. The abundance of brightly colored flowers among the ephemerals has led to the frequent observation that the desert "blooms" after a rainstorm. Within a month or so, however, the moisture is gone; the flowers have produced their seeds and died; and the desert reverts to its normal sere appearance.

Trees are not found in most desert areas because their moisture requirements are too high. On occasion, though, scattered xerophytic trees do occur, especially near the desert margins. This is especially true in Australia, where occasional drought-resistant eucalyptus trees are found in many desert areas. Other localities where trees can secure enough moisture for growth are *oases*, where the water table intersects the surface to produce a spring, and riverbanks, where dense *riparian* (riverbank) vegetation often flourishes. Riverbanks in parts of the southwestern United States have recently

been invaded by the tamarisk, an imported tree whose dense foliage has choked out the native vegetation.

Tundra Association

The *Tundra* is the most cold-tolerant vegetation association and consequently is found at higher latitudes and elevations than any other association. Because the tundra has been relatively uninfluenced by human activities, it is the least modified of all the major plant groups.

Most tundra areas are located within the northern hemisphere. The three largest regions are northern North America from Alaska to Labrador, the Arctic coast of Eurasia from Norway to the Bering Sea, and the coastal rim of Greenland. These areas all have a Tundra Climate characterized by long cold winters, summers with monthly means below 50° F (10° C) and occasional frosts, and a growing season of only about two months. Only during the summer, extending from late June or early July into early September, does the snow melt and the vegetation appear.

Tundra vegetation consists of a mixture of small plants, most of which are herbaceous perennials (see Figure 9.21). Two layers of plants normally cover most of the surface. The upper layer of grasses, grasslike sedges, flowers, and occasional small shrubs often produces a meadowlike appearance. Grasses are so common in the tundra that some biogeographers consider it a grassland subassociation rather than a separate vegetation association. Several species are common, but perhaps the best known is cotton grass, a tall grass that grows in poorly drained areas and is surmounted by a cottony tuft. Beneath the upper layer of grasses exists a well-developed surface layer of lichens and mosses. Lichens consist of algae and fungi growing together in a symbiotic relationship to produce a plant form unlike either one alone. The best-known example is the so-called "reindeer moss," which grows to heights of 2 to 3 inches (5–10 cm).

Near the southern margin of the tundra, dwarf trees and shrubs are found on better-drained sites. They grow extremely slowly and at maturity may attain heights of only a few feet. The very small size of the trees and shrubs is believed to result largely from the alternate freezing and thawing of the surface, which causes the ground to heave, thus tearing and crushing the roots. In addition, an underlying layer of *permafrost,* or perennially frozen soil, blocks root development.

Localized differences in topography exert perhaps a greater influence on the vegetation of the tundra than upon any other plant group. The effects of permafrost and past glaciation have caused large areas to be poorly drained, and lakes and marshes abound. The marshes are sites where hygrophytic grasses grow. Slightly higher

and drier sites a short distance away are likely to be covered with completely different types of vegetation, including patches of colorful flowers such as buttercups, anemones, and poppies. With the low sun angles that always exist in the high latitude tundra, slope directional orientation is also a crucial factor; south-facing hillsides are sunnier, warmer, and better vegetated.

Proceeding poleward through the tundra, the density, size, and biological sophistication of the plants gradually decline. At its northern edge, the tundra surface consists largely of bare rock, with only widely scattered patches of moss and lichen.

Mountain Vegetation

Within mountainous regions, all the factors that control the distribution of natural vegetation are subject to large local variations. As a result, great differences in natural vegetation are encountered within small distances, especially in higher mountains. The resulting vegetative patterns are too complex to discuss in detail except at the local level. At any specific site, however, within the constraints imposed by topography and soils, the natural vegetation may be expected to be closely associated with the local climate. This correlation again indicates that climate is the most powerful of the natural vegetation distribution controls. Local climatic conditions may cause the vegetation within a highland area to be completely unlike that of the surrounding lowlands. Areas of montane (high mountain) forest can therefore be found in the midst of deserts, and tundra (alpine meadow) can exist at the equator.

Patterns of mountain vegetation are by no means random, however, especially when examined on a broader scale. Because climatic influences are dominant and elevation largely determines mountain climate characteristics, a vertical zonation of plant assemblages representing a compression of latitudinal vegetation gradients is normally well developed (see Figures 7.36 and 9.22). In general, the vegetation of mountain areas consists of assemblages that thrive under cooler and moister conditions than do those of the surrounding lowlands.

Summary

Vegetation covers about 80 percent of the earth's land surface. Through the processes of natural selection and biological mutation, a great variety of plant species

Figure 9.21 Tundra or alpine meadow vegetation above the tree line in the Colorado Rockies. (*Ralph Scott*)

Figure 9.22 A zonation of vegetation types based on elevation is apparent in this view of Wyoming's Grand Teton Mountains. Prairie vegetation in the foreground gives way to coniferous forest, which in turn merges with tundra and, finally, with ice and bare rock. (*Richard Jacobs, © JLM Visuals*)

adapted to a diversity of environmental conditions have evolved. Given a stable environmental setting, a given locality will develop a plant community that will utilize the available space and resources to its maximum advantage. When a plant community is disturbed or removed, vegetation will generally return to the site in an organized successional sequence, eventually reaching ecological stability when a climax plant assemblage is achieved. The combination of human activities and recent climatic changes has prevented stable climax plant communities from currently existing in many areas.

The four chief environmental controls on the distribution of natural vegetation are climate, topography, soil, and biological organisms. Climate is the most important control, largely because it influences moisture availability and temperature. Moisture availability helps determine such vegetative characteristics as plant size, root development, transpiration rates, leaf number and size, growth rates, and regional biomass. Temperature is especially important for its influence on the length of the growing season, growth rates, and leaf shedding. Topographic influences are often localized. In general, areas of steep slopes support a vegetation assemblage of limited complexity and biomass; and poleward-facing slopes support plants adapted to cooler and moister

surface conditions. A number of soil characteristics are important to plants; especially crucial are mineral content and water retention ability. Biological organisms must successfully coexist within a given habitat, or ecosystem alteration will occur. The interaction between the lifeforms inhabiting an ecosystem is highly complex, but an overriding factor is the competition for available resources.

The four major plant groupings, or associations, are forest, grassland, desert, and tundra. Each is associated with a different range of environmental conditions. Forests cover a larger geographical area than do any of the other associations and comprise the great majority of the total biomass of land areas. They generally dominate in areas with at least a five-month period of annual soil moisture surplus and a warmest monthly mean temperature of 50° F (10° C) or higher. Most of the tropics, middle latitudes, and subpolar regions meet these criteria and are forested. Broadleaf evergreen trees dominate near the equator, while broadleaf deciduous trees are dominant in forested portions of the subtropics and middle latitudes. Needleleaf trees are most abundant in subpolar and alpine environments.

Grassland regions are typically encountered on the dry margins of forests. Savanna grasslands, which are tropical grasslands with scattered trees, occupy large portions of the low latitudes that have dry seasons from five to eight months long. Short steppe grasslands are found in semiarid regions of both the low and middle latitudes. Tall prairie grasslands formerly occupied subhumid areas within the middle latitudes, but have largely been removed for agricultural purposes.

Deserts are most widespread in the subtropics, but also exist in portions of the middle latitudes. The driest deserts are unvegetated, but most are vegetated by a variety of xerophytic herbs and shrubs.

Tundra vegetation consists of a mixture of small, cold-tolerant herbaceous perennials, which often exhibit a meadowlike appearance. The plants of the tundra are only visible and biologically active for approximately three months, in summer. They dominate subarctic regions with warm season monthly means between 32° and 50° F (0°–10° C).

Review Questions

1. What is a climax plant community? How does the concept of vegetative succession relate to this concept?

2. List the four major global natural vegetation distribution controls. Which has the greatest effect on world vegetation patterns?

3. What structural differences exist between xerophytes and hygrophytes?

4. What competitive advantage do trees have over other types of vegetation? Why don't trees dominate the vegetation of all areas?

5. Why are the world's forest areas divided into tropical and nontropical segments? Are these two segments connected? Where and why?

6. Through what natural vegetation associations and subassociations would you pass on a trip from New York City to San Francisco? Briefly explain the environmental factors responsible for each of the changes in vegetation.

7. What basic similarities and differences exist between the vegetation of the Tropical Rainforest and the Mid Latitude Deciduous and Mixed Forest?

8. In what respects is the Northern Coniferous Forest different from all the other forest subassociations?

9. Describe the basic environmental characteristics that favor the dominance of each of the four natural vegetation associations.

10. Describe the appearance of the Tropical Savanna. In what ways does it differ from the grasslands of the middle latitudes?

11. What are some common misconceptions regarding the nature of desert vegetation? Which of the general categories of desert plants is most widespread?

12. Discuss the major environmental limitations of the tundra and explain how they have influenced the tundra vegetation.

13. Imagine a supercontinent extending from the equator to the North Pole. Starting with the Tropical Rainforest, list in order the hypothetical sequence of vegetation groups that a person traveling poleward would encounter.

14. In what natural vegetation association or subassociation is your own area located? Explain how local environmental conditions have resulted in the dominance of this type of vegetation.

Key Terms

Photosynthesis
Biogeography
Ecosystem
Vegetative succession
Climax community
Plant association
Xerophyte
Hygrophyte
Mesophyte

Tropophyte
Deciduous
Evergreen
Annual
Perennial
Broadleaf
Needleleaf
Jungle

CASE STUDY

The Destruction of the World's Tropical Rainforests

Reference was frequently made in this chapter to the impact of human activities on global vegetation patterns. In some areas, this impact has taken place over the course of many centuries, and little now remains of the original surface cover. In other areas, only minor modifications have occurred, and the original vegetation remains largely intact. In at least one instance—that of the Tropical Rainforests—human beings are currently destroying a major vegetation assemblage. This destruction is proceeding with unprecedented speed and is likely to produce major adverse consequences for the tropics and perhaps even for the entire planet.

Causes and Extent of Deforestation

Until a few decades ago, most portions of the Tropical Rainforest were untouched and largely unexplored. Little lumbering took place because of the lack of roads, the long distances to markets, and the great diversity of tree species. Beginning in the 1950s and 1960s, however, several factors combined to alter the economic conditions that had previously protected the rainforests. This initiated the large-scale forest destruction that continues at an accelerating pace in the present. One

of the chief factors in this turnabout was population pressure caused by the tremendous growth rates of the developing countries that contain tropical rainforests. Many landless peasants in Asia, Africa, and South America are willing to settle and farm rainforest areas despite the land's low fertility and are burning rainforest to clear land for agriculture.

Commercial lumbering is increasingly profitable, despite the inherent difficulties already mentioned (see Figure 9.23). Modern machinery has greatly increased the efficiency of operations, and uses have been found for more and more tree species. Clearcutting methods are often employed, and even if they are not, uncut species are frequently damaged and later die from insect infestation and disease. Settlers also need wood for lumber and fuel. Tropical Rainforest lumber sales, largely to Japan, total several billion dollars a year. This income is desperately needed by the developing countries, most of which are in dire economic straits. Other important causes of rainforest destruction include cattle ranching, mining operations, and the flooding of vast areas by dams built for the generation of hydroelectric power.

In 1990, the remaining area of Tropical Rainforest totaled some 3 million square miles (7.8 million km²), distributed among 70 countries. This is an area nearly the size of Europe. Each year, approximately 44,000 square miles (113,000 km²) of rainforest are destroyed by land clearing for crop production, fuelwood gathering,

and cattle ranching, and an additional 17,000 square miles (45,000 km²) are damaged or destroyed by commercial lumbering operations. A study released in June, 1990, estimated the current rate of Tropical Rainforest destruction to be 1.5 acres each *second*. In total, an area slightly larger than New York and Vermont combined is lost each year. Within a decade or two, only two large rainforest areas will remain, in the western Amazon Basin and in the central Congo Basin. The destruction of the Tropical Rainforest after some 60 million years of continuous existence is a major event in the earth's biological history and will considerably alter the face of our planet, even as viewed from far out in space. The scope of the clearing operations is larger than the earlier removals of the middle latitude forests in North America and Europe and is being accomplished in a much shorter time span.

Consequences of Deforestation

The removal of the rainforests is likely to produce a number of adverse consequences, the exact nature and magnitude of which are difficult to forecast. One is the permanent deterioration of the vegetative cover. The removal of the trees will not, of course, produce a bare surface for long. Replacement by a dense secondary growth of weeds, shrubs, vines, and saplings is rapid. If undisturbed, the area may eventually return, after several centuries, to its original state. In most cases, though, cutover areas

continued on next page

are settled or grazed, thus inhibiting reforestation. In addition, if cutting is too extensive, few seeds are locally available for the return of the trees. The tendency is for the area to revert to a coarse savanna grassland.

A second important concern is soil deterioration. Soils in the humid tropics are naturally low in fertility because of the leaching of soluble plant nutrients by downward-moving soil water from heavy rains. (This process is discussed further in the next chapter.) The rainforest trees, however, are highly efficient at recycling nutrients by absorbing them into their root systems before they can be leached away. At any one time, a large proportion of the total nutrients are locked up in the trees themselves, to be gradually released as they die and decompose. Lumbering activities therefore directly remove many of these nutrients and also destroy the mechanism for protecting the rest from the leaching process. In addition, erosion is greatly accelerated in these rainy regions by the absence of both a protective can-

opy of leaves and branches and the soil-binding influence of tree roots. Flooding also becomes more common when the trees, with their large water-absorbing capabilities, are gone and the soils become more compact. Finally, some tropical soils tend to bake brick-hard upon exposure to sunlight.

Concern has also been expressed that world temperatures may increase because the rainforests currently absorb more carbon dioxide than the rest of the world's vegetation combined. Without this continuing CO_2 removal and storage in the form of trees, the greenhouse effect may intensify. The rainforests may also become considerably drier because vegetative transpiration provides a substantial proportion of the atmospheric water vapor content in rainforest areas.

The greatest tragedy from a biological standpoint is the impending extinction of countless plant and animal species. The Tropical Rainforest contains hundreds of thousands of species of plants and animals that exist nowhere else on earth and that would

not survive if their habitat were destroyed. Most of the denizens of the rainforest have not yet been studied scientifically; extinction would eliminate them before we had a chance to learn about them or, in some cases, even to become aware of their existence. In addition, these varied life-forms represent a great storehouse of genetic diversity. Their destruction would considerably diminish the world's pool of genetic resources, probably altering future evolutionary history.

Partial Solutions

The fate of the Tropical Rainforest as a vast expanse of virgin forest appears to be sealed. The human transformation of the face of the earth, which is an inevitable consequence of the scientific revolution, is well underway. Sadly, the maintenance of large uninhabited and unutilized wilderness areas such as the rainforest appears to be incompatible with this transformation.

If the rainforest itself cannot survive, at least some of the adverse consequences attending its removal can be minimized. One possible activity is the extensive replanting of trees immediately after logging operations within an area are completed. Replanting would protect the soil and would likely reduce the climatic consequences of deforestation. Little replanting is now taking place. A second action, currently being undertaken by the Brazilian government, is the setting aside of large natural rainforest reserves. The establishment of these reserves should save many plant and animal species from extinction and preserve for future generations at least a segment of a unique ecological system.

Figure 9.23 Appearance of the landscape following rainforest cutting operations in Ecuador. (*J.K. Nakata: Sight & Sound Productions © 1990*)

Chapter Ten

Soils

Focus Questions

1. What physical and chemical characteristics of the soil most greatly affect its ability to support vegetation?
2. What are the five factors in soil development, and what influence does each have on soil characteristics?
3. What are the major world soil types, and what are the basic attributes of each?

S oil is one of our most essential natural resources; without it, we, and all other advanced lifeforms, could not exist. Unlike vegetation, soil is not a living entity; but it is intimately associated with life in several major respects. In the first place, it contains a great deal of life. Soil organisms include a tremendous variety of both plants and animals. Much of this life is microscopic, but some, like the root systems of large trees, dwarfs all but the largest animals living on the surface.

Secondly, the soil supports nearly all plant life; and this in turn supports animal life. Even aquatic plants and animals are largely dependent upon minerals derived from soil particles eroded from the land by water or wind.

A third association with life is the soil's lifelike ability to adjust to its environment. The soil is sensitive, to some degree, to almost every aspect of its surrounding natural environment. Soil characteristics at any given site are therefore determined by the interplay of past and present environmental factors at that site. The soil's response to its environment is dynamic, not static. Not only does the soil develop characteristics that are in equilibrium with the environment in which it forms, it also responds to environmental changes, such as variations in climate or vegetation. Any such change produces a slow alteration of soil characteristics until a new equilibrium is achieved. The geographic distribution of soil types, then, is not random, but is closely related to other components of the environment.

SOIL ORIGIN AND CHARACTERISTICS

The soil is so complex and variable in composition, and means so many different things to different people, that it is difficult to define. In general, though, **soil** can be described as a loose mixture of weathered rock material, organic matter, water, and air that can support plant growth. By volume, a typical soil might contain 45 percent inorganic material, 5 percent organic matter, 25 percent water, and 25 percent gases (see Figure 10.1). These last two components fill the voids, or *pore spaces*, between the solid soil particles. The proportions of the four components vary greatly in different soils, but nearly all soils contain at least a small quantity of each. Each component plays a vital role in the soil's ability to perform its most important function: the support of plant life.

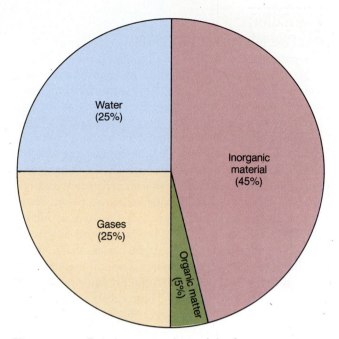

Figure 10.1 Relative proportions of the four components of a typical soil.

The largest proportion of most soils consists of weathered rock material. This material is a major source of the minerals required by plants and animals living in the soil. Weathering, which is discussed in more detail in Chapter 13, refers to the action of physical and chemical processes that break down rock and may eventually convert it to soil-forming materials.

The organic components of the soil consist of the remains of plants and animals. The great majority is derived from vegetation such as leaves, grass, roots, and branches. The organic matter undergoes gradual decomposition, which converts it to a dark-colored substance called humus and slowly releases its mineral content. Many of these organically derived minerals are required by the soil's living population of plants and animals. In addition, chemical reactions involving organic material release nutrients bonded to inorganic mineral particles so that these nutrients become available for plant use.

Soil water and air are both derived primarily from the atmosphere. The proportions of these two components are to some extent inversely related, since both share the soil pore spaces. Following a rainstorm, water floods the pore spaces, expelling much of the air. As the water gradually drains away, evaporates, or is used by plants, air refills the pores.

Soil Color

Soils exist in a great variety of colors (see Figure 10.2). One group varies from white through all shades of gray to black. Other soils range through yellows, reds, and browns. Even greenish and bluish soils are sometimes formed, and colors often differ considerably at varying depths at a single site.

Most soil coloration is imparted by just two substances: humus and iron. As the humus content increases, soils tend to darken through shades of brown to a nearly black color. Oxides of iron are usually responsible for imparting a reddish or yellowish color. These two colors are especially typical of the soils of humid tropical areas, because iron tends to form relatively insoluble compounds with oxygen under warm, moist conditions. In arid environments, the soil may contain whitish flecks or layers due to the presence of calcium carbonate ($CaCO_3$) or salts. These substances are soluble in water and do not collect in the soils of humid climates.

Soil Texture

In contrast to soil color, which is important largely for the information it provides about the composition and condition of the soil, the **soil texture,** or size of the individual soil particles, has important direct effects on the soil. Excluding gravel and larger-sized rocks, soils consist of three basic sizes, or "grades," of particles. The largest or coarsest grade is *sand; silt* particles have an intermediate texture; and *clay* particles are the smallest or finest. Each texture grade has been assigned a specific range of particle diameters, as indicated in Table 10.1.

A factor complicating the description of soil textures is that nearly all soils contain a mixture of soil particles of differing sizes. Various proportions of particles of different sizes can thus produce a variety of texture classes. Soils that are significantly influenced by all three particle grades, and that therefore have an intermediate texture, are referred to as **loam** soils. The classes adopted by the U.S. Department of Agriculture, which are in general use in the United States and many other countries, are shown in Figure 10.3 and are explained in the accompanying legend.

The soil texture in any given area is controlled largely by the climate and by the chemical stability of the material from which the soil was derived. Regions with warm, moist climates typically have deep, fine-textured soils because chemical weathering processes

Figure 10.2 Soils with contrasting colors. The soil on the left is a dark, humus-rich South Dakota mollisol that contains a white layer of calcium carbonate ($CaCO_3$) within the E horizon. The soil on the right is an alfisol from California that has been colored orange by iron and aluminum oxides. *(left: Marbut Memorial Slide Collection, American Society of Agronomy; right: Soil Taxonomy Alt 436 Soil Conservation Service)*

are exceptionally active under these conditions. Conversely, cold, dry areas usually have thin, coarse-textured soils, if a soil cover has developed at all. Likewise, soils developing from chemically stable rock or sediment, especially if it has a high quartz (SiO_2) content, tend to be coarse textured, whereas those developing from parent material with unstable mineral assemblages weather rapidly to produce deeper, finer-textured soils.

The texture of the soil is of practical significance in two major ways. First, it affects the looseness and "workability" of the soil from an agricultural standpoint. Coarse-textured, sandy soils are "light" and easily worked. Plant roots penetrate readily, but the soil may lack the firmness to keep plants from being bent or uprooted during floods or windstorms. Fine-textured soils, in contrast, tend to be dense, sticky when wet, and "heavy." They may be difficult to till, but provide strong physical support for plants.

Secondly, the soil's ability to absorb and retain water is greatly influenced by soil texture. Coarse-textured soils have relatively large intergranular pore spaces because sand-sized particles generally do not fit together as snugly as do silt and clay particles. As a result, sandy

Table 10.1
Soil Texture Grades

Size Grade	Diameter (in)	Diameter (mm)
Gravel	greater than .08	greater than 2.0
Very coarse sand	.04–.08	1.0–2.0
Coarse Sand	.02–.04	.5–1.0
Medium sand	.01–.02	.25–.50
Fine sand	.004–.01	.10–.25
Very fine sand	.002–.004	.05–.10
Silt	.00008–.002	.002–.05
Clay*	below .00008	below .002

*It should be noted that the term "clay" denotes the grade, or size, of the soil particles and does not mean that they consist of clay minerals.

soils have high water-infiltration rates but low water-retention capacities because water passes readily through the pore spaces and relatively little adheres to the soil particles. The situation is reversed for fine-textured soils, which have large water-retention capacities, but which absorb water very slowly. This may lead to conditions of poor drainage and excessive runoff, resulting in rapid erosion and in sediment pollution of streams near cultivated fields. Loamy textures are generally considered most favorable for agricultural purposes because they provide a balance between the problems resulting from excessively fine- and coarse-textured soils.

Soil Structure

Soil particles normally do not remain detached from one another. If they did, fine-textured soils, when dry, would behave like the finest dust, to be easily washed away by rainwater or blown away by the slightest wind. Instead, soil particles tend to adhere to each other, forming larger aggregates called **peds**. The shape, size, and organization of these peds constitute the soil's **structure**.

The peds are held together largely by the clay-sized particles. These fine particles develop strong electromagnetic charges that bind the soil particles together. As a result, coarse-textured soils tend to have weakly developed structures or even a complete lack of structure. As a general rule, the coarser the texture of a soil, the weaker its structure is. The type and strength of the structure are also influenced by the chemical nature of

the clay mineral particles, the type and abundance of soil microorganisms, the frequency and extent of wetting/drying and freezing/thawing cycles, and cultivation practices.

Although a wide variety of specific structural forms can develop, soil structures are divided into four primary groups (see Figure 10.4). *Blocky* structures contain peds that are compact in shape and are relatively flat-sided. Soils with *platy* structures consist of a large number of thin platelet-shaped peds paralleling the surface. *Prismatic* or *columnar* structures are sometimes associated with excessively drained soils, because the pore spaces between the peds offer water a direct path downward through the soil. They are the most massive of the four types. *Granular* or *crumb* structures are common in soils with a moderately high humus content. The degree of structural development and even the type of structure may vary with soil depth.

One of the most important practical effects of a soil's structure is its influence on the absorption and movement of soil air and water. A soil with a well-developed

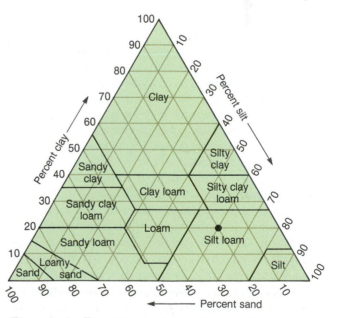

Figure 10.3 The soil texture triangle is used to determine the textural class of the soil, based on the proportions of sand, silt, and clay it contains. Each side of the triangle indicates the percentage content of one of the three texture grades. The three lines that represent the proper proportions of sand, silt, and clay in a soil sample are traced inward to their point of juncture. For example, a soil consisting of twenty percent sand, sixty percent silt, and twenty percent clay is classified as a silt loam. (Note the location of the dot in the diagram for this soil type.)

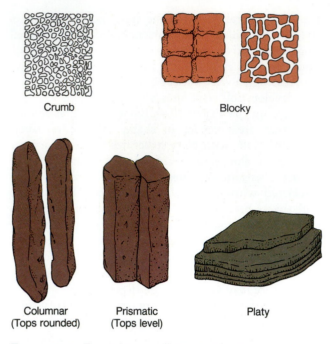

Crumb   Blocky

Columnar   Prismatic   Platy
(Tops rounded)   (Tops level)

Figure 10.4 The primary soil structural groups.

structure is typically less compact and has a greater permeability and porosity than does a structureless soil. Some types of structure, however, may restrict the absorption of water. For example, a soil with a strongly developed platy structure may be poorly drained because each "plate" acts as a barrier to water infiltration.

The strength with which the individual peds are internally bound influences both the soil's resistance to erosion and ease of cultivation. A strong structure holds the soil together and causes it to resist erosion. This same characteristic, though, makes the soil more difficult to plow. Plowing tends to alter and usually to weaken soil structure, and the passage of farm machinery leads to soil compaction. As a result, the soils of agricultural areas typically have poorly developed structures and lower water-infiltration capacities than do nearby unfarmed areas.

Soil Air and Water

Although soil water and air come mostly from the atmosphere, their chemical compositions are modified by their residence in the soil. These chemical changes have a significant effect on the soil biota.

The soil atmosphere is critical to soil organisms because it supplies the oxygen needed for animal respi-

ration and the carbon dioxide required for plant photosynthesis. A lack of oxygen in a saturated soil can suffocate both plants and animals. In addition to air that enters the soil through its pore spaces, the soil also contains gases released from chemical reactions and biological activities. As a result of evaporation and plant respiration, soil air contains slightly less oxygen and nitrogen and somewhat more water vapor and carbon dioxide than does the atmosphere. The constant exchange of soil air with air from the atmosphere, however, keeps the content differences between the two from becoming too pronounced.

Soil water, of course, is derived primarily from precipitation. During a heavy rainfall, large amounts of water typically enter the soil. In response to gravity, most passes downward through the pore spaces toward the water table, but a portion remains within the soil. Water can be held within the soil against the pull of gravity because it is attracted to the surfaces of the soil grains by *capillary tension.* This property results from the electromagnetic charges developed by the soil particles. Because fine-textured soils have a large total soil surface area per unit of volume, they have higher water-retention abilities than do medium- or coarse-textured soils. Plants are able to withdraw much of the soil water for their own use, but finally, at the *wilting point,* the remaining water is bound so tightly to the soil particles that plants are unable to extract it.

Water in the soil dissolves soil minerals, making them available to plants as the water is absorbed by their root systems. As water passes through the soil, dissolved minerals and fine-textured particles are carried, or **translocated,** from one place to another. These materials are often deposited in locations at a considerable distance from their points of origin.

In humid climates, an excess of downward-percolating water removes minerals from the upper soil in a process called **leaching.** Leached minerals may be deposited in a deeper soil layer, but under very wet conditions many of them are carried completely out of the soil. The leaching process is the primary reason for the general lack of fertility of soils in humid climates. Fertilizers may replace leached or otherwise unavailable plant nutrients temporarily, but because these, too, are susceptible to leaching, periodic reapplications are necessary.

In arid climates, the drying of the surface soil causes capillary water to be drawn toward the surface. This water transports minerals dissolved from the parent material or lower soil upward to be deposited near the surface in a sort of reverse leaching process. Although the upper soil's supply of plant nutrients is usually increased by this process, some of the substances brought up are toxic to most plants. A prime example is provided by salts of various types. In many arid areas where water is available for agriculture, the high salinity of the soil makes the raising of crops impossible or impractical. The use of irrigation water for dry land agriculture has in many cases intensified the problem by dissolving large quantities of salts from the lower layers and carrying them to the surface. Approximately one-third of the irrigated farmland in the western United States has been damaged by increasing salinity, and some once-productive land has had to be abandoned. The problem of soil salinization resulting from irrigation is discussed in the Focus box on page 212. The surfaceward movement of water may also deposit substances that form a resistant **duricrust** layer on or just beneath the surface (see Figure 10.5). When well developed, a duricrust layer may resemble a buried layer of concrete; and under extreme conditions dynamite has been employed to break it up.

As a general rule, the most naturally fertile soils exist in subhumid or semiarid climates, where precipitation approximately equals evaporation. Under these conditions, vertical translocations of soil water and dissolved minerals in either direction are minimized.

Figure 10.5 A duricrust layer appears between the 1- and 2-foot levels in this aridisol from southern New Mexico. The horizontal structure of this layer contrasts markedly with the vertical prismatic structure of the overlying A horizon. *(Marbut Memorial Slide Collection, American Society of Agronomy)*

Organic Matter in the Soil

Organic matter is incorporated into most soils in the form of humus. **Humus** is finely divided, decomposed organic matter with a brownish black color. The remains of both plants and animals serve as a source of humus, but most humus has a vegetative origin, being largely derived from leaves, branches, grass, and roots. Humus forms from the slow oxidation of organic litter by bacterial action. It is a relatively stable residue remaining after the bulk of the decomposition has already occurred.

The amount of humus present in the soil is determined by the balance between the rates of humus formation and consumption. Because humus forms primarily from plant matter, significant quantities exist only in places where vegetation grows in abundance. In most cases, these are humid or subhumid forest or grassland environments. The plants and animals from which humus is derived live mostly near or above the soil surface, so the humus content of most soils is greatest in the near-surface layers. Excessive bacterial decay rates can cause humus to be consumed as rapidly as it is formed. As the temperature increases, the quantity and activity levels of bacteria also increase; and net humus production declines. For example, a forest soil in an area with an annual mean temperature of 40°F (5°C) typically contains about 8 percent humus, by volume. Under similar vegetative conditions and an annual mean temperature of 70°F (21°C), the humus content is likely to be only about 1 percent.

Humus has profound physical, chemical, and biological effects on the soil. It is a major source of plant nutrients and therefore of soil fertility. The spongy consistency of humus increases the capacity of the soil to hold water, along with its dissolved minerals. As discussed in the next section, humus also plays an essential role in the process of cation exchange, whereby plant nutrients are transferred to plants from the soil. The presence of humus aids the soil in developing and maintaining a structure favorable for plant growth and for cultivation. A high humus content also gives the soil a looseness and high aeration capacity that are favorable to vegetative growth.

Cation Exchange and Soil pH

The fertility of a soil is determined by its plant nutrient content and by the availability of these nutrients to plants. The nutrients are derived almost entirely from the weathering of parent material and from the decomposition of humus. By what mechanisms, though, are nutrients transferred from the soil to the plants? The answer to this question lies in the properties and activities of the soil *clay-humus complex*. This complex consists of colloidal combinations of tightly bound clay and humus particles. Soil colloids develop negative electrical charges and therefore become capable of attracting and holding positively charged ions (cations). The cations are formed by the separation or dissociation of molecules when they are dissolved in water—in this case, soil water. The substances ionized in this fashion are mostly bases such as calcium, magnesium, phosphorus, and potassium, which plants require and must be able to withdraw from the soil. The cations are electrically attracted to the negatively charged clay-humus colloids and are therefore kept from leaching away. Plants gain needed cations by exchanging other cations for them; the most commonly exchanged cations are hydrogen and aluminum (see Figure 10.6).

The ability of a soil to exchange cations is called its **cation exchange capacity** (CEC). The CEC of the soil is determined not only by its supply of exchangeable base cations, but also by the quantity of hydrogen ions (H + ions) that can be traded by the base cations. The quantity of available H + ions is measured by the pH scale, which indicates the logarithmic concentration of H + ions (see Figure 10.7). The pH scale is numbered so that a pH of 7, which means that each 1000 cm^3 of solution contains 10 grams of hydrogen ions, is chemically neutral. Each whole number below 7 means that there are 10 times as many H + ions, making the soil increasingly *acidic*. Each number above 7 means that there are only one-tenth as many H + ions as for the previous number, making the soil increasingly basic, or *alkaline*.

Soils range in pH between approximately 3 and 10. In general, a soil pH between 5 and 7 is best for most agricultural crops, although different soil flora and fauna have developed greatly different pH tolerances. A highly acidic soil contains an excess of H + ions and probably does not have enough exchangeable base cations. On the other hand, a highly alkaline soil is generally infertile because it lacks H + ions to exchange for bases and possibly contains toxic concentrations of sodium.

Soils of humid climates tend to be acidic due to an abundance of H + ions produced when water (H_2O) molecules dissociate, or break apart. In addition, rainwater is slightly acidic and has recently been increasing in acidity because of the widespread burning of fossil fuels by human beings. Arid and semiarid climates, conversely, generally have alkaline soils. Because most forests grow in humid environments, forest soils are usually somewhat acidic. The soils of grasslands and especially of desert environments tend to be alkaline.

Soil Horizons

Although soils vary greatly in different areas, one may have to travel a considerable distance before encountering major changes in soil type. Vertically, however, major differences in soil characteristics may exist over distances of only a few inches or, at most, a few feet. In the soil, as in the atmosphere, distinctive horizontal layers tend to form as a result of different environmental controls operating at differing levels. In the soil, these layers are termed horizons.

Soil horizons are layers, roughly paralleling the surface, that differ in their physical, chemical, and biological characteristics. Any of the soil characteristics, including color, texture, structure, consistency, physical and chemical maturity, numbers and types of organisms, humus content, and pH, may vary between horizons.

Soils develop horizons for two basic reasons. First, the soil is sensitive to its environment; and nearly all

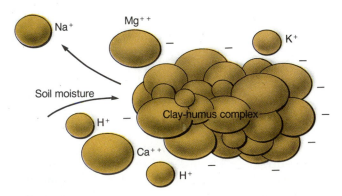

Figure 10.6 The cation exchange process. Plants gain needed base cations from the clay-humus colloids in the soil by exchanging other cations such as hydrogen and aluminum for them.

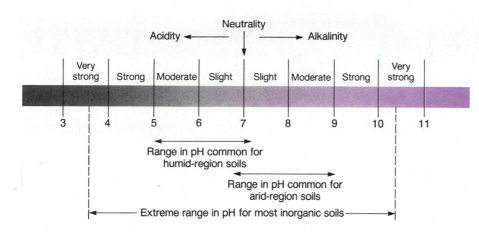

Figure 10.7 The pH scale is a logarithmic scale for measuring the degree of acidity or alkalinity of the soil.

soil-forming environmental controls vary rapidly in a vertical direction. Second, a vertical translocation of materials occurs in all soils because of the movements of soil water. As a result, selected materials are removed from some parts of the soil and accumulated in others. The removal of solid or dissolved materials from one horizon is called *eluviation,* while their deposition in another horizon is known as *illuviation.*

The soil horizons are identified by a lettering system (see Figure 10.8). From the surface downward, five major horizons, designated the O, A, E, B, and C

horizons, are recognized. In order to provide greater precision, transitional horizons are often identified. For example, the subhorizons EB and BE are sometimes recognized between the E and B horizons. The characteristics and extent of development of horizons vary greatly in differing pedogenic (soil-forming) environments, and some of the subhorizons, or even horizons, may not be present in a given soil sample.

The uppermost horizon, referred to as the O horizon, consists of fresh or partially decomposed organic matter. This horizon develops only where there is a significant

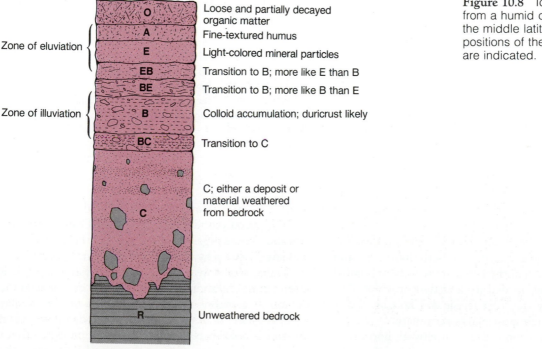

Figure 10.8 Idealized soil profile from a humid climate region in the middle latitudes. The relative positions of the various horizons are indicated.

Soil Salinization—A Growing Problem in Arid Lands

The soils in many desert regions are inherently fertile, and the use of irrigation water can make these regions "bloom," creating some of the most productive agricultural lands on earth. Irrigated cropland, which constitutes about 15 percent of all cultivated land, produces up to 30 percent of the world's food supplies. The extensive irrigation of California's Imperial Valley by water from the Colorado River, for example, has helped make that state the nation's leader in value of agricultural production. Unfortunately, the irrigation of arid lands often causes a gradual increase in soil salinity. This problem, unless controlled, can eventually ruin even the most productive agricultural lands. Approximately one-fourth of the world's irrigated land suffers from salinity to some extent, and that figure is increasing.

Salinization of soils caused by irrigation is probably the most ancient of all the factors promoting desertification (see the Case Study at the end of Chapter 16). In the fourth millennium B.C., the Sumerians occupied the southern Tigris-Euphrates Valley and built a prosperous civilization based on irrigation agriculture. By the second millennium B.C., this region, often referred to as one of the "cradles of civilization" in world history texts, was so badly salinized that agriculture was largely abandoned. Even today, many parts of this valley are semidesert wastelands.

The rapid spread of irrigation agriculture has greatly increased the geographical scope of the salinization problem during the past century. For example, extensive areas of Pakistan have been affected, as have large portions of the cotton-producing lands of the southwestern Soviet Union. Salinization is becoming a problem in Egypt's Nile Valley now that the Aswan High Dam has ended the annual flooding that once flushed the salts from the soil. Extensive portions of the American West also suffer from this problem, and some areas have already been abandoned (see Figure 10.9).

The Salinization Process

The problem of soil salinization exists because salts in large quantities are toxic to most crops. Common table salt ($NaCl$) is the worst offender. As it dissolves, the sodium cations further increase the pH of the already alkaline soil. Sodium attacks the soil structure, disintegrating the structural peds that provide channels for the percolation of moisture. Precipitated salts also clog the soil pores, making it impervious to air and water and producing a white crustal layer.

When fields in dry climate areas are flooded with irrigation water, the excess water percolates down to the water table. If natural ground water drainage is poor and no artificial drainage system is provided, the water table rises. The rising water carries with it dissolved salts from both the lower soil and from the irrigation water (which in arid regions is often slightly saline). As capillary action

continued on next page

cover of natural vegetation; it is therefore absent in arid and some tropical regions.

The second, or A horizon, is often referred to as the *topsoil.* It contains a mixture of both inorganic and organic matter (much of the latter in the form of humus) and is typically low in density and dark in color. The topsoil is generally the most fertile soil horizon and is the horizon in which most plants germinate.

Below the A horizon is the E, or eluvial, horizon. It is a mineral horizon that is lighter in color than the overlying horizons, and is best developed in high rainfall areas, especially under a forest cover. Clay-sized particles and minute organic substances have been washed or eluviated from it by downward percolating water.

The underlying B horizon, termed the *subsoil,* is often a zone of illuviation. It is typically denser than the A and E horizons and may develop zones of clay accumulation. In some cases, iron oxides give the B horizon a reddish coloration. Unless humus has been translocated downward from the A and E horizons, the

Figure 10.9 This abandoned, salt-encrusted farmland in California's San Joaquin Valley has been ruined by salinization resulting from the surfaceward movement of irrigation water. *(Howard Wilshire)*

causes the water to approach the surface, the water evaporates and the salts are left behind. Salinization and waterlogging increase in severity until further farming becomes uneconomical and the land is abandoned.

Possible Solutions

There are a number of existing and potential solutions to the salinization problem that provide a measure of optimism for the future. In areas where fresh water is abundant and the water table is not too high, the application of large quantities of water can flush the accumulated salts away. If the irrigation water itself contains a significant quantity of dissolved salts, it can be desalinated before use. Unfortunately, this is a costly process because of the large amounts of energy required.

Irrigated fields in some regions now employ "tile" drainage: a subsurface system of perforated plastic pipes that conducts away excess irrigation water and dissolved salts. This system is expensive to install, but is extensively used in California's Imperial Valley and in Uzbekistan Region of the southwestern Soviet Union.

The process of drip irrigation, pioneered by the Israelis in the 1960s, is currently considered one of the most promising methods of conserving water and combating salinity in arid lands. It employs a system of aboveground perforated plastic pipes that drip water into the soil directly beside each plant.

Potentially, the most promising solution of all involves the genetic engineering of crops to produce salt-tolerant plants. Irrigation water containing up to 3000 ppm (parts per million) of dissolved salts is now being routinely used in some areas for a wide variety of salt-resistant crops, including wheat, cotton, potatoes, and sugar beets. Some crops are now being irrigated with water containing up to 10,000 ppm dissolved salts—close to one-third the salinity of seawater. Researchers dream of developing strains of crops that can be irrigated with seawater itself, thereby assuring humanity of an unlimited supply of irrigation water. This dream may someday become a reality; already, salt-resistant varieties of barley have been grown with seawater irrigation under laboratory conditions.

B horizon contains little organic material. The A, E, and B horizons together comprise what is sometimes referred to as the "true soil," or *solum.*

The C horizon consists of weathered rock material that, as yet, has been little affected by pedogenic processes. This material is undergoing changes that will eventually convert it to soil, but has not yet crossed the threshold that separates soil from weathered rock material or regolith. Rapid weathering in the tropics can form a C horizon more than 150 feet (50 m) thick. In the middle latitudes, thicknesses of 3 to 10 feet (1–3 m) are typical, but much greater thicknesses occur in some places. Below the C horizon lies unweathered bedrock or other parent material, sometimes labeled the R horizon, that represents the raw material for future soil.

Considerable time is required for soil horizons to develop. For this reason, the presence of well-developed horizons indicates that the soil is mineralogically mature and that environmental conditions have remained relatively constant over a prolonged period.

Pedologists (soil scientists) study soils by examining **soil profiles,** which are vertical slices of the soil in which the various horizons are exposed. Each soil exhibits a different profile which can be used to identify it. In this respect, a soil profile is much like a fingerprint, no two of which are exactly alike. A person's fingerprint, though, tells little or nothing about the characteristics of its owner, whereas a soil profile can provide a great deal of information about a soil's physical and chemical characteristics. In many places, soil profiles have been destroyed by human activities such as plowing and the construction of buildings and highways.

A mature soil provides a good example of a system in *dynamic equilibrium* with its environment. The soil is dynamic because it undergoes a constant progression of soil-forming activities. Each soil mineral particle passes, in turn, through a sequence of horizons, beginning as unweathered R horizon material and eventually being eroded at the surface as A horizon material. On the other hand, the soil is capable of achieving and maintaining an equilibrium condition with its environment. If the environmental controls in a given area remain constant for a long enough period, the soil type will remain the same even though different material is constantly passing through the soil "assembly line" from initial weathering to erosion.

FACTORS IN SOIL DEVELOPMENT

To this point, the discussion has centered on the physical and chemical characteristics of soils. Although the reasons, or controls, for some of these characteristics have been alluded to briefly, this subject has not been covered in any systematic fashion. It is the geographic distribution of the pedogenic controls, however, that has determined the global pattern of world soil types. Before proceeding to a discussion of world soil types, then, the five major factors controlling the type of soil that will develop in any given locality need to be examined. These five factors can be summarized by the so-called "clorpt equation," written as follows:

$$\text{Soil} = f\,(\text{CL, O, R, P, T})$$

This equation indicates that the soil at any given site is a function of climate, organisms, relief, parent material, and time.

Parent Material

The first factor to influence a newly developing soil is its **parent material,** which may be defined as the material from which the soil develops. The parent material is the "raw material" of the soil, which undergoes progressive physical and chemical alterations as the soil-forming processes continue. As might be expected, the parent material has the greatest influence on recently formed soils. Its influence diminishes, but does not entirely disappear, as the soil matures. Pedogenic processes operating in differing environments can cause greatly differing soils to form from the same parent material.

Nearly all parent materials have formed from combinations of the most abundant elements in the earth's crust. Table 10.2 lists the eight elements that individually comprise 2 percent or more, by weight, of the crust. Differing parent materials vary greatly in the numbers and proportions of the elements they contain. In addition, some of these elements are essential for plants, while some are not. As a result, the type of parent material is an important factor in determining soil fertility. For example, soils formed from sandstone or quartzite are generally less fertile than those formed from limestone or basalt because the latter two rock types contain more of the minerals that provide important plant nutrients.

From the standpoint of origin, soil parent materials are often designated as either residual or transported. *Residual* parent materials consist of weathered or unweathered bedrock that remains in place before and during the pedogenic processes. *Transported* parent materials, which are geographically widespread, have been brought in from other sites through the actions of gravity, water, ice, or wind.

Relief

Topographic relief, like parent material, is a pedogenic control that can vary greatly over a very short distance, leading to a variety of localized soil types. Most differences in soil characteristics relating to slope steepness occur because of the slope angle's effects on soil erosion and drainage. Because erosion becomes increasingly effective as slope angles increase, soils on steep slopes tend to be thin, coarse textured, and poorly developed. Soil depths decrease as steepness increases, until, at slope angles of about 45° or more, loose material washes, slides, or falls off slopes as soon as it forms, leaving bare rock surfaces. Steep slope angles are largely responsible

Table 10.2
Abundant Elements of the Earth's Crust

Element	Percentage of Crust by Weight
Oxygen	46.6%
Silicon	27.7%
Aluminum	8.1%
Iron	5.0%
Calcium	3.6%
Sodium	2.8%
Potassium	2.6%
Magnesium	2.1%
Total	98.5%

The quantity and biomass of soil plants generally far exceed that of soil animals. By far the smallest and most numerous of the plants living in the soil are the *bacteria*. Under favorable conditions, several million of these tiny, single-celled plants can inhabit each cubic inch of soil. The bacteria, more than any other organisms, enable rock or other parent material to undergo the gradual transformation to soil. Some bacteria produce organic acids that directly attack soil parent material, breaking it down and releasing plant nutrients. Others decompose organic litter to form humus.

Higher vegetation plays several vital roles with respect to the soil. Trees, grass, and other large plants supply the bulk of the soil's humus content. The minerals released as these plants decompose constitute an important nutrient source for succeeding generations of plants as well as for other soil organisms. In addition, trees in particular are able to reach and recycle nutrients located deep within the soil, returning them to the surface (see Figure 10.10). Finally, plants perform the vital function of slowing runoff and holding the soil in place with their root systems, thus combating erosion. The accelerated erosion that often accompanies agricultural use of sloping land is principally caused by the removal of its protective cover of natural vegetation.

The faunal counterparts of the bacteria are the *protozoa*. These single-celled organisms are the most numerous representatives of the animal kingdom and, like the bacteria, can sometimes number in the millions for each cubic inch of soil. Protozoa feed on organic matter and hasten its decomposition.

Of the larger soil animals, the earthworm is probably the most important. Up to a million earthworms, with a total body weight exceeding 1000 pounds (450 kg), may inhabit an acre of soil under exceptionally favorable conditions. Earthworms ingest large quantities of soil, which is altered chemically as it passes through their digestive systems to be excreted as casts. The casts form a high-quality natural fertilizer. In addition, earthworms mix soil both vertically and horizontally, improving aeration and drainage.

for the stark, rocky surfaces of the mountains of much of the western United States, as opposed to the soil-covered and forested Appalachians. Globally, the best developed and most agriculturally important soils are located in low relief areas.

Slope aspect affects soil for the same reasons that it affects vegetation. North-facing slopes in the northern hemisphere and south-facing slopes in the southern hemisphere are inclined away from the sun. As a result, they develop under conditions of lower temperature and greater moisture (resulting from decreased evapotranspiration rates) than do slopes oriented toward the sun. East- and west-facing slopes, as well as all slopes in the tropics, are not so greatly influenced, because the directional orientation of sunlight is less constant in these settings.

Organisms

Organisms play an essential role in soil formation. The numerous soil flora and fauna release minerals from the parent material, supply organic matter, aid in the translocation and aeration of the soil, and help protect the soil from erosion. The types of soil organisms present have a vital influence on the soil's physical and chemical characteristics. In fact, for mature soils in many parts of the world, the predominant type of natural vegetation is considered the most important direct influence on soil characteristics. For this reason, the description of a soil as a "prairie soil" or "tundra soil" conveys to the pedologist a great deal of information about the soil's attributes.

Climate

While natural vegetation, in a direct sense, may be considered the most important single control in determining the characteristics of a mature soil, climate is certainly the most important control in an indirect or ultimate

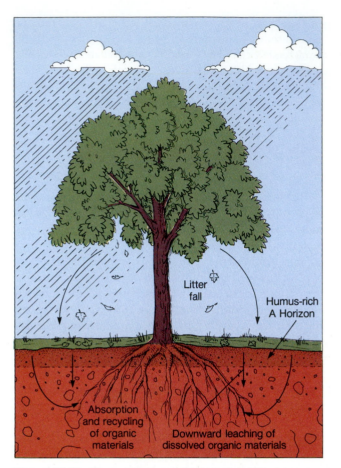

Figure 10.10 Trees help to maintain soil fertility by drawing up nutrients from the deeper soil layers through their root systems. These materials are subsequently added to the surface soil through the fall of organic litter.

available to support biological activity and humus formation. Conversely, an excess of precipitation may lead to conditions of poor drainage, low aeration, leaching, and excessive erosion. An intermediate amount of precipitation therefore is most favorable for the formation of fertile soils. As a general rule, the most fertile soils are located in areas where precipitation amounts roughly equal evapotranspiration rates. When this balance occurs, no large-scale translocations of soil water and minerals take place; and neither significant leaching nor duricrust formation occurs. Table 10.3 summarizes the differences in soil characteristics between excessively wet and dry environments. It should be kept in mind that the most productive soils develop under moisture conditions approximately midway between those represented in the table.

In the absence of inhibiting factors such as insufficient moisture, the rate of soil formation is positively correlated with the mean temperature of an area. The tropics therefore contain the deepest soils, and soil depths generally decrease poleward. In fact, soil formation rates in the humid tropics average about 9 or 10 times those of the subpolar regions because the rate of chemical weathering is greatly accelerated by warm temperatures and ceases entirely when the soil is frozen. The soils of warm and humid environments also tend to display an increased rate of differentiation from the characteristics of their parent materials. This is because they pass through a greater variety of chemical weathering reactions in a shorter time span than do the soils of colder climates.

Another important difference between the soils of warm and cold environments is that soil microorganisms are much more numerous and active in warm environments. This factor, in turn, affects the rate of humus formation and consumption. In cold, heavily vegetated regions such as the taiga, humus forms relatively slowly. Once formed, though, it may remain for many years before being consumed by microorganisms. As a result, humus in the high latitudes can accumulate to excess, leading to highly acidic soil conditions. Within the tropics, conversely, organic materials on or in the soil are rapidly consumed by microorganisms, producing soils deficient in humus. Once again, intermediate temperatures produce soil conditions that are generally most favorable for agriculture. As at least a partial consequence of this fact, the most fertile soils are largely concentrated in the middle latitudes. Table 10.4 summarizes the major characteristics of soils developing in warm and cold environments.

sense. This is because climate is the major controlling factor for the world distribution of natural vegetation. Indeed, climate has at least some effect on all of the soil development factors discussed in this section. The same two climatic elements that most strongly affect human activities—temperature and precipitation—exert the greatest effect on the soil. Extremes of either element have an unfavorable effect on soil characteristics.

Soil formation begins when minerals in the parent material combine chemically with water from precipitation. If this water is not available, as in an arid region, most physical and chemical weathering reactions cannot occur; and a mature soil will not form. In addition, a lack of precipitation means that little moisture will be

Table 10.3
Differences between Soils in Wet and Dry Environments

Wet	Dry
1. Deep (unless formation inhibited by standing water)	1. Shallow
2. Intermediate to fine texture	2. Coarse texture, stony
3. Chemical weathering predominant	3. Physical weathering predominant
4. Leached of soluble minerals	4. Near-surface accumulation of soluble minerals
5. Acidic condition	5. Alkaline condition
6. Variable humus content (depending on temperature and abundance of vegetation)	6. Low humus content due to lack of vegetation

Time

Each of the four soil formation factors just discussed needs time to influence the soil. Therefore, a fifth and final essential factor determining the characteristics of the soil in any locality is the amount of time that has elapsed since soil formation began.

The amount of time needed for soil formation and maturation varies geographically, making it impossible to cite precise figures that will hold true everywhere. In general, soil development is fastest in areas where the climate is warm and moist; soil microorganisms are abundant, and the soil parent material is composed primarily of chemically unstable minerals that weather quickly. Specifically, average rates of development are most rapid in the humid tropics and slowest in polar and arid environments.

A rough approximation of 1 inch (2.5 cm) per century as a world average rate of soil formation is sometimes given, but actual rates vary so greatly in different areas that such a figure is of limited value. Even at a particular site, rates of soil formation and development do not remain constant. A recently formed soil is not in equilibrium with many of the soil-forming factors in its environment. Change therefore proceeds at a relatively rapid pace. As the soil matures, it gradually approaches a condition of equilibrium with its environment, and the pace of development slows. Eventually, a steady-state condition may be reached, in which soil characteristics remain essentially unchanged with time. This condition of equilibrium can be attained only if all factors affecting the soil remain constant for a prolonged period. In the face of major recent geologic, climatic, and especially human influences, few places on earth currently have the necessary degree of environmental stability for this situation to exist.

SOIL CLASSIFICATION AND WORLD SOIL TYPES

The geographic pattern of world soil types is highly complex. As is true of all natural earth phenomena, however, the pattern is not random; it displays an organization that is closely related to the distribution of the pedogenic factors just discussed. Because the two most important pedogenic controls are climate and natural vegetation, the pattern of world soil types is more closely correlated with these factors than with any other pedogenic influences. It should be stressed, though, that each of the five factors in soil development is capable, under certain conditions, of exerting a dominating influence on local soil characteristics and that each will exert at least some influence on the soil at any given site.

Table 10.4
Differences between Soils in Warm and Cold Environments

Warm	Cold
1. Deep	1. Shallow
2. Fine textured	2. Coarse textured
3. Chemical weathering predominant	3. Physical weathering predominant
4. Rapid rate of soil formation	4. Slow rate of soil formation
5. High degree of mineralogic maturity	5. Low degree of mineralogic maturity
6. Soil microorganisms abundant	6. Soil microorganisms sparse
7. Humus lacking	7. Humus abundant
8. Less acidic	8. More acidic

Soil Classification Systems

Differences in soil types have been recognized from antiquity. For example, there is evidence that the Chinese divided soils into different types over 4000 years ago. Early soil classification systems were primarily descriptive and differentiated soils on the basis of physical characteristics such as color, structure, and depth. Russian soil scientists first developed modern concepts of soil formation and classification in the late 1800s. They realized that the soil is not a static medium produced merely by the weathering of parent material, but that it evolves through time as the result of energy and material gains and losses from climate and biological activity. Emphasis on soil classification therefore shifted from parent material to the process of soil development and evolution, as determined by the analysis of soil profiles. Further work, especially by pedologists in the U.S. Soil Survey of the Department of Agriculture, led to the publication in 1938 of a genetic (formation-oriented) soil classification system that was in official use in the United States for 27 years.

The 7th Approximation System

During the 1950s, growing dissatisfaction with the Russian-American genetic system led the U.S. Soil Conservation Service to devise a completely new soil classification system for official use. This system, called the "U.S. Comprehensive Soil Classification System," was first published in 1960 and was adopted for official use in the United States in 1965. Often referred to as the *"7th Approximation,"* this system has been expanded and refined and remains in current use. It is the system used in the inventory of world soil types appearing later in this section. It should be noted that although the 7th Approximation system is used by a number of countries, other countries, including Canada, have their own national soil classification systems.

The major characteristics of the 7th Approximation system are as follows:

1. Soils are classified on the basis of their present characteristics rather than by their method of formation. It is thus a descriptive system rather than a genetic system.
2. Soils modified by human activities are classified along with "natural" soils.
3. The names of soil types convey exact information concerning their characteristics. Each syllable has a

meaning, so the name describes the soil in almost the manner of a chemical formula.

The system is organized into a hierarchy of six levels of classification, with individual soil types forming a seventh level, as follows:

> Orders
> > Suborders
> > > Great groups
> > > > Families
> > > > > Series
> > > > > > Individual soil types

Ten soil orders exist in the system. The number of entries increases at each lower level, until, at the Series level, 14,000 kinds are recognized in the United States alone. Several times as many Series exist for the entire world, but they are as yet incompletely classified. The chief characteristics of the ten soil orders are described in the next section. Their geographical distribution is mapped in Figure 10.11. An examination of this map will reveal that some of the soil orders are much more geographically extensive and important than others.

World Soil Types
Entisols

The **entisols** (*ent* "recent") are poorly developed soils with little or no profile development (see Figure 10.12). As their name implies, most have formed very recently.

Entisols develop in a number of highly dissimilar and often localized environments; consequently, they vary greatly in their physical and chemical characteristics. Their most widespread areas of occurrence are in sandy deserts and in areas of regolith in dry environments. They also exist in tundra areas, in areas that have been recently glaciated, and in areas recently covered by alluvium (deposits of river sediments) or volcanic ash. Soils at construction sites as well as other sites at which the soil has been extensively eroded or disturbed by human activities are also classified as entisols.

Inceptisols

The **inceptisols** (*incept* "inception") are immature soils that have some profile development, but contain weatherable minerals. Although recently formed, they are generally somewhat more mature and better developed than the entisols. They show little indication of illuvial layers or of the vertical translocation of materials (see Figure 10.13).

The inceptisols are found largely in three types of environmental settings, each of which lacks the physical stability for the development of a mature soil. The first is in moderately steep mountain areas, such as the Appalachians, where erosion and landslides are relatively common. The second is in tundra areas, where soil development is inhibited by cold temperatures, poor drainage, and permanently frozen subsoils (permafrost). The third setting, and the most important from a human standpoint, is in floodplains or deltas. River-deposited soils, or *alluvial soils,* typically have a loamy texture, a high humus content, and a high cation exchange capacity. The alluvial soils of large floodplains, such as those of the Mississippi, Nile, and Ganges Rivers, are intensively farmed and highly productive.

Aridisols

The **aridisols** are soils lacking in available water for extended periods. They occupy large portions of the low and middle latitudes that receive less than about 10 inches (25 cm) of liquid precipitation annually. In all, they cover an estimated 17 percent of the earth's land surface, largely within the subtropical deserts.

Chemical weathering processes are considerably hampered by a lack of water in arid regions. As a result, the aridisols are shallow, stony, and mineralogically immature, with poorly developed horizons (see Figure 10.14). Soil textures are generally coarse and sandy, leading to a low water-retention ability even when water is available. The humus content is low to completely absent because of the sparseness of vegetation.

An important characteristic of the aridisols is their high alkalinity. Evapotranspiration exceeds precipitation, producing a surfaceward movement of ground water and dissolved minerals. Well-drained soils typically experience an accumulation of calcium carbonate and other soluble bases at the site of water evaporation, normally a few inches below the surface. Frequently, this produces a duricrust layer. In poorly drained depressions, the salinization process results in conditions toxic for most vegetation.

Despite their shortcomings, aridisols are often potentially fertile and have proven highly productive under certain conditions. The chief problem is a lack of water. Irrigation is essential not only to support plant growth, but often initially to flush excess salts from the soil. Most agricultural activity occurs in enclosed basins where water is available, the surface is flat, and soils are deep enough for farming.

Mollisols

The **mollisols** (*molli* "soft") are soils that contain a thick, dark-colored, humus-rich A horizon and that have a soft consistency when dry (see Figure 10.15). They exhibit a high cation exchange capacity and a B horizon accumulation of calcium carbonate ($CaCO_3$). Mollisols are associated chiefly with the prairie and steppe grasslands of the middle latitudes. They form in regions that have a rough balance between precipitation and evapotranspiration. As a result, there is little translocation of soil materials and no significant leaching or accumulation of soluble minerals.

Climatic and vegetative conditions combine to allow the mollisols to maintain a high humus content. Water from winter snowmelt and spring rains promotes rapid early luxuriant grass growth. Drier, hotter conditions in middle and late summer as well as cold winter temperatures greatly reduce vegetative decomposition during these periods, allowing humus to accumulate. The soils of the tall grass prairies have the highest humus content, which is generally well mixed throughout a thick, black-colored A horizon by the upper soil fauna. In the drier steppe areas, the humus content is reduced by the smaller quantity of available organic matter, and the soils assume a lighter brown color. Soil depths also decrease toward drier areas because less water is available for chemical weathering reactions.

Soil pH values are conducive to the growth of a wide variety of vegetation. The better-watered areas are almost neutral or even slightly acidic, but pH values increase with dryness, producing soils that are moderately alkaline and causing a subsoil concentration of calcium carbonate to develop (see Figure 10.2 left side).

The mollisols have a high degree of natural fertility and are associated with some of the world's most productive agricultural regions. The better-watered areas support major grain-producing economies. The American Midwest, South American Pampas, Soviet Ukraine, and a part of Chinese Manchuria, which are considered the major "breadbasket" regions of the world, produce huge surpluses of corn, wheat, and oats. The drier steppe areas generally receive too little precipitation to support intense cash crop economies and are used primarily for livestock raising.

Figure 10.11 Global distribution of the 10 soil orders of the U.S. Comprehensive Soil Classification System (7th Approximation).

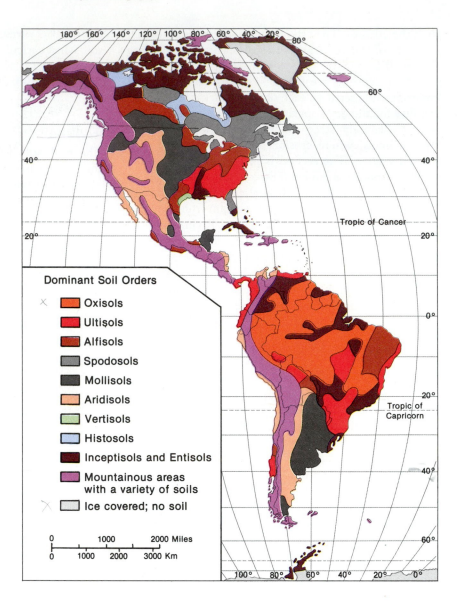

Dominant Soil Orders

- Oxisols
- Ultisols
- Alfisols
- Spodosols
- Mollisols
- Aridisols
- Vertisols
- Histosols
- Inceptisols and Entisols
- Mountainous areas with a variety of soils
- Ice covered; no soil

Spodosols

The **spodosols** (*spodos* "wood ash") are acidic soils with a light-colored eluvial E horizon and an illuvial B horizon (see Figure 10.16). Spodosols generally develop from a quartz-rich and often sandy parent material in a cool, humid climate. They usually occur in association with a needleleaf evergreen forest cover; the most extensive areas containing this soil type are occupied by the Northern Coniferous forests of North America and Eurasia.

The acidity of the spodosols results from the presence of abundant organic matter in the A horizon and a surplus moisture content. The soil pH is especially low in areas of needleleaf forest because these trees have low nutrient requirements and do not cycle many bases from the subsoil. The leaching of soluble constituents is promoted by the abundant precipitation and by the acidity of the soil. Soil acidity restricts the removal of silica in solution while promoting the leaching of iron and aluminum compounds. As a result, silica is the primary mineral constituent in the E horizon, giving it a

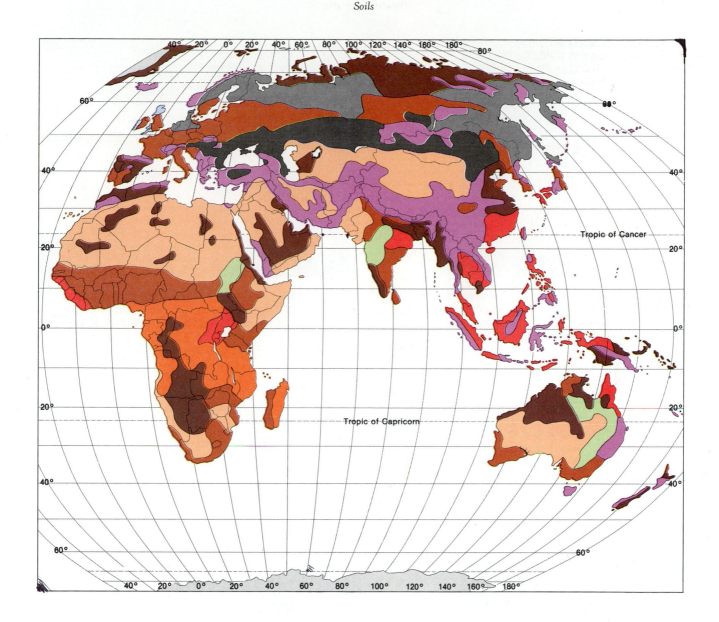

pale gray, ashy appearance. This characteristic led to the name given this soil type. Dissolved minerals and clay-sized particles, as well as some humus, are deposited in the B horizon. The B horizon is therefore an illuvial horizon that is denser, finer in texture, and higher in soluble bases than the E horizon.

The combination of low natural fertility, high acidity, a short growing season, and sometimes a lack of soil depth presents serious drawbacks to agricultural development in regions containing spodosols. These regions are most noted for the production of root crops, such as potatoes, that have modest soil nutrient requirements and a tolerance for acidic conditions.

Alfisols

The **alfisols** (*alf* "aluminum and iron") are fine-textured soils with a yellowish-brown A horizon that has been colored by iron and aluminum compounds (hence the name given this soil type) and that is low in humus. They have a clay-rich B horizon with a moderately high cation exchange capacity (see Figure 10.17).

Figure 10.12a This southwestern Wisconsin entisol developed on a river floodplain and contains many horizons. The development of each horizon was interrupted by an overlying deposit of fresh alluvium during a flood. *(Marbut Memorial Slide Collection American Society of Agronomy)*

Figure 10.12b Sea oats and beach grass partially cover an entisol formed from wave-deposited sand at Cape Hatteras, North Carolina. *(Breck P. Kent, © JLM Visuals)*

The alfisols are well-developed soils that are widely distributed throughout the low and middle latitudes. They tend to develop in areas that contain clay-forming parent materials and that have a season in which evapotranspiration exceeds precipitation. They are therefore especially typical of forest/grassland transitional environments. In the United States, they occur mostly in the southern Great Lakes, the Ohio Valley, Texas, and the Mississippi River Valley. Other middle latitude sites include eastern Europe, the central Asiatic Soviet Union along the taiga/steppe transition zone, and southern Australia. Within the tropics, these soils are especially common in southern and eastern Africa and in the African Sahel, as well as in large areas of India and Pakistan.

Their widespread geographical distribution and moderate-to-high natural fertility allow the alfisols to rival the mollisols in worldwide agricultural potential.

Figure 10.13a A New Zealand inceptisol that has developed on loess deposits. *(Soil Taxonomy Alt 436, Soil Conservation Service)*

Figure 10.13b This glacially derived bog in eastern Wisconsin is a typical site for inceptisol development. *(Carl Harns, © JLM Visuals)*

Figure 10.14a A stony Utah aridisol is covered by semidesert vegetation dominated by sagebrush. *(Marbut Memorial Slide Collection, American Society of Agronomy)*

Figure 10.14b A sparsely vegetated aridisol in the Atacama Desert of Peru. *(Breck P. Kent, © JLM Visuals)*

The wide variety of crops raised in these soils constitute a significant proportion of total world food production.

Ultisols

The **ultisols** (*ult* "ultimate") are highly chemically weathered soils that have developed under warm temperatures and relatively abundant moisture. They are somewhat similar to the alfisols, but display a higher degree of mineral alteration and are more thoroughly leached of bases.

Ultisols have a distinctly reddish A horizon that has been colored by iron and aluminum compounds and is low in humus because of a rapid rate of bacterial decomposition. The illuvial B horizon has a considerable accumulation of clay minerals, but is composed of highly weathered minerals with a relatively low cation exchange capacity (see Figure 10.18).

Figure 10.15a This mollisol, developed on glacial drift in central Iowa, has a thick, humus-rich A horizon and a granular structure. *(Marbut Memorial Slide Collection, American Society of Agronomy)*

Figure 10.15b Recently plowed mollisol near Macomb, Illinois. This soil, derived from weathered glacial till and developed under a prairie grass cover, is among the most naturally fertile in the world. *(Richard Jacobs, © JLM Visuals)*

Figure 10.16a This spodosol, developed on sandy glacial outwash in northern New York, has a well-defined whitish silica-rich E horizon. *(Soil Taxonomy Alt 436, Soil Conservation Service)*

Figure 10.16b Spodosol supporting a spruce forest in Quebec, Canada. *(Breck P. Kent, © JLM Visuals)*

Most ultisols have developed under a broadleaf forest cover where the trees maintain some soil fertility by recycling soluble bases from the B horizon to the A horizon. The clearing of the forests in many areas, though, has led to the rapid leaching of these plant nutrients. The fine texture of the soil also inhibits water infiltration and has resulted in severe erosion in many areas cleared for agriculture.

The ultisols are especially associated with Humid Subtropical Climate regions, but also occur in portions of the humid tropics. They are the dominant soil type of the southeastern United States, where in many areas

Figure 10.17a An alfisol developed under an oak-hickory forest in southern Michigan. *(Marbut Memorial Slide Collection, American Society of Agronomy)*

Figure 10.17b This rolling farmland in the Finger Lakes district of upstate New York contains alfisols that are used for the raising of corn, small grains, and forage crops. *(Marbut Memorial Slide Collection, American Society of Agronomy)*

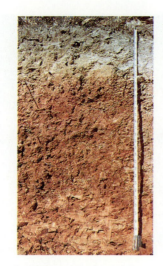

Figure 10.18a An ultisol developed on metamorphosed sandstone in the Ouachita Mountains of western Arkansas. *(Marbut Memorial Slide Collection, American Society of Agronomy)*

Figure 10.18b A typical view of rural farm and forest land in a portion of the North Georgia piedmont containing ultisols. *(Marbut Memorial Slide Collection, American Society of Agronomy)*

they have suffered from nutrient depletion and erosion resulting from past excessive cotton cultivation. They are also common in much of China, Uruguay, southern Brazil, the northern Andes, West Africa, northeastern Australia, and parts of the East Indies.

Oxisols

The **oxisols** (*oxi* "oxide") are the deepest and most chemically altered of the soil types. Almost all minerals other than quartz have been weathered to form oxides and hydroxides within the oxisols, and virtually all soluble bases have been leached from the upper soil layers. These soils occur on ancient, geologically stable surfaces within the humid tropics, especially in South America and Africa.

The intense leaching that is characteristic of these soils results from heavy precipitation, which at least seasonally greatly exceeds evapotranspiration. Not only dissolved soil constituents are transported downward; even clay-sized particles are frequently removed as they are flushed through the pore spaces separating the larger soil grains. Much of the leached soil material is carried completely through the soil profile and is permanently lost to the soil. Because there is no soil layer in which any substantial illuviation takes place, soil horizons are weakly developed and difficult to recognize (see Figure 10.19).

Intensely leached soils are usually highly acidic, but the acidity of the oxisols is kept in check by their lack of humus and its associated organic acids. An only slightly acidic pH, usually in the 5 to 7 range, results. At this acidity level, silica, which is usually a highly stable and inert soil component, enters into weathering reactions that result in its solution. Large amounts of silica are therefore leached from the soil. Iron and aluminum, conversely, weather to form insoluble oxides that accumulate in the soil. They typically impart an orange-red color that is one of the most striking characteristics of the soils of the humid tropics.

The low natural fertility of the oxisols presents imposing problems for agricultural use. Despite the presence of year-round warm temperatures and often abundant moisture, the lack of humus and base nutrients in these soils necessitates the use of large amounts of fertilizer for any sustained crop raising. The removal of the trees from forested areas often causes a further deterioration of soil quality. Trees draw up a modest supply of soluble bases, which are placed on the surface as the plants die and decompose. A nutrient-cycling system is thus established, and a limited amount of soil fertility is maintained. Removal of the trees causes these bases to be leached, increases the likelihood of erosion, and in some places encourages the formation of a rock-hard subsurface layer of iron oxide called *laterite*. The current large-scale removal of tropical forests,

Figure 10.19a An oxisol developed from weathered basalt on the island of Kauai, Hawaii. *(Marbut Memorial Slide Collection, American Society of Agronomy)*

Figure 10.19b A roadcut through an oxisol on the island of Fiji displays the depth as well as the rust-red color typical of this highly weathered soil type. *(Rodman Snead, © JLM Visuals)*

discussed in the Case Study at the end of Chapter 9, is causing a general deterioration of the soils over large areas.

Vertisols

The **vertisols** (*vert* "invert") contain large amounts of montmorillonite clay that swells when wetted and shrinks when it dries. These soils are located in areas that experience alternating (often seasonal) periods of moisture surplus and deficiency. During dry periods, vertisols shrink and develop cracks that may be as much as an inch (2.5 cm) wide and 30 inches (75 cm) deep. Surface soil crumbles and is blown or washed into the cracks. When moistened, the soil swells, closing the cracks and covering the soil that has entered the cracks from above. The vertisols thus possess the unique ability to mix themselves and even, to some extent, to turn themselves "inside out." As a result of this natural churning action, they are almost lacking in horizons. Vertisols generally develop in semiarid regions under a grass cover, have a moderately high cation exchange capacity and organic matter content, and form fertile and agriculturally productive soils (see Figure 10.20).

The vertisols are relatively restricted in geographical extent. Most are located in areas with a Tropical Wet and Dry Climate in western India, eastern Australia, and parts of northeastern Africa. They occur to a lesser degree in portions of the southern and western United States, notably in Texas and California.

Histosols

The **histosols** (*histos* "tissue") are dark-colored organic soils of poorly drained areas. They consist primarily of decomposed vegetation, often forming a peat or muck. A lack of oxygen inhibits organic decay, so these soils can attain considerable depth through the continued addition of vegetative matter at their surface. Most histosols are highly acidic and low in fertility (see Figure 10.21).

The histosols are the least extensive and least important of the 10 soil orders. They can occur in almost any non-arid environment where drainage is poor, but are most common in the tundra regions of North America and Eurasia, where the underlying permafrost prevents adequate drainage. They are also found in lowland areas to the south that were once covered by continental glaciers. Glacial erosion and deposition in these areas disrupted the previously existing drainage systems. Significant sites of histosol formation in the coterminous United States include the Florida Everglades and the Mississippi River Delta in Louisiana.

The artificial drainage of histosol areas leads to improved aeration and often to a major temporary improvement in their fertility and agricultural potential. Oxidation of the organic matter, though, eventually leads to compaction and subsidence and to fire danger in dry periods. In recent years, smoky fires in the Everglades, resulting from a decline in the region's water table, have burned over extensive areas and have caused severe air pollution problems in coastal cities.

Figure 10.20a A vertisol consisting largely of montmorillonite clay in the Lajas Valley of Puerto Rico. *(Marbut Memorial Slide Collection, American Society of Agronomy)*

Figure 10.20b A South Dakota vertisol. This area displays gilgai relief: the ridges, composed of calcium-rich soil, alternate with depressions underlain by soil low in calcium. *(Marbut Memorial Slide Collection, American Society of Agronomy)*

Summary

The soil is the loose mixture of weathered rock material, organic matter, air, and water that mantles most of the land surface of the earth. It is the medium in which most plants and many animals live, and it provides the water and nutrients they need in order to survive.

Soils are highly diverse in their physical and chemical characteristics, and the type of soil that exists at any given site is determined by the interplay of a variety of environmental controls. The soil characteristics of greatest importance to vegetation include texture, structure, organic matter content, and base saturation. The soil's texture is determined by its mix of sand, silt, and

Figure 10.21a A histosol occupying a poorly drained depression in southern Michigan. *(Marbut Memorial Slide Collection, American Society of Agronomy)*

Figure 10.21b This swampy area in the Mississippi River delta near Houma, Louisiana, is a prime site for histosol formation. *(Richard Jacobs, © JLM Visuals)*

clay-sized particles. Texture especially affects the soil's density and ease of cultivation, as well as its ability to absorb and retain water. The soil structure is produced by the tendency of individual soil particles to adhere to one another, forming peds of distinctive sizes and shapes in differing environments. The structure notably influences the soil's ability to absorb water and to resist erosion. Soil organic matter supplies plant nutrients, plays an essential role in the transfer of the nutrients to plants, and improves the soil's water-retention ability. The base saturation of the soil relates to its ability to exchange base cations that are essential for plants. It is therefore a quantifiable measure of soil fertility. The soil's base cation exchange capacity (CEC) is influenced by a variety of factors, including its humus content and its pH.

The five factors in soil development collectively determine the type of soil that will develop at a given site. An unlimited number of combinations of these five factors is possible. The first factor to influence the soil is its parent material, which may consist of bedrock, sediment, organic matter, or a combination of these materials. A second factor is the topographic relief of the land. Steep slopes are usually adversely affected by erosion, and their soil cover is typically shallow, coarse textured, and often relatively infertile. A third factor is biological organisms. Microscopic plants and animals, notably bacteria and protozoa, are crucial in decomposing soil organic matter to form humus. Larger plants and animals help to loosen and aerate the soil and add organic matter. A fourth factor, the climate, plays both a direct and an indirect role in soil characteristics because of its influence on vegetation. The two climatic elements with the greatest direct influence are moisture and temperature, and moderate conditions with respect to each are generally associated with the most fertile soils. Excessively moist conditions promote poor aeration and leaching, while dry conditions are associated with shallow and immature soils, alkaline conditions, and a lack of humus. Soils developing in high-temperature regions often lack organic matter, while low-temperature regions often have shallow, acidic soils. The final factor in soil development is time. All soils need time to develop, and all gradually evolve with time, so a variety of different soils will eventually form from the same parent material in diverse environments.

Pedologists have long been concerned with developing ways of effectively classifying the great variety of existing soils. The so-called 7th Approximation, which divides all soils into 10 major soil orders, is currently in official use in the United States and was used in this chapter.

Two of the soil orders, the entisols and inceptisols, are immature soils with poorly developed profiles. They have developed in areas where pedogenic processes are disturbed by natural or human-induced erosional or depositional processes. The aridisols develop in arid regions and are usually shallow, mineralogically immature, and alkaline. The mollisols typically develop in the middle latitude grassland regions. They have a high cation exchange capacity and a thick A horizon that is rich in humus. They are often considered the most inherently fertile of the 10 soil orders. The spodosols are acidic soils with a silica-rich eluvial E horizon and an illuvial B horizon. They are relatively low in fertility and typically develop under a needleleaf forest cover in the higher latitudes. The alfisols are mature soils that are fine textured and relatively low in humus and that have a high aluminum and iron content. They are usually fertile, are typical of forest/grassland transitional areas, and are widespread in the United States. The ultisols are reddish soils that are highly chemically weathered, have a high aluminum and iron content and limited natural fertility, and are typically associated with humid subtropical climatic conditions. The oxisols are the deepest and most chemically weathered of the soil types. They are especially associated with the humid tropics. Intense leaching has removed most of their plant nutrients and has left concentrations of insoluble oxides of iron and aluminum that give these soils a distinctive red color. The vertisols are clay-rich soils that shrink and crack when dry, and swell when wet. Vertisols are of limited global extent, but most are associated with grassland areas. Histosols are composed largely of organic matter that has accumulated in swampy areas. In their natural state, most are highly acidic and infertile; but they can be productive when drained and treated for acidity.

Review Questions

1. From what four constituents is the soil formed? What proportion of each comprises a typical soil? What alterations must the soil's two solid constituents undergo before they can be incorporated into the soil?

2. What is soil texture? What practical effects does it have on the soil?

3. What is soil structure? What relationship exists

between texture and structure? What practical influences does structure have on the soil?

4. What is humus? What favorable effects does humus have on the soil? What environmental conditions maximize humus formation?

5. Why do most soils develop horizons? In what major respects does the A horizon differ from the B horizon?

6. What are the basic differences in the characteristics of soils that have developed in wet versus dry environments? In warm versus cold environments?

7. The entisols, inceptisols, aridisols, and histosols are all considered as poorly or incompletely developed soils. Explain how the environmental settings of each of these soil types contribute to its lack of development.

8. Which of the 10 major soil types is generally considered to have the highest natural fertility? What characteristics of the soil produce this fertility?

9. Which of the 10 major soil types is the deepest? Why? Explain what physical and chemical

characteristics limit its agricultural potential and what environmental factors have resulted in the development of these characteristics.

10. Which of the 10 soil types dominates in your local area? Describe the soil's basic characteristics and agricultural potential in a short paragraph.

Key Terms

Soil	Parent material
Soil texture	7th Approximation
Loam	Entisol
Ped	Inceptisol
Soil structure	Aridisol
Translocation	Mollisol
Leaching	Spodosol
Duricrust	Alfisol
Humus	Ultisol
Cation exchange	Oxisol
capacity	Vertisol
Soil horizons	Histosol
Soil profile	

CASE STUDY

American Agriculture and Accelerated Soil Erosion

The United States has been endowed with vast expanses of fertile soils. The presence of these soils, coupled with the enterprise of the American farmer and the availability of modern technology, has allowed American agricultural output to far surpass that of any other country. The successes of American agriculture have not been without cost, however. The greatest problem—one that raises serious questions about the continued expansion of our agricultural production—is the alarming rate of soil erosion from American agricultural land (see Figure 10.22).

Soil erosion is a natural and unavoidable phenomenon that becomes detrimental when it occurs more rapidly than new soil can form. Such an unbalanced situation results in a net loss of soil and, generally, of agricultural potential. Most human activities that change the landscape disrupt the delicate natural balance between soil formation and removal, and greatly increase the rate of erosion.

By far the leading cause of accelerated erosion throughout the world is the clearing of land for agriculture. Removal of the protective vegetation cover exposes the soil to rapid erosion by water in wet periods and by wind in dry periods. Both agents of erosion remove topsoil, which in most cases represents the most fertile segment of

the soil. In the United States, the estimated average rate of soil formation is one inch (2.5 cm) per century, or approximately 1.5 tons (1360 kg) per acre per year. Estimates of the annual loss of topsoil from agricultural land in the United States range from 6 to 14 tons (5500-12,700 kg) per acre per year. Our farmland is therefore being eroded at several times the rate at which it is being formed. As a result, the amount of sediment carried by rivers flowing into the Atlantic is estimated to be 4 or 5 times greater than before European settlement.

Most intensive land-use practices increase the susceptibility of the soil to erosion. Unfortunately, farming is a highly intensive land-use practice. At present, approximately 20 percent of the land area of the United States is under cultivation. Yet, about 75 percent of the 5 billion tons (4.5×10^{12} kg) of soil lost annually to water and wind erosion in the 48 contiguous states is derived from agricultural land. During the past two centuries, at least one-third, and perhaps as much as one-half, of the topsoil on U.S. farmland has been lost. For every ton of grain produced in the fertile Mississippi River Valley, 4 tons of soil are eroded from the land. Of the 4 billion tons (3.6×10^{12} kg) of American topsoil removed annually by water erosion, 1 billion tons (0.9×10^{12} kg) is carried to the sea, where it is irretrievably lost. The remaining 3 billion tons (2.7×10^{12} kg) settles in lakes, reservoirs, and river floodplains, where it causes serious sedimentation problems.

Sedimentation can be viewed as the "other side of the coin" in the soil

Figure 10.22 Rill and incipient gully erosion on agricultural land following a rainstorm in the central Coast Ranges of California. *(J.K. Nakata: Sight & Sound Productions © 1990)*

erosion problem. In contrast to erosion, where the effects are often gradual and sometimes unnoticeable over the short term, the results of greatly accelerated sedimentation are often painfully apparent. For example, river channels can be raised so greatly by deposition that they become much more susceptible to flooding. In addition, the rivers frequently develop wide, shallow, and constantly changing braided flow patterns, instead of stable single-channel patterns. Reservoirs are filled so rapidly by siltation

continued on next page

that their useful life spans are often reduced to only a few decades. Harbors and river channels must be dredged constantly in order to permit continued use by shipping.

Numerous factors affect the severity of soil erosion. The most important is slope steepness, while others include length of slope; soil texture, structure, and organic content; cultivation practices; types of crops grown; and climatic characteristics, especially the frequency of heavy rains and strong winds. The U.S. Department of Agriculture estimates that approximately 70 percent of American farmland has experienced moderate-to-severe erosion, while 30 percent—usually on very flat surfaces—has experienced little or no erosion. Figure 10.23 shows the pattern of soil erosion in the United States resulting from the combined actions of water and wind. It can be seen that the most severe problems are occurring in the Southeast, the western Midwest, the Great Plains, and parts of the Southwest. These areas include much of the most intensively farmed and agriculturally productive areas of the country. Within portions of these areas, considerable farmland has been abandoned because of erosion and gullying. Erosion, coupled with rapid urbanization and highway construction, has removed some 200 million acres of land from agricultural availability. This is over half the total acreage still being farmed in the United States.

Some reduction of soil erosion rates in the United States has occurred over the past few decades, although erosion still remains at unacceptably high levels. Conservation measures such as contour plowing, minimum tillage, strip cropping, and mulching, as well as government incentives and technical assistance, are all helping to reduce the problem. Unfortunately, the implementation of many of these conservation practices results in at least a short-term reduction in income for the farmer, who often finds it difficult to appreciate potential benefits that may be years in coming. Many farmers are presently in a precarious financial situation and simply cannot afford to sacrifice immediate income for long-term gains. It is also difficult to argue with the continued success of American agriculture.

continued on next page

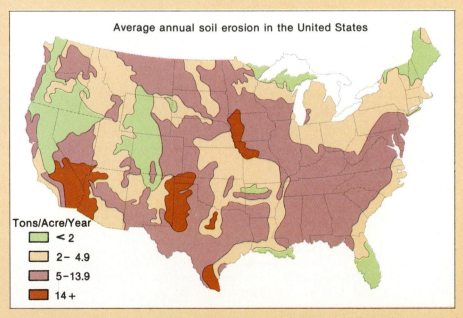

Figure 10.23 Soil erosion severity map of the United States. (Information provided courtesy of the U.S. Department of Agriculture.)

Many of the farmers' economic problems stem from too much, rather than too little production, and agricultural yields continue to increase. This increase is not, however, due to any improvement in the soil. Rather, it is a result of the continued impact of the "Green Revolution"—the highly successful application of modern technology to agriculture—which includes such factors as hybrid crops, modern pesticides, increased irrigation, modern marketing facilities, and, above all, the extensive and increasing use of fertilizers. It is tempting, but dangerous, to place faith in the continued ability of science to offset soil deterioration. An increased commitment to provide for the welfare of one of America's most vital natural resources—its soil—is of the utmost importance.

Chapter Eleven

Landforms and the Earth's Interior

Outline

Focus Questions

1. What are the two basic landform-producing forces, and where do they originate?
2. What are the characteristics and locations of the internal layers of the earth?
3. What are the major categories of rocks and minerals comprising the earth's land surface?
4. How do plate tectonic movements occur, and how do they create and destroy crustal material?

The last section of this book examines the lithosphere—the massive, predominantly solid spheroid of rock and metal that comprises the heart of the earth system. Because geography is concerned chiefly with the earth's surface, the focus of this section is primarily on the outermost boundary of the lithosphere. It is here that the lithosphere, atmosphere, hydrosphere, and biosphere all meet and interact; and it is here that humanity resides.

BASIC TERMINOLOGY AND CONCEPTS

Despite the leveling influence of gravity, the earth's surface is highly irregular, and a great variety of surface features exist. Individual features are referred to as *landforms*. Geographers have an interest not only in the characteristics of landforms, but also in their surface distribution. The combination of landform characteristics and distributions within a region is referred to as its topography. Maps are usually employed when landforms are studied, and those displaying both the vertical dimensions of the landforms and the spatial relationships existing among them are termed *topographic maps* (see Appendix B). Geomorphology (from the Greek *geos* "earth," *morphe* "form," and *logos* "description") is the systematic study of landforms, including their origin, characteristics, and distribution.

Like climate, vegetation, and soils, the landforms of an area have multiple controls. A number of these controls are dynamic, causing the topography to change through time. Some changes occur rapidly and can be witnessed readily by human beings. Examples include changes in the course of a river during a flood, alterations in the size or composition of a beach as the result of a storm, or the eruption of a volcano. Most changes, though, occur very slowly, and centuries or millennia may pass before the landscape is significantly altered.

A region's topography, like its climate, vegetation, and soils, strongly influences the number, distribution, and economic characteristics of its human inhabitants. Lowlands, especially those near large rivers or the sea, often support large populations and contain few, if any, uninhabited areas. In contrast, rugged mountainous regions are typified by localized population clusters in favored sites, separated by large areas with few, if any, inhabitants.

It will be recalled that climate, natural vegetation, and soils, despite their variations from place to place,

display large-scale patterns that are repeated to a greater or lesser extent on the continents of both hemispheres. The tendency is for latitudinal zones of similar conditions to form; solar energy powers all three systems, and its gradual reduction in intensity with latitude is ultimately responsible for the progression of changes that occur in soil, climate, and vegetation. The sun is also a basic energy source for the production of landforms. Its influence results in a general latitudinal zonation of those features resulting primarily from solar-powered processes.

There exists, however, a second group of geomorphic processes powered by the atomic fission of certain radioactive elements in the interior of the earth. These are the tectonic processes. No organized latitudinal zonation exists with regard to these forces and their resulting topographic features. As a consequence, nearly all large regions contain an assemblage of landforms, some produced by solar-powered processes, others by internal earth processes, and many others initially produced by one of these groups and subsequently modified by the other. The net result is the production of topographic patterns considerably more complex, at least on the large scale, than the global patterns of climate, vegetation, or soils.

Because surface features are generally very slow to change, the existing assemblage of landforms within any area is the result of both present and past geomorphic processes. In some instances, the controlling processes have changed greatly, and the landscape is dominated by *relict* features produced by environmental processes no longer active within the region. An important group of relict landforms are the glacial features that formed during the Pleistocene epoch over much of northern North America and northwestern Europe (see Chapter 15). At any given time, however, the topography is actively evolving as it is modified by existing conditions. Relict landforms tend to be removed gradually and to be replaced by others in equilibrium with the current environment. Large-scale landform changes occur so slowly, and environmental processes that can influence landforms vary so frequently, that continuous topographic changes are occurring nearly everywhere.

LANDFORM CLASSIFICATION

Despite the complexity of landforms in terms of origin, form, and distribution, all terrestrial landforms are produced by understandable natural processes. The earth's surface can be likened to a wonderfully complicated jigsaw puzzle that has already been assembled. It is the task of earth scientists, especially geomorphologists, to understand the processes by which the pieces were constructed, how they were joined together, and why they exist in their current positions.

Physical geographers have long been confronted with a choice of two differing conceptual approaches to the study of landforms. One approach is to categorize landforms on the basis of physical appearance (see Figure 11.1). This is often referred to as the "descriptive approach" to landform study. A shortcoming of this approach is that features of similar general appearance, such as mountain ranges, may have completely different origins. The alternative approach is to categorize landforms by their method of formation, or, more precisely, by the geomorphic processes responsible for their production. This is sometimes referred to as the "genetic approach" to landform study. Difficulties are also inherent in this approach. First, virtually all landforms are polygenetic in origin; that is, they have been produced by the combined influence of a number of processes. A second problem is that features that differ greatly in physical form may have been produced by the same geomorphic processes and will therefore be grouped together. In recent years, most physical geographers have employed the genetic approach to the study of landforms, and this approach is stressed in the following chapters.

Genetic Approach to Landform Study

The genetic approach to the study of landforms emphasizes the causal factors in the development of the topography. The chief subdivisions within this system are the various landform-producing forces outlined in Table 11.1. As indicated by this table and the preceding text, two basic groups of geomorphic forces exist. Each is powered by a different energy source; one influences the earth's surface from below and the other from above (see Figure 11.2).

The first group of forces, the tectonic forces, originate within the earth. They are powered primarily by the nuclear decay of unstable isotopes of the elements uranium and thorium, and their overall effect is to increase the elevation and relief of the surface. Examined from a global standpoint, the tectonic forces are largely responsible for raising land areas above sea level and for initially producing mountains and plateaus.

Table 11.1
Landform-Producing Forces

I. Tectonic forces
  A. Diastrophism
    1. Folding
    2. Faulting
  B. Volcanism
    1. Intrusive
    2. Extrusive
II. Gradational forces
  A. Flowing water
  B. Ice
    1. Expansionary forces
    2. Glacier movement
  C. Wind
  D. Ocean waves and currents
  E. Gravity

The tectonic forces are, in turn, divided into two groups. Diastrophism involves solid-state movements of the earth's crust. This displacement may occur either through the *folding* (bending) of the rock material or through *faulting*, which involves rock fracturing, followed by the movement of the two sides of the fracture relative to one another. Volcanism involves the transfer of molten rock material either from one point to another beneath the surface or, more rarely but spectacularly, its expulsion onto the surface. The mechanisms by which tectonic forces are generated and the nature and distribution of the resulting diastrophic and volcanic landforms will be examined shortly.

The second group of forces, the gradational forces, are external forces powered by solar energy and given a downward directional component by gravity. They tend to lower the surface and to reduce topographic relief. The application of these forces is very uneven, however, so that a sequence of changing features is produced during the gradational process. The ultimate effect of gradation would be to reduce all land surfaces to smooth plains approaching sea level in elevation.

Three different substances accomplish most of the gradational reduction of the surface. These are water, ice, and air (as wind), often collectively referred to as the "three tools of gradation." It is interesting to note that one is a liquid, one a solid, and one a gas. They accomplish their work by being heated, evaporated, or both, largely by solar energy. This causes them to

Figure 11.1 Global distribution of major landform types. Plains are areas of flat-to-gently rolling topography that are generally situated at a low elevation. Plateaus are relatively level upland surfaces frequently bounded on at least one side by an abrupt descent to lower elevations. Areas with hills and mountains have rugged surfaces with considerable local differences in elevation. High mountain areas are produced by lithospheric plate collisions.

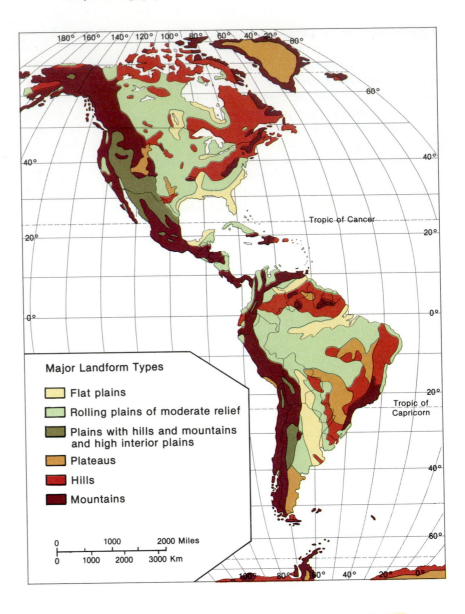

Major Landform Types

- Flat plains
- Rolling plains of moderate relief
- Plains with hills and mountains and high interior plains
- Plateaus
- Hills
- Mountains

expand, to become more buoyant, and to rise into the atmosphere. When the energy is later lost, gravity returns them to the surface in the forms of liquid or solid precipitation, or wind. Earlier chapters have described the processes by which this is accomplished. Once on the surface, they tend to flow downhill or, in the case of air, to areas of lower pressure. The frictional drag exerted on the surface by this movement accomplishes the work of gradation. The force of gravity acting alone can also transport materials downhill in areas where loose or weak surface materials are present.

Examined in a temporal sequence, there are four steps in the gradational process, regardless of which "tool" is involved. The first step is **weathering**. Weathering is actually a preliminary process that makes gradation possible. It involves the breakup of solid rock or large rock fragments into pieces small enough to be moved by the three tools of gradation. This is necessary because the forces of gradation are not as powerful as the tectonic forces. For example, although a mountain range may be raised tectonically as a single unit, it must be carried off by water, ice, or wind piece by piece.

Once weathering (discussed further in Chapter 13) has prepared the surface, the weathered materials are in relatively quick succession picked up or *eroded,* transported, and finally deposited by one or more of the three

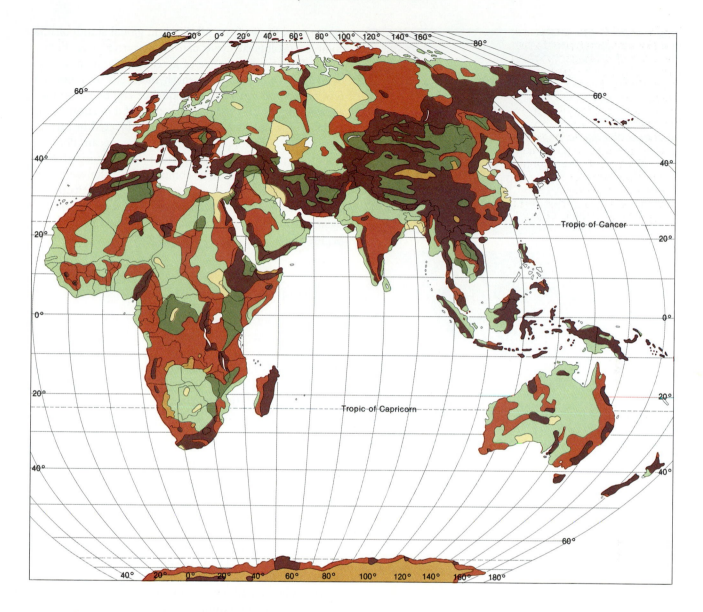

tools of gradation. Because erosion involves the removal of material, it results in a lowering of the surface. Deposition, conversely, produces an increase in surface elevations. Erosion occurs at sites where the available tools of gradation have excess energy and are able to pick up and transport weathered material. Deposition usually occurs at relatively low elevations and in areas of low relief. It takes place at sites where the gradational tools lack sufficient energy to continue to transport the materials they have been carrying. For the world as a whole, erosion must in the long run equal deposition, since everything that is picked up must eventually be put back down. The geographical distributions of the

sites in which erosion and deposition are dominant, though, differ greatly, because the environmental conditions that favor these processes directly oppose one another. As a basic geographical generalization, erosion exceeds deposition on land areas, resulting in their net gradational reduction, while deposition exceeds erosion on the ocean floor.

The tectonic and gradational forces are in essence directly opposed to one another in terms of their effects on the surface, and can be envisioned as being engaged in a constant struggle for a controlling influence on the topography. As soon as tectonic forces produce surface uplift, forming hills, mountains, and plateaus, gradational

Figure 11.2 Landscapes illustrating the effect of tectonic and gradational processes. The top photo shows a peak in the Canadian Rockies that has been uplifted and complexly folded by diastrophic forces. The bottom photo of an Alaskan glacier illustrates the effectiveness of ice in eroding and transporting rock material (the dark bands in the glacier) downhill under the influence of gravity. *(top: Richard Jacobs, © JLM Visuals; bottom: John S. Shelton)*

forces attack these uplands, carrying them bit by bit to lowlands or to the sea. The tectonic forces, however, occur very sporadically, with periods of intense activity followed by long interludes of quiescence. A good example was provided by the violent eruption in 1980 of Mount St. Helens, after 123 years of dormancy. The gradational sequence, conversely, is generally slower in its operation, but it is continuous and inexorable. As soon as a tectonic feature is produced—even as it is being uplifted—gradational forces attack it and begin to wear it down.

If one looks at the workings of the tectonic and gradational forces from a long-term global standpoint, a complex geological cycle can be discerned (see Figure 11.3). The same sequence of activities occurs over and over again, and the same earth materials are continuously used and reused. The sequence can be envisioned as beginning with the tectonic uplift of a portion of the surface. The uplifted region is then subjected in turn to weathering, erosion, transport to the ocean floor or some other basin, and deposition. The deposited sediment is eventually buried, compacted by the weight of the overlying materials, and gradually reconstituted as rock. At some later time this rock is once more uplifted by another episode of tectonic activity, and the sequence begins anew.

EARTH'S INTERIOR

The geographer is concerned primarily with the earth's surface. The internal composition and structure of our planet is of relevance to physical geography largely because it sheds light on surface processes and features. Events deep within the earth produce the tectonic activity that forms the continents, the oceans, and the atmosphere. All knowledge of the earth's interior must be obtained indirectly because human beings have barely scratched the surface in their mining and drilling activities. Most information on the composition and structure of the interior has been gained through the analysis of shock waves produced by earthquakes and explosives, as recorded by instruments called *seismographs*.

The earth's interior has an orderly layered structure produced by density differences resulting from variations in composition and temperature. It is dominated by a small group of chemical elements consisting mostly of iron, silicon, and oxygen, along with lesser quantities of magnesium, aluminum, and calcium. The interior may be separated into an iron-dominated central region called the core, and a larger silicon- and oxygen-dominated periphery containing the mantle and crust (see Figure 11.4).

The development of a layered interior apparently occurred during the formation of the earth. The process of planetary accretion was associated with very high temperatures, which, it is believed, were sufficient to partially melt the developing planet. The melting of the earth's constituents allowed them to form the present layers of the interior by settling at different levels according to their respective masses. Fortunately, the separation process was not complete; or none of the

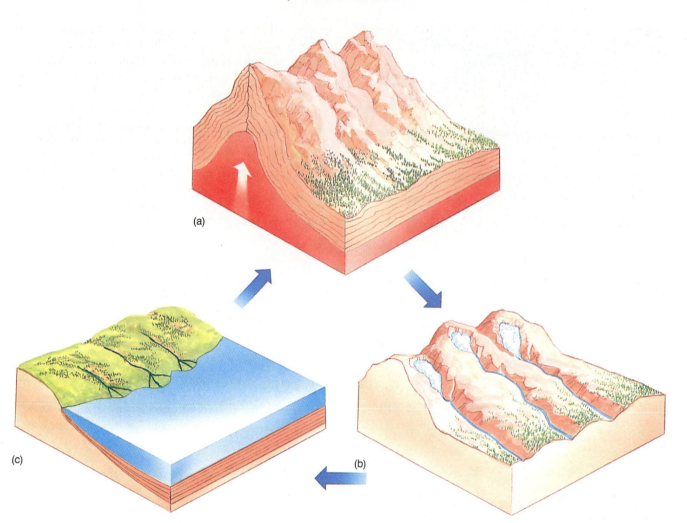

(a)

(c)

(b)

Figure 11.3 The geological cycle. In the top view, a rugged mountainous surface is formed as a result of tectonic activity. The gradational forces of water, ice, and wind, in conjunction with gravity, gradually erode the mountains (right view). The eroded material is transported, largely by rivers, to the ocean, where it is deposited to form layers of sedimentary rock. At some later time, renewed tectonic activity lifts these rock layers to form a new mountainous landmass; and the cycle is repeated.

heavy elements that are so essential to us today would exist at the surface.

The Core

The core is the innermost layer of the earth and occupies a fifth of our planet's total volume. It forms a dense metallic sphere with a radius of about 2160 miles (3460 km). Temperatures and pressures within the core are higher than anywhere else inside the earth. All values are estimates, but temperatures are believed to range between 7200° F and 9000° F (4000-5000° C), and pressures range between 20 million and 50 million pounds per square inch (1.4 × 10^6 and 3.5 × 10^6 kg/cm^2). The composition of the core has not yet been ascertained, but, because of its mass, it is thought to consist largely of iron, perhaps alloyed with a small quantity of nickel.

Seismic data indicate that the core consists of two distinct layers. The larger *outer core* is apparently liquid, since it cannot transmit the seismic s-waves that pass only through solids. It is the only major portion of the earth's interior to exist in the liquid state. Inside the

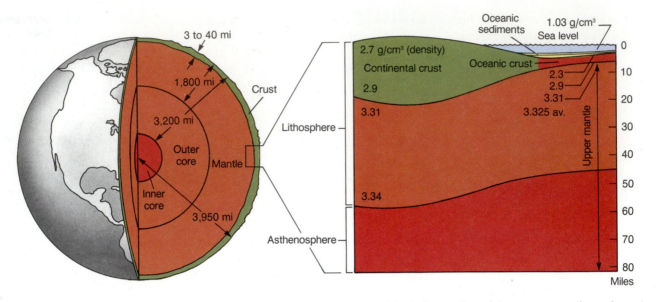

Figure 11.4 The internal layers of the earth. An expanded view is provided of a portion of the upper mantle and crust, with representative rock densities indicated.

outer core is a solid *inner core* with a radius of approximately 750 miles (1200 km). Temperatures here are undoubtedly even hotter than those of the outer core, but pressures are so high that melting cannot occur.

The Mantle

The mantle forms an extensive layer of rocky material extending from the top of the outer core almost to the surface of the earth. It is nearly 1800 miles (2900 km) thick and comprises about four-fifths of the earth's volume. In contrast to the core, it is composed primarily of rock. The rock type is apparently peridotite, which is dense and dark colored, and which consists largely of oxides of silicon, iron, and magnesium.

The mantle can also be subdivided on the basis of its physical state. The lower mantle, below a depth of 450 miles (700 km) beneath the surface, consists of immobile crystalline rock. Most of the upper mantle, in contrast, seems to be in a plastic state and is capable of undergoing a slow creeping flow in response to forces acting upon it. This plastic layer, termed the asthenosphere, plays a critical role in tectonic processes at the earth's surface. The extreme upper mantle, above the asthenosphere, is a rigid solid. This uppermost portion of the mantle and the overlying crust to which it is attached are sometimes collectively termed the **lithosphere**.[1]

The Crust

The crust is the outermost layer of the earth. It consists of a relatively thin "skin" of rigid low-density rock that covers the underlying mantle to a depth of between 3 and 40 miles (5-70 km). Nearly three-fourths of the crust is formed from the elements oxygen and silicon. The crust, like the core and mantle, is divided into two segments. The lowermost segment is the oceanic crust, a thin, continuous layer of relatively dense basaltic rock that reaches the surface beneath the ocean basins. Above this, and forming the continental landmasses, is a layer of very low density granitic rock referred to as the continental crust.

The oceanic crust is a continuous layer that generally varies from 3 to 5 miles (5-8 km) in thickness. It has a mean specific gravity of about 3.0 grams per cubic centimeter, a value some 10 percent greater than that of the continental crust.[2] In addition to oxygen and silicon,

---

1. Note that this use of the term "lithosphere" is much more restrictive than its earlier application to denote the entire interior of the earth.

2. *Specific gravity*, commonly stated in grams per cubic centimeter, is a frequently used measure of density. Fresh water has a specific gravity of exactly 1.0 g/cm³; therefore, a given volume of oceanic crust weighs three times as much as the same volume of water. By comparison, the mean specific gravity of the earth as a whole is about 5.5 g/cm³; and that of the inner core averages 12.5 g/cm³.

the oceanic crust contains a higher proportion of the heavier elements magnesium, iron, and calcium than does the continental crust. Its thinness and high density cause it to rest at a relatively low level on the underlying mantle; for this reason, it forms the ocean basin floors.

The continental crust is the uppermost of the solid layers of the earth and is the only layer that is discontinuous. It covers approximately 35 percent of the earth. In most places where it exists, the continental crust extends from the surface to a depth of between 6 and 45 miles (10-70 km), making it much thicker than the oceanic crust (see Figure 11.5). The continental crust contains a relatively high proportion of the elements aluminum, potassium, and sodium, which often combine with oxygen and silicon to form the rock called granite.

Because its thickness and low density (about 2.7 g/cm^3) give the continental crust enough buoyancy to form the highest portions of the earth's surface, it is the material of which the continents are composed. The oceanic crust sags elastically downward beneath the weight of the continental crust. The thickest areas of continental crust therefore underlie the highest mountains while the thinnest areas are located beneath shallow ocean water, usually along the continental margins. Ocean depths increase dramatically past the tapered margins of the continental crust (see Figure 11.6).

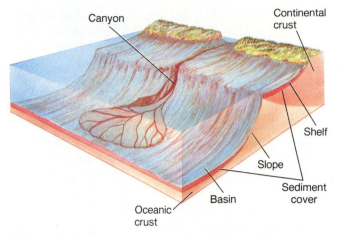

Figure 11.6 The tapered margins of the continental crust are overlain by the shallow ocean water covering the continental shelves. As the continental crust ends completely, ocean depths increase dramatically.

ROCKS AND MINERALS OF THE EARTH'S CRUST

The solid surface of the earth is composed of rocks and of weathered materials derived from them. Rocks in different regions vary greatly in origin, composition, and appearance, and in their physical and chemical characteristics. The distribution of different rock types over the earth's surface is in itself an important aspect of our environment. Two consequences of this distribution that are of great practical importance to humanity are the global patterns of soil types and of mineral deposits. Rock types and structures play a crucial role in determining the characteristics of landforms. In general, high-standing portions of the surface have either been uplifted recently by tectonic forces or are composed of physically strong and chemically stable rock types that are resistant to gradational processes. In contrast, lowlands are frequently underlain by rocks that are highly susceptible to weathering and erosion. A basic understanding of the major categories of surface rock types is therefore an important prerequisite to the study of landforms.

The earth's crust contains only a small number of chemical elements in abundance (see Table 10.2). Oxygen and silicon are about three times as abundant as all the rest combined. These two elements usually occur in chemical combination with metals to form the **silicate minerals,** the dominant mineral group of the crust and mantle. A **mineral** is a naturally occurring, solid,

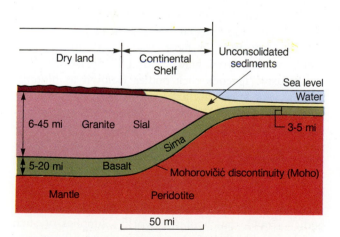

Figure 11.5 Cross-sectional view of the margin of a continental plate. The oceanic crust is shown in green and the continental crust in violet.

inorganic compound. It has a specific chemical composition and usually exhibits a crystalline structure that results from a distinctive atomic arrangement of its constituent elements. The more than 2000 known minerals are divided into several major mineral groups dominated by the silicates, which comprise 92 percent of the crust.

Minerals can be considered the building blocks of rocks. A rock is simply a solid aggregate of minerals. Most rocks contain several different minerals, the individual crystals of which are sometimes readily visible (see Figure 11.7). In most areas, the solid surface rock, or bedrock, is overlain by comparatively thin layers of weathered rock and soil. The majority of the continental crust is composed of rocks with a high content of silicon and aluminum. These rocks are normally low in density and light in color. The oceanic crust, in contrast, consists almost entirely of denser, dark-colored rock containing a greater abundance of iron, magnesium, and calcium. This rock is sometimes brought to the surface by volcanic eruptions or is exposed by the long-continued erosion of a landmass.

Major Categories of Rocks

Rocks are generally categorized by their method of formation. Three major categories are recognized: igneous, sedimentary, and metamorphic. Within each category, a large variety of individual rock types exists. Any rock type in any one of the three major rock categories discussed below can be converted by earth processes to one or more types in either of the other two categories.

Igneous Rocks   *impt. class in earth's crust*

Igneous rocks (from the Latin *igneous* "from fire") are formed directly from the solidification of molten rock material and are produced by tectonic activity. If the earth were at one time in a molten state, as it apparently was during the period of planetary formation, it must have consisted entirely of igneous rock when it solidified. This means that the constituents of all rocks must originally have been derived from igneous rock. The great majority of the earth, including all of the mantle, the oceanic crust, and even most of the continental crust, still consists of igneous rock. Only within a few miles of the surface are sedimentary and metamorphic rocks found.

Figure 11.7 This closeup photo of a granite specimen clearly indicates its crystalline structure. (*John S. Shelton*)

Most igneous rocks, including many currently located at the surface, formed in the lower crust or upper mantle. Although both of these zones of the earth's interior consist primarily of solid rock, localized magma bodies form in the vicinity of zones of weakness where pressures are not high enough to keep the rock from expanding and melting. Such masses of magma may be intruded into overlying rock layers before solidification occurs, forming bodies of intrusive igneous rocks.

The most obvious characteristic of intrusive igneous rocks is the large size of their constituent mineral crystals. These crystals had time to form because of the very slow cooling and solidification of intrusive igneous rock bodies. The best-known intrusive igneous rock type is *granite*, which is composed chiefly of the minerals quartz, mica, and feldspar (see Figure 11.7). Granite is relatively resistant to most gradational processes. It therefore often forms high-standing landmasses.

Intrusive igneous rock bodies (known as *plutons*) serve as sites for the formation of veins of economically important metallic minerals. These veins form along fractures or other zones of weakness in the slowly cooling plutons. Mineral-bearing fluids enter and move along the fractures, depositing their dissolved mineral content because the reduction of pressure in these zones of weakness reduces the ability of the fluids to carry minerals in solution. Richly mineralized igneous rock bodies provide much of the world's supplies of such metals as gold, silver, copper, and tin.

On occasion, magma makes its way to the earth's surface, where it flows out as *lava* or is thrown out as *volcanic ejecta* (see Chapter 12). The solidification of this material forms what is termed **extrusive igneous rock** because it is extruded onto the surface while still in a molten state. Because of the speed with which it cools, extrusive igneous rock generally has microscopic mineral crystals and, unless it contains air pockets, a relatively smooth texture. *Basalt,* a dense, black-colored rock containing substantial quantities of iron and magnesium, is the best-known extrusive igneous rock type. The Hawaiian Islands and the Columbia Plateau of the northwestern United States consist largely of basalt (see Figure 11.8).

Figure 11.9 depicts the global distribution of surface rock types. It can be seen that intrusive and extrusive igneous rocks together cover about 30 percent of the earth's land surface and that the intrusive rocks are the more common of the two groups. The distributional pattern of intrusive igneous rocks is complex, but the largest areas consist of the cores of ancient plutons, such as those in Scandinavia, Africa, and eastern Canada. Only the erosion of great thicknesses of overlying materials has allowed these ancient rocks, which originally formed at great depth, to appear at the surface.

Extrusive igneous rocks are more commonly associated with oceanic crust than with continental crust. On land, they occur mostly within active volcanic zones located along the margins of the continents. They are quite common, for example, all along the borderlands of the Pacific. Because volcanic eruptions are usually rather localized phenomena, the resulting extrusive igneous rocks are scattered in numerous patches, many of which are too small to appear in Figure 11.9. In a few instances, though, large-scale fissure outpourings of lava have covered extensive areas. Notable in this regard are the Columbia Plateau of the northwestern United States, the Paraná Plateau of southern Brazil, and the Deccan Plateau of India. These eruptions are discussed further in the next chapter.

Sedimentary Rocks

Weathering processes produce loose surface materials sufficiently small and light to be readily picked up and transported to other sites by the gradational agents. Any such solid materials, of either organic or inorganic origin, that have been transported by water, ice, wind, or gravity are collectively termed **sediments**. Familiar examples include mud or gravel deposits left by streams, beach sand and shells washed ashore by the ocean,

accumulations of unsorted rock debris deposited by retreating glaciers, and windblown leaves and dust.

These sediments are deposited in sites where the energy of the transporting agents is insufficient to carry them farther. Because water is by far the most effective gradational tool, most depositional sites are associated with bodies of slow-moving or standing water. Examples include valley floors, lake bottoms, and especially the ocean floor. The ocean may be thought of as the ultimate sediment trap. Once sediments reach it, as they do in tremendous quantities, there is normally no way for them to be removed. The great majority of land-derived sediments reaches the ocean by way of rivers, so the greatest rates of sediment accumulation are along the continental margins, especially near the mouths of large rivers.

Where conditions remain favorable for the deposition of sediments over extended periods of time, they gradually accumulate to form horizontal layers termed **strata.** If accumulation is rapid, great thicknesses of material may eventually collect, resulting in the deep burial and compaction of the lowermost sediments.

Figure 11.8   The margin of a geologically recent basaltic lava flow on the Snake River Plain of southern Idaho. (*John S. Shelton*)

Figure 11.9 World distribution of rock types on land. Sedimentary rocks cover approximately two-thirds of the earth's land surface.

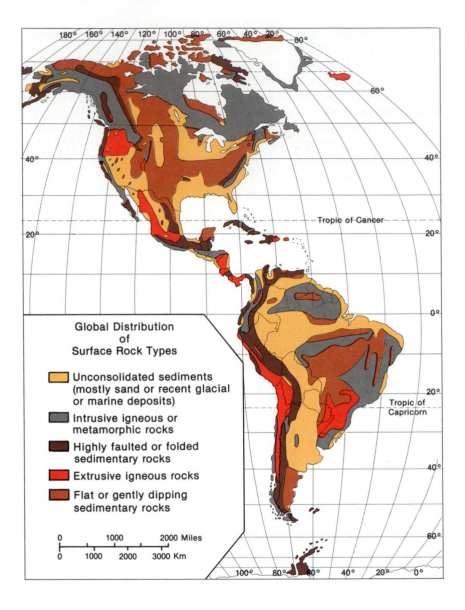

Global Distribution
of
Surface Rock Types

- Unconsolidated sediments (mostly sand or recent glacial or marine deposits)
- Intrusive igneous or metamorphic rocks
- Highly faulted or folded sedimentary rocks
- Extrusive igneous rocks
- Flat or gently dipping sedimentary rocks

Frequently, water circulating through the compacted sediments deposits chemical precipitates that cement the deposits together. The processes of compaction and cementation, along with chemical changes in the sediments themselves, result in their gradual conversion to **sedimentary rock.**

Sedimentary rocks are categorized by their method of formation. Three basic categories are recognized: clastic, organic, and chemical. *Clastic sedimentary rocks* are composed of particles of preexisting rocks. These particles were weathered from the parent rock and were subsequently transported, deposited, and lithified (converted to rock). Individual clastic sedimentary rock types are differentiated by the sizes of their constituent rock particles. *Conglomerate* contains the largest particles, which are gravel-sized or larger. *Sandstone,* one of the most common and important sedimentary rock types, consists of sand-sized particles (see Figure 11.11). *Siltstone* and *shale* are composed, respectively, of silt-sized and clay-sized particles. Shale is the chief sedimentary rock produced on the deep sea floor and comprises approximately half of all sedimentary rock.

Organic sedimentary rocks are produced from the remains of plants and animals. *Limestone* ($CaCO_3$) is by far the most abundant rock type within this category. It forms on seafloors beneath clear tropical water where

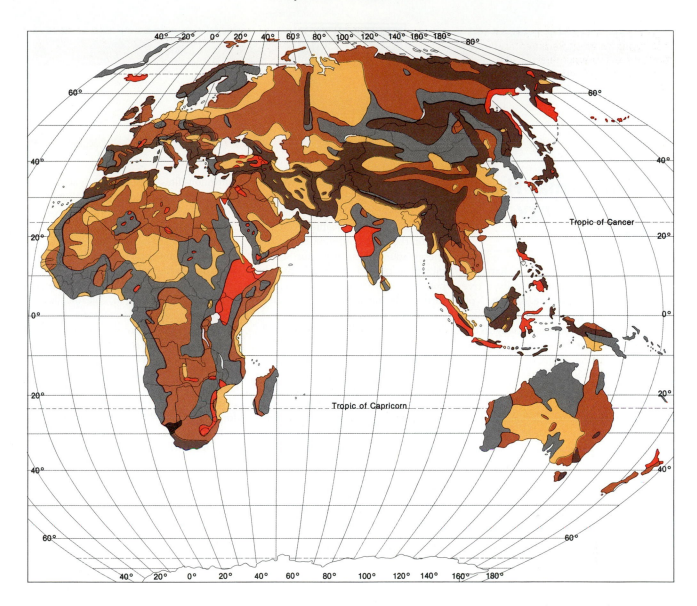

marine animals synthesize the calcium carbonate they need for their shells or exoskeletons. When these animals die, their remains accumulate on the seafloor and are buried, compacted, and lithified into limestone. Another important organic sedimentary rock is *coal*. Composed almost entirely of carbon, this important economic mineral is formed from the remains of ancient plants.

Chemical sedimentary rocks are formed from the precipitation of chemicals in water. It was already noted that chemical precipitates play an important role in the formation of the coarser-grained clastic rocks, especially conglomerate and sandstone, by serving as the cement holding the rock particles together. Under certain conditions, rocks are formed largely or entirely of these precipitates. The most common chemical precipitate is calcium carbonate ($CaCO_3$), which can accumulate beneath tropical waters to form chemical limestone deposits. Another important chemical sedimentary rock is *halite*, or rock salt (NaCl).

Sedimentary rocks cover nearly all of the ocean floor and about two-thirds of the continents, making them the most widely distributed of the three rock categories at the earth's surface (see Figure 11.9). As a group, they are physically the weakest of the rock categories, so they are easily affected by tectonic and gradational forces.

245

Geothermal Heat as an Energy Source

An important geological characteristic of many tectonically active regions is the presence of large volumes of hot ground water. In some areas, this water makes its way to the surface to form such natural phenomena as geysers, fumaroles, and hot springs. In addition to their scientific interest and economic value as tourist attractions, hot springs have been used from at least the time of the Roman Empire for heating, recreation, and health. During the twentieth century, however, geothermally heated water has become increasingly important for the production of electricity.

The chief advantage of geothermal power is that it is, when used within reason, a renewable resource with almost unlimited potential for development. The ultimate source of energy is the decay of fissionable materials within the earth's interior. So far, this energy source has lasted for several billion years, and apparently has the potential for several billion more. Unfortunately, technological limitations presently restrict the areas where geothermal energy can be economically utilized.

There are currently several site requirements for the development of a geothermal field to produce electric power. A large subsurface reservoir of superheated water must be present. Most geothermal electric power plants tap water with temperatures between 450° and 570° F (230° and 300° C). The water must be present at relatively shallow depths of 10,000 feet (3,000 m) or less and must have sufficient recharge ability to allow for a continuous flow of hot water or steam for many years. Finally, the water must be relatively free of corrosive chemicals. In addition, since the transportability of pressurized hot water is limited, the power plant must be built near the geothermal field.

Most currently developed geothermal fields are located in volcanically active areas, where hot igneous rock exists near the surface. The first geothermal electric power plant was constructed in Larderello, Italy, in 1904. Since then, plants have been built in Japan, Iceland, the Philippines, New Zealand, the Soviet Union, El Salvador, Nicaragua, Mexico, the United States, and several other countries (see Figure 11.10). The field with the largest electrical output is The Geysers, in California; but a new plant currently under construction in the

continued on next page

The widespread presence of sedimentary rocks of marine origin in continental settings indicates that the land in these areas must have risen above sea level since the rocks were formed. Marine sedimentary rocks, for example, cover most of the southern and midwestern United States (see Figure 11.12). Most of the high mountain ranges of the world, including the Himalayas, are composed of sedimentary rocks that formed on the ocean floor. In places, great thicknesses of sedimentary strata were deposited over lengthy geologic time spans. These rocks, which often contain the fossil remains of plants and animals, are invaluable in reconstructing the earth's natural history. They bear eloquent testimony to the great age of our planet and to the inconstancy of surface environmental conditions.

Metamorphic Rocks

Metamorphic rocks (from the Greek *metamorphosis* "transformation") are rocks altered from their original form by heat, pressure, and/or chemical activity, while remaining in the solid state. Metamorphic changes in rocks produce textures, structures, mineral compositions, and general appearances that differ from those of the original rocks. The constituent chemical elements

Figure 11.10 Steam collection pipe system at a geothermal power plant in Warrakei, New Zealand. (*J.N.A. Lott, McMaster Univ./Biological Photo Service*)

and environmental problems. Geothermal plants presently cost more to build than conventional plants. There is also concern that overzealous development could damage or destroy natural geothermal phenomena in such places as Yellowstone National Park, Hawaii, and New Zealand, and deplete the geothermal potential of individual fields. Obnoxious gases, such as hydrogen sulfide, are released to the surrounding air in geothermal fields; and the spent geothermal water, because of its dissolved salt content, may be a source of thermal and chemical pollution if released to surface lakes and streams. All in all, however, the potential benefits of geothermal power seem to far outweigh the problems associated with the continued development of this natural resource; and it appears to have a bright future.

Kamchatka Peninsula of the Soviet Union may eventually surpass it in production. Recent major discoveries of geothermal reservoirs in Central America make this area one of the most promising for development in the near future.

Despite its future potential, geothermal electric power production is associated with a number of economic

remain the same, but they are rearranged into different mineral combinations. Metamorphism results from the high temperatures and pressures generated well beneath the surface by tectonic processes. Most metamorphic rocks are produced by large-scale or *regional metamorphism* that occurs during mountain-building activity. These rocks, formed in the root zones of mountain systems, are exposed at the surface only after long periods of erosion.

All types of rock are subject to metamorphism with sufficient applications of heat and pressure; this includes igneous, sedimentary, and even metamorphic varieties. Each rock type has one or more metamorphic equiva-

lents. Varying degrees of metamorphism exist, from slight alterations to complete recrystallization (see Figure 11.13). The type of metamorphic rock produced at any given site depends on the original type of rock and on the nature, intensity, and duration of the metamorphic processes involved. Most metamorphic rocks are classified as one of seven important types, listed in Table 11.2.

Metamorphic rocks are by far the least abundant of the three major surface rock categories. They are much more common deep within the crust, where they have been produced by tectonic movements. Because erosion must remove great thicknesses of overlying materials before rocks subjected to regional metamorphism can

Figure 11.11 The cross-bedding displayed by this sandstone in Apache County, Arizona, indicates that it was originally deposited as sand dunes. (*E.D. McKee, USGS*)

appear at the surface, these rocks are exposed mostly in ancient plutons.

PLATE TECTONICS

The tectonic and gradational forces act in concert to produce the surface features of the earth. Each year, gradational forces (ignoring human influences), erode some 11 billion tons (1.0×10^{13} kg) of material from the land and transport it to the oceans. If sediment removal proceeded unopposed at this rate, the continents would be reduced to sea level in only 40 million years. A comparable mass of material, however, is annually lifted above the oceans by tectonic activity, so the total mass of land remains relatively constant.

Although the net gains and losses of mass from the earth's land areas resulting from tectonic and gradational forces may be roughly equal, the geographical distributions of these opposing forces and the types of surface features they produce differ greatly. In a basic sense, the tectonic forces are primarily responsible for the large-scale patterns of surface features, such as the arrangements of the continents and ocean basins and the locations of the major mountain ranges. Gradational forces, on the other hand, can only operate where, and when, the tectonic forces have first created topographic features to erode.

The sizes, shapes, and locations of the continents, ocean basins, and mountain ranges are among the most

basic and important aspects of our planet's physical geography. They form the largest units of the earth's surface features. In addition, as a result of the complex interactions that exist among the various earth surface phenomena, they exert a vital influence on global patterns of climate, plant and animal distributions, and human activities. We will now examine the mechanisms by which these earth-shaping forces operate.

Evidence for Continental Drift

Many of us have been struck by the thought that a map of the earth resembles a jigsaw puzzle, with the continents representing puzzle pieces separated by areas of ocean. On closer inspection, a number of the continental "pieces" do seem to fit together remarkably well. Especially striking is the match between the east coast of South America and the west coast of southern Africa. Scholars as early as the sixteenth century noticed these coastal similarities and wondered if they could be coincidental or if, indeed, the land masses had somehow separated and had moved long distances over the surface of the planet.

The matter remained largely speculative until 1912, when German geophysicist Alfred Wegener presented a detailed theory for *continental drift* that not only traced past positions and subsequent motions of the continents, but that for the first time suggested a means by which this movement could have been accomplished. According to Wegener, all the larger landmasses were at one

Figure 11.12 A sequence of marine sedimentary rocks hundreds of millions of years old is exposed on the south wall of the Grand Canyon in Arizona. (*John S. Shelton*)

Figure 11.13 Hand specimens of banded gneiss. The application of heat and pressure during tectonic activity caused the alignment of the mineral crystals in these metamorphic rocks. (*Richard Jacobs, © JLM Visuals*)

time assembled into a single supercontinent that he called Pangaea. This supercontinent later separated into a number of smaller blocks that subsequently drifted to their present positions. However, Wegener's theory that the continental crust moved about on the underlying oceanic crust was proven impossible. This basic shortcoming caused the idea of continental drift to fall into disrepute for a time.

A revival of research into continental drift took place following World War II. By this time, more sophisticated geologic instrumentation was being developed, and earth scientists soon amassed additional evidence in support of large-scale crustal movements. It was discovered, for example, that certain rock sequences extending to the coast of Brazil reappeared on the African coast precisely where they would be expected to be if the two continents were once joined. Fossil remains indicated that similar, or identical, plants and animals appeared during the same geological periods on landmasses currently located great distances from one another. Evidence of a past ice age was unearthed in currently subtropical portions of South Africa and South America, and fossil beds of tropical plants were discovered in Antarctica and Britain. In addition, the current locations of active volcanoes, earthquake foci, and the major mountain belts are all strongly aligned with the apparent boundaries of the crustal plates.

Some of the strongest evidence for continental drift came from analysis of the crystal orientation of the

mineral magnetite (Fe_3O_4) in oceanic basalts. These crystals, when free to move, act like tiny compass needles by aligning themselves along the axis of the earth's magnetic field. When lava containing magnetite crystals solidifies, they are frozen into place, so that their orientation provides a record of the direction toward the North and South Magnetic Poles at the time of solidification.

It was discovered in the 1960s that the North and South Magnetic Poles occasionally "flip," or reverse polarity. This causes the North Magnetic Pole to become the South Magnetic Pole temporarily, and vice versa. Any magnetite crystals located in basalt that solidified during a period of reversed polarity have a reverse alignment from crystals aligned with the present polarity. The earth's magnetic field has reversed polarity dozens of times over the past few tens of millions of years, and the latest reversal occurred approximately 700,000 years ago.

In 1964, aircraft equipped with magnetic sensing devices flew over the Mid-Atlantic Ridge and discovered that the basalt on both sides exhibited a mirror-image pattern of stripes of normal and reversed magnetic polarity (see Figure 11.14). The Mid-Atlantic Ridge is part of a system of oceanic ridges extending through the Atlantic, Pacific, and Indian Oceans and reaching heights of thousands of feet above the surrounding deep ocean floors. These ridges are the sites of active and nearly continuous submarine volcanism, which extrudes great quantities of lava that solidifies into basalt containing magnetite crystals. The inescapable conclusion was that these ridges are the boundaries between newly forming portions of two crustal plates that are diverging at rates of a few inches a year. The movement

Table 11.2
Major Metamorphic Rocks and Parent Rock Types

Metamorphic Rock	Parent Rock(s)
Quartzite	Quartz conglomerate, sandstone, or siltstone
Slate	Shale
Phyllite	Shale, slate
Schist	Shale, slate, phyllite
Gneiss	Granite, diorite
Marble	Limestone, dolomite
Anthracite coal	Bituminous coal

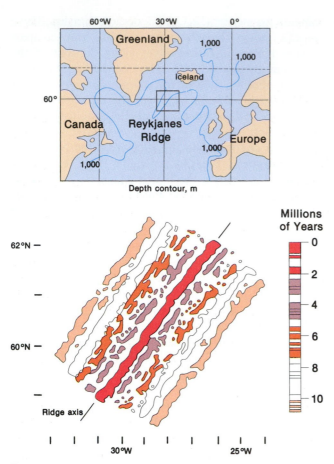

Figure 11.14 Magnetic rock patterns along a portion of the Mid-Atlantic Ridge southwest of Iceland. The shaded areas display normal polarity; they are separated by areas of reversed polarity. Note the symmetry and the continually increasing age of the magnetic bands outward from the ridge axis. (*Robert A. Phinney, ed.,* The History of the Earth's Crust. *Copyright © 1968 by Princeton University Press. Fig 6, p. 83, reprinted with permission of Princeton University Press.*)

is much like that of two belts located on adjacent rollers rotating in opposite directions. Because most new crustal material seems to form and diverge along these oceanic ridges, continental drift appears to be largely attributable to the process of **seafloor spreading**.

The accumulation of evidence caused the idea of continental drift to gain widespread acceptance during the late 1950s and the 1960s. The entire process of plate formation, movement, and destruction, now known as **plate tectonics,** has been worked out in considerable detail over the past few decades. The development of plate tectonic theory has been the landmark occurrence

in modern earth science. Plate tectonic theory has shown that the earth's surface does not consist merely of a random assortment of land and water bodies, but that it forms an integrated system intimately related to materials and processes deep within the earth's interior.

Plate Tectonics Processes

It is now known that the moving plates are not composed exclusively of crustal material, but that they also include the rigid uppermost portion of the mantle. Taken together, the crust and uppermost mantle comprise the lithosphere, which averages about 60 miles (100 km) in thickness and which rests on the asthenosphere. It is therefore more accurate to refer to these plates as **lithospheric plates** than to use the older term "crustal plates." The asthenosphere is the mobile layer upon which the lithospheric plates float.

Heat from the decay of radioactive elements in both the core and mantle is conducted very slowly surfaceward through the mantle. When it reaches the asthenosphere, this heat generates slow convection currents within the rock material. The radioactive elements and the heat they produce are distributed unevenly within the earth, much as they are at the surface. In places where they are relatively abundant, more heat is generated, and the mobile areas within or above them in the asthenosphere expand, increase in buoyancy, and become the sites of rising convection currents. In places where radioactive elements are not as abundant, less heat is produced, and the affected mobile areas are likely to be the sites of descending currents of somewhat cooler and denser material. A continuous circulation system is thus established (see Figure 11.15). The regions of mantle upwelling and sinking are not fixed; instead, they slowly shift in various directions at mean rates of a few inches a year. As the plates float on this moving layer, they are slowly carried by its currents like ice on a stream. Atomic energy, manifested in the form of convection currents in the asthenosphere, therefore provides the long-sought mechanism for continental drift. Continental drift, in turn, results in collisions and separations between adjacent lithospheric plates; and these occurrences produce tectonic landforms.

The lithosphere is divided into six major plates and a number of smaller plates, as shown in Figure 11.16. The plates are moving in various directions at speeds of up to several inches a year. Movement in each case is

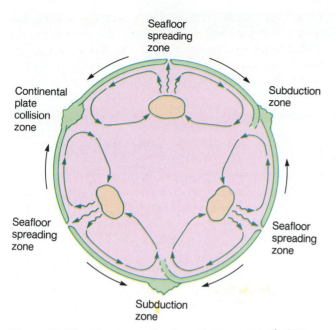

Seafloor spreading zone

Continental plate collision zone

Subduction zone

Seafloor spreading zone

Seafloor spreading zone

Subduction zone

Figure 11.15 Diagrammatic view of the cause of plate tectonic movements. Convection currents in the asthenosphere, heated by radioactive decay in "hot spots" in the mantle, rise to the base of the lithosphere and then spread laterally, dragging the overlying lithospheric plates with them. The resulting plate collisions produce mountain ranges, while seafloor spreading zones in areas of plate divergence are the sites of new oceanic crust formation.

away from the oceanic ridge along which the plate is currently being manufactured. For example, the American Plate, which contains the continents of North and South America as well as the western half of the Atlantic Ocean, is being manufactured along the Mid-Atlantic Ridge. As new crustal material is extruded along the ridge, the plate is carried westward by convection currents in the underlying asthenosphere. This westward motion causes the American Plate to collide with the Pacific and the Nazca Plates, which are located to its west. This collision has crumpled the forward edge of the American Plate to produce the Rocky Mountains in North America and the Andes in South America. The movement is also gradually increasing the width of the Atlantic Ocean, which began forming about 150 million years ago, when North and South America first began to separate from Europe and Africa during the breakup of Pangaea. In similar fashion, the northeastward motions of the African and Indian-Australian Plates are thrusting up a great mountain belt extending from the East Indies westward to

North Africa and Spain as they collide with the Eurasian Plate (see Figure 11.18).

Plate Formation and Movement

The convection currents of the asthenosphere are responsible not only for the motions of the lithospheric plates, but for their formation and destruction as well. Internal heat seems to be carried surfaceward largely by upwelling "plumes" that form in various locations within the asthenosphere. As these plumes approach the upper margin of the asthenosphere, they gradually change direction from vertical to horizontal. Over 120 active plumes have been tentatively identified. They are most numerous along and near the oceanic ridges, but a large number are scattered in seemingly random locations beneath other parts of the oceans and the continents. They are usually associated with surface volcanic activity, and most mid-oceanic volcanic islands or island groups such as the Hawaiian Islands, the Azores, Iceland, and Réunion have formed on the sites of active plumes.

In places where several nearby plumes break through the crust, they may weaken a plate so much that it splits apart to produce two smaller plates. The splitting of a plate substantially reduces the pressure on the underlying asthenosphere. The reduced pressure lowers the melting point of the rock comprising the upper mantle, allowing the selective melting of its lighter constituents, which have the lowest melting points. Eventually, part or all of this molten rock material makes its way to the surface through a system of vertical passageways lying along the fracture and is expelled onto the ocean floor during submarine volcanic eruptions. As it solidifies, it forms new oceanic crust of basaltic composition. The combination of upward expansion due to heat and the expulsion of magma produces an oceanic ridge and apparently forces the newly forming lithospheric plates on either side to diverge. Three major zones of plate formation and divergence marked by oceanic ridges presently exist. They are the Mid-Atlantic Ridge, the East Pacific Rise and Pacific-Antarctic Ridge system, and the Y-shaped Indian Ocean Ridge system (see Figure 11.16). They total some 47,000 miles (75,000 km) in length.

The rate of seafloor spreading varies considerably in different areas. Along the Mid-Atlantic Ridge it averages about 1.5 inches (4 cm) a year. The fastest rate of spreading is along the East Pacific Rise, where it

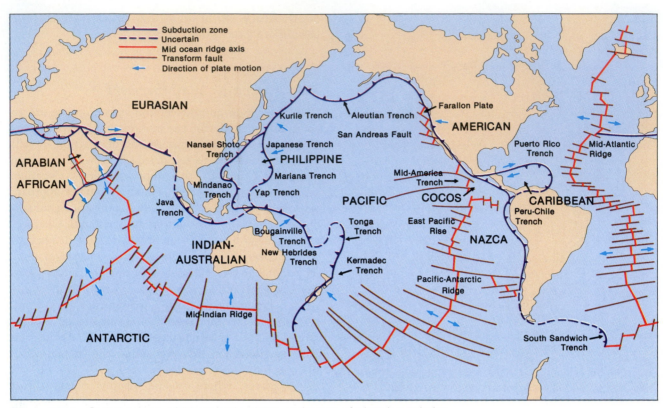

Figure 11.16 Global pattern of lithospheric plates and types of plate boundaries.

averages about 6 inches (15 cm) a year. Even though spreading rates of a few inches a year may seem very slow, these movements can add up to thousands of miles in a few tens of millions of years. They are therefore capable of completely rearranging the surface of our planet over relatively short geologic time spans.

Plate Collisions and Destruction

Based on their relative motions, lithospheric plate boundaries are classified as divergent, transform, or convergent. Along divergent boundaries, as already noted, the adjacent plates move away from one another; and new oceanic crust is formed. Along transform boundaries, the motions are lateral, so that the two plates slide past each other while moving in opposite directions. Neither of these two types of boundary movements directly produces a large proportion of the earth's continental landforms.

Convergent plate motions most greatly influence the production and distribution of continental tectonic features. These movements actually form continental crust while destroying oceanic crust. Without the convergence of lithospheric plates, no major mechanism would exist for the production of light, buoyant, and therefore high-standing masses of continental granitic rock; and the earth would be almost completely covered by ocean.

The nature of the surface features produced along convergent plate boundaries depends primarily on whether the plates are covered by oceanic or continental crust. Three types of convergent boundaries exist; the first is formed by the collision of two oceanic plates; the second, by a collision between an oceanic and a continental plate; and the third, by a collision of two continental plates.

When two oceanic plates collide, one of the plates is bent downward by the other so that it is gradually pressed into the asthenosphere at an angle of 30° or more in a process called *subduction*. The subducted plate, upon reaching the base of the asthenosphere at depths exceeding 400 miles (600 km), is heated until it is eventually destroyed. During this process, the lighter rock-forming material of the subducted plate melts and rises toward the surface as magma. This magma may either solidify beneath the surface to form large granitic plutons known as *batholiths* (see Chapter 12) or reach the surface in volcanic eruptions. In either case, the resulting igneous rock material becomes continental crust.

The initial eruptions of magma derived from the subducted plate take place on the seafloor. As these eruptions continue, however, a sufficient quantity of relatively low density igneous rock may accumulate to protrude above sea level, forming an *island arc* near the margin of the overriding plate. Examples of such island arcs include the Marianas of the western Pacific, the Aleutian Islands of Alaska, and the Lesser Antilles that separate the Atlantic from the eastern Caribbean Sea.

The gradual accumulation of lighter rock from both surface volcanism and subsurface granitic intrusions eventually forms continents. This process is accomplished largely by the collision and joining of numerous island arcs or other small segments of continental crust, each of which is termed a *terrane*, by the sweeping action of the lithospheric plate movements. The west coast of North America from California to Alaska, for example, has been formed by the accretion of numerous terranes onto the western margin of the American Plate.

When an oceanic plate collides with a continental plate, the denser oceanic plate is always subducted (see Figure 11.17). A marginal volcanic zone forms as magma from the subducted oceanic plate rises into and some-times through the adjacent margin of the overriding continental plate to form new continental crust. This category of plate collision is the most common of the three and is primarily responsible for the fact that the majority of the world's actively forming mountain belts are located along the continental margins. The Pacific Plate is currently being subducted by continental plates on all sides, producing its surrounding ring of volcanoes.

When two continental plates collide, neither one is subducted; and little volcanism normally occurs. Instead, a broad zone of compression and diastrophism, often hundreds of miles wide, is formed. Frequently, one of the plates is thrust for long distances over the other. This produces a double thickness of continental crust, which, in combination with its light weight and buoyancy, results in the formation of a high, rugged mountain system. Continued convergence eventually results in the fusion, or **suturing,** of the two plates into a single larger plate. This mechanism balances the process of plate separation and keeps the crust from being divided into progressively smaller units.

The premier example of a continental plate collision is the one occurring between the Indian subcontinent and southern Asia. This collision, which began 50 million years ago because of the northward movement of the Indian-Australian Plate, has produced the Himalayas, the world's loftiest mountain range (see Figure 11.18). Asia has proven to be the weaker of the two

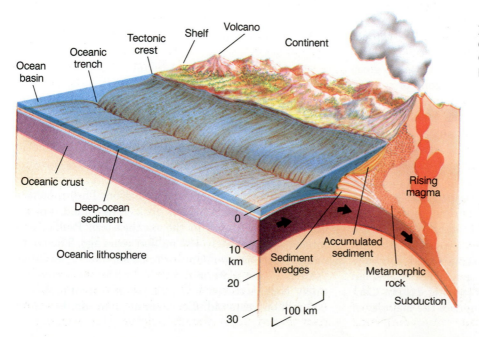

Figure 11.17 Diagrammatic view of the collision between a continental plate and an oceanic plate.

Figure 11.18 View of the Himalayas. These mountains are produced by the northward movement of the Indian subcontinent into the Eurasian Plate. Mt. Everest, the world's highest peak, is on the right. (*Gary J. James/Biological Photo Service*)

continental participants and has deformed the most. Faulting in Asia extends nearly 3000 miles (4800 km) north of the collision zone, and the continent has been shortened latitudinally and stretched longitudinally by the collision. A similar collision at an earlier stage is occurring to the west, where the northward-moving African Plate is running into the European portion of the Eurasian Plate. The intervening Mediterranean Sea is the last remnant of an ocean that once separated these two plates, and continuing plate movement should eventually compress it into oblivion. These plate collisions are major generators of earthquakes—a topic explored in Chapter 12.

The process of continental plate collision forms great thicknesses of continental crust in the zone of convergence. For this reason, ancient plate boundaries, which now are generally located within the interiors of sutured lithospheric plates, are exceptionally strong and resistant to further deformation. These areas are termed *shields* because of their strength and their broadly convex surface profiles. Shields form the stable core areas of continents. They have remained near or above sea level for long periods of geologic time, causing them to be subjected to slow erosion that has exposed the granitic and often metamorphosed "roots" of the ancient mountain systems. At present, shields underlie more than a third of the continents. The Canadian Shield, for example, extends over most of eastern Canada, while the African Shield comprises the majority of

that large continent. The locations of the major shield areas are depicted in Figure 11.19.

Isostatic Uplift

The lithospheric plates, which float on the asthenosphere to just the depth needed to support their weight, are said to be in isostatic equilibrium (from the Greek *isos* "equal" and *stasia* "standing") with their surroundings. When erosional processes remove material from the surface, the weight of the plate in that location is reduced. The reduction in weight results in the **isostatic uplift** of the region. This is a broad, gentle upwarping that, like the gradual rise of the underwater portion of an iceberg when melting occurs at its top, allows the plate to remain in a state of isostatic equilibrium.

Isostatic uplift occurs most rapidly in mountainous regions, where the reduction of mass by the erosional processes is more rapid than it is on other types of surfaces. As erosion proceeds, then, the surface lowering of a region, produced by the continuing removal of weathered materials, is largely offset by its gradual uplift because of its loss of weight. This allows old mountain systems in tectonically stable regions like the Appalachians to maintain themselves for a much longer time than they could otherwise. Isostatic uplift also allows shield areas, with their great underlying thicknesses of buoyant continental crust, to remain above sea level for lengthy geologic time spans despite constant slow erosion.

Future of Plate Tectonics

Two hypothesized long-term trends in the earth's plate tectonic activity are the gradual increase in the volume and areal extent of continental crust at the expense of the oceanic crust and the slow worldwide decline in tectonic activity. Because only oceanic plates are subducted, they are destroyed within a relatively brief geologic time span after their formation. They therefore consist of relatively recent rock. The oldest known oceanic rock, located in the northeastern Pacific, was formed approximately 180 million years ago. The continental plates are too buoyant to be subducted, and they "ride out" the plate movements on the surface indefinitely. For this reason, many consist of ancient rock over 600 million years old. Continental plates do, however,

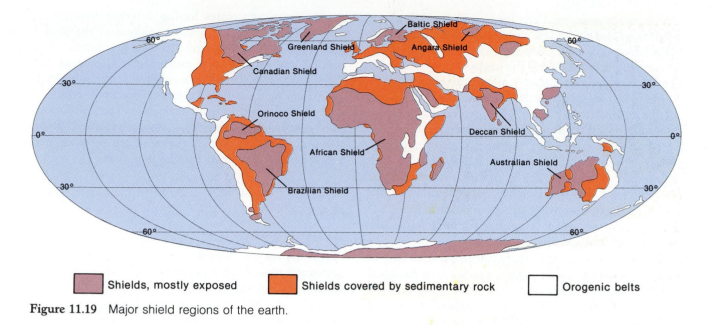

Figure 11.19   Major shield regions of the earth.

lose surface material to the erosional action of water, ice, and wind; and a current area of investigation concerns the relative rates of formation and destruction of continental crustal material. Many researchers believe that the total mass of continental crust is continuing to increase gradually with time.

The radioactive isotopes of uranium and thorium that power the movements of the lithospheric plates are gradually declining in abundance as they are converted to stable elements through the process of atomic decay. In the earth's early history, the output of heat from this source was perhaps three times the present amount. It is estimated that the earth is currently losing heat at about twice the rate that it is being produced. This means that the interior must be slowly cooling.

In the future, the interior of the earth will continue to cool slowly. The lithospheric plates will thicken, and eventually the asthenosphere will become rigid, so that the upward transport of heat by convection currents will cease. This will allow no further tectonic activity to occur. In that distant future, probably several billion years from now, the contest between the tectonic and gradational forces will finally be decided, as gradation achieves the ultimate victory. The earth in its old age will have its landmasses eroded to low plains. These will eventually be covered by shallow seas, as the waves accomplish the final step in the destruction of the continents.

Summary

The earth's land surface contains a great diversity of landforms, produced by a variety of forces acting on numerous types of earth materials. Large-scale surface features form and evolve very slowly; as a result, most landscapes contain features of widely varying age and often contain landforms produced by forces no longer regionally active.

Landform-producing forces may be divided into two major groups. The first group, the tectonic forces, originate from heat released by the nuclear decay of radioactive elements deep within the earth. The general tendency of the tectonic forces is to produce surface uplift and to increase topographic relief. As a result of tectonic activity, new landmasses are produced and mountain ranges are formed. The second group of landform-producing forces is the gradational forces. Powered largely by solar energy and given a downward directional component by gravity, these forces lower surface elevations and generally reduce topographic relief. Gradation operates chiefly through the erosive action of water, ice, and wind.

The earth is divided into three concentric internal layers on the basis of composition. The innermost layer, the core, is a dense metallic sphere composed largely of iron. The inner core is believed to be solid, while the outer core is liquid. The surrounding mantle is an extensive layer of dense rock containing large amounts of iron and magnesium. The lower and extreme upper portions of the mantle are rigid solids, but an intermediate portion, the asthenosphere, is plastic in nature. The earth's surface layer, the crust, consists of low-density rock composed largely of oxygen and silicon. A continuous lower portion, the oceanic crust, reaches the surface in the ocean basins. The uppermost portion is discontinuous and forms the continents.

Most rocks comprising the earth's crust consist of silicate minerals, which are composed largely of oxygen and silicon. This is especially true of the continental crust, which has a predominantly granitic composition. The oceanic crust consists largely of basaltic rock; it is denser than the continental crust and contains a greater proportion of ferromagnesian minerals.

Rocks are divided into three categories according to their mode of formation. Igneous rocks have solidified directly from a molten state. Intrusive igneous rocks form below the surface and are exposed by the erosion of overlying materials. Extrusive igneous rocks are formed at the surface by volcanic eruptions or fissure outpourings of lava. The second rock category, sedimentary rocks, are formed from the accumulation and lithification of sediments. Clastic sedimentary rocks, including conglomerate, sandstone, and shale, consist largely of particles of preexisting rocks. Organic sedimentary rocks, including limestone and coal, are composed of plant and animal remains. Chemical sedimentary rocks, including salt and some limestones, result from chemical precipitation. The final rock category, metamorphic rocks, are altered from their original states by mountain-building tectonic stresses. Familiar examples include slate, marble, and gneiss.

Tectonic forces originating deep within the interior are responsible for the continual alteration of t'e earth's surface. New crustal material is manufactured, while old material is destroyed. At the same time, the positions of the continents and oceans are constantly rearranged by the slow creep of huge volumes of plastic rock in the mantle. These motions are responsible for the formation of the continents and for the uplift of mountain ranges.

The movements of the lithospheric plates are caused by the frictional drag of convection currents in the underlying asthenosphere. These currents are produced by heat released by the nuclear decay of radioactive elements in the mantle and core. Upwelling mantle plumes result in the formation and divergence of new oceanic crust along the oceanic ridge systems. A colli-

sion between two plates having differing densities results in the subduction and destruction of the denser plate as it is forced downward into the mantle. The melting and surfaceward movement of some of the lighter and more volatile components of the subducted plate form new continental crustal material, which is volcanically erupted onto the surface. The collision of two continental plates typically produces a high-standing mountain belt composed of greatly thickened and complexly deformed continental crust.

Review Questions

1. What are the two basic approaches to the systematic classification of landforms? What are the advantages and shortcomings of each approach?

2. What are the two groups of geomorphic forces, and what is the ultimate energy source for each? How are the two groups of forces opposed to each other in terms of their influence on the topography?

3. Name the internal layers of the earth and briefly describe their physical and chemical characteristics.

4. Explain how the continental and oceanic crusts differ in terms of composition, thickness, density, age, and surface distribution. How does this distribution relate to the current global distribution of continents and ocean basins?

5. List and define the three major categories of rocks. Describe the environmental settings in which each is formed. What types of rocks dominate in the area in which your college or university is located?

6. Discuss several types of evidence that support the concept of continental drift.

7. What are lithospheric plates? Describe the current motion of the one on which you live and relate its movement to the major topographic features of the plate.

8. Explain how and why plate tectonic movements occur.

9. Explain the process of subduction. Why is oceanic crust destroyed by the subduction process, while continental crust is not?

10. How does the subduction of oceanic crust form new continental crust?

11. What will be the ultimate fate of plate tectonic movements, and what impact will this have on the earth's surface?

Key Terms

Topography
Geomorphology
Tectonic forces
Diastrophism
Volcanism
Gradational forces
Weathering
Core
Mantle
Crust
Asthenosphere
Lithosphere
Oceanic crust
Continental crust
Silicate mineral
Mineral

Rock
Intrusive igneous rock
Extrusive igneous rock
Sediment
Strata
Sedimentary rock
Metamorphic rock
Plate tectonics
Seafloor spreading
Lithospheric plates
Subduction
Suturing
Isostatic uplift

I. IGNEOUS ROCKS — solid — molten state

   A. INTRUSIVE - below surface

   B. EXTRUSIVE - outpourings of lava

II. SEDIMENTARY ROCKS — accumulation & lithufication of sediments

   A. CLASTIC — preexisting rocks — conglomerate, sandstone, shale

   B. ORGANIC — plant + animal remains — limestone + coal

   C. CHEMICAL — chem. precip. salt + some limestone

III. METAMORPHIC ROCKS

   — altered

   — slate, marble, gneiss

Chapter Twelve

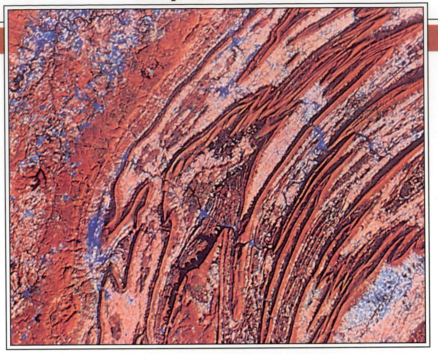

Diastrophic and Volcanic Landforms

Outline

Focus Questions

1. What causes rocks to fold, and what topographic features result from folding?
2. What are the major types of fault motions and their resulting topographic features?
3. What causes earthquakes? What controls the geographic distribution of earthquake hazards?
4. What causes volcanic eruptions, and what types of surface features do they produce?

Tectonic processes, as we saw in Chapter 11, control the formation and distribution of the earth's large-scale surface features, including the continents, ocean basins, and mountain belts. This chapter examines at much closer range the surface features produced by tectonic activity. The emphasis is on localized crustal motions and the types and distribution of the resulting tectonic landforms. Two basic categories of tectonic activity are recognized: diastrophism, or crustal displacements resulting from solid-state motions; and volcanism, or crustal displacements involving the transfer of molten rock-forming material.

DIASTROPHISM

Diastrophism (from the Greek *diastrophe* "twisted or distorted") is the solid-state deformation of the earth's crust resulting from internal pressures. Deformation can occur either through the folding of rock material or through faulting, which involves the breaking and subsequent displacement of the rock. The nature of the crustal response to tectonic forces depends on both the characteristics of the rock material and on the nature and extent of the stresses involved.

Linear Folding—Processes and Features

We tend to think of the rocks at the earth's surface as possessing great strength. Strength, though, is a relative term; and the tectonic forces are easily capable of deforming great thicknesses of crustal rock. Most folding takes place in response to lateral compression resulting directly or indirectly from lithospheric plate collisions. The rocks are buckled and folded, and the entire area is generally uplifted as the relatively buoyant layers of crustal rock are compressed into a smaller horizontal area and increased in thickness.

Folding usually involves sedimentary rocks, which, as a group, are softer and more flexible than either igneous or metamorphic rocks. The ability of these rocks to deform elastically without breaking is further increased by the pressure and heat resulting from their burial at depths of often several miles at the time of folding and by the fact that the compression is normally applied at maximum rates of only a few inches a year.

The compression of rock strata produces a linear pattern of folding. Folded rock structures are highly similar in both cause and appearance to the pattern of folds produced when a carpet is pushed against a wall. Direct compression of this type produces a fold pattern with its long axis or *strike* perpendicular to the axis of compression (see Figure 12.1). This explains, for example, why the westward movement of the American Plate has resulted in the north-south orientation of the Rockies and Andes, and why the northward movement of India into Asia has produced the east-west strike of the Himalayas.

The areal extent and intensity of folding are controlled largely by the degree of compression and by the strength and thickness of rock strata involved. With moderate compression, a wavelike series of open folds is produced. Each upfold is termed an **anticline**, and each downfold a **syncline.** Both the vertical dimensions and the slope angle, or *dip*, of the flanks of the folds increase as the amount of compression increases. If compression is intense, the folds may be squeezed tightly in on one another.

The relationship between fold patterns and topographic features might be expected to be straightforward with upfolded areas producing mountain ridges and downfolded areas producing valleys. This situation rarely exists in reality because of the effects of **differential erosion**, or erosion occurring at differing rates. In few regions has folding occurred recently enough, and on a large enough scale, for folding to control the present pattern of surface features directly. Most such regions have dry climates, and the lack of surface water reduces the effectiveness of the gradational processes. Included are parts of North America, the Middle East, New Zealand, and limited portions of western North and South America. Within these regions, the surface is dominated by anticlinal ridges and synclinal valleys produced directly by folding. An *anticlinal ridge* is a ridge composed of upfolded rock strata, while a *synclinal valley* is a valley developed upon downfolded strata. It should be stressed, though, that in most areas ridges are not produced directly by upfolding; nor are most valleys produced by downfolding. How then, are ridges and valleys in folded topography formed?

Over extensive areas, folding has acted in concert with the forces of gradation to exert a dual control on the topography. Folding initially deforms the rock strata so that they vary greatly in their angle of inclination, or dip. As erosion lowers the surface, rock strata that vary considerably in their susceptibility to weathering are exposed. Erosion occurs at differential rates in these strata, with layers of resistant rock eventually forming ridges or other uplands and weaker rock layers forming

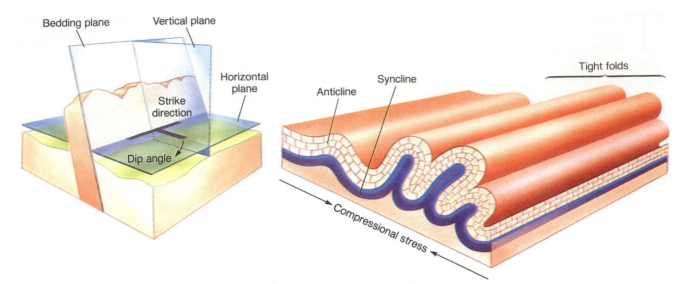

Figure 12.1 *Diagrams*—Linear fold terminology and features. *Photo*—Tight folding in limestone strata exposed in a railroad cut in the central Appalachians near Cumberland, Maryland. (*John M. Morgan, III*)

valleys or lowlands. The resulting topography is therefore controlled by differences in erosional rates between stronger and weaker rock strata whose surface patterns have resulted from folding.

The stretching and compression of rock strata during the folding process may also influence their susceptibility to erosion. Upfolding stretches the rock, weakening it and often producing stress fractures that facilitate subsequent weathering and erosion. Conversely, downfolding compresses the rock, generally increasing its strength. Other factors being equal, then, anticlines contain somewhat weaker rocks than do synclines. After folding occurs, therefore, the initial anticlinal ridges tend to be removed rapidly by erosion; and, as time goes by, the greater weakness of the upfolded rock

strata results in the formation of *anticlinal valleys* in their places (see Figure 12.2). The more resistant downfolded rock strata, meanwhile, are being eroded more slowly; so they eventually extend above their surroundings in the form of *synclinal ridges*. At this time, an inverse relationship exists between surface features and rock structure, and a **reversal of topography** is said to have occurred since the time of initial folding. Because of the influence of folding on rock strength, a reversal of topography is more common than an accordance between structure and topography.

In many folded areas, the erosional resistance of the rock strata is influenced more by rock type than by fold structures. Where differing rock types exist, some are likely to be more resistant to erosion than others because

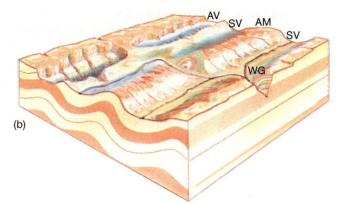

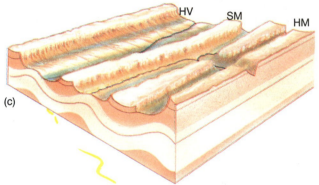

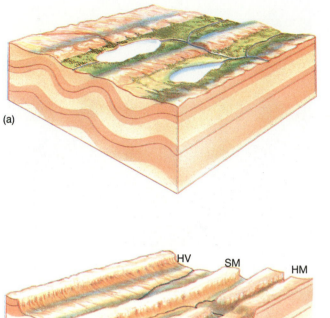

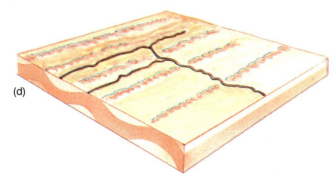

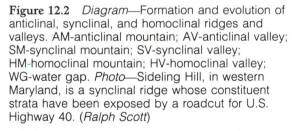

Figure 12.2 *Diagram*—Formation and evolution of anticlinal, synclinal, and homoclinal ridges and valleys. AM-anticlinal mountain; AV-anticlinal valley; SM-synclinal mountain; SV-synclinal valley; HM-homoclinal mountain; HV-homoclinal valley; WG-water gap. *Photo*—Sideling Hill, in western Maryland, is a synclinal ridge whose constituent strata have been exposed by a roadcut for U.S. Highway 40. (*Ralph Scott*)

of their physical hardness and/or their chemical stability. In areas containing folded or tilted rock strata of greatly differing resistances to erosion, the positions of the ridges and valleys will be controlled by the surface distribution of strong and weak rocks. Resistant strata will undergo reduced erosion rates and become ridges, and nonresistant strata will erode rapidly to become valleys. As time goes by and the erosional reduction of the land continues, the positions of the ridges and valleys will continually shift to coincide with the changing patterns of resistant and nonresistant rock types exposed at the surface. The ridges and valleys of the Appalachians are a good example of the results of this process (see Figure 12.3 and chapter opener profile).

Regardless of the types of fold structures, if the folding occurred geologically recently and resulted from substantial crustal compression, the affected area will normally be uplifted to a high mean elevation. Erosion should therefore be vigorous, and, if the rock layers within the area differ considerably in their erosional resistance, a high relief surface will result. Conversely, ancient folded rock areas that have been geologically stable for long periods usually display low relief surfaces because even the resistant strata have been worn down.

The major mountain belts associated with active lithospheric plate boundaries are the primary global sites of recent folding; active folding is still occurring within many portions of these belts. Evidence suggests that average maximum rates of uplift within portions of these belts are approximately 0.25 inches (6–7 mm) per year. Older mountain systems such as the Appalachians, the Urals, the Great Dividing Range of Australia, and the highlands of the British Isles contain highly folded strata but are not currently undergoing active folding. They are located on now-sutured lithospheric plate boundaries and owe their survival as low mountain systems to sporadic isostatic uplift of the crust. The ancient shield regions also contain complexly folded, faulted, and metamorphosed rock strata produced by plate collisions that occurred hundreds of millions or even billions of years ago. In many areas they are covered by more recent undeformed strata. Taken together, most land areas are underlain, at least at depth, with rocks that have been folded by compressional forces associated with lithospheric plate collisions.

Domes and Basins

Linear folding is largely a product of horizontal compressionary forces resulting from lithospheric plate collisions. Folding, however, can also be produced by the application of vertical forces. These forces consist of upward pressures from the earth's interior and the downward pull of gravity. If the upward pressures exceed the gravitational pull, the affected area is arched upward to produce a rounded anticlinal structure known as a **dome** (see Figure 12.4). If, on the other hand, internal pressures become insufficient to counter-

Figure 12.3 The Susquehanna River has cut water gaps through these linear Pennsylvania ridges composed of resistant sandstone (*John S. Shelton*)

act gravity, the surface rock layers sag, producing a synclinal **basin.**

Structural domes and basins underlie extensive portions of the earth's surface. They display great variability in size, shape, age, and extent of folding. Some very large domes and basins cover areas of tens of thousands of square miles. These structures are all of considerable geologic age, contain very gently arched or downwarped rock strata, and have rather obscure origins. Perhaps the most important practical topographic effect of large dome and basin structures occurs when they involve land near sea level. Doming may in this case generate sufficient uplift to raise a portion of the seafloor and produce a land area, as in the case of the Florida peninsula. Downwarping, conversely, may submerge an area of continental crust, as in the case of Canada's Hudson Bay.

Most domes and basins are old enough that the associated landforms result directly from differential erosion rather than from vertical bedrock movements. A dome or a basin will normally stand topographically higher than its surroundings if the rocks exposed in its center are more resistant to erosion than those outside the structure. One of the best-known examples of a large dome with an elevated, mountainous center caused by the exposure of resistant rocks is the Black Hills of South Dakota (see Figure 12.21). If, on the other hand, weaker rocks are exposed in the center of either a dome or a basin, a lower surface will result. A good example is provided by the Nashville Basin in central Tennessee.

Figure 12.4 The location of an eroded dome near Sinclair, Wyoming, is indicated by the pattern formed by the eroded edges of the relatively resistant rock strata. (*John S. Shelton*)

Although this feature is a structural dome, it is a topographic basin because erosion has stripped away the overlying more resistant sandstone layers, exposing less-resistant limestone in the dome's center.

Faulting

Faulting involves the breaking or fracturing of rock under pressure, followed by movement of the two sides of the fracture relative to one another. While folding is most frequently associated with sedimentary rock, faulting can occur in rocks of any type but is especially associated with the more brittle igneous and metamorphic rocks. Folding and faulting generally occur together in areas subjected to tectonic stress.

Although the stresses producing both fold and fault movements are applied very gradually, a critical difference exists between the two movements. While folding slowly and continuously relieves the stresses, fault movements frequently store stresses for a time and then release them suddenly, producing earthquakes. Individual fault movements are relatively small, ranging from a fraction of an inch up to about 25 feet (7.5 m), but they are repeated periodically when stresses build to the point that they exceed the rock strength on opposing sides of the fault. For a fault that has long been active, the total displacement of the opposing sides over mil-lions of years can attain values of hundreds of miles horizontally and tens of miles vertically.

Fault Terminology and Types of Faults

A fault takes the form of a two-dimensional plane that typically extends from the earth's surface downward to a variable but often considerable depth (see Figure 12.5). The trace of the fault on the surface is termed the *fault line*. Fault lines may extend for hundreds of miles, but lengths of a few tens of miles are more common. Most faults are nearly straight. This linearity, which results from the tendency of rock to fracture along straight lines, contrasts markedly with the irregularity of the features produced by most other geomorphic processes.

Geologists recognize four general categories of faults according to the nature of the displacements that occur; these are termed normal, reverse, transcurrent, and thrust faults. **Normal faults** are the most common of the four fault types. Relative motion is more vertical than horizontal, and an expansionary component is present, so the opposing sides also move apart, resulting in crustal extension (see Figure 12.5a). Normal faults are usually produced by broad regional arching in areas of tectonic stress.

Reverse faults are so-named because the movement of the opposing sides is reversed from that of normal faults. Like normal faults, reverse faults have deeply dipping fault planes and undergo predominantly vertical motion (see Figure 12.5b). Unlike normal faults, though, reverse faults are produced by regional compression. Crustal shortening results, and a net uplift of the surface normally occurs.

Transcurrent faults undergo a predominantly horizontal offsetting of their opposing sides (see Figure 12.5c). They are most frequently located along transform plate boundaries, where the relative motions of the opposing plates are essentially parallel. Most transcurrent faults are located on the floors of oceanic plates and are produced by seafloor spreading movements, but some, like California's San Andreas Fault (see the Focus box on page 270), occur on land.

Thrust faults result from the extreme compression of rock strata produced by lithospheric plate collisions. The relative movement of the opposing sides is similar to that of a reverse fault. Relative motion, however, is predominantly horizontal, as one side is thrust over the other, sometimes for considerable distances (see Figure 12.5d).

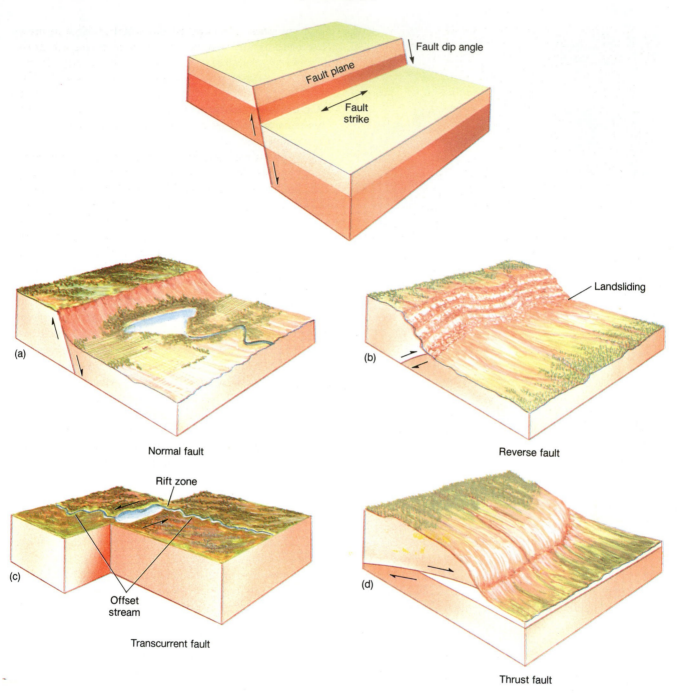

Figure 12.5 Fault terminology and profile views of the four basic fault types.

Landforms Associated with Faulting

Fault movements produce a great variety of topographic features. These range in scale from large mountain systems and fault troughs with many thousands of feet of relief down to microrelief features that may go completely unnoticed on the landscape. It should be stressed that fault features of large size are the net result of many successive fault displacements occurring over extended periods of time. Fault features may be produced directly through the movement of the blocks of

264

Figure 12.6 A portion of the fault scarp that formed in Gallatin County, Montana, as the result of a major earthquake in August 1959. A 19-foot displacement between the two sides was measured. (*J.R. Stacy, USGS*)

land on opposing sides of the fault, or indirectly as a result of differential erosion. Indirectly produced fault features usually result from the fault movements bringing rocks of differing erosional resistances into contact. Gradational forces will subsequently erode the less-resistant rocks at a more rapid rate.

High-angle faults chiefly involve vertical displacements of rock strata; consequently, if these faults are active for a long time, they can produce features with large vertical dimensions. The basic landform feature associated with high-angle faulting is the **fault escarpment** (or **fault scarp**). It is ideally a linear, smooth-sided slope consisting of the exposed surface of the fault plane along the edge of the higher fault block (see Figure 12.6). Repeated movements along a major fault over millions of years can produce fault scarps thousands of feet high extending for hundreds of miles. Most of the earth's loftiest mountain ranges, including the Himalayas, are produced largely by high-angle fault movements.

Topographic features produced by low-angle faulting normally have lower relief and are much less conspicuous on the landscape than are features produced by high-angle faults. Transcurrent faults, especially, may provide little surface indication of their existence in localities where the offsetting of the two sides is essentially parallel. Frequently, though, a zone of crushed rock is produced along the fault line as the opposing sides grind against one another. This forms a zone of weakness that can be exploited by the agents of weathering and erosion to produce a linear depression, perhaps occupied by a stream or lake. As a stream crosses a transcurrent fault line, its course may be offset by subsequent fault movements so that the stream channel is pulled apart, giving it a zigzag course (see Figure 12.7).

California's famed San Andreas Fault is the best-known example of a transcurrent fault. Extending through the state for some 550 miles (880 km) from the Mexican border to Tomales Bay just north of San Francisco, it marks the boundary between the American and Pacific Plates (see Figure 12.8). Geologically, the portion of California situated west of the fault is not a part of North America, but is simply moving northward along the margin of the continent at the present time. The San Andreas Fault has apparently been active for several tens of millions of years, and at least 350 miles (550 km) of offsetting of the two sides has occurred. It is believed that the continuing northward movement of the Pacific Plate will eventually carry western California to a position off the coast of Alaska. Because the plate's movement in this direction averages only a few inches per year, the chief concern for California's residents is the potential destruction from earthquakes resulting from future fault displacements. This topic is examined in the Focus box on page 270.

Figure 12.7 The course of this stream channel has been offset by repeated slippage along the San Andreas Fault. (*John S. Shelton*)

265

Figure 12.8 Map of California showing the location of the San Andreas Fault and its subsidiary faults. The dates and Richter scale magnitudes of major historical earthquakes are given.

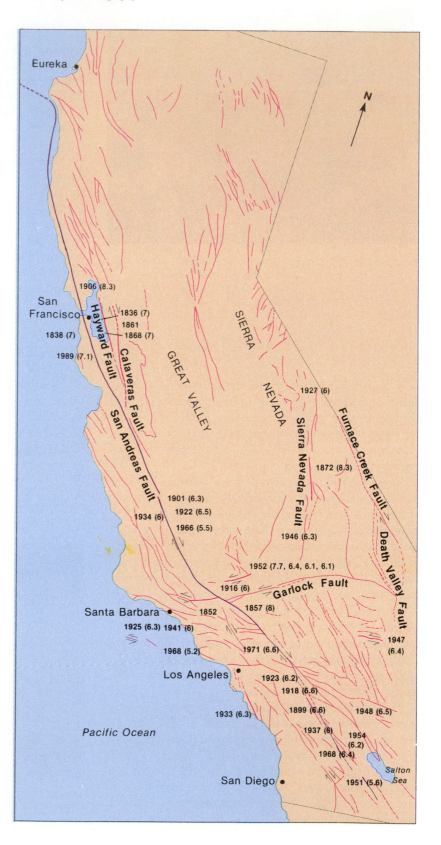

Multiple Faulting

When an area is subjected to tectonic stresses, it is rare for only a single fault to develop; instead, faults, like folds, tend to form in multiples. An area under stress will usually develop a series of faults that parallel one another. The blocks of land between such multiple fault systems are relatively lacking in support and are likely to experience vertical displacement. This may be accompanied by tilting, so a considerable variety of movements, and resulting topographic features, can be produced.

The basic features produced by the movement of blocks of land between parallel faults are horsts and grabens. A **horst** consists of a raised block bounded by two normal faults. A recently formed horst will ideally appear as a linear, steep-sided, flat-topped ridge. The forces of erosion soon modify this form, so it eventually reduced to a range of hills or low mountains. Horsts exist in a variety of sizes; some well-known examples of large horsts include the Palestine Plateau, the Black Forest of Germany, and the Sinai Peninsula. A **graben** is structurally similar to a horst, except that the central block of land is downdropped between two normal faults. Ideally, a graben initially forms as a linear, smooth-sided, flat-bottomed lowland. Once again, the forces of erosion soon modify it so that its original angularity and linearity are lost. Well-known examples of grabens include California's Death Valley, the Rhine Valley separating Germany and France, and the Dead Sea Trough.

If the expansionary stresses that produce grabens continue, the separation between the opposing sides may increase until they assume the dimensions of **rift valleys**. Rift valleys are especially large, elongated grabens whose formation frequently marks an early stage in the separation of two lithospheric plate segments. The outstanding present-day example of this feature is the East African Rift Valley (see Figure 12.9). This immense and complex system, which extends over 2000 miles (3200 km) through East Africa from Mozambique northward to Djibouti, is bounded by fault scarps up to 3000 feet (900 m) high. On its floor lie some of the world's largest lakes, including Lakes Tanganyika, Nyasa, Albert, and Rudolf.

In some parts of the world, large areas of continental crust have been faulted and broadly arched upward by internal pressures that are later relaxed. The faults divide the surface into large blocks that, upon relaxation of pressure from below, drop and tilt to varying degrees.

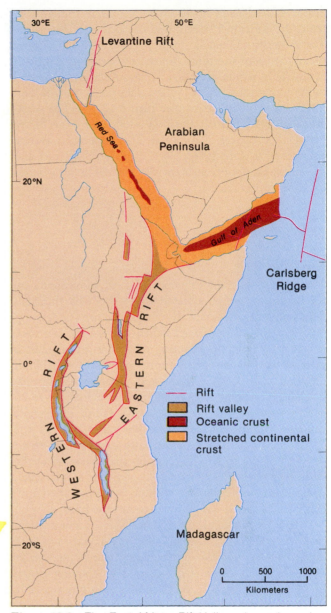

Figure 12.9 The East African Rift Valley, shown here, is apparently being formed by the gradual separation of eastern Africa from the rest of the continent as a result of divergent plate movements. The earlier separation of other blocks formed Madagascar and the Arabian Peninsula.

The higher blocks tend to form steep-sided linear mountain ranges. They are separated by broad, flat lowlands developed on the downdropped blocks. Regions containing multiple fault-block mountains and basins include the Great Basin of the interior western United

States and adjacent northern Mexico, the Iranian Plateau, the Tibetan Plateau, and the Bolivian Altiplano. The American Great Basin provides an excellent example of the basin and range topography developed in such areas. This region contains a large number of normal faults separating block-faulted mountain ranges and intervening broad basins (see Figure 12.10). The western edge of the Great Basin is bounded by the Sierra Nevada of eastern California. This mountain system consists of a giant tilted block of ancient igneous and metamorphic rock that originally formed as an igneous intrusion (or batholith).

Earthquakes

The most important influence of fault movements on the inhabitants of an affected region is the earthquakes they frequently generate. **Earthquakes** are sudden surface movements produced when the slowly accumulated strain along the opposing sides of an active fault is released as the two sides are torn apart. Individual fault motions are relatively small and involve relative displacements ranging from a fraction of an inch up to 20 or 25 feet (6–7.5 m). The same faults experience slippage and produce earthquakes repeatedly as the accumulation of stress periodically exceeds the shear strength of the rock. The rocks on the two sides of the fault deform elastically much like a rubber band, which stretches in response to a buildup in strain, then suddenly snaps when the strain becomes too great.

Most earthquakes are undetectable except to delicate recording instruments called *seismographs*. Thousands of perceptible earthquakes, however, occur somewhere on earth each year. The shaking of an earthquake is actually produced by the passage of a shock wave resulting from the sudden fault movement.

Earthquake magnitudes are commonly measured by use of the *Richter scale,* devised by seismologist Charles F. Richter in 1935. It is a logarithmic scale, with each number having 10 times the earthquake wave amplitude of the preceding number. An earthquake with a Richter number greater than about 2.5 can usually be felt; one above 5 is considered damaging; and one of 8 or above is catastrophic. The most intense earthquake measured since accurate readings have been available occurred in central Chile in 1960 and had a magnitude of 8.9. By way of comparison, the great San Francisco earthquake in 1906 had an estimated magni-

Figure 12.10 The Great Basin of Nevada and western Utah is in an area of crustal extension and consists of block-faulted mountain ranges and broad intervening basins. (*Marbut Memorial Slide Collection, American Society of Agronomy*)

tude of 8.3; the Alaskan earthquake of 1964 was measured at 8.5; and the 1989 California earthquake was measured at 7.1.

The distribution of earthquakes is essentially identical to that of active faults: they are concentrated along lithospheric plate boundaries, especially those where convergence and active subduction are taking place. A map of earthquake epicenter distributions clearly outlines the major plate boundaries (see Figure 12.11). Because these boundaries are frequently the sites of active tectonic uplift, earthquakes on land generally are centered in mountainous regions. Nearly 95 percent of recorded earthquakes occur around the rim of the Pacific and from southern Asia westward through southern Europe. Most other earthquakes take place along generally less active faults located within the lithospheric plates. Some of the movements along these faults can also produce devastating earthquakes on occasion. For instance, much of Charleston, South Carolina, was destroyed by an 1886 earthquake. A series of quakes centered near New Madrid, Missouri, in 1811 and early 1812 sent huge waves rolling down the Mississippi River, changing its course in several locations. Although some regions are notably less prone to earthquakes than others, few if any parts of the world can be considered immune to a major quake (see Figure 12.12).

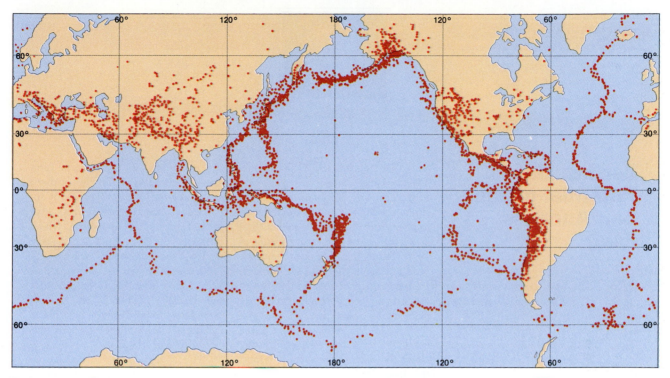

Figure 12.11 Global distribution of the epicenters of earthquakes recorded between 1963 and 1977. Note that the dots collectively outline the lithospheric plate boundaries depicted in Figure 11.16.

Individual earthquakes generally have only a minor impact upon landforms. They may result in the offsetting of surface features by several feet, produce small fault scarplets (see Figure 12.6), and trigger landslides. By far the most important effect, though, is on people, especially those occupying buildings constructed of massive materials at the time of the quake and those living along coasts that may experience earthquake-produced tsunami (giant ocean waves). Major earthquakes rank with hurricanes and volcanic eruptions as producers of the most catastrophic natural disasters in recorded history (see Figure 12.13). Cities are especially vulnerable, since buildings may be shaken to the ground; and ruptured gas pipes and downed electric wires cause fires that cannot be reached to be extinguished. The greatest loss of life resulting from an earthquake in the twentieth century occurred in Tangshan, China, in 1976, where an estimated 655,000 persons lost their lives. A 1923 earthquake in Japan destroyed most of Tokyo and claimed 100,000 victims. As already noted, the growing potential for a major earthquake along the San Andreas Fault is currently of great concern to the residents of California.

VOLCANISM AND VOLCANIC LANDFORMS

Volcanism is the second major category of tectonic activity. In contrast to diastrophism, which involves solid-state movements of the earth's crust, *volcanism* involves the surfaceward movement of fluid magma. Volcanism occurs where the lithosphere contains deep fractures or faults, a situation common along and near plate boundaries. These breaks in the lithosphere serve two essential purposes in promoting volcanic activity. First, they may relieve pressures sufficiently to permit the localized melting of rock that forms the magma; second, they provide pathways for magma's surfaceward movement.

Extrusive Volcanism

Extrusive volcanism consists of processes by which molten rock reaches, and is extruded upon, the earth's surface. Although it occurs less frequently than intrusive volcanism and generally involves smaller quantities of molten rock, extrusive volcanism is much better

269

Earthquake Potential of the San Andreas Fault

The fault best known and most feared for its earthquake potential of any within the United States is California's San Andreas Fault (see Figure 12.8). Minor slippages occur frequently along this great fault and its numerous offshoots, but the last large-scale movement took place in 1906. This involved a 10- to 20-foot (3–6 m) displacement of the opposing sides of the northern portion of the fault. It produced the great San Francisco earthquake and fire that leveled much of the city and claimed 700 human lives.

Much more recently, on October 17, 1989, another slippage on the San Andreas Fault caused the Loma Prieta earthquake. This earthquake, which lasted less than 15 seconds, had its epicenter in the Santa Cruz Mountains about 60 miles (100 km) south of San Francisco. It resulted in 67 deaths

and 3500 injuries, left 12,000 homeless, and caused property damage estimated at nearly $6 billion. The Loma Prieta earthquake had a Richter scale intensity of 7.1, less than one-tenth as powerful as the great earthquake of 1906. Future earthquakes along the fault are inevitable, and Californians are increasingly concerned that one of catastrophic proportions may take place in the near future.

Two lines of evidence point to the possibility of a powerful earthquake along the San Andreas Fault within the next few decades. In the first place, over 85 years have now passed since the San Francisco earthquake of 1906. Prior to that time, major slippages had been taking place at approximately 50-year intervals, so the next movement is now overdue. A second line of evidence comes from measuring instruments and geological observations which indicate that a large amount of strain has developed between the opposing sides of the fault in several areas. One ominous indication of this strain is the so-called "Palmdale Bulge," centered on the fault about 35 miles (56 km) northeast of Los Angeles, where the surface of

the earth has in recent years risen about 12 inches (0.3 m). The southern section of the San Andreas Fault apparently became tightly locked after its last large-scale movement, which produced the Fort Tajon earthquake of 1857; to date, it has resisted any further movements of consequence. Many experts currently believe that the southern portion of the fault is more likely to produce the next major earthquake than is the northern portion near San Francisco.

The tremendous population growth in western California during the twentieth century has greatly increased the potential for severe loss of life and property. Construction standards for new buildings have been improved substantially to make them more earthquake resistant, and civil defense authorities have long been preparing to cope with the consequences of a major earthquake. All that most Californians can do at present, though, is to wait and hope that the fault will provide prior warning of its next movement.

known, more spectacular, and has a more direct influence on the topography.

Two basic groups of extrusive volcanic features exist: volcanoes and lava flows. The difference between the two is determined largely by the fluidity of the magma and the force with which it is expelled onto the surface. A comparison of volcanoes and lava flows illustrates the fact that features with similar origins may differ greatly in their topographic expressions.

Volcanoes

A **volcano** is a hill or mountain constructed from material that has been ejected under pressure from a vent (see Figure 12.14). Large volcanic peaks are one of the most beautiful and distinctive types of landforms. They typically have a conical and sometimes nearly perfectly symmetrical shape and may extend to great heights above the surrounding landscape. They are often surmounted by a cap of snow and ice and, if

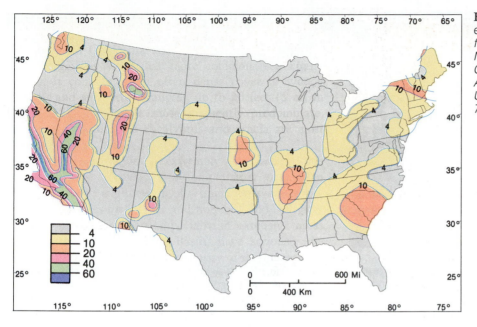

Figure 12.12 Map of U.S. earthquake potential. (*Modified from "A Probabilistic Estimate of Maximum Acceleration in the Contiguous United States." S. T. Algermissen and D. M. Perkins U.S. Geol. Surv. Open-File Rept. 76-416, July 1976.*)

active, may extrude steam, smoke, and sometimes ash or lava. A volcanic eruption, especially when viewed at night, may be the most spectacular and awe-inspiring of all geomorphic events. The fascination of volcanoes is increased by their unpredictable nature and demonstrated potential for dealing death and destruction on a large scale.

At present, approximately 500 active volcanoes are scattered throughout the world, and their eruptions have taken over 250,000 human lives during the past four centuries. The greatest known loss of life during that period occurred in the 1815 eruption of Tamboro Volcano in the East Indies, which killed 56,000 people. Most eruptions occur from existing volcanoes, but about 15 or 20 new volcanoes have formed during historic times. Two well-documented cases of volcanic births are of Parícutin in a Mexican cornfield in 1943 and of Surtsey off the coast of Iceland in 1963.

Figure 12.13 Most of the casualties resulting from the Loma Prieta earthquake near San Francisco in October 1989 occurred when the top section of Interstate Highway 880 collapsed onto the bottom section. (*The Bettmann Archive*)

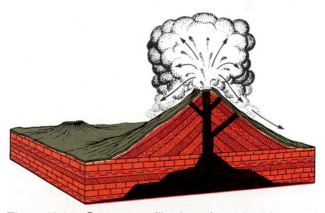

Figure 12.14 Cutaway profile view of a composite volcano, showing the magma chamber, system of vents, and layers of lava and ash comprising the cone.

271

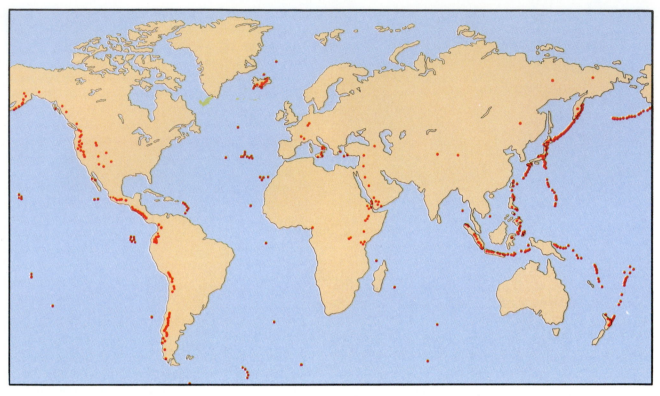

Figure 12.15   World distribution of volcanoes that have erupted in geologically recent times.

The global zones of present-day volcanic activity are closely associated with those areas where folding, faulting, and earthquakes are currently taking place. As a result, most volcanoes are located on or near lithospheric plate boundaries. Nearly four-fifths of the active volcanoes on land are located in the "Ring of Fire" around the circumference of the Pacific. Portions of the Pacific margin where volcanoes are especially numerous include the southern and northern Andes, southern Mexico, the Aleutian Islands, and the entire western Pacific arc from the Soviet Union's Kamchatka Peninsula southward to northern New Zealand (see Figure 12.15). Other notable centers of activity include the Mid-Atlantic Ridge, especially in the vicinity of Iceland; the East African Rift Valley, the Lesser Antilles, the Mediterranean Sea, the islands of Java and Sumatra in the East Indies; and scattered mid-oceanic "hot spots," such as Hawaii, Tahiti, the Galápagos, and Réunion, situated above mantle plumes.

The nature of the volcanic activity within any region is controlled largely by the type of magma involved in the eruption. Eruptions involving fluid magmas are relatively nonviolent and are characterized by the outpouring of large quantities of lava. At present, most such eruptions are occurring on the ocean floor or on mid-oceanic islands situated above mantle plumes, such as Hawaii. The basaltic rock formed from the solidification of these eruptions has produced the oceanic crust.

Volcanoes situated above areas of plate convergence and subduction typically erupt magma that is cooler and considerably more chemically acidic, or rich in silicon dioxide. The result is a much more viscous or "pasty" type of magma. The high viscosity of this magma inhibits the release of its compressed gases. This usually results in violent volcanic eruptions powered by the explosive escape of the pent-up gases. This factor, for example, caused the great violence of the initial eruption of Mount St. Helens in May 1980 (see Figure 12.16).

The type and quantity of available magma, the pressure under which it is extruded, and the persistence of subsequent eruptions at the same site all affect the nature of the surface features produced by volcanic eruptions. The most abundant, most widespread, and generally the smallest of the major types of volcanoes are

Figure 12.16 Aerial view of the Mount St. Helens eruption of May 18, 1980. (*USGS, JLM Visuals*)

cinder cones. They are steep-sided conical hills or low mountains constructed of loose fragmental material thrown into the air from a central vent (see Figure 12.17). The central vent forms a depression, or crater, which is filled by loose volcanic rock. Cinder cones exist in a variety of sizes, but most are only a few hundred feet high. They generally form very rapidly, and most become extinct after only one period of eruption. Their composition of loose, easily weathered material allows cinder cones to be eroded rapidly by gradational forces, so they are short-lived features on the landscape. Geologically recent cinder cones are found in several portions of the western United States, notably in central Arizona and in Idaho's Snake River Plain.

Shield volcanoes, in contrast to cinder cones, have broad, gently sloping surfaces and are constructed primarily of solid basaltic rock rather than loose materials (see Figure 12.18). They are located mostly in the ocean basins at "hot spots" above mantle plumes and form volcanic islands or island groups. Examples of shield volcanoes are Mauna Loa and Mauna Kea in Hawaii, Mt. Hekla in Iceland, and Mt. Etna in Sicily, as well as the volcanoes that have formed the islands of Réunion, Mauritius, Tahiti, Samoa, and the Azores. The eruptions of shield volcanoes are relatively quiet and consist largely of outpourings of great quantities of fluid lava. Successive eruptions of individual shield volcanoes may extend over long geologic time spans before these volcanoes become extinct, so mountains of great size are eventually produced.

Most of the largest volcanic peaks are constructed of alternating layers of fragmental materials and lava. Because their surficial and eruptive characteristics are in some respects transitional between cinder cones and shield volcanoes, they are known as **composite volcanoes.** Nearly all are located within mountain belts or island arcs situated above convergent lithospheric plates. Some of the better-known examples are Mt. Fuji

Figure 12.17 Sunset Crater, near Flagstaff, Arizona, is a geologically recent volcanic cinder cone with a basal lava flow. Its only eruption occurred in A.D. 1065. (*John S. Shelton*)

Figure 12.18 The gently sloping profile of Mauna Loa volcano as seen from the north coast of Hawaii. (*B.F. Molnia*)

in Japan; Mounts Shasta, Hood, Rainier, and St. Helens in the western United States; Mt. Vesuvius in Italy; Mt. Kilimanjaro in Tanzania; Mt. Mayon in the Philippines; and Mt. Egmont in New Zealand. Composite volcanoes combine the steepness and distinctiveness of cinder cones with the large size and geological persistence of shield volcanoes. Unlike cinder cones, composite volcanoes frequently have multiple craters and are asymmetrical in shape.

The sudden eruptions of composite volcanoes have been responsible for most major historic volcanic disasters. Eruptions are sometimes exceedingly violent because extreme gas pressures can build up when hardened lava from earlier eruptions has blocked the volcano's vent. When these pent-up gases are finally released, explosive eruptions capable of releasing prodigious quantities of ash occur. (The Case Study of the eruption of Krakatau at the end of the chapter provides a good illustration.) Heavy ashfalls can bury the surrounding countryside to depths of tens or even hundreds of feet, producing ash-covered plains or ash-filled basins. Fine ash can be injected high into the stratosphere, reducing solar energy sufficiently to produce a detectable lowering of worldwide temperatures over several years. Associated lava flows are also capable of burying the towns and agricultural lands that often surround volcanoes.

In various parts of the world, large, roughly circular depressions of volcanic origin known as **calderas** are found. Most are believed to be formed by the collapse of a volcano into its subsurface magma chamber after the chamber was emptied by an eruption. Diameters of calderas range from about 2 miles (3.2 km) up to 10 miles (16 km) or more. The best-known caldera in the United States is that formed by the prehistoric eruption and collapse of Mt. Mazama in Oregon. This 5-mile (8 km) wide caldera is now occupied by Crater Lake, and its natural beauty and unique origin have caused the United States government to preserve it within a national park (see Figures 12.19 and 12.20). The violent eruption of Krakatau in the East Indies in 1883, described in the Case Study, resulted in the formation of an underwater caldera.

Lava Flows

Lava flows occur when subsurface gas pressures are insufficient to throw molten rock material into the air during an eruption; instead, it wells out of the ground in

Figure 12.19 Aerial view of Crater Lake, in Oregon. Wizard Island, a cinder cone, is visible on the far side of the lake. (*John S. Shelton*)

a fluid state. This may occur during the latter stages of a volcanic eruption, due to the reduction in gas pressures at this time. Lava flows, sometimes on a massive scale, can also occur where magmas are so fluid that they offer little resistance to their dissolved gases, allowing them to escape readily to the surface.

At some sites beneath divergent plate boundaries and above mantle plumes, temperature and pressure conditions allow the partial melting of large quantities of rock within the upper mantle, producing reservoirs of basaltic magmas. If these magmas reach the surface, extensive outpourings of highly fluid *flood basalts* are produced. The greatest quantities of flood basalts, by far, occur on the ocean floor along the seafloor-spreading centers of the oceanic ridges. The movement of the lithospheric plates away from these ridges has spread their eruption products over the earth to produce the oceanic crust.

Of greater direct significance to the topography of the continents are outpourings of flood basalts on land above "hot spots" that are apparently associated with mantle plumes. These outpourings occur as **fissure eruptions** along a major fracture or fault rather than from a single vent, as in a volcano. Huge outflows of flood basalts have occurred in a number of widely separated locations, including the Columbia River Plateau and Snake River Plain of the northwestern United States, the Deccan Plateau of northern India, the Paraná Plateau of southern Brazil, and the Drakensberg Plateau of South Africa, as well as in portions of

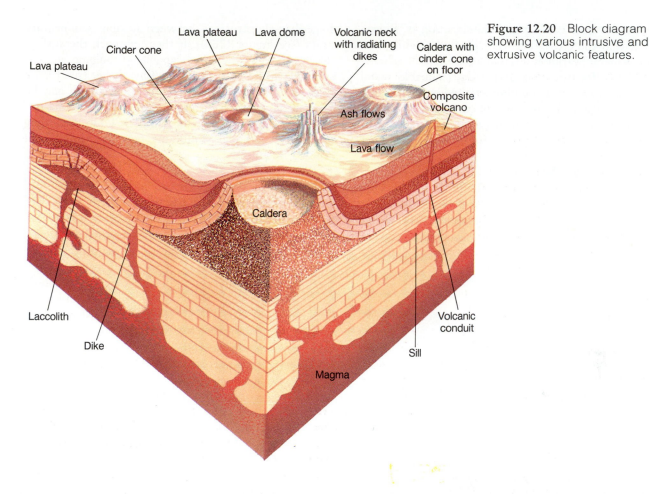

Lava plateau
Cinder cone
Lava plateau
Lava dome
Volcanic neck with radiating dikes
Caldera with cinder cone on floor
Composite volcano
Ash flows
Lava flow
Caldera
Laccolith
Dike
Volcanic conduit
Sill
Magma

Figure 12.20 Block diagram showing various intrusive and extrusive volcanic features.

Antarctica, Bolivia, Ethiopia, Iceland, the Soviet Union, and northwestern Arabia.

Fissure eruptions may largely or completely bury the original surface, producing a gently rolling landscape that, depending on its elevation, may be termed a **lava plain** or a **lava plateau.** Successive flows may accumulate to depths of thousands of feet. The Columbia Plateau, for instance, consists of 50,000 square miles (130,000 km²) of lava-covered surface, and in some locations the lava reaches depths of 10,000 feet (3000 m). The lavas of the Columbia Plateau have been extruded geologically recently in a large number of separate eruptions. The eruptive cycle is probably not yet completed, and future outpourings of flood basalts may again cover large portions of this region.

Intrusive Volcanism

Intrusive volcanism occurs when surfaceward-moving magma solidifies into igneous rock before reaching the surface. If solidification takes place far below ground, any effect on the surface is indirect and occurs at a much later geologic period when the deep-seated pluton is exposed by the erosional removal of the overlying rock. Near-surface intrusions as well generally have their greatest influence on the topography when exposed by erosion. Unlike plutonic intrusions, however, near-surface intrusions may also have a direct and immediate effect if they deform the overlying rock layers during their upward movement. The surrounding rock may be domed up and may also be metamorphosed by the heat and pressure of the molten rock. When intrusive igneous rock masses do reach the surface, their effect on landforms results chiefly from differential erosion rates. In most, but not all cases, the igneous rock is more resistant to erosion than the surrounding rock and produces high-standing landscape features.

By far the largest of the plutonic rock features are **batholiths.** They are huge plutons with diameters of at least several tens of miles and indeterminate depths that form by the slow cooling and solidification of magma deep within the crust. The very gradual solidification allows mineral-bearing fluids to deposit veins of metallic minerals in fractures, so that today valuable metal deposits are being mined within exposed batholiths in a number of regions. Batholiths are currently exposed in the cores of many mountain ranges, and it is believed that they may underlie, at depth, all major mountain regions. Within the United States, batholiths are exposed in the Sierra Nevada, the Salmon River Mountains of central Idaho, the Front Range of the Colorado Rockies, the Adirondacks of upstate New York, and the Black Hills of South Dakota (see Figure 12.21).

Laccoliths are smaller bodies of magma that work their way surfaceward through vents and solidify as

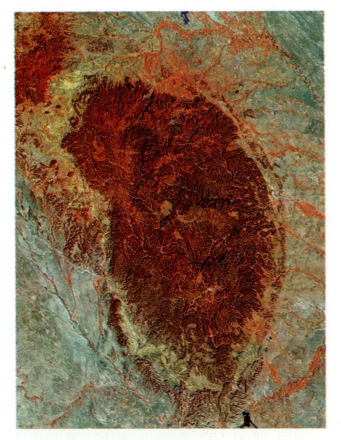

Figure 12.21 EROS satellite photograph of the Black Hills of South Dakota and Wyoming. The Black Hills formed from a batholithic dome whose igneous core has been partially uncovered by the erosional removal of the overlying sedimentary strata. (*Department of the Interior, USGS*)

dome-shaped intrusions between flexible sedimentary strata (see Figure 12.20). If they approach the surface closely enough, they are capable of bulging up the overlying strata to create *laccolithic domes* (see Figure 12.4). In time, the igneous core of the laccolith may be exposed, forming a central mass of hills or mountains. Several mountain masses in southern Utah, including the Henry and LaSal Mountains, represent the cores of exposed laccoliths.

Sills consist of sheets of magma injected between sedimentary rock strata relatively near the surface. They are rather similar to laccoliths, but are flatter because of the greater fluidity of the magma involved. As a result, they do not noticeably dome the overlying rock layers. Sills vary greatly in horizontal extent and in thickness, depending on the quantity of magma involved. Exposures of the edges of large sills often display the columnar jointing typical of basalt, as in the well-known examples of the Giant's Causeway on the Irish coast and the Palisades along the lower Hudson River in New York.

Dikes are igneous rock masses formed from sheets of magma that were injected nearly vertically through preexisting rock fractures. They are very much like sills, except that dikes cut across the structures into which they are intruded, while sills parallel the structures. Dikes are extremely narrow compared to their linear extent. They may extend for several miles, but are usually only a few inches to several feet in thickness. They are widespread in their occurrence, being found in "swarms" beneath some volcanoes or volcanic remnants.

Volcanic Erosion and Remnants

Individual volcanic peaks are relatively short-lived features on the landscape. Once eruptions cease, the forces of gradation quickly lower and eventually remove them completely. Most symmetrical volcanoes currently on the earth's surface have been active within the last few thousand years. A volcano composed primarily of loose rock material may leave as its last erosional remnant a **volcanic neck**. This is a solid spire of hardened lava that was in the vent of the volcano at the end of its final eruption. The rapid removal of the unconsolidated material that once surrounded it and that formed the volcanic cone may leave the volcanic neck standing conspicuously alone above the neighboring landscape. A well-known example of a volcanic neck in the western

Figure 12.22 Ship Rock is a volcanic neck standing conspicuously above the surrounding New Mexico desert. (*E.D. McKee, USGS*)

United States is Ship Rock, in New Mexico (see Figure 12.22). On occasion, volcanic necks may have exposed *dike ridges* radiating from them and appearing like stone walls. The degree of topographic expression of such volcanic remnants depends largely upon their relative resistance to erosion compared to that of the surrounding rocks. They stand out most conspicuously in areas where they have been intruded into weak sedimentary strata.

Flood basalts involve much more massive quantities of volcanic materials than do individual volcanic peaks and remain as landscape features for correspondingly longer time spans. Their basaltic composition, however, causes them to weather rapidly at the surface, so in most areas they soon become covered by fertile soils. If the accumulation of basalt has formed a high-standing plateau, over time it is likely to become increasingly dissected by steep-sided river valleys or canyons. In arid regions, especially, the margins of lava plateaus are frequently marked by retreating escarpments.

Summary

Tectonic processes are divided into two general categories based on the physical state of the rock material undergoing movement and deformation. Diastrophism involves solid-state motions and encompasses the processes of folding and faulting. Volcanism involves the movement of magma within the earth and its expulsion onto the surface. Most tectonic processes and their resulting landforms occur along and near the boundaries between lithospheric plates.

Folding is caused chiefly by the linear compression of sedimentary rock strata. This results in the formation of elongated anticlinal and synclinal rock structures. Subsequent weathering and erosion of rocks of differing erosional resistance will typically produce shifting patterns of ridges and valleys. Rounded fold structures in the form of domes and basins are caused by the subsurface injection or removal of plastic or fluid substances such as magma or salt. Domal structures vary greatly in their size, age, and extent of rock deformation and are widely distributed over the earth.

Faulting involves the fracturing of rock, followed by the subsequent offsetting of the two sides. Individual fault displacements are small, but they may be repeated at frequent intervals, eventually producing major topographic features. Two of the four basic fault types are high-angle faults that experience primarily vertical offsetting and may eventually produce mountain ranges or large escarpments. These include normal faults, which have an expansionary component, and reverse faults, which result from compression. Faults primarily involving horizontal offsetting typically produce lower relief topographic features, but are feared for their earthquake potential. Low angle faults include transcurrent faults, which involve lateral motion of the two sides of the fault, and thrust faults, which involve extreme compression resulting in the overlapping of their two sides. Multiple faults often occur in areas subjected to tectonic stresses, and the intervening blocks of land may be raised, lowered, or tilted.

Earthquakes are sudden surface movements produced by fault displacements. They occur as the rocks on opposing sides of a fault are torn apart and rebound elastically after being subjected to gradually increasing strain. Earthquakes vary greatly in magnitude as a result of such factors as the depth of the focal point, the type of rock, and the amount and nature of the stresses involved. They can be devastating events in densely populated regions, especially those containing large stone structures such as buildings and dams. The distribution of earthquakes closely coincides with that of the active lithospheric plate boundaries. The most severe earthquakes are experienced in regions where the opposing plates are tightly "locked" together, so that a great deal of strain must be developed before displacement occurs.

Volcanic eruptions involve the surface expulsion of molten rock either in the form of lava or fragmental material. Surfaceward movement of magma occurs because of its buoyancy, and the violence of the eruption itself results largely from pressurized gases originally dissolved in the rock. Ideally, a volcano takes the form of a symmetrical conical peak with a central vent. Three basic types of volcanoes are recognized. Cinder cones are typically small and steep and are composed primarily of loose material. Shield volcanoes, conversely, are very large, gently sloping, and composed of accumulated flows of basaltic lava. Composite volcanoes consist of alternating layers of ash or other loose material and of lava. Lava flows associated with volcanic eruptions are usually of limited extent, but extensive fissure outpourings of fluid basaltic lava in the geologic past have covered large regions.

On occasion, magma will solidify before reaching the surface, resulting in the formation of intrusive volcanic features such as batholiths, laccoliths, sills, or dikes. These may later be uncovered by erosion, and they often form high-standing surface features because of their resistance to erosion.

Review Questions

1. Under what tectonic and geologic conditions does folding take place? Give three global examples of folding and relate them to particular lithospheric plate movements.

2. In what ways do the characteristics of domes and basins differ from those of linear folds? What are some of the causes of dome and basin formation? What determines whether the center of a dome or basin will be topographically higher or lower than its surroundings?

3. What is faulting? What factors determine whether folding or faulting will take place in response to tectonic stresses?

4. List and describe each of the four basic fault types. Describe the types of lithospheric plate boundaries with which each type is normally associated.

5. What causes earthquakes? What factors determine their frequency and intensity? How are earthquake magnitudes measured?

6. Why do many experts expect a major earthquake in western California in the near future? Where will this earthquake most likely occur? Why there?

7. What is volcanism? What is the source of the magma involved in volcanic activity, and how does it reach the surface?

8. Describe, in general terms, the global distribution of active volcanism and relate this distribution to the global patterns of folding and faulting. Explain the distributional relationship between these patterns.

9. Describe the general characteristics of each of the three major types of volcanoes, and give two examples of each type. Which type has produced the most catastrophic eruptions from a human standpoint, and why?

10. Explain how a plutonic igneous intrusion such as a batholith can influence the surface topography. Cite one or more actual examples of such an influence.

Key Terms

Diastrophism	Earthquake
Anticline	Extrusive volcanism
Syncline	Volcano
Differential erosion	Cinder cone
Reversal of topography	Shield volcano
Dome	Composite volcano
Basin	Caldera
Fault	Lava flow
Normal fault	Fissure eruption
Reverse fault	Lava plain
Transcurrent fault	Intrusive volcanism
Thrust fault	Batholith
Fault escarpment (scarp)	Laccolith
Horst	Sill
Graben	Dike
Rift valley	Volcanic neck

CASE STUDY

The Eruption of Krakatau Volcano

The spectacular eruption of Krakatau Volcano, in the East Indies, provides a graphic illustration of the awesome powers locked within the earth. Although not the largest volcanic eruption on record, it was the most violent and certainly one of the most celebrated eruptions to have occurred during recent historical times. The eruption took place over a century ago—from May through August 1883—but investigations are ongoing, and only in recent years is a full understanding of this event beginning to emerge.

Among the characteristics of Krakatau's eruption that have resulted in its notoriety are the following: (1) Its climactic explosion on August 27 produced the loudest sound ever known. It was heard as far away as the Philippines, Sri Lanka, central Australia, and Rodriguez Island in the Indian Ocean. Rodriguez Island is located over 3100 miles (5000 km) from the site of the eruption. If Mount St. Helens had erupted as loudly in 1980, it would have been heard in New York City and Miami, Florida. (2) The eruption destroyed most of Krakatau Island and ejected several cubic miles of ash and pumice into the air. Fine ash that reached the stratosphere produced numerous optical phenomena, including spectacular sunrises and sunsets around the world for months following the eruption. (3) The eruption produced a series of tsunamis (ocean waves) that drowned an esti-

mated 36,000 people. This is the second greatest loss of life recorded for any single historical eruption.

Krakatau in 1883 was a large uninhabited island located in the Sunda Strait between Sumatra and Java (see Figure 12.23). It consisted of three overlapping volcanic cones, the highest of which reached an elevation of 2667 feet (813 m). The island was located at the juncture of two intersecting submarine grabens in an area of crustal extension along the border of the Indo-Australian and Eurasian Plates. The most recent previous eruption of one or more of the island's volcanoes is believed to have occurred in 1680.

The 1883 eruption began rather suddenly on May 20, preceded by a period of frequent minor earthquakes apparently caused by the surfaceward

movement of magma. The May eruptions were accompanied by loud explosions heard over 90 miles (150 km) away in Jakarta, on Java. Intermittent eruptions continued through June, July, and most of August. These eruptions were not unusually intense, and their main effect was to deposit some 20 inches (50 cm) of ash on the island.

At approximately 1 P.M. on August 26, the character of the eruption suddenly changed. Intense explosions occurring at about 10-minute intervals produced a cloud of smoke and ash that rose well into the stratosphere and reached a peak altitude of about 16 miles (25 km). Heavy ashfalls of up to 60 feet (20 m) were experienced on the island and by nearby ships. At

continued on next page

Figure 12.23 Krakatau's location at the "hinge" between the islands of Java and Sumatra.

about 5:30 the following morning, the eruption increased greatly in violence, culminating in an enormous explosion at 9:58 A.M. that destroyed the island. This was the explosion heard over a radius of thousands of miles. Several cubic miles of ash and pumice were hurled high into the air, and the eruption cloud reached an altitude of 25 miles (40 km). The denser portion of this cloud of incandescent gases and volcanic fragments quickly fell back to the surface. Much of the material fell into the sea, but a considerable portion rushed off laterally above the water surface. The finer ash stayed in the atmosphere, plunging the neighboring islands into darkness.

The momentum gained by its surfaceward fall of several miles imparted tremendous speed to the ash cloud as it rushed away from the shattered remnants of Krakatau and out over the adjacent sea. As it traveled, a winnowing action took place, and a thick deposit of ash and pumice was left on nearby islands. Deposition also produced floating islands of pumice on the sea surface of the Sunda Strait, temporarily blocking shipping. The lighter ash and gases traveled at least 25 miles (40 km) to the northeast and eventually reached the coast of Sumatra. As they blew up from beneath the floorboards of raised houses, they were still hot enough to cause burns to the residents of the town of Kalimbang. The final eruption also produced a devastating tsunami that reached an estimated height of 130 feet (40 m) and obliterated coastal settlements on the nearby islands, taking some 36,000 lives.

The explosion produced a caldera up to 950 feet (290 m) deep on the site where the island had been (see Figure 12.24). The caldera resulted primarily from the collapse of the island into the magma chamber that fed the eruption. Field investigations show that approximately 95 percent of the erupted material came from the magma chamber and 5 percent was torn from the island by the explosion.

Two theories exist for the origin of the tremendous quantities of gas that were necessary to produce the explosion that destroyed Krakatau. One long-held idea is that seawater penetrated into the magma chamber, forming great volumes of water vapor that "blew the lid" off the chamber and hurled its contents into the air. Recent analyses of the ash produced by the eruption, however, have led to a different hypothesis. The ash, rather than being of uniform composition, consists of a mixture of silica-rich (rhyolitic) and silica-poor (basaltic) fragments. It is hypothesized that most of the eruption involved rhyolitic materials, but that late in the eruption a body of gas-rich basaltic magma was injected into the base of the magma chamber. Pressures within the chamber were low enough to allow huge quantities of gases to come rapidly out of solution from this magma, causing the explosion and subsequent collapse of the chamber's roof.[1]

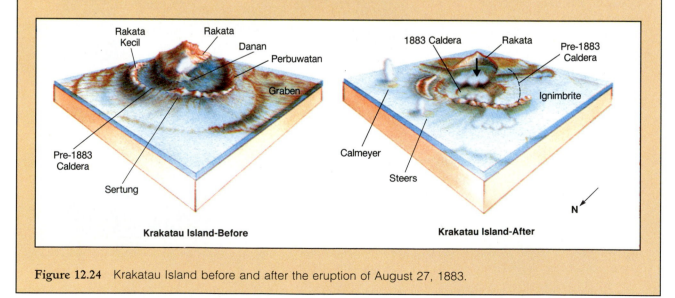

Figure 12.24 Krakatau Island before and after the eruption of August 27, 1883.

Differing theories concerning the origin of the tsunami produced during the final eruption also exist. The wave may have resulted from the collapse of the magma chamber, but it may also have been caused by the splashdown of the vast quantities of material hurled into the air by the final explosion.

Currently, the remnant of Krakatau consists of the bisected cone of one of the three volcanic peaks that comprised the original island. The fertile soils and humid tropical climate have allowed a rapid return of life to the island. The area has been set aside by the Indonesian government as a nature preserve, and it is the last home of the nearly extinct Javan rhinoceros. Sporadic eruptions have continued within the adjacent caldera formed by the 1883 eruption, and in 1928 a new island emerged that was named Anak Krakatau ("Child of Krakatau"). This island is slowly increasing in size as new eruptions add to its mass, and it seems to be gradually taking on the shape of the original island. Krakatau may thus eventually be resurrected, perhaps at some future time to be destroyed in another volcanic cataclysm.

---

1. Peter Francis and Stephen Self. "The Eruption of Krakatau." *Scientific American* 249, no. 5 (November 1983): 172–87.

Chapter Thirteen

Weathering, Mass Wasting, and Karst Topography

Outline

Focus Questions

1. How do the weathering processes operate, and in what environmental settings are they most effective?
2. What causes mass wasting, and what types of mass wasting are most widespread and important?
3. What are the causes and characteristics of karst topography, and how is it distributed around the world?

Tectonic uplift is accompanied, and followed, by the gradational processes that operate under the influence of gravity to lower the land surface. Gradation, unlike tectonism, is continuously active everywhere on land. As a consequence, a much larger total surface area is dominated by gradational than by tectonic features. Gradation primarily involves the transport of materials from higher to lower elevations by the three "tools of gradation": water, ice, and wind. Chapters 14 through 17 are devoted to the actions of these gradational agents and to the landform features they produce. In order for gradation to be effective, however, surface materials must first be reduced in size and erosional resistance through weathering processes. Earth materials, especially if weakened by prior weathering, do not always require transportation by water, ice, or wind, but can be eroded directly by the force of gravity. In addition, differential weathering processes can selectively dissolve some rock types to produce a distinctive landscape assemblage called "karst." The three topics of weathering, gravity movements, and karst are covered in this chapter.

WEATHERING

Weathering may be defined as the combined action of physical and chemical processes that disintegrate and decompose bedrock. The term is used because agencies associated with the weather—specifically, air and water derived from precipitation—perform weathering operations. Weathering processes are of vital importance because they produce essential weathering products and because they greatly increase the effectiveness of subsequent gradational processes. Weathered rock material is the major component of most soils, a commodity without which advanced plant and animal life on land would never have evolved. Weathering also produces loose sediments that are later converted to the sedimentary rocks that now comprise the majority of the earth's surface rock cover.

Weathering is an essential prerequisite for gradation because it reduces the size or alters the chemical composition of rock materials in such a way that erosional agents can remove and transport them much more readily. The tools of erosion, even ice, are largely ineffective when operating on solid bedrock. The gradational process therefore would be slowed immeasurably if bedrock were not first prepared for removal by prior weathering activities. Weathering sets the pace of gradation; because its progress is slow and steady, the gradational processes also operate in a slow and steady manner.

Because weathering involves contact between rock materials and the constituents of the atmosphere, it acts most effectively on near-surface materials. In general, the nearer rock materials lie to the surface, the more they have been altered by weathering. Weathering at any given site should not be thought of as a single process, but rather as a series of processes, each of which further alters the rock and generally weakens or disassembles it. Waves of sequential weathering processes work their way slowly downward through the rock material; the position of the lowermost wave is sometimes called the *weathering front*.

Types of Weathering

For purposes of classification and discussion, weathering activities are commonly divided into the two categories of physical weathering and chemical weathering. Although in this section the various processes in each category are discussed one at a time, it should be realized that several processes normally act in conjunction or in sequence at a given site.

Physical Weathering

Physical weathering involves processes that reduce the size of rock masses without altering their chemical composition. Often described as producing rock disintegration, physical weathering converts larger masses of rock, into smaller pieces of the same kind of rock allowing them to be more readily removed by erosional agents. The most important physical weathering processes are joint and fracture formation, frost wedging, salt crystal growth, thermal expansion and contraction, and biological forces.

JOINT AND FRACTURE FORMATION The first weathering process to affect many rocks, and the deepest acting, is the formation of joints or fractures. Such breaks are often produced by tectonic stresses or result from cooling contractions. The formation of joints or fractures facilitates subsequent weathering activities by greatly increasing the amount of surface exposed to physical and chemical attack.

Horizontal fracture systems may also be produced by the process of **unloading**. This refers to rock breakup caused by the gradual decrease in gravitational pressure as erosion removes the overlying material. The resulting vertical expansion causes the rock to **exfoliate,** or to separate into exfoliation sheets (from the Latin *folium* "leaf") paralleling the surface. Exfoliation sheets may vary in thickness from less than an inch to many feet; when the rock is exposed at the surface, the sheets tend to peel or break off somewhat like the layers of an onion. When exfoliation operates on a large scale, mountain-sized *exfoliation domes* can be produced, especially in areas of exposed granite (see Figure 13.1). Half Dome in Yosemite National Park, California, and Stone Mountain near Atlanta, Georgia, are well-known examples of exfoliation domes.

FROST WEDGING Frost wedging results from the growth of ice crystals, derived either from precipitation or sublimation, within rock fractures or hollows. This process can generate pressures of hundreds of pounds per square inch of rock surface. The force concentrates at the apex of the fracture and often causes further splitting and widening.

Frost wedging can be highly effective in regions where daily freeze and thaw cycles occur during a large part of the year and where little soil cover exists to insulate the solid bedrock from the effects of atmospheric temperature fluctuations. In particular, locations within the subpolar regions and at high elevations above the tree line but below the zone of perennial ice and snow are susceptible to this process.

Figure 13.2 Talus cones lie beneath towering peaks in the Canadian Rockies near Moraine Lake in Banff National Park, Alberta. (*Ralph Scott*)

Figure 13.1 Enchanted Rock, Texas, is a classic exfoliation dome. (*David Butler*)

Active frost wedging in mountainous regions can litter the surface with angular rock fragments of all sizes, producing *boulder fields*. Most of the rocks have broken off nearby cliffs. The lower portions of large, steep-sided cliffs may be buried beneath cone-shaped accumulations of such rocks. These accumulations, known as *talus cones*, are common in mountainous portions of western North America (see Figure 13.2).

SALT CRYSTAL GROWTH In arid regions, saline ground water is drawn surfaceward by capillary tension, and the salt is eventually deposited at or near the surface. Runoff from precipitation can then dissolve the salt and carry it into bedrock fractures, intergranular voids, or other openings. The subsequent evaporation of the water allows salt crystals to form and grow, generating expansionary pressures similar to those associated with frost wedging. Fractures can thus be enlarged and rocks eventually split apart.

THERMAL EXPANSION AND CONTRACTION   The diurnal heating and cooling of bedrock exposed to sunlight at the surface exerts stresses on the rock material because thermally produced changes in volume occur. Thermal expansion and contraction was once thought to be the primary cause of the granular disintegration of exposed bedrock often observed in arid regions. Recently, its effectiveness has been questioned because laboratory experiments subjecting rocks to repeated heating and cooling have produced no perceptible effects. However, repeated heating and cooling, when combined with the presence of air and water over a prolonged period in an outdoor environment, may well play a significant role in the weathering of exposed rock surfaces.

BIOLOGICAL FORCES   Biological forces are of limited significance with respect to physical weathering. Probably the most important is the growth of plant roots, especially those of trees, in rock crevices. The buckling of concrete sidewalks by tree roots is a familiar illustration of the power that growing plants can exert on rock material.

If the effects of human beings are included within the category of physical weathering, biological force attains substantially more importance. Humans, especially during the present century, have developed the technological capacity to alter landforms on a major scale (see Figure 13.3). Examples of human activities resulting in the mechanical breakup of bedrock include drilling and blasting associated with mining and quarrying, building and road construction, and warfare.

Chemical Weathering

<mark>Chemical weathering</mark> consists of weathering processes involving chemical changes in the rock material that result in the formation of new substances. In general, the new substances are softer, more finely divided (i.e., they consist of smaller pieces), more soluble, and larger in volume than the original rock. These changes make the rock material more susceptible to further weathering and erosion. The breakup of rock by chemical weathering is often referred to as rock decomposition.

Chemical weathering often, but not always, follows physical weathering processes, such as joint formation

Figure 13.3   Bingham Copper Mine, in Utah, is one of the world's largest human excavations. (*U.S. Department of Agriculture, ASCS Western Aerial Photo Lab, Salt Lake City, Utah*)

and frost wedging, that open the way for chemical attack. As a result, the weathering reactions advance most rapidly along fractures or other lines of weakness such as bedding planes. In contrast to the angularity of rocks that have been influenced primarily by physical weathering, those dominated by chemical weathering processes are typically more smoothly rounded. Rounding occurs because rock corners and edges have large surface areas in proportion to their masses, allowing chemical attack to alter them rapidly.

As in the case of physical weathering, several important categories of chemical weathering activities exist. Most involve chemical reactions with water or with gases dissolved in water; consequently, the availability of water is critical to both physical and chemical weathering processes. The five most important categories of chemical weathering are hydration, hydrolysis, oxidation, carbonation, and solution.

Hydration involves the absorption, or adhesion, of water molecules to the molecules of a mineral. It is not a permanent chemical combination and can be reversed with the application of sufficient heat. An important example is the conversion of anhydrite ($CaSO_4$) in the presence of water to gypsum ($CaSO_4 \cdot 2H_2O$). Hydration increases the mass and volume of a mineral and softens it. In some rocks it produces the exfoliation of thin sheets, and in others it aids in granular disintegration (see Figure 13.4).

285

Figure 13.4 Granite exposed in this roadcut in California is undergoing granular disintegration. (*Ralph Scott*)

Hydrolysis is a permanent chemical combination with water. It is especially effective on igneous rocks, because hydrolysis particularly affects feldspars and micas, which these rocks contain in quantity. The hydrolysis of igneous rocks by ground water, especially in the humid tropics, can occur at considerable depths.

Oxidation involves the chemical combination of minerals with oxygen that normally has first been dissolved in water. It is most effective on iron compounds, especially in warm climates, and primarily affects igneous rocks containing iron. The rusting of automobiles, iron railings, or other objects when exposed to moisture is a familiar example of oxidation. The presence of large quantities of iron oxides in the soils of the humid tropics is responsible for their distinctive red coloration.

Carbonation consists of chemical reactions of rock material with carbonic acid (H_2CO_3), which is produced when carbon dioxide is dissolved in water. The carbon dioxide involved is generally produced by plant decay, so carbonation is most active in humid, well-vegetated environments. Carbonation results in the conversion of oxides to more soluble carbonates. It is highly effective on the feldspar minerals that comprise a large proportion of most igneous rocks. Carbonation is also the process by which limestone and dolomite are converted to calcium bicarbonate, a physically weak and highly soluble mineral that can be readily removed by ground water flow.

Solution refers to the dissolving of solid rock material in water. In effect, the rock is converted from a solid to a dispersed molecular state and therefore loses all resistance to erosion. Solution is not necessarily accompanied by permanent chemical change because the dissolved substances may eventually be precipitated back into their original chemical forms. All substances are soluble in water to some degree, especially if the water has the proper pH value. Solution is most effective by far, however, on soluble rock types like limestone and dolomite. In areas where these rock types dominate, they may be removed wholesale by erosion following the carbonation/solution process to produce a karst landscape. This subject is covered later in the chapter.

Global Distribution of Weathering Activities

Chemical and physical weathering activities differ considerably in the climatic and lithologic conditions under which they operate most effectively. As a result, major differences exist in the relative importance of these two groups of processes in different parts of the world.

For the earth as a whole, chemical weathering is more active than physical weathering. This is because chemical weathering operates at a much faster rate than physical weathering, under favorable conditions, and because it most completely transforms the rock material. The end product of physical weathering is merely smaller rock particles of the same composition as the original rock. The end products of most chemical weathering processes, on the other hand, are completely new substances. The complete chemical weathering of rock-forming silicate minerals forms the clay minerals that are essential components of most soils.

Because chemical weathering processes generally operate most efficiently under conditions of warm temperatures and abundant moisture, the most rapid chemical weathering occurs in the humid tropics (see Figure 13.5). Here, the deepest and most highly weathered soils exist, and the fastest topographic modification takes place. Conversely, chemical weathering processes are least active in polar and arid regions, and landscapes in these environments evolve relatively slowly. Chemical weathering is also aided by the presence of fragmented, soluble, or chemically unstable rocks, so that geology as well as climate is important in controlling the rate of chemical weathering within a region.

Physical weathering occurs most rapidly in regions that favor the operation of its primary mechanism: frost wedging. It is therefore most vigorous in the subpolar regions and in high mountain areas in the low and

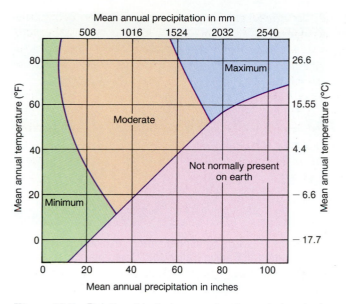

Figure 13.5 Relationship between climate and chemical weathering rates.

middle latitudes (see Figure 13.6). Physical weathering also attains relatively great significance in deserts, largely because the lack of water inhibits chemical weathering activities. Desert regions are frequently littered with fragmented rocks resulting from the effects of salt crystal growth, frost wedging, and thermal expansion and contraction. Physical weathering processes are also aided by favorable lithologic characteristics. These include large surface exposures of brittle, fractured, thinly bedded, or physically soft rock. In most regions, physical and chemical weathering processes work together so intimately that it is difficult and rather arbitrary to separate them; and the existing landscape represents their collective contributions.

Regional variations in the rates, types, and products of weathering are largely responsible for differences in the geomorphic characteristics of regions with differing climates. For example, humid tropical environments dominated by chemical weathering processes typically display a deep soil cover and rounded topographic features. This is in sharp contrast to the stark, angular features that result in many arid regions from the lack of a soil and vegetation cover and from the dominance of physical weathering activities (see Chapter 16). The correlation between climate and topography would be still greater were it not for the topographic variability imposed by differing rock types and especially by variations in the nature and degree of activity of the tectonic processes. Both of these topographic controls

operate independently of climatic conditions, so a newly formed volcano, for example, looks much the same in Alaska as it does in Mexico.

In concluding this section, it should again be stressed that weathering is not a landscape producer in itself, but is merely the necessary preparatory step in the gradational process. Differential weathering is important because it facilitates differential erosion by water, ice, wind, and gravity. In general, high-standing gradational landforms are composed of materials that are relatively resistant to weathering; and low-standing features are composed of easily weathered materials. An example is provided by the folded Appalachians. Here, the ridges are composed of physically hard and chemically resistant sandstone, while the valleys are composed of soluble limestone or physically weak shale.

MASS WASTING

Once weathering processes have weakened surface materials, these materials become much more susceptible to the influence of outside forces, including gravity. Gravity is the direct motivating force in the process of gradation. Most gradation is accomplished by water, ice, and wind, which, impelled by gravity, transport surface materials downslope. A vital role in the gradational process, though, is performed by gravity acting alone.

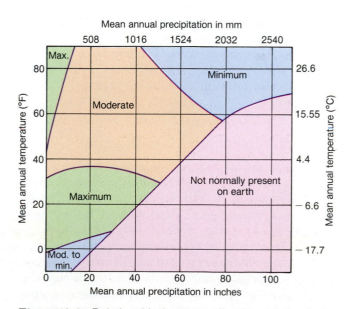

Figure 13.6 Relationship between climate and physical weathering rates.

The downslope movement of surficial material under the direct influence of gravity is termed **mass wasting**. Although no medium such as water or ice serves as a transporting agent, one or both of these substances are present and serve as destabilizing factors in most individual mass movements.

Mass wasting has a much greater impact on the earth's surface than is often realized. It performs the vital role of transferring the products of weathering from their original sites, primarily on hillsides or other upland surfaces, to the valleys or lowland sites where they can be picked up for long-distance transport by the erosional tools of water and ice.

Mass movements exhibit a tremendous range of physical attributes. They vary in size from the motion of a single grain of sand to the massive displacement of several cubic miles of rock. The materials involved may consist of unweathered rock, soil, vegetation, water, ice, or, more commonly, some combination of these ingredients. The speed at which mass movements travel varies from the imperceptible to velocities exceeding 100 miles per hour (45 m/sec). Distances involved range from fractions of an inch to many miles.

The larger and faster types of mass movements are not only awesome phenomena that are capable of sweeping away everything in their path; they are also likely to occur suddenly and unexpectedly. As a result, they can produce great property destruction and loss of life. In fact, mass movements comprise one of the major categories of natural hazards that have afflicted humanity for millennia. Even minor mass movements are quite capable of cracking road surfaces and building foundations, breaking water mains, bending trees and telephone poles, and littering highways with debris following rainstorms in mountainous terrain.

Factors Influencing Mass Wasting

Gravity is the motivating force that initiates mass wasting; therefore, as a general rule, the steeper a slope, the less stable it is. The steepest angle that can be maintained on any given slope without failure (mass movement) of the slope materials is called the **angle of repose.** Its exact value depends upon the type, amount, and condition of the slope materials. The angle of repose for slopes covered by unconsolidated materials normally ranges between 25° and 40°. At progressively steeper slope angles, the thickness of the weathered mantle generally declines as mass wasting and other erosional

Figure 13.7   Large rotational slump resulting from wave undercutting at Pacific Palisades, California. (*John S. Shelton*)

processes become more efficient in transporting it downslope. Slopes steeper than 40° usually have solid rock surfaces.

A number of processes may increase slope steepness past the angle of repose. Probably the most important is undercutting by water, so mass wasting along stream banks is a common phenomenon. Likewise, wave undercutting, especially during storms, often produces mass movements along oceanic coasts or the shores of large lakes. The California coastline and portions of the Great Lakes shorelines are noted for such occurrences (see Figure 13.7). Oversteepening leading to slope failure may also result from valley glaciation within mountainous regions and from artificial roadcuts that have been made at too steep an angle. The latter factor is largely responsible for the frequent mass movements that occur along roads constructed through mountainous terrain.

Weak or loose slope materials are much more prone to mass wasting than is solid bedrock. Vertical or even overhanging cliffs can be maintained in a relatively stable condition in massive, solid rock such as granite, limestone, or basalt. Zones of weakness such as fractures or steeply inclined bedding planes, though, can greatly reduce the angle of repose. Poorly consolidated rocks, such as shale and many sandstones, are also unable to maintain steep slopes. In particular, a strong "caprock" (such as basalt) overlying a much weaker rock (such as shale) may lead to unstable slope conditions, since the more rapid erosion of the weak rock will likely under-

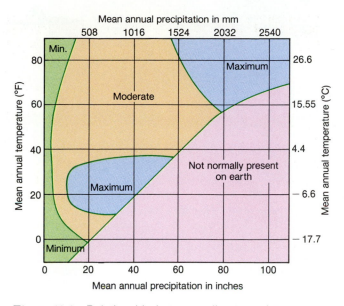

Figure 13.8 Relationship between climate and mass wasting potential.

movements often result from earthquakes, which are relatively common in tectonically unstable mountainous regions prone to mass movements. If a slope is unstable enough, a lesser disturbance such as a loud clap of thunder may initiate slope failure. *Avalanches,* which are rapid movements of masses of snow and ice down steep mountain slopes, may be triggered by the sound of a gunshot or even a shout.

Large quantities of water that may produce slope failure are usually provided by heavy storms. The water adds weight to the slope and lubricates and buoys the individual slope particles so that their cohesion is reduced. Water is especially effective on slopes with a high clay content, since clay is not only capable of holding large quantities of water, but also consists of platy minerals that readily slide over one another when wet. (This accounts for the slickness of mud, as opposed to the firmness of dry soil.) Several types of mass movements occur only when the water content of the slope materials is unusually high.

mine the caprock. Many steep slopes in the western United States are destabilized by these conditions.

The presence of a deep weathering mantle also favors mass movements. Many mass movements, in fact, involve only the displacement of weathered materials; the underlying bedrock is largely or entirely unaffected. This factor brings into consideration the indirect influence of climate, which is crucial in determining the types and rates of weathering processes (see Figure 13.8). The humid tropics are particularly susceptible to mass wasting of chemically weathered materials. Mass movements involving physical weathering products, conversely, achieve great geomorphic significance in alpine and high latitude environments.

Triggering Mechanisms

The slope instability factors just described are considered "passive" factors because over time they tend to make a slope gradually more susceptible to mass movement. Most mass movements are actually triggered by an "active" factor that temporarily reduces the resistance of the slope materials below the critical angle of repose. Although no clear-cut triggering mechanism exists for some mass movements, most are initiated either by strong vibrations or by the temporary presence of large amounts of water. Vibrations that trigger mass

Types of Mass Wasting

Mass movements, like most other natural phenomena, exist in a continuum of types, and, with distance, one type often grades into another. Although this means that no mass movement classification system can fully describe the existing spectrum of types, a number of such systems have been devised. Mass movements are usually classified according to whether their motion consists primarily of sliding, falling, or flowing; whether the material involved is predominantly rock, earth, or debris (a mixture of soil, rock, and vegetation); or whether their motion is relatively slow or rapid. We shall use the last of these criteria to divide mass movements into two groups and will briefly describe the major types of mass movements included within each (see Figure 13.9).

Rapid Mass Movements

Rapid mass movements are those in which perceptible motion occurs. As a group, they are much briefer, more destructive, and more potentially dangerous, but less geomorphologically significant than the slow mass movements. They tend to occur on relatively steep slopes and are likely to contain unweathered rock as at least a component of their material. Rapid mass movements in which the primary motion is sliding are

Figure 13.9 Some major types of mass movements. (a) *Rockfall* The relatively free falling of a newly detached mass of bedrock of any size from a cliff or steep slope. (b) *Rockslide* The usually rapid sliding of newly detached masses of bedrock down a steep bedrock slope. (c) *Slump* The usually slow downward slippage of a coherent body of rock or regolith along a curved slip-plane. (d) *Earth flow* The slow flowage of a mass of nearly saturated soil or regolith down a moderate-to-steep hillside. (e) *Mudflow* The perceptible and often rapid downhill flow of mud, usually mixed with rocks and other debris, along a linear track.

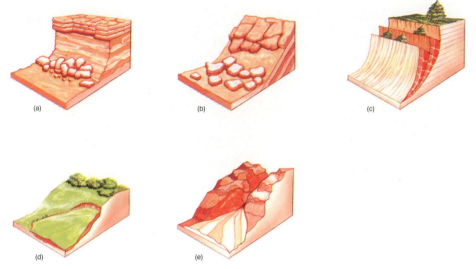

sometimes collectively referred to as "landslides." This term is not employed here because it lacks specificity.

ROCKFALLS *Rockfalls* are among the smallest, most common, and most rapid mass movements. They are produced simply by the falling of individual rocks from nearly vertical cliffs. Rockfalls usually result from frost wedging in alpine environments and are common phenomena in mountainous portions of western North America. Accumulations of angular rock fragments from innumerable individual rockfalls form the talus cones described earlier in the chapter.

ROCKSLIDES Rockslides are produced by the sliding of masses of detached rock down moderate-to-steep slopes (see Figure 13.10). They vary tremendously in size, ranging from individual rocks to entire mountainsides. Large rockslides are usually triggered by earthquakes. They may begin as coherent masses but quickly become fragmented by the violence of their motion. On long, steep slopes in high mountain areas such as the Andes or Himalayas, they can attain speeds well in excess of 100 miles per hour (160 km/hr).

The world's largest known single mass movement of any kind was the prehistoric Saidmarreh rockslide, which occurred in southwestern Iran. It consisted of approximately 5 cubic miles (21 km³) of limestone and marl that became detached from an anticlinal ridge and slid into the adjacent valley and well up the ridge on the valley's opposite side. It left a scar some 9 miles (14 km)

long, and the resulting debris covered an area of 64 square miles (166 km²).

DEBRIS SLIDES, DEBRIS AVALANCHES, AND DEBRIS FLOWS These mass movements all involve the rapid transport of a mixture of soil, weathered and unweathered rock material, and vegetation down moderate-to-steep slopes. These movements differ from one another primarily in their water content and their consequent fluidity of motion. The debris slide is the driest of the

Figure 13.10 The Kelly rockslide of 1925 in western Wyoming blocked the Gros Ventre River, forming a lake. (*John S. Shelton*)

three, while the debris flow is the most fluid. They are all similar in overall shape, as they tend to be elongated and rather narrow, and to typically follow hillside depressions or "hollows" (see Figure 13.22). When they reach the base of the slope, they generally spread out in fan-shaped deposits. All three are commonly triggered by exceptionally heavy rainstorms and are capable of inflicting considerable destruction.

Recently, a major natural disaster resulted from a huge debris flow that began when an earthquake triggered an avalanche on Mt. Huascarán, Peru's highest peak, in 1970. Accelerating down the steep mountainside and picking up debris as it went, the flow may have attained speeds as great as 200 miles per hour (90 m/sec). It overwhelmed and destroyed several towns near the base of the mountain, taking an estimated 70,000 lives. The Case Study at the end of the chapter discusses the causes and consequences of a large group of debris avalanches in central Virginia that resulted from torrential rainfalls associated with Hurricane Camille in 1969.

Figure 13.11 Earth flows that occurred on a grassy hillslope near San Francisco following heavy rains. (*Martin Miller,* © *JLM Visuals*)

MUDFLOWS **Mudflows** are the most fluid of the major types of mass movements. They consist of a perceptible and often rapid downhill flow of mud, usually mixed with rocks and other debris. Mudflows generally follow preexisting channels and are long and relatively narrow. Most occur in arid and semiarid regions, where they are triggered by heavy thunderstorms; but they may also form in alpine areas and in ash-covered volcanic regions. In all three of these environments, the surface is covered by loose, fine-textured materials that are not stabilized by vegetation.

In interior portions of the western United States, mudflows caused by mountain thunderstorms may suddenly and unexpectedly debouch from canyons onto flat farmland in adjacent basins, burying large areas in mud. In southern California, slopes denuded of vegetation by summer brushfires often become the sources of destructive mudflows following the onset of the winter rains.

Slow Mass Movements

Slow mass movements advance at a visually imperceptible rate and may take days, weeks, or even years to become evident. Although they are not nearly as spectacular or dangerous as the rapid mass movements, they are more widespread and are responsible for the downhill transport of a greater total volume of weathered materials.

SLUMP A **slump** is the intermittent movement of a mass of earth or rock along a curved slip-plane. It is characterized by the backward rotation of the slump block so that its surface may eventually tilt in the direction opposite the slope from which it became detached (see Figure 13.9c). Slumps are most likely to occur on steep slopes with deep, clay-rich soils after a period of saturation by heavy rains. The movement generally takes place over a period of days or weeks and is nearly impossible to control or halt once it has begun.

Slumps are common along the California coast, where slopes have frequently been oversteepened by wave undercutting (see Figure 13.7). They also occur along the side-slopes of river gorges in various parts of the western United States. Small slumps capable of blocking traffic frequently occur on steep roadcuts following heavy rains.

EARTH FLOW *Earth flows* involve the slow flowage of nearly saturated soil down moderate-to-steep slopes. These movements are more fluid, shallower, and usually smaller in size than slumps. An earth flow typically has a depressed upper portion and a bulging "toe" at its base where the material has come to rest, perhaps beneath a mat of vegetation (see Figure 13.11). They commonly occur in humid climates on grassy, soil-covered slopes following heavy rains. Under favorable conditions, large numbers of earth flows may form on a single hillside.

Land Subsidence— Its Causes and Environmental Impact

Land subsidence refers to the vertical settling or sinking of the ground resulting from compaction or the removal of supporting materials. Although most forms of subsidence are slow and relatively unspectacular, it exists in many forms and has numerous causes. Some forms of subsidence are of natural origin, but many are caused or abetted by human activities. Human-induced subsidence occurs in at least 38 states within the United States, and the total area affected exceeds 15,500 square miles (40,000 km²). Conservative estimates of its annual costs within the United States exceed $100 million.

Four categories of human activities result in significant subsidence in portions of the United States. Probably the oldest of these is underground mining, especially of coal. Approximately 12,000 square miles (32,000 km²) of land have been undermined by mining, and a third of this land has already experienced subsidence. Annual costs, largely resulting from building damage due to settling and cracking, may run as high as $30 million.

The withdrawal of ground water and petroleum from unconsolidated sediments has caused well-publicized problems in a number of major cities around the world, including Venice, Mexico City, and Bangkok. Within the United States, petroleum removal from the Wilmington oil field produced as much as 27 feet (8.8 m) of subsidence in the city of Long Beach, California, between the early 1940s and the mid-1960s.

The drainage of excess water from organic soils prior to agricultural and suburban development is probably the most costly cause of subsidence. Drainage not only leads to the compaction of the soil, but often allows its organic materials to be consumed rapidly by aerobic bacteria. Resulting subsidence rates can exceed 3 inches (8 cm) a year. In the Sacramento Delta area of California, some 300 square miles (775 km²) of drained land have subsided below sea level, necessitating the construction of an expensive levee system.

continued on next page

CREEP

Creep is the slowest and least noticeable, but the most widespread and geomorphologically important of all the categories of mass movements. It consists of the very slow downhill movement of soil or rock material over a period of years. Creep involves the entire hillside and probably occurs, to some extent, on any sloping, soil-covered surface. Because the near-surface layers move fastest, creep is commonly responsible for the downhill tilting of buried or partially buried objects such as trees, telephone poles, stone walls, and inclined rock strata.

Creep is caused by any disturbance to the surface mantle, since such disturbances almost invariably result in a net downhill displacement of surface particles. Some disturbances contributing to creep include wetting and drying, heating and cooling, plant root growth, raindrop impact, and trampling by livestock. Creep is especially active on slopes covered by deep soils with a high water content and is therefore much more significant in humid regions than in arid or semiarid ones. It is thought to be largely responsible for the generally rounded topographic features of humid landscapes. The subdued profiles of the Appalachian Mountains, for example, stand in stark contrast to the angular features of mountain ranges in arid portions of the West. Creep may be the chief mechanism by which weathered materials are transported down hillsides in soil-covered areas to the valleys of the streams that ultimately carry these materials to the sea.

KARST TOPOGRAPHY

In some areas underlain by thick sequences of limestone and dolomite, the weathering processes of carbonation

The most spectacular form of subsidence is associated with the sudden appearance of collapse sinkholes in areas of limestone. Sinkhole development, as discussed in this chapter, is a natural occurrence in limestone regions; but water extraction for irrigation or other purposes sometimes lowers the water table sufficiently to trigger surface collapse. The development and rapid growth of a large sinkhole during a period of dry weather between May 8 and 9, 1981, in Winter Park, Florida, made headline news around the country (see Figure 13.12). This sinkhole eventually attained a diameter of more than 330 feet (100 m) and a depth of 115 feet (35 m). It swallowed up a home, six vehicles, part of a city swimming pool, and portions of two streets.

Figure 13.12   This collapse sinkhole formed in May 1981 in Winter Park, Florida, as a result of a drought that lowered water table levels. It grew to be 330 feet wide and 115 feet deep, and swallowed up several buildings and vehicles. (*USGS*)

and solution can be considered responsible for the major characteristics of the topography. Even here, a transporting agent (in the form of ground water) is needed to remove the weathered materials from their original sites. Because these materials have first been converted to a molecular state, however, carbonation and solution are often considered to be the primary geomorphic mechanisms in the development of the resulting topography.

Cause and Global Distribution of Karst Topography

Ground water in humid regions tends to become somewhat acidic from the solution of atmospheric carbon dioxide as well as from the acquisition of organic acids from the soil. It becomes, in effect, a weak carbonic acid solution. Carbonic acid reacts with carbonate rocks to produce calcium bicarbonate, which is highly soluble in water and is readily transported away by ground water flow. An area with landform features produced largely by the differential solution and subsequent erosion of limestone and dolomite is called a **karst** area. The term is derived from the Kras region of Yugoslavia near the Adriatic Coast, where some of the earliest research on this topographic assemblage was conducted.

Karst topography is geographically widespread, although individual sites are sometimes highly localized. Conditions in humid portions of the middle latitudes are generally favorable for karst development. It reaches optimum development, however, in the humid tropics, where heavy precipitation and organic acids produced by the decomposition of the dense vegetation combine to produce the rapid solution of limestone and dolomite.

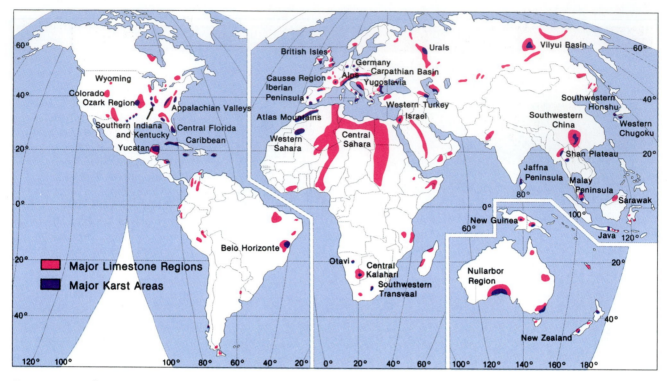

Figure 13.13   Global distribution of major limestone and karst regions.

Major karst areas are depicted in Figure 13.13. The United States probably contains more karst than does any other country. Important areas include Kentucky and Indiana, the central Appalachians, Florida, the Ozark Plateau, western Texas and eastern New Mexico, and the Black Hills of South Dakota. Karst is also well developed in southern Europe, especially in the southern Alps and in a number of regions bordering the Mediterranean Sea. Australia contains an extensive but little-publicized karst region in the Nullarbor Plain, located in the south-central part of the continent. The most important areas of tropical karst include South China and adjacent portions of Vietnam, Thailand, and Burma; parts of the East Indies; a large area in eastern Brazil; Mexico's Yucatán Peninsula; and the West Indian islands of Jamaica, Cuba, and Puerto Rico.

Karst Features

Different karst regions vary greatly in topographic characteristics. Some may be characterized as plains, and some are mountainous, but most frequently they are hilly. The larger relief features tend to be very abrupt and steep-sided, giving the topography a picturesque and sometimes spectacular appearance, but typically reducing accessibility and land-use potential.

The solution rates of limestone and dolomite can vary greatly over very small distances because of fracture and bedding patterns, slight variations in chemical composition, and access to acidic ground water. As a result, even in areas of low relief, differential removal produces an extremely rough and intricately sculpted surface topography on exposed bedrock outcrops (see Figure 13.14). Rock outcrops are common and frequently extensive in karst regions, and they typically exhibit gullies, grooves, solution pits, and other small-scale differential solution features in abundance. In addition, loose stones and boulders often litter the surface. The reddish, clay-rich soils of karst areas are fertile but discontinuous and often quite thin, sometimes existing only in patches within low-lying sites.

Sinkholes, also termed "sinks" or dolines, are the most widespread of the major karst features. They are rounded depressions formed by subsurface solution. Most develop on bedrock joint intersections, are rather small in size, and are produced by the gradual subsid-

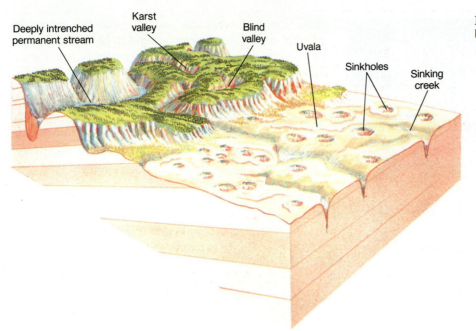

Figure 13.14 Important surface karst features.

Deeply intrenched permanent stream
Karst valley
Blind valley
Uvala
Sinkholes
Sinking creek

ence of the weathered surface mantle into an enlarging solution hollow (see Figure 13.15). Sinkholes can also form rather suddenly by the collapse of the surface into an underground void. Such *collapse sinkholes* tend to be larger, deeper, and more steep-sized than the more common *subsidence sinkholes* (see Figure 13.12). Sinkholes commonly range up to 100 feet (30 m) in diameter and up to 30 feet (10 m) in depth, but most are considerably smaller and shallower than this. Sinkholes are typically dry if their bottoms are situated above the water table, but are partially filled with water if they extend below it. The rounded lakes that characterize

the Central Florida Lake District, for example, occupy large sinkholes in an area with a high water table. Sinkholes are most abundant and conspicuous in low-relief areas. *Karst plains,* such as those in portions of Indiana and Kentucky, are plains covered with sinks that may number in the hundreds per square mile (see Figure 13.16).

Well-developed karst areas display an almost complete lack of organized surface drainage. The underlying carbonate rocks are honeycombed with solution channels that receive surface water through *swallow holes* typically located at the bottoms of sinkholes. The

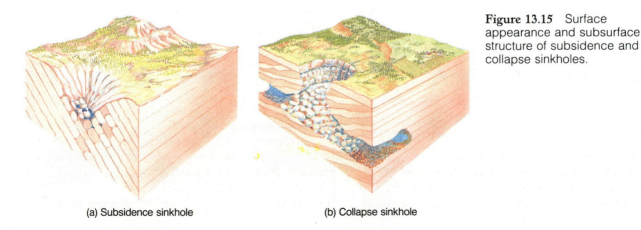

(a) Subsidence sinkhole

(b) Collapse sinkhole

Figure 13.15 Surface appearance and subsurface structure of subsidence and collapse sinkholes.

Figure 13.16 This karst plain in western Illinois contains numerous water-filled sinkholes. (*B.F. Molnia*)

surface is almost like a sieve, and runoff cannot travel far before it enters a sinkhole and disappears underground. One consequence of the thin soil cover and rapid underground diversion of surface water is the development of a drought-prone surface incapable of retaining moisture for long after a rainfall.

In areas where sinks and swallow holes are somewhat less numerous, surface streams may form, only to disappear suddenly below ground to become **sinking creeks** as the water enters large swallow holes in the streambed. A sinking creek may flow beneath the surface for a considerable distance—perhaps several miles—before suddenly reappearing as a spring. Some springs in karst areas have very large volumes of flow. A well-known example is Silver Springs, located near Ocala, Florida. It has a flow volume of approximately 500 cubic feet (15 m^3) of water per second and has been developed as a tourist attraction.

Caves are produced by a variety of geomorphic processes, but they are most numerous and complex in karst areas, where they are formed by subsurface solution and erosion. A **cave** is any natural cavity within or beneath the earth's surface. It may range from a small bedrock hollow set into a cliff to an elaborate cave system consisting of an interconnected maze of rooms and passageways. Although normally predominantly horizontal in extent, caves may be developed on several vertical levels.

Within karst areas, caves are formed largely by ground water solution along joints, fractures, or bedding planes located at, or just below, the water table. Because

these zones of weakness usually exhibit a geometric pattern, cave systems likewise tend to be organized into distinctive, and often rectilinear, patterns (see Figure 13.17). As solution progresses and the water table falls, the original water-filled caves are gradually drained, while new levels may be formed below them. In this way, a multiple-level cave system is believed to develop from the upper levels downward. The world's largest known cave system, Mammoth Cave, in Kentucky, contains over 200 miles (350 km) of interconnecting passageways at several levels.

Adding to the mystery and beauty of limestone caves is the great variety of *travertine* features that often develop within them (see Figure 13.18). These features consist of calcium carbonate deposits left by the slow dripping of water from the roof and walls of the cave. The most important travertine features include the icicle-like *stalactites* that hang from the cave roofs, upward-growing *stalagmites* on the cave floors, pillars or columns formed by the coalescence of stalactites and stalagmites, and travertine curtains or ribbons formed

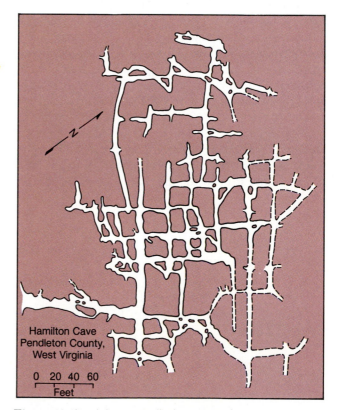

Hamilton Cave
Pendleton County,
West Virginia

0 20 40 60
Feet

Figure 13.17 Joint-controlled system of passageways in Hamilton Cave, West Virginia.

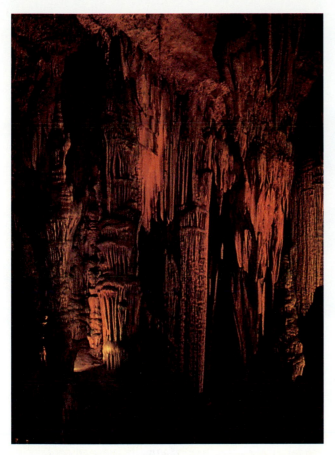

Figure 13.18 A view of various travertine features within Luray Cavern, Virginia. (*Ralph Scott*)

by water dripping along an incline. All these features come in an infinite variety of sizes and shapes, and they may be stained various hues by chemical impurities in the water.

Although cave systems are located in most areas underlain by extensive limestone and dolomite deposits, they are readily accessible only if they are currently located above the water table. As a consequence, most well-known cave systems are associated with hilly or mountainous regions situated well above sea level and with deep water tables. Important regions include the Kentucky-Indiana area, the Ozarks, the central Appalachians, and the Black Hills of the United States (see Figure 13.19); the Dinaric Alps of Yugoslavia; Mexico's Yucatán Peninsula; and the mountains of south-central China.

Cave systems are inherently unstable and geologically short-lived natural phenomena because their continued growth eventually leads to loss of roof support and collapse. The collapse of cave systems usually occurs in segments rather than all at once. Uncollapsed portions may form *natural tunnels,* if broad, or *natural bridges,* if relatively narrow. Probably the most famous limestone bridge in the world is Natural Bridge, in Rockbridge County, Virginia.

Another type of erosion remnant consists of hills that remain on a karst surface as indicators of a former higher topographic level. These hills vary in steepness

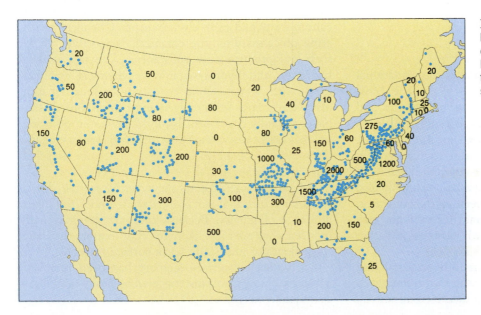

Figure 13.19 Distribution of karst cave areas in the coterminous United States. Numbers indicate the estimated total number of caves in each state.

Figure 13.20 Tower karst in the Li River Valley of southern China. (*R. Krubner/H. Armstrong Roberts*)

and height in different areas, but reach optimal development in the humid tropics. They are known regionally by a variety of terms, including *haystack hills*—an allusion to their steep-sided and highly rounded forms. In parts of Southeast Asia, such remnant limestone peaks attain mountainous proportions, producing a spectacular and highly distinctive landscape referred to as **tower karst** (see Figure 13.20). This landscape has inspired Chinese artists for many centuries.

Summary

This chapter begins our examination of the gradational processes and their effect on the earth's surface. Most gradational landforms are produced by erosion and deposition by water, ice, and wind. Before these agents can operate effectively, however, the surface rock must be acted upon by weathering processes, which greatly reduce its resistance to erosion. Gravity is the chief force that powers the three tools of gradation, and it also produces a variety of mass movements of the surface on its own. Another set of gradational features not normally produced by the "traditional" tools of gradation are the karst features of limestone regions.

Weathering involves the actions of physical and chemical processes that reduce the size of rock materials and usually produce softer, more erodible substances.

Physical weathering involves only a physical reduction in rock size and is most important in cold and dry environments. The chief physical weathering processes include the formation of joints and fractures, frost wedging caused by ice crystal growth, the growth of salt crystals in rock fractures, thermally induced changes in volume, and the mining and construction operations of humans.

Chemical weathering has greater overall geomorphic importance than physical weathering; it is most significant in warm and moist environments. Chemical weathering processes alter the chemical composition of the rock and produce new substances. The processes of hydration and hydrolysis involve chemical combinations with water, while oxidation and carbonation involve reactions with oxygen and carbon dioxide that have generally first been dissolved in water. Solution, another important chemical weathering process, causes the molecules of the rock material to become detached from one another and to be suspended in water.

Mass wasting, or mass movements, are the downward movements of surface materials under the direct influence of gravity. Because they are gravitationally induced, they are most likely to occur in areas that have steep slopes or physically weak surface materials, or in places where subsurface support has been removed. Most mass movements are triggered by heavy rainfalls or earthquake vibrations.

Rapid mass movements such as rockfalls, rockslides, and debris avalanches are short in duration, spectacular, and often highly destructive. They are most common in geologically youthful mountain regions along or near lithospheric plate boundaries. Slow mass movements, especially creep, have a greater total geomorphic impact because they are more widespread and collectively involve the downward transport of a greater volume of surface material.

Karst topography results from the differential solution and erosion of soluble rocks, notably the carbonate rocks limestone and dolomite. Carbonate rocks are widely distributed around the world, but are especially susceptible to solution processes in areas of heavy precipitation and abundant vegetation. These conditions favor the production of large quantities of acidic surface water. Most important karst regions are located in humid portions of the tropics and middle latitudes.

The karst process involves three steps. The rock is first attacked by the chemical weathering process of carbonation, which changes the carbonate minerals to highly soluble bicarbonates. The bicarbonates are then dissolved by acidic water and are removed by surface or ground water flow. A karst landscape is therefore characterized by a variety of differential solution features. Small-scale surface features include grooves, solution pits, and loose rocks. The most important larger surface features are circular depressions known as sinkholes. Subsurface solution features are also very prominent. Most drainage takes place through underground routes, and the frequent result is the formation of cave systems containing a variety of travertine features.

Review Questions

1. What is weathering? Explain the role that weathering plays in the gradational process.
2. What is the difference between physical weathering and chemical weathering in terms of their effects on rock? Give two examples of each weathering category.
3. How do the global distributions of physical and chemical weathering rates vary? Why do they do so? For the world as a whole, which of the two categories of weathering processes has the greatest geomorphic significance? Why?
4. What is mass wasting? Discuss the importance of the role that mass wasting plays in the gradational process.
5. Explain the significance of the "angle of repose" in mass wasting. Describe several ways by which human activities can cause slope angles to exceed the angle of repose and initiate mass movements.
6. Describe the environmental conditions that favor the development of karst topography. Why is karst more widespread in the humid tropics and middle latitudes than in the subtropics and high latitudes?
7. Describe how the drainage in a karst area differs from the drainage in a "normal" area. What particular karst features develop as a result of these drainage characteristics?

Key Terms

Physical weathering	Mudflow
Unloading	Slump
Exfoliation	Creep
Frost wedging	Karst
Chemical weathering	Sinkhole
Mass wasting	Sinking creek
Angle of repose	Cave
Rockslide	Tower karst

CASE STUDY

Debris Avalanches Produced by Hurricane Camille

Under some conditions, mass movements can occur in large numbers within a restricted area, resulting in considerable destruction and even loss of life. A major example of such an occurrence was the formation of several hundred debris avalanches in western Virginia as a result of torrential rains associated with Hurricane Camille in 1969.

The Storm

The development and path of Hurricane Camille were highlighted in the Case Study at the end of Chapter 6. This intense storm, which came ashore near Gulfport, Mississippi, on the night of August 17, 1969, was responsible for over $1 billion in property damage and the loss of 171 lives. By the evening of August 19, as the weakened remnants of the storm were crossing the central Appalachians, its rainfalls briefly flared up to unprecedented levels, producing widespread flooding and debris avalanches in several Appalachian counties of West Virginia and Virginia.

The center of heaviest rainfall was located in Nelson County, Virginia (see Figure 13.21). It is a rural county of about 10,000 inhabitants situated in the west-central part of the state on the eastern flanks of the Blue Ridge Mountains. Excessively heavy rain began falling in Nelson County about 9:30 P.M. and continued until about 3:30 A.M. on the twentieth—a period of six hours. The National Weather Service later conducted bucket surveys (field surveys of water levels in containers known to be empty preceding the storm) and in one location confirmed a storm rainfall total of 31.5 inches (80 cm). Unofficial indications showed that in a few sites the storm total may have exceeded the area's annual precipitation mean of 40 inches (100 cm). These figures are far larger for a six-hour time span than any officially recorded totals anywhere in the United States.

The cause of the sudden and rather localized increase in storm rainfall intensities is apparently related to

continued on next page

Figure 13.21 Track of Hurricane Camille and location of Nelson County, Virginia.

a combination of four circumstances. First, the entire region was covered by an unusually moist maritime tropical air mass. Second, even before the onset of the storm, soils in the area were very moist from one of the wettest summers on record. Third, the center of the storm passed just south of the Nelson County area, producing an easterly wind flow. Because Nelson County is located on the easternmost flanks of the Appalachians, a significant orographic lifting factor was produced. Finally, an area of pre-frontal thunderstorms from an approaching cold front reached the area at the same time as the remnants of Hurricane Camille.

The Debris Avalanches

The torrential rainfall initiated small roadbank slumps by 10 P.M. and the first debris avalanches by about 11 P.M. By 1 A.M., debris avalanches were occurring in large numbers, and new ones continued to form until about 4 A.M.

The sites on which the debris avalanches occurred were the mountainside hollows that serve as conduits for subsurface drainage. The highly porous soils on these slopes normally absorb water as rapidly as it is received and direct it as ground water into the hollow systems. This water eventually surfaces at the foot of the slopes to form small streams. The debris avalanches were apparently initiated by the coalescence of extraordinarily large amounts of subsurface water into the hollow systems. This water may have literally lifted the regolith in each hollow from the underlying bedrock, allowing the entire detached mass to slide and flow down the hollow and into the stream valley below.

Witnesses reported that the debris avalanches occurred in groups. The noise and vibration from one avalanche apparently triggered several others nearby in a chain reaction. In all, over 275 avalanches occurred, including many with multiple branches totaling several thousand feet in length. The debris avalanches left narrow, elongated scars on the hillsides, while the debris—a chaotic jumble of soil, rocks, giant boulders, and trees—was deposited in the adjacent flat stream valleys (see Figure 13.22). Many houses in the affected area were swept away by either the debris avalanches or the flooded streams, and approximately 100 people died in Nelson County alone.

In the years following the storm, the debris avalanches have been healing rapidly. The debris itself was quickly cleared, with government assistance, from the agriculturally productive valleys. The hillside scars remain, but are much less conspicuous because they are being slowly recovered by soil that is washing and creeping onto them from upslope. Vegetation in this humid temperate setting returns quickly once a soil cover is present, and most scars already contain small trees.

Debris avalanches are undoubtedly recurrent phenomena within these mountainside hollow systems. Creep and debris avalanching apparently play complementary roles in transporting weathered materials down the steep slopes to the stream valleys below. Creep initially fills the hollow systems with weathered materials from upslope. These materials are then periodically flushed down the hollow systems by debris avalanches during exceptionally heavy storms.

What may at first seem to be a single, random occurrence thus becomes a step within an organized gradational sequence. It is another indication that the various earth phenomena we have been studying are organized into complex interrelated systems. A basic research goal in physical geography is to discover the existence of such systems, how and where they operate, and the role they play in the development and evolution of the earth's surface features.

Figure 13.22 Debris from the three debris avalanche scars on the hillslope has been deposited in the foreground. This photo was taken in Nelson County, Virginia, a few months after the area was devastated by torrential rains from the remnants of Hurricane Camille. (*Donald Poole*)

Chapter Fourteen

Fluvial Processes and Landforms

Focus Questions

1. How do drainage systems develop, and what geographical patterns do they exhibit?
2. How do streams erode, and what are the most important landforms produced by stream erosion?
3. Why do streams deposit sediments, and what are the most important landforms produced by stream deposition?

The surface features of the land are more the work of streams than of any other geomorphic agent. In Chapter 8, the discussion centered on the hydrologic aspects of streams, especially on their widespread distribution and their role in the hydrologic cycle. In the present chapter, we examine the physical characteristics of streams and stream systems, with particular emphasis on their role as shapers of landforms. The term **fluvial** (from the Latin *fluvius* "river"), which is commonly used to identify stream-related processes and features, appears frequently during the course of the discussion.

Streams are a dynamic component of the physical environment and are capable of undergoing major changes over relatively short periods of time. They perform two vital natural functions. First, they remove excess water from the land. The total quantity of water transported by streams to the ocean comprises about one-fifth of all the precipitation that falls on the earth's land surface. By flowing in response to gravity, streams gain enough kinetic energy to perform a second function. This is the erosion, downhill transport, and eventual deposition of surface weathering products supplied to the streams from adjacent slopes largely through mass wasting processes. The redistribution of these materials by streams produces most fluvial landforms.

The profound impact of streams on human activities, though largely beyond the scope of this text, can be mentioned briefly. Streams have served vital functions as transportation routes and irrigation sources since the dawn of recorded history. They have formed the fertile agricultural floodplains on which we raise a large proportion of our food crops. From a broader perspective, fluvial processes acting over lengthy geologic time spans are responsible for most of the plains on which the majority of our planet's human inhabitants reside. Two examples of human interactions with stream systems are provided in the Case Study at the end of the chapter and in the Focus box on Rio Grande boundary changes.

Satellite imagery indicates that fluvial landform processes dominate perhaps three-fourths of the earth's land surface (see Figure 14.1). This geomorphic dominance results from the nearly global distribution as well as the great gradational power of stream systems. Only the 10 percent of the land surface covered by continental glaciers does not contain at least intermittently active streams. In addition, certain nonglaciated regions contain some fluvial features, but are dominated by landforms produced by other geomorphic processes. Chief among these are the coastal margins (discussed in

Figure 14.1 Even in desert regions, fluvial features often dominate the landscape. Here, dendritic stream channels cover the surface of a portion of the Mojave Desert in southeastern California. Vegetation in the moister channels serves to accentuate the drainage pattern. (*John S. Shelton*)

Chapter 17), some tectonically active areas (Chapter 12), and areas covered by shifting sands (Chapter 16). All three are restricted in geographical extent.

The geomorphic effect of flowing water, though important nearly everywhere, displays significant spatial variations in its nature and intensity. These variations are related chiefly to climatic and vegetative patterns and are influenced strongly by local soil and bedrock characteristics. One might expect fluvial effects on the landscape to be least important in arid regions and to increase directly with precipitation averages, making the wettest regions the areas with the best-developed and fastest forming fluvial features. Figure 14.2 indicates that this is not the case. Flowing water is actually most effective as a landscape modifier in semiarid and subhumid regions. Very dry regions are minimally influenced because of a lack of water, but fluvial features in humid regions with abundant precipitation have only a moderate rate of development. This results largely from the protective influence of the vegetation cover, which generally increases in density with increased precipitation. The climatic zone of maximum fluvial effectiveness receives enough rainfall to produce considerable storm runoff, but is dry enough for the year as a whole to support only a limited cover of protective vegetation.

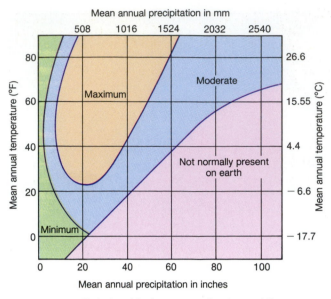

Figure 14.2 Relationship between climate and the effectiveness of fluvial geomorphic processes.

Although most surface topography results primarily from fluvial processes, streams at any given time occupy and actively modify only a very small proportion of the total surface area. All stream action and energy are concentrated in a small area, producing features that stand in marked contrast to the remainder of the surface. The majority of the surface in most areas consists of interfluves, or areas between streams.

This chapter is divided into two sections. The first discusses the causes and characteristics of stream flow: how and why streams develop, the mechanics of fluvial erosion and deposition, and the forms assumed by individual streams and the drainage systems of which they are a part. The second section examines the major landforms produced by fluvial processes.

CAUSES AND CHARACTERISTICS OF STREAM FLOW

Streams are formed by water that reaches the earth's land surface and cannot be absorbed. The surface in each locality has a maximum rate at which it can absorb water. This is known as the infiltration capacity and is related primarily to the texture, degree of compaction, and existing water content of the surface materials. If precipitation exceeds the infiltration capacity, water begins to collect on the surface. Depressions are filled

first, and then, assuming that some slope exists, the excess water begins to flow downhill. In the absence of existing channels, the initial flow of water is disorganized and spreads over the surface in a nonchannelized form known as *overland flow*. Because this flow is not concentrated, it is very shallow, and friction with the underlying surface is great. As a result, even on relatively steep slopes, its speed of flow is slow, and its capacity for eroding and transporting sediments or debris of any kind is highly limited. Overland flow dominates near the crests of interfluves, producing zones of no erosion.

Proceeding downhill, the flow increases in volume as the total upslope area being drained becomes progressively greater. Channelization, and the beginning of a stream, occurs when the water depth and speed suffice to overcome the cohesion of the soil particles. On undisturbed, well-vegetated slopes the soil is highly absorbent and is held firmly in place by the vegetation mat. In this case, water from even heavy storms is typically absorbed as fast as it is received. This water travels downslope as thoughflow (see Chapter 8) until it surfaces as a spring or seeps out along the side of a stream bank at a point where the surface intersects the water table. Deeper ground water flow travels much more slowly than either surface water flow or throughflow. Thus, in a humid climate, a stream supplied by ground water is likely to maintain a stable base flow even during dry periods.

Drainage Systems and Patterns

Streams are not separate, independent entities, but are organized into drainage systems consisting of numerous interconnected stream segments. In form and function a drainage system is much like a tree. It normally consists of a master or trunk stream at its base that branches uphill into a progressively larger number of increasingly small segments, or *tributaries*. The tributaries extend into all portions of the drainage basin. Also like a tree, the branches of a drainage system conduct water and other materials (in this case, stream sediments) from one part of the system to another.

Around the peripheries of drainage basins of any size are highland rims that serve as *drainage divides;* these divides separate the flow and direct it toward adjacent drainage systems. In some areas, the divides are low and indistinct, while in others they form high mountain systems. The largest and most important drainage divide

International Boundary Changes Along the Rio Grande

Rivers have long served an important political function as boundaries between nations. The tendency of some rivers to shift their courses, however, can cause major diplomatic problems for the countries involved. A good example is provided by the Rio Grande, which forms the international boundary between the United States and Mexico over a distance of roughly 1000 miles (1600 km).

Soon after the 1848 Treaty of Guadalupe Hidalgo established the middle course of the Rio Grande as the international boundary, it became evident that the river occasionally changed its course during floods. This raised a sticky political problem: should these changes cause the international boundary to shift, or should the original boundary be maintained, leaving increasingly numerous enclaves of each country on the "wrong" side of the river?

Most changes in the Rio Grande's course during the nineteenth and early twentieth centuries were more of a nuisance than a major problem, because the lands involved were barren and unpopulated. In 1905, Mexico and the United States signed the Banco Treaty in which they agreed to exchange approximately equal parcels of land from time to time in an attempt to offset, as much as possible, shifts in the river's course.

A more serious situation had developed, however, in the densely populated area where the Rio Grande separated El Paso, Texas, from the neighboring Mexican city of Ciudad Juárez. Southward shifts of the river occurred in 1852 and again in 1864, placing tracts of Mexican land on the American side of the river. Beginning in the late 1800s, Mexico repeatedly requested that the lands involved be returned to her. The U.S. government's position at the time was that, since the shifts in the river's course had resulted from natural processes of erosion and deposition, the lands involved now belonged to the United States.

In 1910, the United States finally agreed to settle the problem by arbitration, and a joint Mexican-American commission was appointed to decide the matter. The commission eventually decided that Mexican land lost by the 1852 shift should remain as part of the United States, but that the larger "El Camizal" tract lost in 1864 should be returned to Mexico. The United States long refused to accept this, although it had originally agreed to abide by the commission's decision. The transfer of the land finally took place in 1963 as the result of an act signed by President John Kennedy and Mexican president Lopez Mateos.

in the United States is the Continental Divide, which separates runoff flowing eastward into the Mississippi River system and ultimately into the Gulf of Mexico from that flowing westward through the Colorado and Columbia River systems to the Pacific (see Figure 14.3).

Drainage Patterns

The drainage pattern is the surface pattern collectively formed by the streams in a drainage system. All the stream systems in a given region typically exhibit essentially the same type of pattern. The existing pattern is largely a response to differences in rock types and structures as well as to preexisting surface features. It therefore provides an indication of the region's geologic and geomorphic history as well as its present characteristics. A large number of drainage pattern types exist, and, as might be expected, some regions exhibit patterns that are poorly defined, are intermediate between two types, or are complex because of multiple influences. The types described here and illustrated in Figure 14.4 are among the most widespread in their distribution.

The dendritic pattern is the most widely distributed of the drainage pattern types. It develops on gently sloping surfaces that are homogeneous in their resistance to erosion and that therefore exert no significant

Figure 14.3 The Continental Divide (in red) separates streams flowing into the Pacific from those flowing into the Atlantic via the Gulf of Mexico.

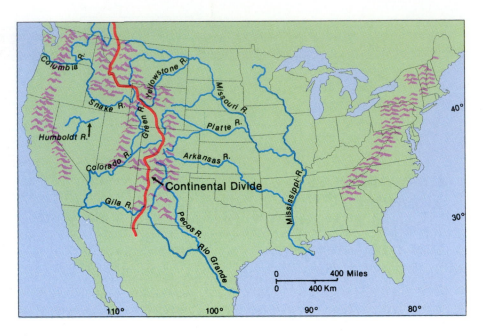

structural control on stream location. It is especially likely to occur in areas of horizontally bedded sedimentary rocks and glacial deposits. The dendritic pattern is treelike in form, with a large degree of variability in the orientation of tributaries and with stream junctures normally forming angles well under 90°.

The <mark>trellis pattern</mark> typically develops in regions of parallel <mark>ridges and valley</mark>s where resistant and nonresistant sedimentary strata alternate. It consists of a parallel pattern of major streams that occupy valleys eroded into the weaker rocks, while much smaller and shorter tributaries flow down the steep ridge flanks to join the major streams at nearly right angles. In the United States, trellis drainage patterns are well developed in the Ridge and Valley portion of the central Appalachians and in the Ouachita Mountains of Arkansas and Oklahoma.

The <mark>rectangular pattern</mark> is usually produced when the drainage pattern is controlled by intersecting fault or fracture systems in areas of crystalline rocks such as granite. Streams alternate between unusually straight segments, where they flow along one fault or fracture line, and sudden right-angled bends, where they are redirected onto an intersecting line. A rectangular pattern is well developed in portions of the Adirondack Mountains of upstate New York.

The <mark>radial pattern</mark> is perhaps the simplest of the basic drainage pattern types. It consists of the outward flow of streams in all directions from a central peak or upland. It is especially well developed in the vicinity of large, isolated volcanic peaks such as those in the northern Cascades of Oregon and Washington. Conversely, a converging flow of streams toward a central basin produces a *centripetal pattern*. Centripetal patterns are typical of both karst regions and arid regions with interior drainage; consequently, they are widespread in occurrence.

The <mark>deranged pattern</mark> displays a high degree of spatial disorganization. Streams are very irregular in their directions of flow, sometimes exhibiting sharp, V-shaped bends. In addition, they do not flow smoothly downhill, but pass alternately through depressions occupied by lakes and swamps, and steep segments marked by rapids and waterfalls. The existence of a deranged pattern indicates that it is of geologically recent origin and that an organized drainage pattern has not yet developed. It is common in glacially scoured regions such as the Canadian Shield and Scandinavia, where the previous drainage systems were erased by the ice.

Stream Erosion, Transportation, and Deposition

It is through the ability of streams to erode, transport, and deposit sediment that fluvial landforms are developed. Because the characteristics of fluvial features are determined by the nature of the activities that produce

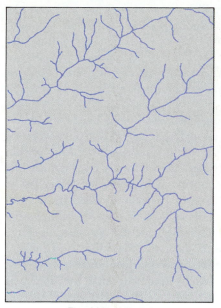

Dendritic pattern
Virginia, Ill., quadrangle

Trellis pattern
Monterey, Va., quadrangle

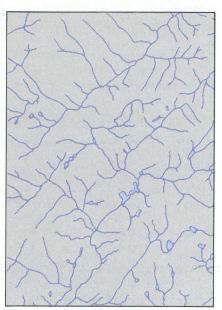

Rectangular pattern
Elizabethtown, N.Y., quadrangle

Scale

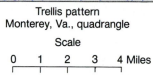

0   1   2   3   4 Miles

Radial pattern
Katahdin, Me., quadrangle

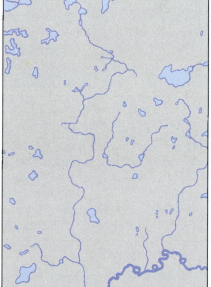

Deranged pattern
Galesburg, Mich., quadrangle

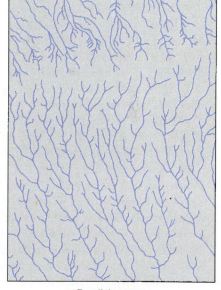

Parallel pattern
Mesa Verde Nat. Park, Colo.

Scale

0   1   2   3   4 Miles

Figure 14.4 Map depictions of major drainage pattern types.

them, we now examine the processes by which stream erosion, transportation, and deposition occur.

Erosional Processes

Stream erosion results from solution and from the friction and shear forces generated between the flowing water (including any sediments being transported by the flow) and the nonmoving materials with which it comes into contact. If the stresses brought to bear by this action exceed the cohesive strength of the materials, these materials will be eroded and transported downstream. Approximately 95 percent of the gravitational energy that causes stream flow is dissipated through friction between the water molecules of the stream and is converted to heat; the critical remaining 5 percent is used for sediment erosion and transport and therefore serves to shape fluvial landforms.

The erosive energy of a given stream is related to its velocity of flow, its volume of flow, and the amount of friction generated with its *banks* (sides) and *bed* (bottom). An increase in any of these factors will increase the erosive power of the stream. At the same time, a change in any of the three factors will affect the other two, since they are all interrelated. A fourth factor that must be acknowledged when assessing a stream's ability to perform erosion is the erosional resistance of the surface on which it flows. If the stream banks and bed consist of solid rock, the erosional process is normally very slow. If the stream flows within accumulations of unconsolidated materials, however, it can effect very rapid erosional modifications.

The speed at which the water flows and the amount of friction it encounters are critical factors in determining the *turbulence* of the flow. Turbulence refers to the tumbling eddy motions of small parcels of water within the stream that are superimposed on the main downhill direction of the current. This phenomenon also influences wind-flow characteristics (Chapter 4). Just as friction between the atmosphere and the earth's surface produces turbulence, which causes the wind to be gusty and to erode and transport sediments, so too the friction of a stream with its banks and bed causes it to flow in a turbulent manner and gives it the ability to erode and transport sediments with a higher specific gravity than water. Water flows at a much slower average speed than the wind, but it is a much denser medium and has correspondingly greater erosional and transportational capabilities. An increase in either streambed roughness

or speed of flow will increase turbulence and give the stream a greater erosional ability. In fact, it has been determined that the erosive power of a stream increases as approximately the cube of its increase in speed of flow. This means that if a stream's speed of flow doubles, its capacity to erode sediments increases by 2^3, or eight times.

In order for any stream erosion to occur, the water must flow rapidly enough that the friction of the moving water, coupled with the lift produced by eddies within the flow, can dislodge the materials over which the water passes. The minimum stream velocity needed to erode a given streambed particle is called its **critical erosion velocity.** This velocity is lowest for sand-sized particles. Smaller and lighter silt and clay particles have higher critical erosion velocities than does sand because of their cohesiveness and because they collectively produce such a smooth streambed that turbulence is reduced. Larger particles, on the other hand, require higher flow speeds for erosion because of their weight.

Although small streambed particles are difficult to erode, once eroded they can continue to be transported at very low flow speeds. Conversely, high-velocity flow is required for both the erosion and transport of coarse materials. Most streams do not maintain a high flow velocity for very long; as a result, the great majority of transported stream sediments consists of sand-sized or finer particles.

During times of high flow volumes, streams have the capacity to erode and transport a much greater sediment load than at times of normal flow. This occurs for several reasons: the quantity of water available to erode and transport sediments is greater; the total surface area covered by the stream increases, giving the stream access to more sediments; and, most important, the speed of flow increases significantly. As a result, more stream erosion and fluvial landscape modification in general occur during floods than during the much longer periods when normal or low flow conditions prevail.

The actual process of sediment erosion by streams is accomplished in three different ways. **Hydraulic action** consists of the direct sweeping away of loose materials by the friction and turbulence generated by the moving water. Although very effective on unconsolidated materials, it has a negligible effect on solid rock. **Abrasion** involves the scraping of water-carried particles against the stationary materials of the banks and bed. This process accomplishes slow erosion of even solid rock if the stream-borne particles are harder than the rock they are scraping. Abrasion also occurs within the flow itself as the transported sediments are repeatedly scraped or knocked together by turbulence. This process results in the gradual rounding of the sediments. *Solution* is the dissolving of rock materials so that they enter the flow in a molecular state.

Most streams simultaneously erode both vertically and laterally. The relative proportion of erosion that takes place in each of these directions is critical to the nature of the resulting topography, since a predominance of vertical erosion will produce a high relief surface containing deep, steep-sided valleys, while a predominance of lateral erosion will result in the formation of broad, gently sloping valleys. In general, although factors such as volume of flow, channel form, and surface erosional resistance are important, vertical erosion is dominant where the downhill gradient of the stream is steep, and lateral erosion dominates where it is gentle. A gentle gradient reduces the stream's velocity, thereby lessening the rate of downcutting and enabling valley-widening processes such as sheet flow, tributary inflow, and mass wasting to keep up with deepening and to maintain a gentle valley side-slope.

Variations in stream velocity, turbulence, and resulting erosional capacity can exist over short distances. Erosion therefore does not occur simultaneously everywhere along the course of a stream, but is at any given time concentrated at certain points. These points are located where active currents come in contact with bed and bank materials, especially if these materials are weak or unconsolidated.

Transportation

The same water turbulence that lifts sediments from the streambed keeps them from settling back to the bed, allowing the current to transport them downstream. As long as the stream velocity and turbulence that eroded the sediments are maintained, stream sediments will be transported downstream without being deposited.

Streams transport sediments in a number of ways, depending upon the sediment sizes and the physical state of the flow (see Figure 14.5). A substantial proportion of the total sediment load is carried in solution. The dissolved load tends to be especially large in relation to the total load in areas of relatively soluble rock such as limestone, in areas where the velocity of flow and resulting turbulence are low, and in humid areas where vegetation is abundant and the water is mildly acidic. Most dissolved materials are transported

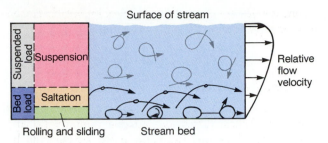

Figure 14.5 Methods of stream sediment transport. The velocity profile at right indicates that water flows fastest near the surface of a stream and most slowly along its bed.

entirely through the stream system to the sea or to another standing water body that the stream enters. On the average, about 30 percent of the total sediment load of streams is transported in solution.

Over half the total sediment load collectively transported by streams travels as suspended load. Suspended materials consist of generally fine-textured solid particles that are carried within the flow itself. They are kept from settling by the upward component of the turbulent eddies generated by friction. Streams flowing through areas in which unconsolidated materials are plentiful carry the greatest volumes of suspended sediments. The greatest quantities of such materials are picked up and transported during floods (see Figure 14.6).

The coarsest particles that a stream is capable of moving are carried as bed load. As the term implies, these sediments are moved along the streambed because they are too heavy for turbulence to transport them within the flow for any great distance. Depending on their size and shape, they may travel by rolling, sliding, or bouncing (*saltation*) along the bed. The bed load, unlike the suspended and dissolved loads, moves intermittently. It is typically transported a short distance downstream, temporarily deposited, and later picked up and carried farther. Many bed load particles are transported only during floods, when the carrying capacity of the stream is greatly augmented temporarily. This accounts for the frequent presence of giant flood-transported boulders on the bed of a small stream that seems utterly unable to move them. As it travels, the bed load is gradually rounded and reduced in size by abrasion.

Streams quickly achieve an adjustment between the quantity and texture of their sediment load and their ability to transport that load. The basic mechanics of

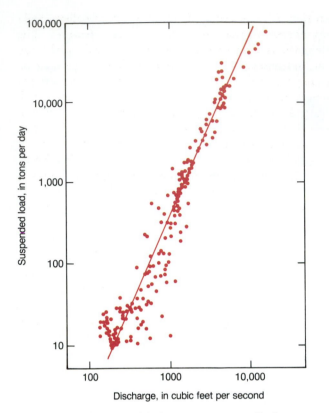

Figure 14.6 Relationship between stream discharge (volume of flow) and suspended sediment load in Brandywine Creek near Wilmington, Delaware. Each dot represents a separate measurement. An increase in discharge can be seen to produce a marked increase in suspended load.

the adjustment process are relatively simple. If the sediment-carrying capacity of a stream exceeds its load, the excess energy is used to erode the stream channel vertically. This, in turn, lowers the elevation of the stream and reduces its downhill gradient, producing a reduction in its speed of flow, its turbulence, and its sediment-carrying capacity. In the absence of any disrupting factors, downcutting will continue until the carrying capacity of the stream is reduced to the point that it becomes just sufficient to transport its sediment load. No further significant downcutting will subsequently occur, because there is no longer a supply of excess stream energy.

By the same token, if the stream's sediment supply exceeds its carrying capacity, the excess sediment will be deposited. This will aggrade (build up) the channel, increasing the stream's downhill gradient, which in turn will increase the stream's flow speed, turbulence, and

sediment-carrying capacity. Deposition and aggradation will continue until the stream's ability to transport sediment has been raised to the point that it can carry all of its load.

A stream that has attained a balance between flow and load is said to have become graded. A **graded stream** typically flows along a gradually declining downhill gradient from head to mouth. Streams can continue to transport their sediment loads at progressively reduced slope angles because their increased downstream volumes of flow result in a gradual reduction in the amount of friction per unit volume of water. In addition, a gradual reduction in the size and settling velocity of the streambed particles occurs in a downstream direction because of abrasion. As an illustration of the influence of discharge on speed of flow, the lower Amazon River, with an extremely gentle gradient of only 3 inches per mile (4.75 cm/km), flows at nearly 8 miles per hour (2.4 m/sec) during times of high water. This is faster than many mountain streams with much steeper gradients.

Deposition

Streams deposit sediments when a reduction in speed, turbulence, and/or volume of flow make them incapable of transporting all of their sediment load. As their carrying capacity declines, they first deposit the coarsest materials, which have the greatest settling velocities, and then deposit progressively finer sediments. The finest sediments, including the stream's dissolved load, are generally transported completely through the stream system to the water body into which the stream empties. This association between stream-transporting ability and the size of the particles being deposited causes fluvial deposits to be *sorted,* or separated by size. A high degree of sorting, in fact, is one of the most important and easily recognizable characteristics of fluvial deposits (see Figure 14.7).

The accumulation of thick deposits of sediments is common in the lower reaches of stream systems, especially at the mouths of large rivers that empty into the sea. Deposition, like erosion, also occurs locally and temporarily at numerous points along the courses of nearly all streams. Sites favoring deposition include shallow stretches along the stream course such as the insides of bends, places where the channel widens, and behind obstructions to flow, such as boulders lying on the streambed. Human actions can greatly influence the

310

Figure 14.7 Closeup view of rounded and sorted fluvial deposits on the bank of a stream near Lafayette, Indiana. (*B.F. Molnia*)

pattern of stream deposition, with dams and reservoirs acting as sediment traps. The effect of channel modification on the depositional pattern of the Mississippi River is discussed in the Case Study at the end of the chapter.

Accumulations of stream-deposited sediments are called *alluvium.* The chief source for the alluvial deposits of most streams is soil eroded from areas upstream. Thick accumulations of alluvium currently cover the lower valleys of most large river systems, and their generally high content of soluble plant nutrients often makes them agriculturally productive.

FLUVIAL LANDFORMS

Because of their gradational power and widespread distribution, streams can be regarded as the leading geomorphic agent in terms of their influence on the earth's surface topography. In many parts of the world, the three processes of weathering, mass wasting, and fluvial erosion act in sequence to erode the surface and produce most gradational landforms. Weathering acts first, producing unconsolidated surface materials; mass movements transport these materials to stream valleys; and streams then transport them, eventually, to the sea. During the last stage of the sequence, a variety of fluvial erosional and depositional features is likely to be formed.

In a direct sense, the dominant fluvial feature is the *valley.* A valley is a linear lowland produced by stream erosion. Most streams, of course, produce valleys, making them common and familiar surface features. The dimensions of a valley are determined largely by the size (or, more precisely, the volume of flow) of the stream that produced it, the rate of stream erosion, and the length of time over which the valley has developed. Valleys contain a variety of both erosional and depositional features, which will be examined shortly. From an indirect standpoint, a second major category of fluvial landforms is erosional hills and mountains. These features are the upland remnants that lie between stream valleys in a fluvially dominated landscape.

In a very general sense, fluvial landforms can be divided into two groups: those formed primarily by erosion and those formed primarily by deposition. It is often rather arbitrary, however, to classify fluvial features as either erosional or depositional because many are composite in nature, produced by both erosional and depositional processes operating in different places or in the same place at different times. Nonetheless, these two groups of processes are used in this section to organize the discussion of fluvial landforms.

Fluvial Erosional Features

Streams erode when their volume, speed, and turbulence of flow enable them to remove and transport a greater quantity of sediment than they already contain. An excess erosional capacity is most common in the upper reaches of drainage systems. In these areas, the downhill gradient is normally considerably steeper than it is farther downstream; the sediment load is usually much smaller in volume; and significant tectonic uplift is more likely to have occurred in the recent geologic past. The two major categories of fluvial erosional features, as already noted, are valleys and erosional hills and mountains.

Erosional Valleys

Nearly all river valleys are produced largely or entirely by fluvial erosion, but the term *erosional valley* as used here refers to a valley still being actively deepened by a river that is downcutting toward base level. **Base level** is the level or elevation at any point along a stream's course that provides the stream with a gradient just sufficient for it to transport its sediment load past that point with no erosion or deposition. It is, in other words, the level of flow/load equilibrium for the stream.

Any erosion or deposition can be viewed as the stream's response to a changing or as yet unattained base level. If base level is achieved all along its course, the entire stream system is considered to be graded.

A temporary base level or equilibrium level may be achieved quickly through the reworking of stream sediments. During floods, for example, a stream gains excess erosional capacity and rapidly erodes accumulations of sediments lying within its channel. The equilibrium level in this case has been temporarily lowered by the stream's increased sediment-carrying capacity, and the stream therefore reduces its gradient in order to restore a balance between flow and load. When the flood subsides, the stream is located at too gentle a gradient to transport its sediment load with the reduced volume of water it now contains. It therefore deposits enough sediment to aggrade the channel back to its original level. Because of nearly constant fluctuations in their flow volumes, most streams are continuously eroding or depositing at various points along their courses in order to adjust to shifting temporary equilibrium levels.

On a long-term basis, an ungraded stream may have to erode through great thicknesses of rock over lengthy geologic time spans in order to reach a stable base level. A well-known example is the Colorado River, which, over many millions of years, has eroded a mile (1.6 km) into the surface of the Colorado Plateau of northern Arizona to produce the Grand Canyon (see Figure 14.8). The river in this area has apparently not yet achieved its base level elevation, partially because tectonic uplift is still taking place, and active downcutting continues. Superimposed on this overall erosional trend, though, the Colorado River locally erodes and deposits sediments at various points along its course in response to short-term variations in its sediment load and volume of flow.

If a stream is downcutting rapidly, assuming that the near-surface rock material is strong enough to maintain steep slopes, the erosional valley will be deep, steep-walled, and relatively narrow. The width of such a valley is restricted because valley-widening processes depend largely upon the slow progress of weathering to weaken valley-side materials so that they may be removed by mass wasting. Valley deepening, on the other hand, can be effectively accomplished by stream particle abrasion on unweathered materials. Even the deepest and most steep-walled stream valleys rarely have depths that exceed their widths. If it is exceptionally deep and narrow, the valley is referred to as a *gorge*. A steep-walled but broader and often flat-bottomed valley is

Figure 14.8 Step and slope topography caused by differing resistances to erosion of rock strata exposed along the rim of the Grand Canyon. The Grand Canyon—a mile deep, approximately 20 miles wide, and over 200 miles long—is one of the most impressive fluvial landform features in the world. (*John S. Shelton*)

called a *canyon*. The side slope characteristics of stream valleys, especially of rapidly expanding erosional valleys, depend largely on the nature of the materials of which they are composed. If the materials are homogeneous, a smooth side slope is produced. If the materials consist of alternating layers of varying erosional resistance, a highly irregular slope may be produced (see Figure 14.8). This slope will normally consist of protruding ledges and cliffs formed on the resistant strata and inset slopes of intermediate steepness formed on the weaker strata.

Erosional valleys are found mostly in uplands such as plateaus or in hilly or mountainous regions. Most large river systems have their headwaters in such regions and eventually enter plains farther downstream. Erosional valleys are especially steep, narrow, and spectacular in arid or semiarid regions (see Figure 14.9). Here they are less masked by vegetation, and much less moisture is available for chemical weathering to facilitate valley-widening processes.

Erosional Hills and Mountains

Erosional hills and mountains are the highland remnants that remain on the landscape following the fluvial dissection of a plateau, fault block, or other raised surface. We have in previous chapters given credit for

Figure 14.9 The geologically youthful gorge of the Colorado River upstream from the Grand Canyon. This gorge is deeply incised into the horizontally bedded sedimentary rocks of the Colorado Plateau. (*John S. Shelton*)

the widespread distribution of such features to the tectonic forces that produced the initial surface uplift, but perhaps equal credit should be given to the gradational forces that have carved these raised blocks into their present intricate patterns of hills or mountains and intervening valleys.

The existing pattern of hills or mountains within a region is determined largely by the type of drainage pattern that develops. The drainage pattern, in turn, is controlled by the nature of the uplift and by the rock type and structures within the area. If, for example, a plateau is formed by the broad uplift of a mass of homogeneous or horizontally bedded rock, a dendritic drainage pattern will develop. The random locations of the stream segments in this pattern will cause the intervening hills or mountains to display a random pattern also. The Appalachian and Ozark Plateaus are good examples of rugged but disorganized regions formed by the development of dendritic drainage patterns in regions of uplifted horizontal sedimentary rocks.

In places where folding or faulting has exposed strata of differing resistance to erosion, streams develop linear drainage patterns, such as the trellis pattern, as they preferentially carve valleys in the weaker rock layers (see Figure 14.10). This produces a parallel system of resistant linear ridges and alternating valleys such as that found in the Appalachian Ridge and Valley region and in portions of the Coast Ranges of California.

Fluvial Depositional Features

Streams deposit materials when their sediment supplies exceed their carrying capacities. Deposition occurs on the largest scale in the lower portions of drainage systems. Here, slopes are normally gentle, the streams and their sediment loads are large in volume, and a stable long-term base level is more likely to have been achieved than farther upstream. The dominant landform resulting from fluvial deposition is the floodplain. Floodplains, in turn, contain a number of important secondary features such as meander deposits, natural levees, deltas, and alluvial terraces, which are examined in this section.

Floodplains

Floodplains are broad, flat-floored valleys covered by alluvium and subject to flooding at times of high water. Floodplains comprise the floors of valleys originally formed by vertical erosion, and they are normally in the process of being slowly widened by lateral erosional processes. In one sense, then, the floodplain, as is true of any stream valley, can be considered more a product of erosion than of deposition. On the other hand, a stream occupying a floodplain has typically reached its long-term base level and has essentially ceased downcutting. Furthermore, nearly all floodplains are covered with stream-deposited alluvium; and the small-scale features

Figure 14.10 The central Appalachians display a trellis drainage pattern of parallel valleys eroded in limestone or shale, separated by resistant sandstone ridges. (*Rodman Snead, © JLM Visuals*)

within the floodplain are alluvial and depositional in origin. A floodplain is therefore a depositional feature located within a larger erosional feature. Because the erosional processes that result in valley formation have already been discussed, the emphasis in this section is on the depositional aspects of floodplains.

Floodplain formation begins when a stream reaches its base level of erosion and active downcutting ceases. This allows the valley floor to be gradually widened by lateral gradational processes. During major floods, the stream temporarily erodes loose deposits and may actively scour the bedrock floor of its valley. As water levels fall and the stream loses its erosional capacity, it deposits a covering of alluvium over the width of the valley, bringing its surface back to base level. Many floodplains are currently so deeply covered by alluvium that bedrock exposure is impossible. In some cases, this has occurred because the floodplains are located in slowly subsiding tectonic basins. This is true, for instance, of the Mississippi, Amazon, and Ganges River Valleys. For example, the Mississippi River Valley near Vicksburg, Mississippi, contains an average of over 150 feet (45 m) of sediment, and in places sediment thicknesses are twice as great.

One of the most important characteristics of a floodplain, from both a geomorphic and a human standpoint, is that it is subject to occasional flooding. Streams flowing within floodplains are not normally incised into the surface, but instead flow close to the level of the surrounding land. During times when they carry large amounts of runoff, they can readily overtop their banks (see Figure 8.7). As a result, overbank floods occur in a typical floodplain section about two years out of three. The relative surface flatness usually causes these floods to be extensive but rather shallow. Large quantities of sediments are transported by a stream during floods and are deposited in a layer over the surface, especially as the water retreats. Such deposits add valuable topsoil eroded from upstream portions of the drainage basin. The short-term economic hardships of floods for floodplain residents must therefore be weighed against the long-term benefits of naturally maintained soil fertility.

Extreme floods, which are fortunately quite rare on any given stream, may cover the floodplain so deeply and with such swiftly flowing water that erosion, rather than deposition, predominates. These floods help to maintain a long-term surface level equilibrium because they offset the slow buildup of sediments from "normal" floods by periodically flushing out the excess.

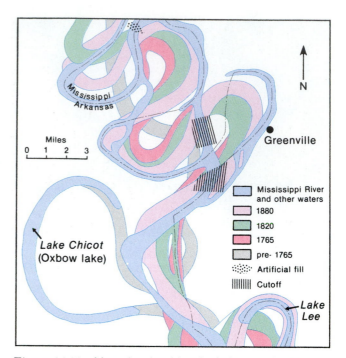

Figure 14.11 Map showing historical changes in the course of a portion of the Mississippi River on the Arkansas-Mississippi border. The boundary between these two states no longer follows the river's course because the channel has changed several times since the boundary was established in the early twentieth century.

Because a stream occupying a floodplain flows near the surface in unconsolidated alluvium, it can shift its course much more readily than can a stream occupying a narrow erosional valley. Over long periods of time, then, the stream is likely to flow over all portions of its floodplain in turn (see Figure 14.11). On occasion, it will impinge on portions of the sidewalls of the floodplain, producing bank undercutting that widens the floodplain. Fluvial features are therefore not restricted to the immediate vicinity of the present channel, but are distributed widely over the floodplain surface. Most of the modest surface relief found within floodplains is associated with these features.

MEANDERS AND ASSOCIATED FEATURES   The development and migration of broad looping bends, or meanders, in the courses of streams occupying floodplains are responsible for a large proportion of floodplain features (see Figures 14.12 and 14.13). Stream meanders

Figure 14.12 A variety of active and abandoned meander features is evident in this view of a river in central Alaska. (*B.F. Molnia*)

speed and depth of flow causes the deposition of sediments, which form a series of **point bars.** Centrifugal force, acting on the current as it passes through the meander, causes the sides of the meander to converge gradually to form a narrowing *meander neck.* Eventually, the sides intersect to produce a *meander cutoff.* When this happens, the meander is abandoned by the active current as the river takes the shorter and steeper route through the cutoff.

The abandoned meander is eventually separated from the stream by bank deposits along its new course so that it becomes an arcuate lake called an **oxbow lake** (see Figure 14.15).[1] Lakes are relatively ephemeral features on the landscape, and an oxbow lake gradually fills with fine sediments and organic matter to form first a swamp and eventually dry land. The meander scar may long be visible on the landscape, especially from the air, because its vegetation typically differs from that of its surroundings.

NATURAL LEVEES When rivers within floodplains spill out of their channels during floods, their water is subjected to a sudden increase in friction, and its speed of flow is consequently greatly diminished. This reduces its sediment-transporting capacity and results in the deposition of the sediment excess on the floodplain surface. The coarsest materials, which have the largest settling velocities, are sometimes deposited atop the stream banks, where they accumulate to form natural levees.

Natural levees take the form of low, often broad ridges of fine sand and coarse silt that parallel each side of the stream channel. They are sometimes the highest natural features within the entire floodplain and, as such, increase channel stability by helping to keep the stream within its established channel. Natural levees along the lower Mississippi River, for example, reach heights of up to 15 feet (5 m) above the main floodplain surface. If the river is aggrading its channel, the presence of natural levees may result in the water surface standing higher than the level of the surrounding floodplain. This is an unstable situation because a breach in any portion of the levee will allow the river to abandon its channel and form a new channel at a lower level in another portion of the floodplain. Human

display a distinctive cycle of formation, growth, and disappearance. A meander begins as a slight bend in the channel, perhaps initiated by differing degrees of cohesiveness in the materials on the opposing stream banks or by the inflow of water from a tributary. The diversion of flow results in the erosion of the unconsolidated material on the outside of the bend, allowing the bend to become more pronounced (see Figure 14.14). As the bend increases in amplitude, centrifugal force furthers its growth by deflecting the main current toward the outer bank, where erosion occurs. As it grows outward, the meander slowly migrates down-valley. On the inner side of the developing meander, the reduction in the

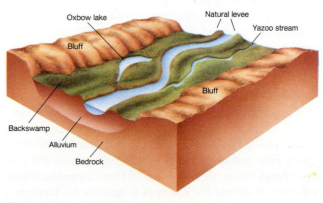

Figure 14.13 Major floodplain features.

Oxbow lake · Natural levee · Yazoo stream · Bluff · Bluff · Backswamp · Alluvium · Bedrock

---

1. An oxbow, which gives this type of lake its name, is the U-shaped part of an ox yoke that passes under and around the animal's neck and helps attach it to a cart or plow.

Figure 14.14 Stages in the "life cycle" of a stream meander.

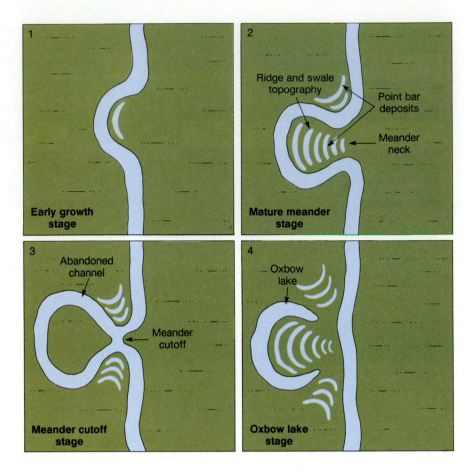

1 Early growth stage

2 Mature meander stage
Ridge and swale topography
Point bar deposits
Meander neck

3 Meander cutoff stage
Abandoned channel
Meander cutoff

4 Oxbow lake stage
Oxbow lake

beings, who usually have a vested interest in keeping rivers within their existing channels, may reinforce natural levees in an attempt to avoid such breaches, which usually occur during floods (see the Case Study at the end of the chapter). In places, China's Huang Ho River flows over 33 feet (10 m) above the level of its surrounding floodplain because a combination of natural and artificial levees have long kept it from shifting its course.

The tendency of rivers in floodplains to meander and to form natural levees therefore leads to the development of a variety of relatively short-lived features on the floodplain surface. In general, these features are most abundant and recently formed near the current river courses. Their presence gives the relatively flat floodplain floor a complex microrelief that is often clearly visible on aerial photographs because of the vegetative patterns these features produce (see Figure 14.15).

Deltas

Deltas are deposits of alluvium formed when streams enter standing bodies of water. They exist on a wide variety of scales, from that of a small brook entering a pond to that of a giant river such as the Amazon or Mississippi entering the sea. Deposition takes place because the sudden cessation of flow as the stream enters the standing water body causes the stream to lose its sediment-carrying capacity. This causes all but dissolved sediments and clay-sized particles to settle quickly to the bottom.

In general, deltas are proportional in size to the sizes and sediment loads of the streams that produce them. All of the world's major deltas are produced by large rivers that transport heavy sediment loads to the sea. Some large rivers, though, carry little sediment and have not formed deltas. For example, Canada's St. Lawrence River cannot obtain the sediment load necessary for

Figure 14.15 An oxbow lake, meander scars, and ridge and swale topography produced by the Tallahatchie River in Mississippi. (*Department of the Interior, U.S. Geological Survey*)

delta formation because the Great Lakes, which it drains, act as settling basins. In contrast, many of the large rivers of Asia carry heavy sediment loads and have developed extensive deltas. The Ganges delta, with a surface area of 30,000 square miles (78,000 sq km), is the world's largest.

Delta formation is also greatly facilitated if the standing water body that the river enters is shallow and has weak wave and current action. If the water is very deep, great quantities of sediment must accumulate before the deposits reach the surface. This is one reason why few significant deltas occur along the west coasts of North and South America. Similarly, if wave and current action are strong, stream deposits will be carried away or spread out along an extensive stretch of coastline rather than accumulating in one place to produce a delta. For this reason, most of the best-developed deltas are formed in bays, partially enclosed seas, or other protected waters.

The term "delta" was taken from the Greek letter delta (△), which was originally used to describe the shape of the "classic" delta of the Nile River (see Figure 14.16). Other deltas, like those of the Ganges or Niger Rivers, have a more fanlike shape. A few deltas, most notably that of the Mississippi River, contain long

fingerlike projections of land. These projections result from a combination of very weak currents that are unable to smooth off the outer perimeter of the delta and a slow subsidence of the surface that causes only the crests of natural levees along the major delta channels to remain above the water level at present. This form has been termed a *bird's-foot delta*. Finally, a number of deltas are located in semi-enclosed water bodies (*estuaries*) either in drowned river valleys or in submerged tectonic troughs. Examples of estuarine deltas include the deltas of the Yangtze, Susquehanna, and Colorado Rivers.

Deltas generally form seaward extensions of existing floodplains. Like floodplains, delta surfaces are low, flat, and often swampy and are composed of fine-textured sediments that normally contain a large proportion of organic matter. When adequately drained, they may form highly productive agricultural land and support dense populations.

Terraces

It will be recalled that the formation of a floodplain indicates that the stream has reached a stable base level of erosion and that downcutting is no longer significant. If base level is subsequently lowered for any reason, a stream occupying a floodplain will renew downcutting and may eventually carve a new and narrower floodplain within the old floodplain surface (see Figure 14.17). The now-elevated remnants of the original floodplain form terraces that usually end abruptly in steep slopes well back from the current position of the river channel. In most cases a pair of terraces will be present, one on each side of the valley. The height of the terraces depends upon the extent to which base level has been lowered. The width of the new floodplain level, and therefore the distance that separates the opposing terraces, is related to the amount of time that has passed since the river eroded to its new base level. If the new level is maintained long enough, the floodplain will eventually be widened sufficiently by erosional processes to destroy the terraces. In some cases, multiple terrace levels exist within a valley. This indicates that base levels are being sporadically lowered, with intervening periods of stability.

The primary cause of terrace formation is an increase through time in the stream gradient. This can be accomplished by tectonic or isostatic uplift of the surface or by a decline in sea level. River terraces are

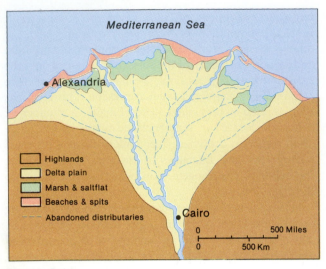

(a) Nile Delta

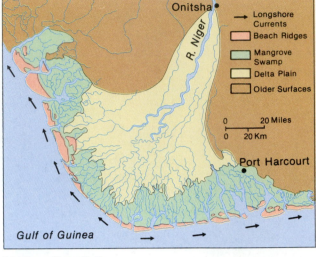

(c) Niger River Delta

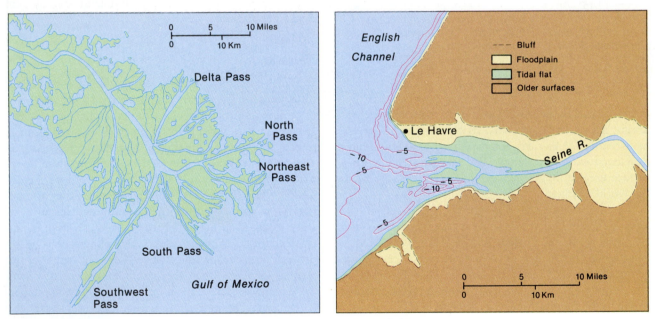

(b) Mississippi Delta

(d) Seine Delta

Figure 14.16 Maps of four deltas with differing forms. Delta forms are controlled by such factors as the rate of sediment deposition, the bottom topography of the water body in which the delta is being formed, and the patterns of coastal currents. The Nile and Niger Rivers have produced "classic" deltas, while the Mississippi has produced a bird's-foot delta and the Seine an estuarine delta.

common in areas currently undergoing tectonic uplift, such as portions of California and the Alps. Another important cause of terrace formation is a reduction in the sediment load supplied to a stream. During the waning stages of the Pleistocene period, streams drain-

ing the melting ice sheets of North America and Eurasia received tremendous influxes of glacial sediments. Unable to transport all of this load, the streams deposited much of it within their valleys to form thick alluvial fills. When the glaciers disappeared, the sediment sup-

Figure 14.17 Terraces have formed along the margins of the Inn River Valley, in Austria, as a result of periodic tectonic uplift. (*Jerry Scott*)

ply was greatly reduced, and the rivers actively eroded into their alluvial deposits, leaving the remnants standing as terraces well above the present valley floors.

River terraces have become favored settlement sites because of their relatively flat surfaces, fertile alluvial soils, and proximity to water. An important additional advantage is that their high-standing position makes them immune to the floods that periodically inundate the lower floodplain.

Summary

Streams perform two vital natural functions. The first is the removal of excess water from the land. The second is the removal of surface weathering products supplied to streams largely from adjacent slopes by mass wasting processes. Because of the gradational power and the nearly global distribution of streams, the majority of the earth's land surface is dominated by fluvial landform features.

Streams develop where there is a sufficient downhill flow of water to erode a channel. Each stream channel conducts water from a drainage basin, and the coalescence of stream channels forms a drainage system. Drainage systems display spatial patterns of organization dictated largely by existing patterns of rock types and structures as well as surface features. Several basic drainage pattern types were discussed in the chapter. The most widespread is the dendritic pattern, which

develops on homogeneous surfaces that exert no strong topographic or structural control on the locations of the stream segments.

Stream erosion takes place through the processes of solution, hydraulic action, and abrasion. All rock materials are soluble to some degree, and a mean of about 30 percent of the sediment load of streams is carried in solution. Hydraulic action, resulting from stresses generated by the force of the flowing water, effectively erodes the unconsolidated materials that form the banks and beds of most streams. Abrasion, the scraping action of stream-borne particles, is capable of slowly eroding solid rock. Once sediments have been entrained into the flow of a stream, they are transported as long as the turbulence generated by the flow keeps them from settling to the streambed.

The primary fluvial erosional feature is the valley. A stream with a large amount of erosive energy will downcut rapidly, producing a deep, narrow valley. Downcutting, however, reduces the slope angle and therefore the gravitational energy available to the stream. Eventually, valley-widening processes become dominant in tectonically inactive areas. The fluvial erosion of uplifted blocks of land also carves them into hills and mountains. The topographic pattern displayed by these upland features depends largely upon the drainage pattern that develops, while their relief is determined by the rate of downcutting.

A stream deposits sediments when a reduction in its speed or volume of flow, or a change in the geometry of its bed, decreases the turbulence of the flowing water so much that the stream can no longer transport its entire sediment load. When this occurs, the coarsest sediments are deposited first, followed by progressively finer sediments as the stream's carrying capacity diminishes. As a result, fluvial sediments are sorted, or separated by size. The primary fluvial depositional feature is the floodplain, which is a broad, flat-floored valley covered by alluvium and subject to occasional flooding. Floodplains exhibit a variety of depositional features formed largely by lateral stream migration. These include bars, natural levees, and various meander features such as oxbow lakes and meander scars. When a stream deposits sediments in a standing body of water, a delta is often constructed. Deltas vary greatly in size and shape, but most display a fanlike pattern.

When a stream occupying a floodplain renews downcutting, it incises itself into the floor of the floodplain. The remnants of the old floodplain are often preserved as terraces along the sides of the valley. Some

streams have formed multiple terraces at varying levels above the present valley floor.

Review Questions

1. What parts of the world are, and are not, dominated by fluvial landforms?
2. Briefly describe the likely general appearance of the surrounding landscapes of areas that have each of the following drainage pattern types: radial, trellis, deranged, dendritic.
3. Describe each of the three methods by which streams transport sediments. Which moves the largest total volume of sediments worldwide?
4. What causes streams to deposit sediments? Describe the physical appearance of stream deposits.
5. What is the major landform feature produced by fluvial erosion? Describe the characteristics of such a feature in which vertical erosion is occurring rapidly.
6. What is a floodplain? How does it differ from an erosional valley? List several important floodplain features.

7. Explain why deltas form. Why are bird's-foot deltas like that of the Mississippi River relatively rare?
8. What does the development of a series of alluvial terraces within a valley indicate about the recent geomorphic history of the valley? Explain.

Key Terms

Fluvial	Suspended load
Interfluve	Bed load
Infiltration capacity	Meanders
Drainage system	Graded stream
Dendritic pattern	Base level
Trellis pattern	Floodplain
Rectangular pattern	Point bars
Radial pattern	Oxbow lake
Deranged pattern	Natural levee
Critical erosion velocity	Delta
Hydraulic action	Terrace
Abrasion	

CASE STUDY

Will the Mississippi Change Its Course?

Human beings have been altering the flow of rivers for thousands of years in an effort to increase their usefulness and to reduce the danger of flooding. Most early alterations involved small-scale diversions of flow for irrigation purposes, but modern technology has greatly expanded the magnitude and variety of alterations that can be undertaken. A few examples of modern alterations of rivers include channel straightening and the construction of dams, reservoirs, locks, and levees.

Because all the components of the environment are complexly interrelated, purposeful human alterations occasionally produce unfavorable and unforeseen consequences. It would seem that our technological abilities sometimes surpass our understanding of the more subtle interactions between the earth's physical systems. The history of efforts to control the Mississippi River provides a good example of how desirable human adjustments to the flow of a major river have produced a highly unstable situation that offers the prospect of eventual disaster for much of southern Louisiana.

The Mississippi River system, one of the world's largest and most important, drains 41 percent of the United States as well as a portion of Canada. Its drainage basin is exceeded in size

only by those of the Amazon, Nile, and Congo Rivers. The heavily farmed Mississippi Basin is the agricultural heartland of the nation, and the river system provides the means for the transport of a tremendous tonnage of agricultural, industrial, and petrochemical products.

The task of developing and controlling the Mississippi River has been assigned largely to the U.S. Army Corps of Engineers. The two major responsibilities of the Corps have been to reduce the danger of destructive floods and to develop the system's navigation potential. On the whole, they have been very effective in accomplishing these goals. The Corps maintains the river throughout its length by means of a complex system of locks, dams, diversion channels, and artificial levees. Channels have been straightened and shortened in various places, and some channels have even been paved in order to increase their stability. The environmental consequences of these activities have been modest in the upper portion of the drainage basin, but are far more profound in the Lower Valley, where deposition predominates.

The Lower Mississippi is an alluvial river. It presently transports 64,000 tons of sediment a day (23,500,000 tons per year). Some of this material is deposited within the river channel, but most is carried into the Gulf of Mexico. Accelerated erosion from agricultural activities upstream has so augmented the sediment load that it is currently twice as large as it was 50 years ago. In the past, the Mississippi repeatedly changed its

course due to channel aggradation; and the deposits were thus spread over a large area to form an extensive delta. Because of the artificial stabilization of the river's channel, however, widespread sediment distribution can no longer occur.

The major problem resulting from channel stabilization is that the present Mississippi channel has aggraded itself well above the level of the surrounding floodplain. This has caused the river to display an increasing tendency to change its course to a more westerly route that is lower in elevation and shorter in its path to the Gulf. Continuing alluviation within the present channel, which is occurring at a substantial rate because of the large influx of sediment from upstream agricultural sources, is building the river ever higher. This makes the situation progressively more unstable and necessitates the constant heightening and strengthening of the levees along the river's current channel.

The site at which the Mississippi River would abandon its present course, if it had its own way, is known. It is located at the juncture of the Mississippi with the Red and Atchafalaya Rivers near the southwestern corner of the state of Mississippi about 50 miles (80 km) northwest of Baton Rouge (see Figure 14.18). The Atchafalaya is a distributary river that carries excess water from the Mississippi southward to the Gulf along a route located well to the west of the course of the Mississippi River itself. In 1839, a logjam on the upper Atchafalaya was cleared to open this river to navigation. The

continued on next page

resulting increase in flow caused the Atchafalaya to become gradually deeper and wider and to slowly divert increasing amounts of water from the Mississippi. During this century the process accelerated, and in 1953 the Corps of Engineers notified Congress that, unless stopped, the Mississippi's flow would be completely diverted into the Atchafalaya by 1990. The consequences of such a diversion would be enormous. Baton Rouge, New Orleans, and many other cities and towns along a stretch of the Mississippi more than 300 miles (480 km) long would be left high and dry, while towns along the Atchafalaya—notably Morgan City, Louisiana—would be flooded out of existence.

In order to prevent such an occurrence, Congress hurriedly authorized the Old River Control Project at an initial cost of $70 million. This project had as its goal the regulation of water flow from the Mississippi into the Atchafalaya through a system of 84 floodgates. The project was completed in 1963 and seemed at first to be highly successful. In 1973, however, a major flood caused the collapse of part of the structure and nearly destroyed it. At present, such an imbalance has developed between the two channels that the undermining of the Old River Control Structure (ORCS) by scour has become a very real danger. Three times each week, a series of several hundred water depth soundings are taken to ascertain whether new scour channels are forming; and a backup control structure is being built. Even if a future flood does not destroy the ORCS and permanently shift the Mississippi's course, floodwaters would be diverted down the Atchafalaya in order to save Baton Rouge and New Orleans. This places the future of

rapidly growing Morgan City in grave doubt.

Another major problem has been created by the tight control of the course of the Mississippi River. In the past, a network of distributary rivers and swamps spread vast quantities of sediments over a broad area of southeastern Louisiana. These deposits offset the slow tectonic subsidence of the

area and allowed for the maintenance and slow growth of the Mississippi delta. With the flow now tightly confined to a channel 1 mile (1.6 km) wide, most sediments are presently shot out into the Gulf of Mexico. The deposition that does occur takes place only within the existing channel, where it is not wanted. The lack of

continued on next page

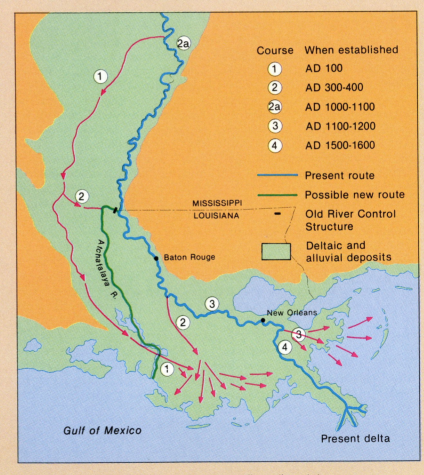

Figure 14.18 Map showing past courses (in red) and the present channel of the Mississippi River. The location of the Old River Control Structure (ORCS) is indicated, as well as the possible new course of the river down the Atchafalaya.

sediment is causing the remainder of the delta to slowly subside into the Gulf. Each year, about 50 square miles (130 sq km) of land are lost. The intrusion of salt water into other portions of the gradually submerging delta is killing vegetation and fresh water fish and is polluting domestic drinking water supplies.

Nature's price for channel stability has thus been the threat of a catastrophic shift in the course of the Mississippi River and the progressive loss of large amounts of valuable land from the lower Mississippi delta. These events provide eloquent testimony that environmental manipulations performed without adequate knowledge of possible long-term consequences may come back to haunt us.

Chapter Fifteen

Glacial Processes and Landforms

Focus Questions

1. What are glaciers, and why do they form?
2. What two major categories of glaciers exist, and where are they found?
3. In what ways do glaciers modify the landscape?
4. What was the Pleistocene, and why did it occur?

Ice is the most powerful of the "tools of erosion." Its gradational effectiveness for the earth as a whole, though, is greatly limited at present by its restricted areal extent. Ice currently covers only a tenth of the land area of the world, and it can directly influence the surface only in those areas. During the very recent geologic past, however, in the epoch called the *Pleistocene* (popularly known as the "Ice Ages"), the global coverage of ice was as much as three times that of today. The effects of this ice on the landscape were profound, and the Pleistocene glaciation occurred so recently that, as yet, many of the landforms it produced have been modified relatively little. The importance of ice as a geomorphic agent is therefore much greater than its current global distribution would indicate, and its significance is increased by the possibility that someday the glaciers may form again.

In this chapter, we examine the geomorphic influences of ice on the land and discuss the surface characteristics of landscapes dominated or strongly modified by glacial ice. The chapter is divided into three major sections. The first provides an overview of glacial geomorphology, stressing the current distribution, mechanics, and surface features produced by glaciers. The second section covers the subject of mountain or alpine glaciers. The final section examines continental glaciers as well as the causes and global impact of the Pleistocene.

GLACIER FORMATION, DISTRIBUTION, AND FEATURES

A **glacier** is a mass of freshwater ice, formed on land, that is or has been in motion. Like rivers, glaciers move largely in response to gravity and perform the basic hydrologic function of removing excess precipitation from the land. The chief source of glacial ice is compacted snow; in order for a glacier to develop, the accumulation of snow must exceed losses by melting and sublimation over an extended period. Therefore, not only cold temperatures, but also significant snowfalls are required for glacier formation. Because of a lack of snowfall, some very cold but dry areas, such as parts of Alaska and much of northeastern Asia, have apparently never been glaciated.

When snowfall accumulation rates exceed losses over a prolonged period, the accumulated excess is gradually converted to glacial ice. A fresh snowfall contains mostly air and has a specific gravity of approximately 0.10, a tenth that of water. After a short period, the snow changes form to a much more compact granular state, called "firn," that has a specific gravity of 0.55 or more. Further compaction, melting, and refreezing result in the expulsion of the remaining air and a slow conversion to glacial ice, with a specific gravity of about 0.85.

Present Extent of Glaciation

Glaciers are divided into two general categories on the basis of location, size, and shape. **Continental glaciers,** as the name implies, cover extensive land areas and form in the high latitudes. **Alpine glaciers,** also called "mountain glaciers," form at high elevations in mountainous regions.

The great majority of the earth's glaciated surface is covered by the two surviving continental glaciers, or ice sheets: the Antarctic and Greenland ice sheets (see Figures 15.1 and 15.2). The Antarctic ice sheet is by far the largest; it covers 4.9 million square miles (12.5 million sq km), an area larger than the United States and Mexico combined. It has a volume of approximately 5,750,000 cubic miles (24 million km^3) and a maximum thickness of 13,000 feet (4000 m or 2.5 miles).

The Greenland ice sheet is considerably smaller, with an area of 696,000 square miles (1.8 million sq km), a volume of 550,000 cubic miles (2.3 million km^3), and a maximum thickness of 11,000 feet (3350 m). These two ice sheets, if melted, would release enough water to supply all the world's rivers at their present discharge rates for nearly 900 years, raising sea levels at least 150 feet (45 m).

In contrast to continental glaciers, alpine glaciers are of limited areal extent, with a total world surface coverage of approximately 196,000 square miles (508,000 sq km) and a combined volume of 19,200 cubic miles (80,000 km^3). Several thousand are scattered throughout the higher mountain ranges of the world. The more significant of these glaciers occur in Alaska, the Canadian Rockies, the Canadian Arctic Islands, the Andes, the European Alps, and several mountain ranges of Asia, including the Himalayas, Karakoram, Tien Shan, and Caucasus (see Table 15.1). A few small glaciers exist in the southern Rockies of the United States, the Cascades, and in East Africa. Australia is the only continent with no glaciers.

Figure 15.1 The Antarctic ice sheet. The continent of Antarctica is almost completely buried by an ice sheet that averages approximately 7090 feet (2160 m) in thickness. It reaches a maximum thickness of 13,000 feet (4000 m) at a location about 300 miles (480 km) inland from the Soviet base of Mirny. Contour lines give surface elevations in meters.

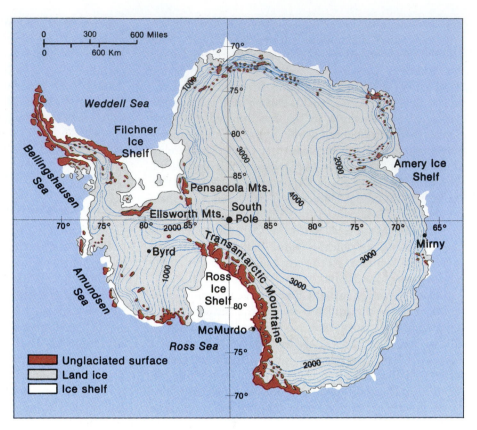

Table 15.1
Present-day Ice-Covered Areas (mi²)

Antarctic	4,859,000
Greenland	696,000
Northeast Canada	59,100
Central Asian ranges	44,400
Spitsbergen group	22,400
Soviet Arctic islands	21,500
Alaska	19,900
South American ranges	10,200
West Canadian ranges	9,600
Iceland	4,700
Scandinavia	1,500
Alps	1,400
Caucasus	700
New Zealand	400
USA (excluding Alaska)	200
Others	about 40
Total volume of present ice: 7 to 8 million mi³	

Source: C. Embleton and C.A. King, *Glacial Geomorphology*, Halsted Press, 1975.

Glacier Motion

Glacier motion is caused by gravitational forces, which compact, deform, and may locally melt the ice as they draw it constantly downward. Unlike streams, the motion is not always downhill; rather it is down-ice—that is, in the direction the ice surface slopes.

In discussing glacial movement, a distinction should be made between absolute and relative motion. The absolute motion is the actual forward movement of the ice particles in the glacier. This motion is nearly always visually imperceptible and usually totals less than 1000 feet (300 m) per year. Stagnant or "dead" glaciers, located in areas where rapid losses by melting, or *ablation,* are occurring, do not move at all, but decay in place. The absolute motion of a glacier may be far from steady. Most glaciers experience surges, or sudden increases in their rates of advance. The causes of these surges are not always apparent, but are sometimes associated with one or more seasons of heavy snowfall in the ice accumulation zone.

The relative motion of a glacier refers to the advance or retreat of the ice front. It is determined by the relationship between the absolute motion of the glacial ice and its rate of ablation. A glacier will advance if its absolute motion exceeds the ablation rate, will retreat if its ablation rate exceeds the absolute motion, and will remain stationary if these two factors are equal. Most glaciers have been retreating over the past few centuries, apparently as a result of a gradual global warming trend, even though the ice itself is still advancing in an absolute sense. Glacier motions are watched carefully, not only because of their geomorphic significance, but also because they are considered to be sensitive climatic indicators.

Every glacier has a *zone of accumulation* and a *zone of ablation*. If the glacier is to maintain a constant mass, the formation of new ice in the former zone must exactly balance the losses in the latter. Separating the two zones is a so-called equilibrium line, where gains and losses are exactly equal (see Figure 15.3). In general, a size correlation exists between the zones of accumulation and ablation, and larger glaciers, with more extensive zones of accumulation, typically extend to lower elevations than do smaller neighboring glaciers.

The exact mechanisms of glacial movement are complex, but may be divided into two major components: basal sliding and internal flow. Basal sliding involves a true sliding on the bedrock surface underlying the glacier. Sliding occurs only with *warm ice glaciers,* which have an underlying thin film of meltwater acting as a lubricant to aid slippage. Most alpine glaciers are of this type.

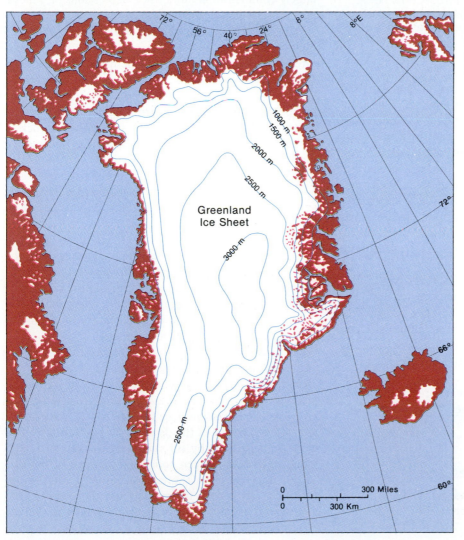

Figure 15.2 The Greenland ice sheet occupies the interior of the world's largest island and has a maximum thickness of approximately 11,000 feet (3350 m).

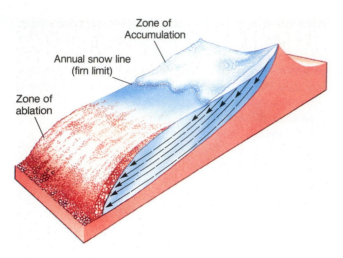

Figure 15.3 Glacial zones of accumulation and ablation. Ice and rock debris are progressively buried by newly forming ice in the zone of accumulation, but eventually resurface in the zone of ablation due to the melting of the overlying ice.

The second component of glacier motion, internal flow, occurs primarily as a smooth flowage of horizontally oriented ice crystals that shear over the ice crystals below. This type of motion, unlike basal sliding, occurs within the ice and has little or no effect on the underlying surface. Internal flow is the only mechanism of movement employed by *cold ice glaciers,* which are frozen fast to the underlying bedrock. Cold ice glaciers exist largely in the colder high latitude environments where continental glaciers form. At any point on its surface, the absolute motion of a glacier is the sum of the velocities of basal sliding and internal flow.

Because the nonmoving rock floor and walls of a glacier (if it is laterally confined) impede ice flow, the surface ice moves fastest. In addition, if the glacier takes the form of a stream of ice, as do most alpine glaciers, flow rates are greatest in the center. In these respects, the flow pattern of a glacier is very similar to that of a river. Major differences are the much slower flow velocities of glaciers and the glacial predominance of laminar flow (smooth flow) rather than of turbulent flow.

Glacier Erosional Processes and Features

The great power of glaciers as eroders and transporters of regolith stems from several factors. One is the fact that, unlike water and wind, glaciers exist in the solid state. This produces the physical hardness and firmness needed to dislodge vast quantities of weathered rock material effectively. A second factor is the tremendous pressure that several hundred to several thousand feet of ice exerts on the surface. Although water depths and pressures on most ocean floors and in some deep lakes equal or exceed those of glaciers, this bottom water is usually nearly stationary and has little or no erosional capacity. Indeed, the deep ocean basins are major depositional sites. A third factor aiding glacial erosion is the great extent of ice coverage within a glaciated area. While only a tiny proportion of the surface in a fluvially dominated landscape is covered by river water at any one time, continental glaciers bury the surface, modifying the entire landscape. Even alpine glaciers, which are largely confined to mountainside depressions or valleys, may cover a significant proportion of the surface.

On the other hand, glaciers do have some erosional limitations. Probably the most important is the reduced rate of chemical weathering of the underlying bedrock, which is inhibited by cold temperatures and sometimes by a lack of liquid water. The prior weathering of the bedrock is crucial in preparing material for glacial transport. Hard as ice is, most unweathered rock is harder yet and will not be significantly affected by direct glacial erosion. Frequently, however, numerous pieces of bedrock become frozen into the base or sides of a moving glacier. They are then scraped over the surface, abrading it much as the hard grains of sand on sandpaper scrape a piece of wood.

A second factor limiting the effectiveness of glacial erosion is the slow rate of ice movement over the underlying surface. Ice is by far the slowest moving of the tools of erosion. In addition, much of the ice motion occurs internally; it therefore does not affect the underlying surface. Indeed, cold ice glaciers, frozen rigidly to their bases, perform little or no erosion.

Glacial erosion occurs by the processes of scraping and plucking. Scraping is accomplished primarily by rocks frozen into the base of the ice and dragged over the bedrock surface. Larger rocks may produce scratches or grooves called *striations* (see Figure 15.4). Striations are valuable indicators of the direction of past ice flow. The abrasion performed by silt-sized particles, on the other hand, smooths and even polishes some rock surfaces.

Plucking involves the excavating of angular, often relatively large, rock fragments from the bedrock. Ice melting on the up-ice side of an obstacle such as a hill or boulder will flow around the obstruction, freeze in rock fractures on the down-ice side, and pluck out rock fragments as the ice moves on. The process is greatly aided by the presence of weathered or highly jointed bedrock. Some previously glaciated areas contain numerous bedrock features with their up-ice sides smoothly rounded and down-ice sides irregularly plucked. These features are sometimes referred to as *roche moutonées,* a French term meaning "rock sheep."

Glaciers are most effective in eroding unconsolidated materials. This most commonly occurs when a glacier advances over terrain that has not been glaciated in the recent past. The hundreds or even thousands of feet of glacial ice involved readily bulldoze the soil and weathered rock or incorporate it into the ice itself, stripping the underlying surface to unweathered bedrock.

Because the sediment-carrying ability of the ice is virtually unlimited, the material is transported in an unsorted state. Warm ice glaciers not bounded by rock walls carry most of their debris along or near their bases. Alpine glaciers contain much basal debris but also transport large amounts of material, commonly dislodged by frost wedging, on their surfaces.

Glacier Depositional Processes and Features

Glaciers transport debris in a conveyor-belt fashion. Even if the forward edge of the ice is retreating, the absolute glacier motion is still down-ice, and this motion does the transporting. The general term for all deposits of glacial origin is glacial *drift.* (The term "drift" is a holdover from when it was believed, for lack of a better explanation, that glacial deposits had been rafted by icebergs to their present sites during the great biblical Flood of the Old Testament.) Drift consists of two general types, depending upon the mode of deposition. Material deposited directly by glacial ice is known as **till.** It is unsorted, and constituent particles can range in size from clay to large boulders. In contrast, *glacio-fluvial sediments* are deposited by glacial meltwaters and are partially stratified, or sorted.

Deposition of till occurs primarily by dumping at the glacier terminus, or *snout,* and by lodging or plastering on the surface beneath the glacier. The deposits produced by either process are termed **moraines.** If the snout remains stationary for an extended period, so much till is conveyed to that point that a ridgelike

Figure 15.4 Closeup of glacial striations on bedrock in Jasper National Park, Alberta. (*Richard Jacobs, © JLM Visuals*)

terminal moraine is produced. This is an accumulation of till with its long axis oriented perpendicular to the direction of ice flow (see Figure 15.5). It is often arcuate, because most glacier snouts are convex in shape. If the glacier margin is retreating, a series of such moraines may form, with each moraine marking the site of a temporary halt in the glacier's retreat. In this case, only the outermost moraine is referred to as the terminal moraine; the others are termed *recessional moraines.* Terminal and recessional moraines, where undisturbed, provide precise information on past glacial positions and are invaluable to researchers reconstructing the chronology of events in a previously glaciated area. Between these moraines, a thinner covering of *ground moraine* is deposited by the more rapidly retreating ice. Although continental glaciers may transport some rock debris for hundreds of miles, most glacial deposits are derived from relatively nearby sources.

Glacio-fluvial materials are normally deposited at or beyond the ice margins, although some subglacial deposits may occur. In general, the farther these deposits, called *outwash,* have been transported by water, the better sorted they are. Glacial meltwater streams are usually overloaded with sediment and frequently aggrade their channels, forcing them to periodically shift their courses. If multiple streams issue from the ice front, a broad *outwash plain* may form. If only one fairly channelized stream emerges, a single long *valley train* of deposits is produced, perhaps extending from the ice front for many miles.

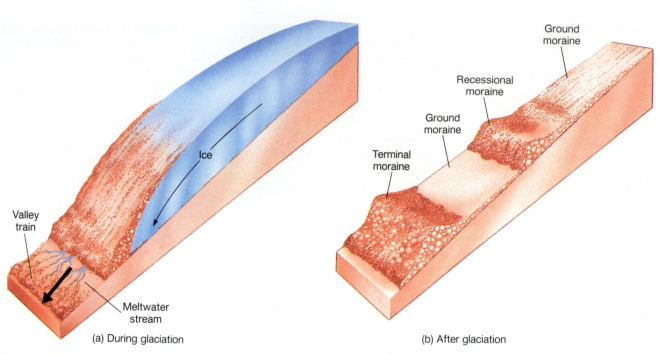

Figure 15.5 Block diagrams depicting the formation of a terminal moraine, as well as the appearance and relative positions of terminal, recessional, and ground moraines after the melting of the glacier that formed them.

ALPINE GLACIERS AND ASSOCIATED LANDFORMS

Alpine glaciers form in mountainous regions at high elevations and flow downhill under the influence of gravity. The lowest elevation at which they form increases as temperatures become warmer equatorward; for this reason, alpine glaciers become progressively less numerous with decreasing latitude. The lower limit of ice accumulation lies essentially at sea level within 10 degrees of the poles and increases to elevations of 15,000 to 16,000 feet (4600 to 5000 m), near the equator in the central Andes and the Ruwenzori Range of East Africa. The zones of ablation of well-fed glaciers can extend significantly below these limits, though, so the lower limit of glaciation depends not only on temperature, but also on snowfall amounts and on the elevation and size of the glacial accumulation area.

Characteristics of Alpine Glaciers

Most alpine glaciers originate in relatively small mountainside hollows called *cirques*. If conditions are marginal for ice accumulation, the glacier may remain as a small, rounded *cirque glacier*. If conditions favor ice accumulation, the glacier will outgrow its cirque and flow down the mountainside as a long, narrow tongue-like mass of ice (see Figure 15.6). When the glacier reaches the base of the mountainside, continued expansion will usually cause the ice to advance down a preexisting valley as a *valley glacier*. Large valley glaciers, like those of the Alps, are typically fed by several glacier tongues, so that a dendritic pattern of intersecting glaciers is developed.

The surface features and appearance of an alpine glacier can vary considerably from point to point. In the zone of accumulation, the ice is usually covered by snow and cannot be seen, while in summer, at least, the surface of the glacier in the zone of ablation is visible. Where the glacier passes over convexities on its bed, the stretching of the near-surface ice causes it to fracture, forming parallel open fissures called **crevasses**. The deepest known crevasses extend downward approximately 160 feet (50 m). Below this level, the weight of the overlying ice gives it enough plasticity to close any gaps that may form.

Lateral moraines typically take the form of thick strips of rock debris along the sides of alpine glaciers. This material accumulates both from glacial erosion and from weathering processes, especially frost shattering,

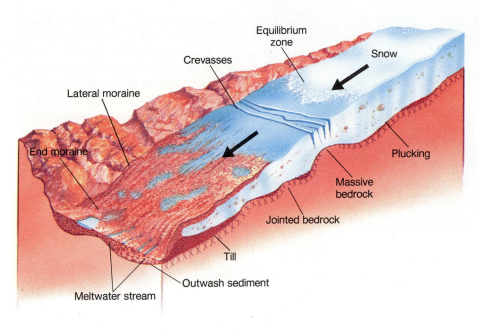

Figure 15.6 Major features of an alpine glacier. The equilibrium zone separates the zone of accumulation from the zone of ablation.

which dislodge rocks from the slopes above. Where two glacial tongues merge, their innermost lateral moraines coalesce to form a single *medial moraine* within the glacier (see Figure 15.7). A large alpine glacier often contains several medial moraines, the total number generally indicating the number of individual valley glaciers that have combined to form the glacier.

Proceeding into the zone of ablation, the glacier appears progressively dirtier and more debris-laden, because the gradual melting of the ice produces a continually increasing proportion of morainic material. Near the glacial snout, debris may completely cover the surface, making it impossible to accurately determine the most forward position of the ice.

Alpine Glacier Erosional Landforms

Erosion by alpine glaciers tends to increase the relief and ruggedness of the topography. Glacial erosion has largely produced the spectacular alpine scenery that annually draws millions of tourists to such highly glacially modified mountain systems as the Alps and the northern Rockies (see Figures 15.7 and 15.8). A number of landform features are produced by alpine glacier erosion, but two of the most important are cirques and glacial troughs.

Cirques are bowl-shaped depressions eroded into the rock faces of mountainsides in glaciated areas (see Figures 15.9 and 15.10). They vary greatly in degree of development and perfection of shape. The precise mechanism by which cirques develop is uncertain, but they are generally believed to form where mountainside snow patches provide water for accelerated physical and chemical weathering. As the hollow enlarges, snow

Figure 15.7 The bands of dark rock in Alaska's Lower Kaskawulsh Glacier are medial moraines formed by the coalescence of two glacier tongues. (*John S. Shelton*)

Figure 15.8 St. Mary Lake occupies a glacial trough in Montana's Glacier National Park. A horn and a knife-edged arête further enhance this area's spectacular alpine scenery. (*Ralph Scott*)

The U-shaped cross-sectional profile of a glacial trough results from the large size of a valley glacier as compared to a river draining a similar area. This size is necessary because glacial ice flows much more slowly than river water. Indeed, a valley glacier usually fills its entire trough, often to a considerable depth. The deepening and steepening of the trough by its glacier considerably increase the local topographic relief because

sufficient to prevent complete summer melting collects, and a cirque glacier forms. The cirque is subsequently enlarged and molded by frost shattering on the headwall and by erosion of the cirque floor by scraping and plucking. Many cirques that formed during the Pleistocene are now devoid of ice and instead contain small lakes.

As growing cirques erode headward from opposite sides of a linear mountain or upland area, they may eventually reduce the intervening mass to a narrow, serrated ridge called an *arête*. Headward cirque growth from three or more sides can produce a sharp-crested pyramidal peak termed a **horn**. The Matterhorn of Switzerland is the most famous example of this process, but most isolated peaks in glaciated mountain regions have formed in essentially the same manner. The cirque growth process is therefore a highly effective erosional mechanism and is largely responsible for the upland features in extensively glaciated mountains.

The valley features in glacially modified mountainous regions result largely from the development of glacial troughs. A **glacial trough** is a deep valley, usually of fluvial origin, that has been subsequently occupied and modified by a valley glacier (see Figure 15.11). Glacial troughs are noted for their broad, rounded U-shapes, as opposed to typical fluvially eroded mountain valleys, which are more V-shaped. In contrast to river valleys, they also possess markedly ungraded down-valley slopes and greater overall linearity.

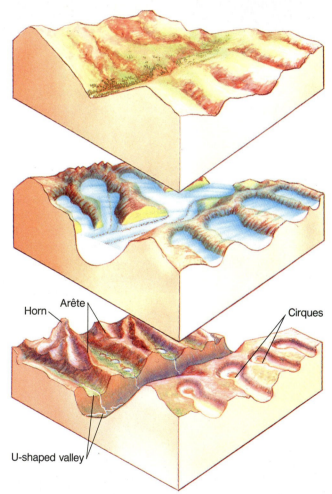

Figure 15.9 The glacial modification of a mountainous landscape is represented in this sequence of diagrams. Figure (a) shows a fluvially dissected landscape before the onset of glaciation. Note the smoothly rounded appearance of the slopes. Figure (b) depicts the same landscape during the height of glaciation. In Figure (c), the glaciers have retreated, leaving in their wake a rugged landscape containing numerous glacial erosional features.

Figure 15.10 Classic cirque development producing "biscuitboard topography" in the Uinta Range of Utah. These cirques are Pleistocene features, and no glaciers currently occupy them. (*John S. Shelton*)

Figure 15.12 Geiranger Fjord on the coast of Norway. (*Geopress/H. Armstrong Roberts*)

nearby mountain peaks have not been similarly reduced in elevation.

Smaller tributary glaciers entering a valley glacier also carve U-shaped valleys, but do not erode nearly as deeply. When the ice melts, their valleys intersect the main glacial trough high on its sides, forming *hanging valleys*. As streams issuing from these hanging valleys cascade into the major trough, spectacular waterfalls

may result. Yosemite Falls, in California's Sierra Nevada, is a famous example.

At higher latitudes, glacial troughs may be eroded below sea level and occupied by ocean water when the ice melts. Such drowned troughs are called **fjords.** They are typically long, winding, steep-walled, and relatively narrow (see Figure 15.12). Fjords formed because the great thickness of the ice enabled it to erode deeply even though it was partially buoyed by the ocean water into which it advanced. Sea level rises subsequent to the melting of the glaciers resulted in further deepening. As a consequence, some fjords in Norway and southern Chile attain maximum depths of over 4000 feet (1300 m). They provide excellent protected routes and harbors for coastal shipping and fishing interests. Fjords are best developed along the coasts of Norway, western Canada, southern Alaska, Greenland, southern New Zealand, and southern Chile.

Alpine Glacier Depositional Landforms

The depositional features of alpine glaciers are typically more poorly defined and less obvious than are the erosional features for several reasons. First, they have

Figure 15.11 Rock Creek Valley, in Wyoming, is a U-shaped glacial trough. (*Carl Harns, © JLM Visuals*)

Alaskan Glaciers—Surging and Retreating

The Coast Mountains of Southeast Alaska and adjacent Northwest Canada, nurtured by copious year-round snowfalls, contain one of the world's most extensive systems of alpine glaciers. These glaciers, which are more accessible than those in many other areas, are of considerable interest to scientists both as climatic indicators and as outdoor laboratories for the study of glacial processes. Although concentrated within a restricted geographical area, the glaciers of Southeast Alaska tend to behave in highly individual fashion. In recent years some have suddenly surged forward, while others have undergone rapid retreat. The recent actions of the nearby Hubbard and Columbia glaciers illustrate both extremes of motion.

For most of the 85 years over which its movements have been recorded, the Hubbard Glacier has flowed into Yakutat Bay, near the northwest tip of the Alaskan Panhandle, at a relatively steady rate of several hundred feet per year. This motion has not been entirely counterbalanced by a loss of ice at the glacier's snout, so the glacier has been advancing slowly into the bay. During the winter of 1986, however, the Hubbard Glacier suddenly began to surge, moving as fast as 46 feet (14 m) a day. A tributary glacier, the Valerie Glacier, also began to surge forward at speeds of up to 112 feet (34 m) per day. This movement soon blocked off Russell Fjord at the head of the bay and turned it into a lake, temporarily trapping numerous seals and other sea animals (see Figure 15.13).

Geomorphologists believe that the surge may be caused by entrapped subglacial meltwater, which provides a sliding plane for the ice and apparently buoys it up from the surface in places.

In support of this theory, it has been noted that some neighboring glaciers, such as the Variegated Glacier, surge at regular intervals. Their surges are preceded by a substantial reduction in meltwater outflow, followed by a flood of meltwater at the end of the surge.

The recent behavior of the Hubbard Glacier contrasts with that of the majority of Alaska's glaciers, which are in retreat. The rapid melting of Alaskan glaciers, in fact, is generally believed to contribute significantly to the gradual global rise in sea levels.

The Columbia Glacier, one of the last of Alaska's great tidewater glaciers and an important tourist attraction, has recently started to retreat at a rate exceeding 1000 feet (300 m) per year. The retreat seems to be taking place because the glacier's snout has now shrunk past a shallow water

continued on next page

much less relief than the erosional features. In addition, they consist of unconsolidated materials and therefore do not have the permanence of the erosional features, which are generally composed of solid bedrock. Finally, because most glacial deposits are located in valleys, the rivers that reoccupy the valleys after the ice departs rework and often largely remove these materials.

The major till features are moraines of various types. Fresh moraines consist of ridges of loose, unsorted rock debris. Lateral moraines form terracelike deposits along valley margins. Those produced by large valley glaciers are sometimes several hundred feet in height and often become progressively larger and better defined down-valley. Medial moraines, which are more likely to have been removed by fluvial processes, may form ridges well within the glacial trough. Terminal and recessional moraines form arcuate ridges, bulging down-valley, that are oriented perpendicular to the lateral and medial moraines. Many formerly extensive alpine glaciers have disappeared completely, and morainic deposits are found from their lower valleys all the way up to their cirque floors.

Glacio-fluvial deposits have already been discussed. Of those produced by alpine glaciers, the most signifi-

ledge into much deeper water, where large chunks of ice can break off. In the late 1970s, the water at the glacier's edge was only 56 feet (17 m) deep. Currently, it is nearly 1000 feet (300 m) deep. It is believed that, over the next few decades, the snout of the Columbia Glacier will leave the sea and retreat several miles inland.

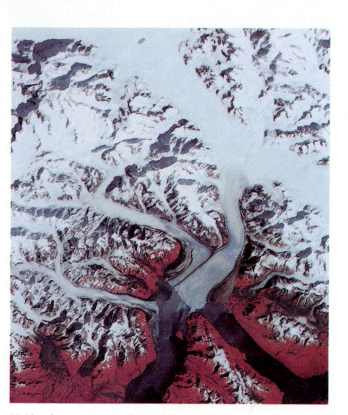

Figure 15.13 A surge in the flow of Alaska's Hubbard Glacier, which occurred in early 1986, temporarily blocked Russell Fjord (the inland water body to the right of the lower portion of the glacier), trapping a large number of marine animals. (*Department of the Interior, USGS*)

cant are the valley trains deposited by debris-laden glacial meltwater streams.

CONTINENTAL GLACIERS AND THE PLEISTOCENE

Continental glaciers differ from alpine glaciers in several major respects. The most important difference is in their size and thickness. Whereas alpine glaciers extend, at most, a few tens of miles and frequently have widths of less than one mile, continental glaciers have horizontal dimensions of hundreds or even thousands of miles.

Alpine glaciers lie cupped in mountainside hollows or in deep valleys, while continental glaciers usually cover the entire surface deeply.

The underlying surface controls continental glacier movement much less than it controls alpine glacier movement. The movement of the entire ice sheet is radial, rather than unidirectional, with the ice spreading outward in all directions from one or more centers of accumulation. The cause of this radial spread is not so much the downhill force of gravity as it is the down-ice pressure exerted by the great weight of the ice itself. Radial expansion gives the glacier a rounded, domelike overall appearance. Examined in more detail, however, the ice front advances as a series of lobes whose positions are dictated by the locations of the ice accumulation centers and by the topography and geology of the underlying surface.

The geographic setting of the ice accumulation zone of alpine and continental glaciers also differs in one major respect. While high elevations are responsible for the cold climates that support alpine glacier growth, high latitudes are responsible for the low temperatures that support the growth of continental glaciers. Since much larger areas of the earth are at high latitudes than at high elevations, continental glaciers are much more extensive than are alpine glaciers.

The Pleistocene Epoch

The geologic record indicates that, for much of its existence, the earth had no significant glaciers. There may always have been alpine glaciers in the highest mountains, but continental glaciers seem to have been rare. Scientists have evidence of only five relatively brief periods of continental glaciation in the past billion years. The first four occurred several hundreds of millions of years ago. The fifth—the **Pleistocene**—began about two million years ago and is still continuing.[1]

Over the past one million years, there have been approximately 10 major glacial periods, or stages, and as many as 40 more minor periods of glacial advance. The Wisconsinan stage (or Würm stage, as it is known in Europe), is the most significant because of its relative recency. The Wisconsinan stage began some 50,000 to 100,000 years ago and is considered to have ended about 10,000 years ago, when the retreating ice left the north-central United States. The last ice may not have melted in central Canada until about 4000 years ago. This period is so recent that only superficial landscape modification has occurred in many of the glaciated areas; as a result, these areas tend to be dominated by relict glacial features.

Between the various glacial stages were lengthy interglacial stages, when the ice sheets generally melted

Figure 15.14 Most of Antarctica consists of a vast, nearly featureless ice plateau covered by a thin layer of drifting snow. (*B.F. Molnia*)

back past their present positions. Evidence suggests that during at least one interglacial stage the ice may have largely or entirely disappeared from Greenland and Antarctica. It is likely that we are currently in an interglacial stage.

Geographic Extent of Pleistocene Glaciation

During the Pleistocene glacial stages, ice covered as much as 30 percent of the earth's present land area. Glacial ice now covers some 5.5 million square miles (14 million sq km), but during the Pleistocene maximum it covered an estimated 17 million square miles (44 million sq km). Then, as now, the Antarctic ice sheet was the largest (see Figure 15.14). During the Pleistocene, it enlarged and spread farther out over the adjacent sea than at present. The Greenland ice sheet was considerably larger than it is today, extending to the coast and probably forming ice shelves in protected coastal waters. The major addition of ice, however, consisted of two giant ice sheets that formed over the northern half of North America and over northern Europe and adjacent northwestern Asia.

The North American or **Laurentide ice sheet** spread outward from two or more centers in interior Canada to cover virtually all of Canada, including Hudson Bay and most of the Canadian Arctic Islands (see Figure 15.15). The United States was covered as far

---

1. Many scientists believe the Pleistocene ended about 10,000 years ago. There is increasing evidence, however, that the cyclical pattern of continental glacier formation and melting will continue in millennia to come. Even if this does not occur, the continuing existence of massive glaciers in Antarctica and Greenland indicates that the present glacial stage is not completed.

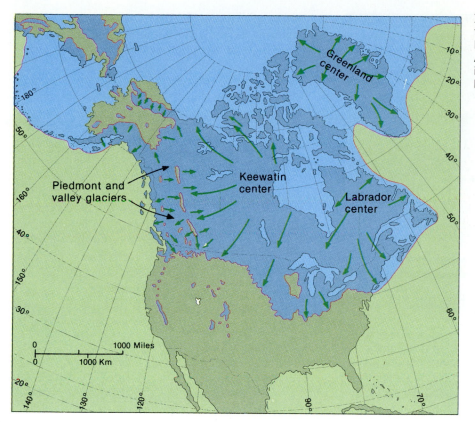

Figure 15.15 Centers of ice formation and maximum extent of Pleistocene glaciation in North America. Areas of both glacial ice and sea ice are shaded in blue.

south as a line extending westward from Long Island through northern Pennsylvania, then along the courses of the Ohio and Missouri Rivers, and finally westward near the international border to northern Washington. Oddly, much of Alaska remained unglaciated, apparently because of a lack of snowfall and the presence of orographic barriers. It should be stressed that the line just described is a composite maximum line of glacial advance. During each of the major glacial stages, the ice advanced to a different boundary. The total area covered by the Laurentide ice sheet was approximately 4 million square miles (10.2 million sq km), about 80 percent as large as the Antarctic ice sheet is today. Its thickness, while unknown, was probably comparable to that of the Antarctic ice sheet and was at least thick enough to cover completely the 6288 foot (1918 m) crest of Mt. Washington, New Hampshire, which was located near the presumably thinner margin of the ice.

The Eurasian or *Fennoscandian ice sheet* radiated from a center in eastern Sweden to cover a maximum of 2 million square miles (5.1 million sq km). Its southernmost boundary extended through southern England, eastward through the Netherlands, then east-southeastward across central Europe to the southern Ukraine, and finally east-northeastward to a point in north-central Siberia (see Figure 15.16). The northern terminus of the ice extended some distance past the present continental margin into the Arctic Ocean. The North Sea and the Baltic Sea were also under glacial ice, which reached far enough westward to cover all of Ireland.

The Pleistocene glacial stages were also periods of formation or expansion for smaller glaciers in many other parts of the world. A subcontinental glacier may have formed over eastern Siberia, which was excessively cold during the Pleistocene but was too dry to support the growth of a more massive glacier. Mountain glaciers in many areas also expanded greatly, and many presently unglaciated mountains supported extensive Pleistocene glaciers. For example, the Canadian Rockies developed a system of coalescent alpine glaciers, sometimes termed the Cordilleran ice sheet, that merged with the Laurentide ice sheet east of the continental divide. The Alps developed an ice cap above which only the higher peaks

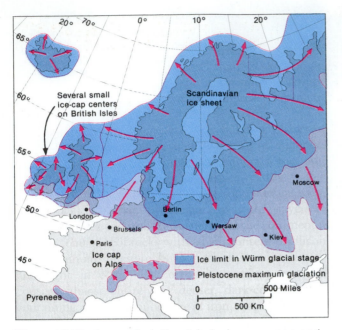

Figure 15.16 Areas of continental glacier coverage and directions of ice flow during the Pleistocene in Europe. Both the maximum extent of the ice sheets and the limits of ice advance during the last (Würm) glacial stage are shown.

protruded. Glaciers formed in the mountains of south-eastern Australia, where none exist today. The southern Rockies, which contain only a few cirque glaciers at present, were extensively glaciated. Glaciers may have descended to 6,500 feet (2000 m) in the Andes at the equator, where today the glacial limit is at least 14,000 feet (4300 m).

Causes of the Pleistocene

The cause of the Pleistocene has produced one of the greatest controversies in geomorphology since nineteenth-century earth scientists confirmed the past existence of an ice age. There are actually two facets to the problem, and their respective solutions may or may not be related.

The first question is why the Pleistocene has occurred at all. More specifically, why, after millions of years of nearly global warmth, did the earth become cold enough within the past two million years to permit the growth of massive continental glaciers? The answer to this question is still unknown, although a number of hypotheses have been developed. These include the

possibility of a recent decrease in the output of solar radiation, an increase in insolation-blocking volcanic dust, a change in the circulation patterns of the ocean currents, and a shift in the location of the North Pole from the Pacific to the Arctic Ocean.

The second question is of greater potential import, because its answer strongly relates to the likelihood of future periods of glaciation. What causes the alternation of glacial and interglacial stages within the Pleistocene? Scientific investigation over the past two decades indicates that the answer apparently relates to variations in the earth's orbital geometry, specifically in (1) the precession of the equinoxes, which determines the times of year that the earth is at aphelion and perihelion, and (2) long-term variations in the tilt of the earth's axis. The influence of these factors is discussed in more detail in the Focus box on page 339. An important implication of these findings is that we are probably now in an interglacial period and that the ice sheets of North America and Eurasia may well re-form in the near geologic future.

Landforms Produced by Continental Glaciers

Continental glaciers produce many of the same types of landforms as alpine glaciers. Important differences exist, however, in the sizes, locations, and relative abundance of most of these features. Continental glacier features can also be categorized as erosional or depositional in origin, although some are produced by both processes. The areas that undergo net erosion by a continental glacier are generally located beneath the central portion of the ice sheet, in the zone of ice accumulation. Surrounding this, in the zone of ablation, are the major areas of deposition.

Continental Glacier Erosional Landforms

A basic difference exists between the erosive actions of alpine and continental glaciers. Alpine glaciers, as already noted, sharpen the topography and usually increase local relief. Continental glaciers have just the opposite effect. Because they cover everything, they subject the entire surface, not just lowland areas, to erosion. Surface features such as hills or mountains are especially susceptible to erosion because they protrude well into the faster moving zones of the ice. Lowland areas, on the other hand, are somewhat protected and

Earth's Orbital Geometry and Climatic Change

Ever since nineteenth century scientists discovered that the earth has been experiencing a recent series of glacial and interglacial periods, a search has been underway for the mechanisms causing these changes. During the 1920s and 1930s, the Serbian mathematician Milutin Milankovitch argued that the glacial and interglacial stages resulted from long-term variations in the earth's orbit. As late as the 1970s, most experts believed that changes in insolation resulting from these orbital variations would be insufficient to produce the necessary climatic changes. In the last two decades, however, knowledge of the chronology of the Pleistocene glacial and interglacial stages has been refined to the point that it has been possible to compare the timing of these stages with those of terrestrial orbital variations. The correlation has been so close that it is now nearly certain that long-term variations in the earth's orbit are the chief cause of the glacial and interglacial stages.

Three orbital variations are involved. First are changes in the tilt of the earth's axis with respect to the plane of the ecliptic. The tilt is currently approximately 23.5°, but it varies between 22.1° and 24.5° over a 41,000-year cycle. An increase in tilt concentrates insolation toward the poles and decreases insolation received at the equator. A decrease in tilt, which is presently taking place, has the opposite effect. A second orbital variation is the precession of the equinoxes. The earth's axis has been found to gyrate or "wobble" like that of a slowly spinning top. In addition, the elliptical orbit of the earth is slowly rotating. The net result of these two independent movements is that the time of year that the earth is at different points in its orbit—at aphelion and at perihelion, for example—gradually changes, completing a full cycle each 22,000 years. A final orbital variation is the change in the eccentricity of the earth's orbit over cycles of 100,000 and 400,000 years. When the eccentricity of the orbit increases, the earth takes a less circular orbit around the sun; consequently, there is a greater annual variation in distance between the earth and sun.

It has been found that, of these motions, the precession of the equinoxes and the changes in axial tilt most greatly influence temperature changes and continental glacier formation and retreat. The influence of the earth's orbital eccentricity is of secondary importance, chiefly influencing the amplitude of the precession cycle. Although human activities may well enter into the picture from this point onward, the net effect of these orbital variations over the next 10,000 or so years would seem to be the winding down of the present relatively warm interglacial stage and the beginning of a new period of continental glaciation.

may even be sites of local deposition within an area of predominant erosion. The surface, as a whole, is therefore smoothed and rounded by the ice.

The extent of actual surface lowering by continental glacier erosion has long been a subject of controversy. The central question has involved how effectively the ice erodes unweathered bedrock. Most researchers now believe that the ice is relatively ineffective on most solid bedrock and that glacial erosion largely removes the soil and regolith down to the weathering front.

The zone of erosion of the Laurentide ice sheet included the Canadian Shield area, the Canadian Maritime Provinces, and northern New England. The Fennoscandian ice sheet zone of erosion included Scandinavia and adjacent areas currently beneath the shallow waters of the North Sea and the Gulf of Bothnia.

For the most part, the surfaces in the zones of erosion of both the Laurentide and Fennoscandian ice sheets were shield areas of ancient crystalline rocks, with predominantly low surface relief. These areas were further smoothed by the ice, and today **ice-scoured plains** are the most extensive features of continental glacier erosion. These areas now consist of vast expanses of irregularly

Figure 15.17 View of the rocky, lake-strewn surface of an ice-scoured plain. Northwest Territories, Canada. (*Alan Kesselheim/Mary Pat Zitzer, © JLM Visuals*)

rolling surface, largely stripped of soil and covered by myriad lakes and swamps (see Figure 15.17). The glaciers originally removed virtually all soil and weathered bedrock from these areas. Most areas today, though, have a thin covering of soil produced either by deposition as the ice melted or by postglacial weathering.

One of the most important features of ice-scoured plains is the great number of lakes and swamps they contain. These features have formed because erosion and deposition by the glaciers obliterated the preglacial drainage network. In addition, areas of deeply weathered rock were scooped out by the ice, forming countless rock basins that have subsequently filled with water. Wandering haphazardly through these plains from lake to lake are stream systems displaying deranged drainage patterns. The presence of glacial lakes has earned the state of Minnesota the nickname "The Land of 10,000 Lakes." The glacially derived lakes of Canada are so large and numerous that they contain over half the world's fresh surface water. The Great Lakes were either formed or greatly enlarged when the ice excavated areas of weathered rock. The extensive glacial erosion of areas such as the Canadian Shield has greatly inhibited their economic potential. The soil that once covered the area was removed, and most weathered rock material was transported southward, where it now enhances the agricultural productivity of the midwestern United States. In addition, much land is agriculturally unusable because of glacially-induced poor drainage. Consequently, most of

interior eastern Canada is an almost uninhabited, rocky, lake-strewn wilderness.

Continental Glacier Depositional Landforms

Continental glaciers produce a variety of depositional features (see Figure 15.18). Some are similar in origin to those produced by alpine glaciers, but occur on a grander scale; others are unique to continental glaciation. Most recognizable depositional features were produced within the past 20,000 years and are the product of the Wisconsinan glacial stage of the Pleistocene.

The most extensive depositional feature is the **till plain,** a thick accumulation of ground moraine covering the preglacial surface over a large area. Till depths often exceed 100 feet (30 m), with thicker accumulations in originally lower areas such as valleys. Most of the till is fine in texture and consists of sand, silt, and clay. Occasional rocks or boulders are mixed in, however, and in some areas are quite numerous. The entire mixture is unsorted, since it was deposited directly by the melting ice. In some localities, the separation of huge ice blocks

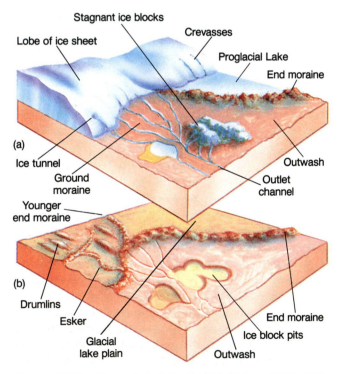

Figure 15.18 Block diagrams of landscapes during and after glaciation, showing the origin of various continental glacier depositional features.

Figure 15.19 The checkerboard pattern of productive farmland attests to the fertility of the soil in this flat till plain area southeast of Aberdeen, South Dakota. (*John S. Shelton*)

from melting glaciers has produced pitted till plains, with the pits termed *kettles*, currently occupied by small lakes.

Till plains cover much of the north-central portions of both the United States and Europe. The general flatness of the terrain, the high mineral content of the still lightly leached till, and the favorable present-day climate have combined to produce fertile soils. The American Midwest and the North European Plain are famous for their agricultural productivity and currently support substantial human populations (see Figure 15.19).

Systems of terminal and recessional moraines form the major relief features on till plains in many areas of both the United States and Europe. These moraines take the form of long belts of low, gently sloping ridges or hills. These belts are arcuate in shape, bulging southward; this reflects the fact that the ice sheets, along whose melting edges the moraines were deposited, were lobate in form. Well-developed moraines are especially numerous near the Great Lakes and to their south and west in Ohio, Indiana, Illinois, Minnesota, and the Dakotas. In Eurasia, a series of moraines extends some 3000 miles (4800 km) from northern Germany through Poland and the eastern Soviet Union to north-central Siberia. Both Long Island, New York, and the Danish peninsula consist largely of morainic deposits.

A smaller feature frequently encountered on till plains is the **drumlin.** Drumlins are smooth, elongated hills composed of till. When well formed, they are oval in shape, resembling an egg that has been cut in half lengthwise and placed with its flat surface downward. They generally occur in groups and are aligned with their tapered ends pointing down-ice (see Figure 15.20). A typical drumlin is perhaps 100 feet (30 m) high and 2000 feet (600 m) long. Some question exists as to their exact method of formation, but many have apparently been produced by the plastering of till around subglacial obstructions lying in the path of the ice, where the ice was overloaded with debris. Especially large groups or "swarms" are located in the Lake Ontario Plain of upstate New York, in southeastern Wisconsin, northern Lower Michigan, the Niagara Peninsula of Ontario, and in Ireland and Scotland.

Till plains, moraines, and drumlins are all till features produced by direct glacial deposition. **Glacio-fluvial features,** deposited by glacial meltwaters, are also of great importance in areas of Pleistocene continental glaciation. Some of these features formed above, within, or beneath the glacier, but most formed beyond the melting edge of the ice.

Foremost among the glacio-fluvial features in importance and areal extent are outwash plains and lacustrine plains. **Outwash plains** may be regarded as the glacio-fluvial counterparts of till plains. They consist of extensive areas covered by stratified drift deposited by melt-

Figure 15.20 These elongated hills east of Rochester, New York, are drumlins. Agricultural fields, which have followed the contour of the land, typically run parallel to the drumlin orientation. (*John S. Shelton*)

waters flowing outward from glaciers. Most outwash plains have rather smooth surfaces, but pitted outwash plains, containing numerous water-filled kettles, have formed where large blocks of ice became detached from the glacier and settled onto the outwash before melting.

Outwash plains are scattered throughout the areas of Pleistocene glacial deposition, primarily in the Great Lakes region and the Upper Midwest of the United States and in the Northern European Plain. Areas of outwash plains are interspersed with complex patterns of moraine systems, till plains, and lacustrine plains because of the numerous positional changes of the ice front. Where the ice front remained stationary for an extended period, however, an orderly sequence of deposits resulted, with till plain deposits beneath the ice, a terminal moraine along the forward edge of the ice, and outwash plain deposits beyond the ice margins.

Outwash plains formed where the surface sloped away from the glacier front; this allowed sediments to be transported and deposited by meltwater streams. Where the surface sloped toward the ice front, drainage was impeded and ice-dammed lakes frequently formed. Very fine sediments accumulated on the floors of these lakes, and their eventual drainage produced *glacial lacustrine plains*. These areas now have exceptionally smooth, level surfaces, and, like the outwash plains,

have been largely converted to productive agricultural land. Many of the Pleistocene lakes were very large, and the lacustrine plains they formed currently cover extensive areas. In North America, the largest lake of all was Lake Agassiz. Late in the Wisconsinan glacial stage it covered over 100,000 square miles (260,000 sq km) in eastern North Dakota, Manitoba, Saskatchewan, and Ontario. Most of this area has now drained, although it is still dotted with thousands of smaller lakes of various sizes. The Great Lakes were much larger during the late Pleistocene, when ice blocked their present outlet through the St. Lawrence Valley (see Figure 15.21). The Great Lakes went through a highly complex sequence of changes during the Wisconsinan stage, but at their maximum extent they coalesced into a single giant inland sea covering nearly twice their present area.

River systems also underwent profound changes as a result of glacial blockage and rerouting. The upper Missouri River, for example, originally flowed northward to Hudson Bay. Glacial blockage diverted it southward and added it to the Mississippi River drainage basin. The present courses of both the Ohio and Missouri Rivers developed roughly parallel to the maximum line of glacial advance and served as lateral drainage routes for glacial meltwaters. During the late Wisconsinan stage, when the St. Lawrence River was

Figure 15.21 This map shows the expansion in areal extent of the Great Lakes during the late Pleistocene, when their present drainage route through the St. Lawrence River was blocked by ice. Major Pleistocene drainage outlets are numbered.

blocked by ice, the Great Lakes drained southward through several routes into the Mississippi River system (see Figure 15.21).

Indirect Pleistocene Effects

The effects of the Pleistocene were not limited solely to those areas covered by glaciers or glacially derived meltwaters; they were worldwide in scope. Probably every part of the earth experienced changes in climate during this period. Pleistocene temperatures over most of the earth are estimated to have averaged some 5 to 10 F° (3–6 C°) lower than at present. These temperature changes were responsible for major redistributions of plants and animals and caused the extinction of numerous species. The cooler world climate also influenced the locations of the global air pressure and wind belts. These belts, with their associated climatic conditions, generally shifted equatorward, giving many areas markedly different precipitation patterns than they have at present.

The cooler, damper Pleistocene climatic conditions produced large lakes in a number of currently dry areas with interior drainage. The arid Great Basin of the western United States, for instance, was much more humid and contained many extensive lakes that have now either dried completely or have shrunk to small, often highly saline remnants (see Figure 15.22). The largest lake of all was Lake Bonneville, which at one point occupied over 20,000 square miles (51,000 sq km) in Utah, Nevada, and Idaho. Its chief remnant is the Great Salt Lake, with an area of 940 square miles (2400 sq km). Lake Bonneville fluctuated considerably in extent, and a series of abandoned elevated terracelike shorelines can be found as much as 1000 feet (300 m) above the present surface of the Great Salt Lake. Numerous other large lakes existed in western Nevada, southern Oregon, and southeastern California, as well as in other parts of the world.

Another Pleistocene effect with worldwide implications was changes in sea levels. The alternative formation and melting of the ice sheets involved the withdrawal and addition of vast quantities of ocean water. During the height of the glacial stages, enough water was locked in glacial ice to lower world sea levels some 300 to 500 feet (90–150 m). As a result, the earth's surface was covered by less water and more land, and the shapes and locations of coastlines differed greatly from those of the present. For example, the

Figure 15.22 Present-day lakes of the western United States (shown in light blue) occupy only a small proportion of the area covered by lakes (dark blue) during the cooler and wetter Pleistocene epoch.

British Isles were connected to mainland Europe, and North America and Eurasia were joined across what is now the Bering Strait. It is believed that the Bering Strait land bridge allowed Asiatic nomads who became the American Indians to migrate from eastern Asia to the Americas.

Conversely, during the interglacial stages, sea levels were substantially higher than at present. A series of seven Pleistocene marine terraces, each representing a past coastline position, has been recognized along the Atlantic and Gulf coastal plain of the United States. The continued slow melting of the world's remaining glaciers has partially caused the gradual rise in ocean levels observed on most present coastal margins.

A related Pleistocene influence is the *isostatic rebound* of previously glaciated areas. The great weight of the continental glaciers depressed the underlying surface much as the pressure of a person's thumb might depress the surface of a rubber ball. When the pressure is removed, the ball springs back to its original shape. Likewise, the recent melting of the Laurentide and Fennoscandian ice sheets has allowed the areas formerly covered by the ice to rebound slowly to their original elevations. This process has continued for thousands of years and is now perhaps two-thirds completed. The rebounding areas are increasing in elevation with respect to sea level, causing coastlines to advance seaward. As a result, countries like Canada and Finland gain land each year. The displaced ocean water, which is relocated southward, has contributed to a slow rise in sea level over much of the world. The resulting widespread problems of coastal flooding and beach erosion are discussed in the Case Study in Chapter 17.

Summary

This chapter has examined the influence of ice on the earth's land surface. Ice, in the form of glaciers, is the most powerful of the three "tools of erosion." However, ice is geographically restricted to low temperature environments at high latitudes and elevations. In the geologically recent past, colder worldwide temperatures prevailed, and ice was of much greater geomorphic importance than at present. Many relict glacial features still exist in previously glaciated areas.

Glaciers are large masses of freshwater ice, derived primarily from the compaction of accumulated snow. They form on land areas where snowfall receipts exceed melting rates over an extended time span.

Glaciers can be divided into two basic types: alpine glaciers and continental glaciers. Alpine glaciers, much smaller than continental glaciers but geographically more widespread, are generally long, relatively narrow tongues of ice occupying depressions in high mountain regions. Continental glaciers, which form over extensive high latitude land areas, are generally domelike in form and move outward in all directions from their centers of accumulation. At present, continental glaciers occupy the continent of Antarctica and the interior portion of Greenland.

Glaciers modify the landscape by eroding and depositing surface materials. Erosion is accomplished by the entrainment or "bulldozing" of weathered surface materials and, to a lesser extent, by the scraping of rocks frozen into the base of the ice against the surface. The chief erosional landforms of alpine glaciers are cirques and glacial troughs. Ice-scoured plains are the chief erosional landform of continental glaciers.

Most glacial deposition occurs after the ice melts sufficiently to become overloaded with sediment. Sediment is transported in a conveyor-belt fashion from the glacial erosional area to the glacial depositional area. The chief depositional features of alpine glaciers are lateral and terminal moraines. Major depositional landforms produced by continental glaciers are terminal and recessional moraines and till plains. In addition, glacial meltwaters can carry large amounts of glacially derived materials away from the ice front to produce such glacio-fluvial features as lacustrine plains, outwash plains, and valley trains.

The Pleistocene was a geologically recent period of colder temperatures and widespread glaciation. Within the past 2 million years, there were approximately 10 stages of extensive glaciation, separated by relatively warm interglacial stages. During the glacial stages, continental glaciers formed over northern North America and northwestern Eurasia, leaving a major imprint still visible on those areas today. Alpine glaciers also increased greatly in numbers and extent. The cyclic alternation of glacial and interglacial stages within the Pleistocene is apparently caused by long-term variations in the earth's orbit around the sun. There is growing evidence that the Pleistocene is not really over and that the earth is currently in an interglacial period.

Review Questions

1. What is a glacier? What conditions are required for the formation of a glacier?

2. What areas of the world are currently covered by continental glaciers? By alpine glaciers?

3. How do glaciers transport and deposit materials? How do the characteristics of direct glacial deposits differ from those of fluvial deposits? What are glacio-fluvial deposits?

4. At what specific sites within a mountainous region do alpine glaciers generally form? What determines the paths they take as they flow?

5. Describe the overall impact of extensive alpine glaciation on a mountainous region such as the Alps. What are some of the more important alpine glacier erosional and depositional landforms?

6. How do the physical characteristics of alpine and continental glaciers differ?

7. What was the Pleistocene? Briefly describe three different hypotheses regarding its origin.

8. Describe the present surface characteristics of the areas subjected to erosion by continental glaciers during the Pleistocene. Where are these areas located?

9. What are till plains? What are the present surface characteristics, geographic locations, and

economic significance of the major till plains produced by the Pleistocene glaciers of North America and Eurasia?

10. What are the two most extensive types of glacio-fluvial features produced by continental glaciers? What are their surface characteristics? Where are these features found in relation to till plains?

11. What changes occurred during the Pleistocene in areas of the low and middle latitudes far removed from the ice sheets?

Key Terms

Glacier
Continental glacier
Alpine glacier
Till
Moraine
Crevasse
Cirque
Horn
Glacial trough

Fjord
Pleistocene
Laurentide ice sheet
Ice-scoured plain
Till plain
Drumlin
Glacio-fluvial features
Outwash plain

CASE STUDY

A Glacial Catastrophe: the Spokane Flood

Geomorphologists and other earth scientists are on the whole committed to the principle of *uniformitarianism,* which is often defined as "The present is the key to the past." The idea is that natural earth phenomena of the geologic past were produced by the same physical processes that operate at present. On occasion, however, landforms are found that differ so greatly in appearance or magnitude from those with which we are familiar that at first they seem to defy the uniformitarian concept. An area containing such landforms exists in the Columbia Basin of southeastern Washington State. After more than 50 years of study, earth scientists have concluded that these features were produced by the largest verified floods in the history of the earth.

The Columbia Basin is a low elevation portion of the Columbia Plateau. The climate is arid, and the topography varies from flat to hilly. Despite its present aridity, however, a large portion of this area shows strong evidence that short-lived floods of gigantic proportions occurred during the Pleistocene. An area of approximately 11,000 square miles (28,500 sq km) is covered by a variety of fluvial erosional and depositional features, some on a scale unknown anywhere else in the world. Among these features are numerous dry canyons (called coulees) containing streamlined mesas and buttes, rock-rimmed basins, low basaltic ridges apparently eroded into segments by rushing floodwaters that overtopped them, dry waterfalls, giant gravel bars, and, perhaps most amazingly, ripple marks 50 feet (15 m) in height and spaced 500 feet (150 m) apart! (see Figure 15.23)

Debate on the origins of this area, called the Channelled Scablands because of the abandoned river canyons scarring its surface, continued for many years following its first description in a 1923 study by J. H. Bretz.* Many geomorphologists originally believed that it resulted from a long-

---

* J. H. Bretz. "The Channelled Scablands of the Columbia River Plateau." *Journal of Geology* 31 (1923): 617–49.

continued flow of glacial meltwater. It was gradually realized, though, that features such as the overtopped drainage divides and giant ripple marks could only result from one or more gigantic, but very brief, floods. The problem was then to determine how and where such floods could have originated. In 1942 an answer was finally found. It had been known for some time that, during the Wisconsinan glacial stage, much of what is now western Montana was covered by a glacially dammed lake that was named Lake Missoula. J. T. Pardee, a geomorphologist employed by the U.S. Geological Survey, hypothesized that the lake was blocked by a lobe of

continued on next page

Figure 15.23 Giant ripple marks produced by the Spokane Flood. (*John S. Shelton*)

ice that extended down the Purcell Trench in northern Idaho.** The melting or breaching of this ice lobe released an estimated 300 cubic miles (1250 km³) of water in Lake Missoula within a period of a few hours or days, creating the catastrophic flood that carved the Channelled Scablands (see Figure 15.24). This origin is now generally accepted.

Recent fieldwork in the area indicates that Lake Missoula drained suddenly and catastrophically on numerous occasions during the late Pleistocene. Geologist R. B. Waite has presented evidence of 40 catastrophic floods during the Late Wisconsinan period (between about 17,000 and 12,000 years ago) alone.*** He believes that, for a time, the lake filled and then suddenly drained at intervals of only a few centuries or even decades. The drainage apparently did not occur due to a warming of the climate, but took place when the water level in Lake Missoula rose to approximately 90 percent of the height of the ice dam (the Pend Oreille Lobe of the Cordilleran ice sheet). At this point, the ice was buoyed up; water began flowing under the dam, and subglacial tunnels were rapidly carved, allowing the lake to drain completely. Each breached ice dam was soon replaced by fresh ice as the glacier continued to move southward, while Lake Missoula, fed by vast quantities of meltwater from the Cordilleran ice sheet to its north, refilled relatively quickly.

Estimates of flow volumes and durations of the floods are imprecise at best, but Lake Missoula may have drained at an approximate rate of 9.5 cubic miles per hour (40 km³/hr) over a period of 40 hours. This is about ten times the present combined flow volume of *all* the world's rivers.

** J. T. Pardee. "Unusual Currents in Glacial Lake Missoula." *Bulletin of the Geological Society of America* 53 (1942): 1569–1600.

*** R. B. Waite, Jr. "About Forty Last-Glacial Lake Missoula Jokulhlaups through Southern Washington." *Journal of Geology* 88 (1980): 653–79.

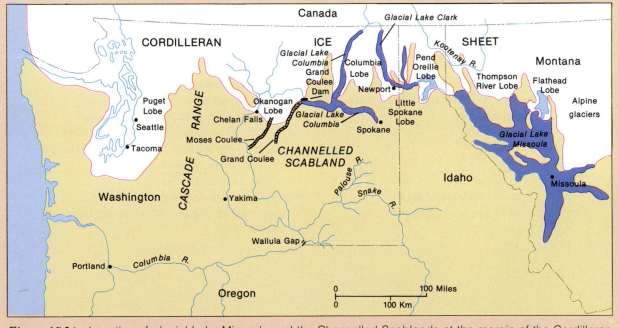

Figure 15.24 Location of glacial Lake Missoula and the Channelled Scablands at the margin of the Cordilleran ice sheet.

Chapter Sixteen

Eolian Processes and Desert Landscapes

Outline

The Wind as a Geomorphic Agent
    Wind Sediment Transport Mechanisms
    Eolian Environments
    Wind Erosion
    Wind Deposition
Desert Landscapes
    Surface Characteristics of Deserts
    Desert Landform Assemblages
Case Study: The Expanding Deserts

Focus Questions

1. In what geomorphic environments are eolian processes most important?
2. What are the primary erosional and depositional features produced by the wind?
3. How do the surface features of deserts differ from those of more humid environments?
4. What are the major types of desert landform assemblages, and what are their characteristics?

This chapter examines the geomorphic aspects of two related topics. We begin with a discussion of the surface effects of the wind. The wind is the weakest of the tools of gradation; it attains importance as a landscape modifier only in areas devoid of a protective vegetation cover. Most such areas have arid or semiarid climates, although sandy beaches and other localized environments are also strongly influenced by the effects of the wind. Geomorphic processes performed by the wind are called **eolian** processes after Aeolus, the Greek god of the winds.

The second section departs from the single geomorphic approach of the last several chapters and examines the landscape characteristics of desert regions. Desert landscapes are not dominated by a single geomorphic process; they are polygenetic in origin, with tectonic, fluvial, and eolian processes especially important. The specific actions and the extent of the contributions of these processes vary from place to place, causing deserts to vary greatly in appearance. As a group, though, the surface characteristics of deserts differ in several basic respects from those of humid landscapes.

Figure 16.1   The awesome front of a dust storm approaching Springfield, Colorado, in the spring of 1937. This storm produced nearly total darkness for a half-hour period. (*AP/Wide World Photos*)

THE WIND AS A GEOMORPHIC AGENT

The ability of desert windstorms to erode and transport enormous quantities of loose sand and dust has long been demonstrated to travelers in arid lands, but the erosive effect of the wind on solid rock is very limited. Its weakness results largely from the gaseous condition of the atmosphere and the air's very low density ratio of about 1:2000 as compared to rock. Water and ice, by contrast, have density ratios of about 1:3 when compared to rock. As a result, the wind is primarily restricted to the redistribution of loose, fine-textured surface materials. Where such materials are plentiful and are not immobilized by moisture or a vegetative cover, the wind can perform substantial erosional and depositional work and produce a variety of distinctive surface features.

Wind Sediment Transport Mechanisms

The wind transports sediments in much the same manner as flowing water, although some important differences exist. Both wind and water are fluids whose flow over the surface generates turbulence. The upward component of the turbulent eddies allows both fluids to pick up loose sediments and to transport them in suspension. Although the wind typically flows at a much greater speed than water, its ability to carry suspended particles is reduced by its lower density. As a result, only silt and clay particles are normally carried in suspension by the wind. These fine particles, however, can be lifted thousands of feet into the air and quickly transported for great distances before being deposited.

Where winds are exceptionally strong and large quantities of loose soil are available, *dust storms* are sometimes produced. These storms can reduce surface visibilities to only a few feet and can last for hours, stripping tremendous volumes of valuable topsoil from agricultural regions. The best-known dust storms in the United States occurred during the 1930s in the "Dust Bowl" region of the southern Great Plains (see Figure 16.1). Although improved agricultural practices and moister conditions have greatly reduced the frequency of such storms within this region, severe dust storms are still common in portions of Africa, Asia, and Australia.

If sediments are too coarse textured to be carried in suspension, both water and wind are capable of transporting them by surface creep or by saltation over the surface. **Surface creep** refers to the slow shifting of surface materials by rolling or sliding; it accounts for approximately one-fifth of all eolian surface sediment transport. **Saltation** is an asymmetrical bouncing of surface grains. It moves materials rapidly and accounts for about four-fifths of all eolian surface sediment transport. It is through saltation that sand grains are aggregated into shifting sand dunes. Because saltation rarely lifts sand grains more than a short distance above the surface, true **sandstorms** consist of a stinging mass of wind-whipped sand grains extending only a few feet above ground level.

Despite the general similarities between the transport of sediments by water and by wind, several important differences exist. Wind transport is much more rapid, but, unlike water transport, it is restricted to fine-textured sediments. In addition, the wind does not carry a dissolved sediment load. Finally, the wind is much less restricted than water in the directions it can travel and in the sites in which it can erode or deposit sediments. Water always transports sediments downhill, while the wind can transport sediments either uphill or downhill (although downhill movement predominates). Fluvial processes are confined largely to the lowland sites where water flow is concentrated. In contrast, the wind flows over the entire surface and is capable of eroding or depositing materials almost anywhere within a suitable region.

Eolian Environments

In order for eolian processes to be effective, surface materials must be fine-textured, dry, and loose. In addition, the affected area must experience at least occasional strong winds. A vegetation cover greatly diminishes the wind's effectiveness by producing a large amount of near-surface friction, by covering the soil, and by holding the soil particles in place. For these reasons, areas subject to significant wind erosion must be at least partially devoid of vegetation.

The most geographically extensive geomorphic environments meeting these criteria are the desert regions. Every continent contains large deserts where the wind plays an important role in landscape development. In most deserts, wind erosion has predominated over deposition. The surface in such areas is covered largely by stones of various sizes because the finer particles have been winnowed out by wind action. Approximately a fourth of the world's deserts, though, have been the sites of overall deposition and are covered partially or completely by sand.

Another well-known but much less areally extensive category of eolian environments consists of the coastlines of large water bodies, including the ocean, seas, and large lakes. Along coastlines, waves and currents supply weathered materials susceptible to wind action, and vegetation is discouraged by storms and tides. The coastal environment is described in detail in Chapter 17.

The valleys of streams draining arid and semiarid regions may also be important sites for eolian processes. These streams support relatively little stream bank vegetation. They also fluctuate greatly in flow volume, often exposing extensive strips of sandy alluvium during periods of low water. Strong winds are able to remove the silt and clay particles and collect the sand grains into dunes. The floodplains of several large rivers of the American Great Plains, including the Missouri, Platte, and Arkansas, are especially affected by these eolian activities.

Over the last few centuries, human activities have unfortunately considerably augmented the global effect of eolian processes. Especially affected are agricultural or pastoral regions with semiarid and seasonally dry climates in portions of Africa and Asia. The activities of the rapidly growing human populations within these regions have led to the extensive removal of the already marginal vegetation cover. The loss of protective vegetation, in turn, has caused the eolian removal of tremendous quantities of irreplaceable soil. This environmental problem is further explored in the Case Study at the end of the chapter.

Wind Erosion

The wind erodes materials from the surface in two different ways. **Abrasion** involves the impact of saltating sand grains. It is often described as a "sandblasting" effect. **Deflation,** on the other hand, involves the removal of loose surface materials directly by the wind.

The effect of wind abrasion on solid rock materials is relatively minor and generally superficial. The wind does not normally produce major bedrock erosional features; it merely modifies their surfaces by pitting, etching, smoothing, or polishing. Sand grains large enough to perform abrasion are transported only a small distance above the surface. As a consequence, features standing more than a few feet above the general surface level are not significantly affected by this process (see Figure 16.2).

Deflation is a much more important mechanism of wind erosion, but it requires the presence of unconsolidated surface materials such as sand or dry soil. One

Figure 16.2 The effect of wind abrasion on solid rock is generally considered to be relatively minor. Repeated sandstorms, though, are sometimes capable of producing "mushroom rock" formations in soft rock outcrops because erosion is most vigorous just above the ground. The example here is in southeastern Iran. (*Rodman Snead, © JLM Visuals*)

widespread category of features resulting from this process is **deflation hollows,** or "blowouts." These are shallow, often circular depressions that can vary greatly in size. The largest are over a mile (1.6 km) in diameter and up to 50 feet (15 m) deep, although most are much smaller. Many thousands of such depressions dot portions of the southern Great Plains of the United States. These features, which may have been formed in part by solution and bison wallowing, become the sites of shallow ponds after heavy rainfalls.

Deflation is the only natural mechanism that can remove sediments from enclosed desert basins. Large desert basins, such as the Qattara Depression in northwestern Egypt and several basins in interior Australia, may owe their origin primarily to deflation; in effect, they may be blowouts of gigantic proportions.

In many arid regions the eolian removal of sand-sized and finer sediments has left behind deposits of pebbles, stones, and boulders that are too large and heavy to be blown away. In time, gravity, salt crystal growth, and ephemeral floodwaters can arrange these deposits into a **desert pavement** of close-fitting pebbles or stones that protects the underlying surface from erosion (see Figure 16.3). Desert pavement surfaces cover large areas in the southwestern United States, the Sahara, and interior Australia.

Wind Deposition

Wind, like water, deposits sediments when its speed of flow and resulting turbulence become insufficient to transport these materials. The transport of sand requires higher wind speeds and greater turbulence than does the transport of finer sediments, and this factor leads to the sorting of eolian deposits on the basis of mass. Although both sand and silt are often derived from the same source region, the coarser sand particles are usually deposited relatively near the source of supply, while the windblown silt, which is commonly termed **loess** (from the German *löss* "loose"), is deposited over a larger region farther downwind. Not only are these two eolian deposits geographically separated from one another, but each has distinctive physical characteristics, landform assemblages, and human implications. They are therefore treated separately in the following discussion.

Topography of Sand-Covered Surfaces

Most eolian sand deposits are composed largely of the stable mineral quartz and have average grain diameters of about 0.04 inches (1 mm). Sand grains, when dry and loose, begin to move over the surface when wind speeds exceed 10 miles per hour (4.5 meters/sec).

Because sand does not normally travel far from its region of origin and can be moved by the wind only when it is dry and loose, it accumulates primarily within desert regions. The impression that many people have of most deserts being blanketed by sand is incorrect,

Figure 16.3 A close-up view of a desert pavement surface in Nevada. The finer surface materials have been removed by the wind. (*John S. Shelton*)

351

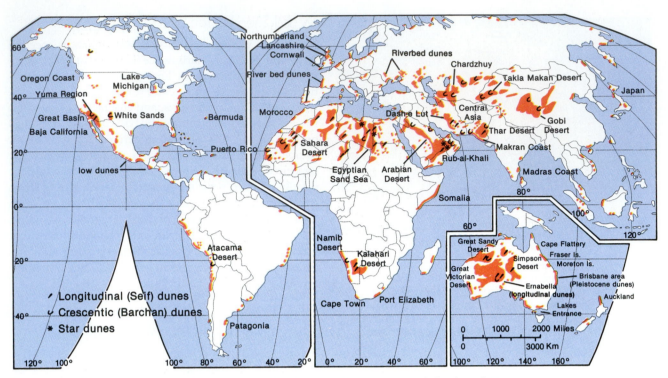

Figure 16.4 Global distribution of sand-covered regions and their characteristic dune forms.

however. Sand does not cover the majority of any large desert. Rather, it has accumulated within certain areas through the action of the wind and, to a lesser extent, of ephemeral streams. Sandy deserts, sometimes referred to as **ergs** (an Arabic term), comprise at most a fourth of the world's deserts. They contain not only local sand, but also sand derived from surrounding nonsandy desert areas. The quantities of sand within such an accumulation area vary greatly. In many instances, sand exists only in scattered patches; in other cases, it may deeply cover the entire surface (see Figure 16.5).

The major sandy deserts of the world are depicted in Figure 16.4. The largest ergs are located in the central Sahara, eastern Saudi Arabia, west-central Australia, southeastern Iran, northwestern India and adjacent Pakistan, and western China. In addition, smaller sandy deserts are scattered throughout other arid lands, including the southwestern United States. A well-known sand accumulation in the United States is located within White Sands National Monument near Alamogordo in southern New Mexico. It is unusual in that it contains snow-white gypsum sand derived from a nearby dry lake bed. Great Sand Dunes National Monument in south-central Colorado is another sand accumulation site that has been preserved as a tourist attraction.

The most characteristic surface features associated with sand-covered regions are dunes. A **dune** is a mound or ridge of wind-deposited sand. Most sand-covered regions exhibit dunal topography. Sand dunes vary greatly in size and shape; small dunes, which are most common, usually range between 10 and 50 feet (3–15 m) in height. Where sand is plentiful and winds are strong, dunes can reach heights of 600 feet (180 m) and extend for miles. Their dimensions and forms are determined chiefly by the quantity of available sand, the speed and directional characteristics of the wind, and the presence and extent of stabilizing vegetation. Areas with winds that are relatively constant in direction generally have the most organized dune patterns.

Dunes basically form and grow because of the frictional resistance they offer to the wind. They are self-generating features because, as they grow, they become more effective barriers to the wind, reducing its speed and causing it to deposit any sand it may be transporting. The sides of most dunes are asymmetrical,

Figure 16.5 Giant sand dunes in the Sahara (Hammada du Guir, Morocco). (*R. Krubner/ H. Armstrong Roberts*)

with a gentler windward-facing *back slope* up which the sand grains are blown, and a steeper leeward-facing *slip face* down which the grains tumble until they land within a niche in which they can lodge (see Figure 16.6). The slip face normally rests at the angle of repose of dry sand, which is approximately 34° from the horizontal. Through the mechanism of individual sand grains saltating up the back slope and masses of sand sliding down the slip face, the dune gradually changes its location and shape.

Dunes are described as "active" if they are not covered by vegetation and are currently moving and changing shape. Conversely, they are "inactive" if they are stabilized by a partial or complete cover of vegetation (see Figure 16.7). Stabilized dunes are often relict features of a past drier climate. For example, the Sand Hills of central Nebraska are sandy Pleistocene outwash deposits reworked by the wind and currently covered by

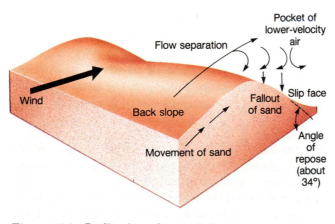

Figure 16.6 Profile view of a sand dune.

Figure 16.7 A sea of stabilized Pleistocene sand dunes in the Palouse region of Washington State. (*H. Armstrong Roberts*)

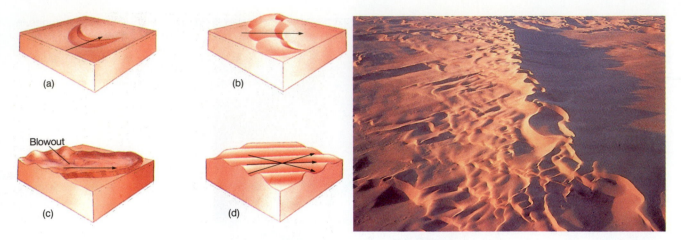

Figure 16.8 *Art:* (a) Barchan dune: A crescent-shaped dune with its horns pointing downwind. Barchans exist on hard, flat floors of deserts where there is constant wind and a limited sand supply. (b) Transverse dune: A dune that forms a wavelike ridge transverse to the wind direction. They exist in areas with abundant sand and little vegetation. (c) Parabolic dune: A U-shaped dune with the open end of the U facing upwind. Some form by the piling of sand along the leeward and lateral margins of a growing blowout. (d) Longitudinal dune: A long, straight, ridge-shaped dune oriented parallel to the wind direction. They exist in deserts with a limited sand supply and strong winds from one prevailing direction. *Photo:* Complex dunal topography in the Namib Desert southwest of the Kuiseb River, Namibia. (*E.D. McKee, USGS*)

a layer of sod and grasses. In addition, large areas of stabilized dunes exist along the equatorward margins of the Sahara and Kalahari Deserts. These dunes provide evidence that the subtropical deserts, at least in Africa, have shifted poleward in post-Pleistocene times.

Local wind systems frequently organize sand dunes into complexes that give a distinctive pattern to the topography. Dunes with similar characteristics tend to occur within a given area because the controlling factors are usually fairly constant. Probably the most common and familiar of the dune forms are the **barchans.** They are relatively small, crescentic in shape, and are oriented with their long axes perpendicular to the prevailing winds and their tapered ends, or "horns," pointing downwind (see Figure 16.8a). When well formed, they are remarkably similar in shape to French crescent rolls. Barchans are found in portions of all large deserts. They develop in areas of relatively constant wind direction and restricted sand supply and are the most mobile of the major dune types.

Longitudinal dunes are long ridges of sand generally aligned parallel to the direction of the prevailing winds (see Figure 16.8d). They may become very large, attaining heights of several hundred feet and lengths of many miles. Longitudinal dunes are especially well developed in central Australia, where they cover nearly a fourth of the continent and are remarkable for their

even spacing. Most of the Australian dunes are stabilized by vegetation at present and apparently formed during a past drier period. Currently active longitudinal dunes cover extensive desert areas in Saudi Arabia, Egypt, and Iran.

Transverse dunes are elongated perpendicular to the prevailing winds. Common in regions of abundant sand, they typically assume a wavelike pattern, with undulating crest lines, and have been likened in appearance to a frozen picture of a storm-tossed sea (see Figure 16.8b). They are much less symmetrical and continuous than longitudinal dunes. Some transverse dunes develop a clearly distinguishable linked barchan form and may separate into individual barchan dunes if they extend into an area containing less sand.

Sand dunes do not always develop the symmetrical forms just described. Complex dunes of almost any shape may form, especially in areas of variable winds. These include the so-called *star dunes* of Egypt and Arabia, which are among the highest in the world. They consist of pyramidal mountains of sand with sharp crest lines on the radiating ridges that form the "points" of the star. At the other extreme, some deserts are covered by flat *sand sheets* that display little or no dunal formation. One of the best examples of these sand sheets covers several thousand square miles of the Libyan desert.

Superimposed on all sand deposits are small *sand ripples* that contain only 1 or 2 inches (2–5 cm) of surface relief. They are highly similar in cause and appearance to the small water ripples that are superimposed on larger wave forms on a windy day. Sand ripples are erased and re-formed during each windy period and develop an undulating pattern elongated perpendicular to the wind direction.

Figure 16.9 This view of a roadcut in Chinese loess illustrates the ability of loess to maintain vertical banks. (*Rodman Snead, © JLM Visuals*)

Topography of Loess-Covered Surfaces

Loess consists of predominantly silt-sized eolian deposits. It is typically buff-colored and is composed of a well-sorted mixture of quartz and calcite ($CaCO_3$) particles. Loess is derived from the same general areas as sand. The three most important loess sources are deserts, Pleistocene outwash deposits, and the floodplains of rivers in semiarid regions.

Despite its general similarity in origin to sand, loess differs markedly in most other geographic and geomorphic characteristics. Because it must be stabilized by moisture and a cover of vegetation in order to accumulate, it is not associated with desert environments. At present, most loess deposits are overlain by both vegetation and a surface soil layer; consequently, they are not as easily recognizable as sand deposits, which are often devoid of both soil and vegetation. Soils derived from loess, in contrast to those developed from sand, are often highly fertile and agriculturally productive.

Uneroded loess deposits usually do not form distinctive topographic features because the wind spreads the loess as a smooth blanket over the surface. As a result, some of the most level terrain on earth occurs in undissected loess regions. These regions become much more topographically distinctive, however, when subjected to stream erosion. Loess particles are angular and tend to interlock and develop a columnar structure. At the same time, the unconsolidated nature of loess readily permits streams to erode into the deposits and form nearly vertical banks and cliffs (see Figure 16.9). As a result, eroded loess areas are typically dominated by deep, steep-walled valleys subject to rapid lateral and headward expansion .

During the Pleistocene, tremendous quantities of fine-textured glacial sediments were transported southward from the margins of the North American and Eurasian ice sheets by glacial meltwaters. These meltwaters were not always organized into discrete streams, but often spread over large areas. They were generally overloaded with sediments, which were deposited over extensive areas. The wind removed great volumes of predominantly silt-sized materials, and the downwind deposition of these materials produced extensive loess deposits.

Loess, predominantly of Pleistocene origin, presently covers approximately 10 percent of the world's land surface and 30 percent of the United States. Depths range from only a few inches to as much as 300 feet (90 m) immediately downwind from a major source area. The major loess-covered regions of the world are shown in Figure 16.10. Among the most important are portions of the North European Plain from Belgium eastward to the Ukraine, parts of the western Asiatic Soviet Union, northern China, the Pampas of Argentina, and the Great Plains, the Midwest, and the Mississippi River Valley in the United States. It is notable that most large deposits occur within the middle latitude semiarid and subhumid areas that collectively comprise most of the world's major grain-producing regions.

Most loess deposits within North America were derived from braided river courses such as that of the Missouri River, from dry lake beds, and from glacial outwash plains at times when they were largely devoid of vegetation. Maximum thicknesses of 50 to 100 feet (15–30 m) occur on the eastern bluffs of the lower Mississippi River and just east of the major Pleistocene

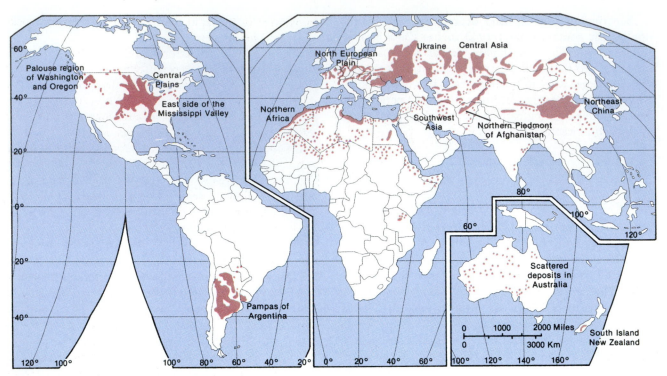

Figure 16.10 Global distribution of major loess-covered regions.

outwash plains in Nebraska, Iowa, and northern Kansas. An important smaller area of loess occurs in the Palouse region of southeastern Washington State.

DESERT LANDSCAPES

A **desert** is an area of considerable size with a surface largely or entirely devoid of vegetation. This broad definition can be taken to include the so-called polar deserts, where plant life is absent because of cold temperatures and a surface cover of ice and snow. In this section, however, the concept is restricted to deserts resulting from aridity.

The causes, climatic characteristics, and global distributions of the world's arid regions were discussed in Chapter 7. It should be recalled that arid regions are generally those in which potential evapotranspiration rates exceed precipitation receipts. Most are located in the subtropics and owe their dryness to the subsidence of air within the subtropical high pressure belts. Additional deserts occur within portions of the middle latitudes. Middle latitude deserts are sometimes described as topographic deserts because their aridity

results largely from the moisture-blocking influences of mountain barriers. In all, some 19 percent of the earth's land surface consists of arid deserts.

Surface Characteristics of Deserts

The first-time visitor to a desert is immediately struck by the great contrast between the desert landscape and the landscapes of more humid environments. Perhaps the most fundamental impression is of the starkness of the land (see Figure 16.11). The surface is not covered by a mantle of soil and vegetation, but instead is likely to consist largely of a loose jumble of rocks of various sizes. The dominant colors are the buffs, grays, and browns of rock in various stages of weathering rather than the soft greens of live vegetation. The topography often appears highly angular, with a prevalence of horizontal and vertical surfaces and a lack of the smoothly rounded slopes typical of many humid landscapes. Even the clear weather and unlimited visibilities that prevail in most desert regions add to the impression that the universe consists predominantly of rock and sky.

The underlying factor controlling all the differences between desert and nondesert regions is the reduced

Figure 16.11 Fault block mountains and salt-encrusted playas are common geomorphic features in the Great Basin of the western United States. (*John S. Shelton*)

availability of surface moisture. The low precipitation totals received in deserts not only discourage the growth of vegetation, they also profoundly affect the processes of landscape modification. Chemical weathering processes, which depend greatly upon the presence of water, progress much more slowly than in humid climates and generally do not reach the clay mineral production stage. Physical weathering activities are less affected, because more of them require either no water or just the short-term presence of water. The most active physical weathering processes in deserts are salt crystal growth, frost wedging in colder areas, and perhaps thermal expansion and contraction related to daily heating and cooling.

The reduced weathering activity causes bedrock to be exposed much more extensively than it is in most humid environments. Based on the nature of their surface materials, about one-fourth of the world's deserts may be classified as ergs, or sandy deserts. Over half are stony deserts, while most of the remainder contain a solid bedrock surface, sometimes covered by scattered boulders. True desert regions generally contain only relict soils, left from a previous more humid period, or, locally, soils developed on alluvial deposits.

The angular topography that characterizes many desert regions has been attributed largely to the virtual absence of creep as a downhill transporter of weathered materials. Instead, most downhill transport of weathering products is performed by surface water during heavy

rainstorms or by rapid mass movements such as rockslides and mudflows.

Despite its restricted availability within desert regions, water is still the most effective erosional agent for deserts as a whole, with the effects of the wind relegated to a distant second place. Because less water is available for weathering and erosional activities, the overall rate of landscape evolution is much slower in arid environments than in humid ones. On the other hand, weathered surface materials within deserts are highly mobile because they are not stabilized by vegetation. This allows flowing water and strong winds to be effective transporting agents and results in the continual rearrangement of surface materials.

A basic fluvial characteristic of desert regions is the small quantity of water normally carried by stream systems. Water tables in desert regions are generally located well below the surface. As a result, ground water outflow is rare, and most streams are fed only by surface runoff. This causes nearly all desert streams to be ephemeral, carrying water for only short periods at infrequent intervals. A major exception to the ephemeral nature of most desert streams is the **exotic rivers** that cross some deserts. These rivers originate in humid regions and happen to pass through deserts on their way to the ocean. Although they have no tributaries and lose large volumes of water to seepage and evaporation as they traverse the desert, the larger rivers have a volume of flow sufficient to survive the passage. A large proportion of the human inhabitants of desert regions reside along the banks of exotic rivers because of the irrigation water, fertile alluvial soils, and means of transport that these rivers provide (see Figure 16.12). Some examples of important exotic rivers are the Colorado and Rio Grande in North America, the Nile and Niger in Africa, and the Indus in Asia.

Except in those desert regions located along exotic rivers or adjacent to coasts, surface runoff does not reach the ocean, but instead flows into enclosed basins where it eventually evaporates. Such desert regions are characterized by **interior drainage.** Runoff usually travels only a short distance before it either sinks into the ground or enters an enclosed basin. Because weathering products are not transported out of the area, depositional features of various kinds—dunes, mass movement debris, and stream deposits—are especially common.

Desert regions are notoriously susceptible to destructive flash floods, which play a vital role in landscape development. Despite the overall paucity of

Figure 16.12 Irrigation water makes intensive agriculture possible in this semidesert area near Bakersfield, California. (*P. Degginger/H. Armstrong Roberts*)

precipitation, desert thunderstorms may produce localized torrential downpours. Flooding is also promoted by the impermeable nature of most desert surfaces. These surfaces are likely to contain a *duricrust* layer formed by the near-surface precipitation of salts, or to consist of a solid bedrock surface beneath a thin cover of sand or stones. No deep soil layer exists to absorb and hold moisture. The lack of vegetation and of soil organisms allows for little working and loosening of the weathered surface materials, so they are typically highly compacted. Finally, major drainage channels capable of holding large volumes of water are few and far between. This lack of surface drainage capacity causes runoff to flow in sheets or shallow rivulets over much of the surface during a heavy storm. Desert storms are typi-

cally very localized and short in duration, though; so floods can subside almost as quickly as they arrive.

Desert Landform Assemblages

There is no predominant desert landscape. Deserts vary greatly in landform assemblages because in different regions they have been subject to differing combinations of geomorphic controls. Factors such as temperature, precipitation amounts and intensities, and wind speeds and directions are all significant. Probably the most critical controls, however, are lithologic factors and recent climatic and tectonic events. Lithologic features such as differing rock strata, bedding planes, and fractures are often strikingly exposed because of the general absence of a soil or vegetation cover. Recent tectonic processes are especially important in deserts because the gradational processes that modify and eventually eliminate tectonic features have been greatly slowed by a lack of water. Because tectonic features are so important, the nature and recency of their actions strongly influence the three categories of desert landform assemblages discussed in this section. These are desert plains, deserts containing plateaus, and mountainous deserts.

Desert Plains

Many deserts have developed in stable continental interiors where little tectonic activity has occurred for long periods of geologic time. These areas have low-to-moderate elevations and limited surface relief, so they are topographically classified as plains. Important examples include the deserts of Australia, Arabia, and Turkestan; the Kalahari Desert; and most of the Sahara.

Desert plains contain some of the most uniformly flat surfaces on earth, chiefly because vertical erosion is restricted by interior drainage. Because no fluvially transported materials can leave the area, local base levels prevail. Sediments are deposited in low-lying sites, eventually filling them to produce an almost perfectly level surface.

If at least some slope is available, the surface is likely to be covered by a braided pattern of shallow channels termed *washes*. The finer textured, damper soils within the washes encourage the growth of vegetation, frequently causing them to become covered or paralleled by growths of small shrubs. These plants accentuate the drainage pattern so that it is readily visible from the air (see Figure 14.1).

Figure 16.13 Monument Valley, in northeastern Arizona, contains a large number of residual sandstone buttes and mesas. Seen here is one of the Mittens. (*E.D. McKee, USGS*)

Surface materials vary considerably in different desert plains. A surface of close-fitting stones is most common, but extensive desert plains are covered by shifting sands. Sandy regions, at least, display highly irregular although usually rather small-scale surface features in the form of dunes. Also in contrast to other desert plains, sand-covered surfaces are usually free of visible drainage channels because of the high porosity and shifting nature of the sand. Still other desert plains have bedrock surfaces or contain stony residual soils.

Deserts Containing Plateaus

Tectonic forces have raised some arid and semiarid regions well above sea level with little deformation so that they now form plateaus. The Colorado Plateau of northern Arizona and southern Utah provides an excellent example of such a region. Other desert plateaus are located in northern Mexico, southern Argentina, the Middle East, western Arabia, Southwest Africa, and western China.

Typically, a desert plateau surface is mostly flat, making it topographically indistinguishable from a desert plain. Higher elevations and some orographic influence often make these surfaces somewhat cooler and less arid than their low elevation counterparts, so plateaus are usually somewhat better vegetated. Most flat upland surfaces are developed on horizontally bedded sedimentary rocks such as sandstone or limestone, which resist weathering and erosion in dry climates.

Among the most notable features of desert plateaus are their steep, clifflike escarpments, usually found either where a river has incised a deep gorge or canyon into the surface or at the eroded margins of the plateau. In either case, the reduced influence of chemical weathering helps to maintain slope steepness. Along escarpment faces, differing erosion rates of resistant and nonresistant strata are often clearly evidenced by the development of a steplike series of alternating cliffs and slopes (see Figure 14.8).

Escarpments in all climates tend to retreat in an irregular fashion, so they have highly uneven margins. Uneroded segments of a retreating scarp frequently become isolated by the erosion of the surrounding rock. These outliers are termed *mesas* if they retain a flat top and *buttes* if erosion has reduced them to narrow spires with no flat upper surface. One or more isolated mesas or buttes may also exist as the last remnants of a plateau that has otherwise been completely consumed by erosion (see Figure 16.13).

Mountainous Deserts

Mountainous deserts contain the greatest variety and complexity of surface features. The major mountainous

The Desert and Off-Road Vehicles

The highly fragile desert environment is easily disturbed by intensive human use. In many areas, this sensitivity presents no major problems, because the desert offers little to induce human occupation. Such a situation, though, is not true of the deserts of southern California, which are located within easy driving distance of a large and highly mobile population. Every weekend, these once-empty lands are invaded by thousands of sightseers, picnickers, fossil and Indian artifact hunters, plant and animal collectors, and recreational vehicle owners. Although all of these groups leave their mark on the land, the recreational vehicle owners are of greatest concern to environmentalists.

In recent years there has been a dramatic increase in the popularity of off-road vehicles (ORVs) of all types, including motorcycles, dune buggies, and especially the relatively inexpensive all-terrain vehicles. Currently, more than 13 million ORVs are in use in the United States, with well over a million of these in southern California alone. The major environmental problems associated with ORV use in the California desert have been soil erosion, changes in hydrology, and damage to plants and animals.

Vehicle tracks made on desert surfaces can last for years, sometimes even for centuries (see Figure 16.14). Passing vehicles destroy the structure binding the thin surface soils, leaving it as loose dust. At the same time, they compact the underlying materials, greatly reducing their ability to absorb water. Runoff from ORV sites is as much as eight times higher than that from adjacent undisturbed sites. Because they compact the soil and reduce its permeability, vehicle tracks, especially those extending downhill, tend to be transformed into gullies by desert rainstorms. What water remains in the soil is more tightly held

and less available to plants. Diurnal soil temperature ranges are also increased, placing additional thermal stresses on plants and animals that must already cope with extremely high daily temperature variations.

The most obvious change in the appearance of many areas frequented by ORVs is the reduction of vegetation. Plants are killed directly by being run over and indirectly by soil erosion, compaction, and desiccation. Animals are also killed or displaced. One 1975 survey by the U. S. Bureau of Land Management following a cross-country motorcycle race between Barstow, California and Las Vegas, Nevada, found a 90 percent reduction in the local population of small mammals.

The ongoing controversy between environmentalists and recreational vehicle owners about use of ORVs in California's deserts boils down to dif-

continued on next page

deserts of the world include the Great Basin of the western United States and adjacent northern Mexico; South America's Atacama Desert; large portions of Southwest Asia, including much of Iran, Afghanistan, and Pakistan; limited areas of the central Sahara and central Australia that contain isolated mountain masses; and portions of western China and Mongolia.

Two gradational activities largely confined to arid regions are of great importance in shaping mountainous desert landscapes; these are back-wasting and interior drainage. Due apparently to the reduced activity of chemical weathering and the virtual absence of creep, the erosional reduction of highland masses in arid regions takes place chiefly by **back-wasting** (see Figure

16.15). The significance of back-wasting is that mountain masses, while being reduced in size and areal extent, remain steep and rugged.

Interior drainage is critically important in mountainous deserts because only weathering products in limited quantities transported by the wind can leave the region; consequently, there is no significant reduction in mean surface elevations. Instead, as highland areas are gradually worn back, the adjacent lowlands that receive the products of erosion are aggraded. This results in a gradual reduction in surface relief unless tectonic activity produces further uplift.

Mountains in most desert regions rise abruptly from a level or gently sloping surface (see Figure 16.11). The

ferences in environmental perception. The environmentalists contend that the desert contains a valuable and fragile ecosystem and that unrestricted vehicular use in these areas should be prohibited. The recreational vehicle owners tend to view the desert as an empty wasteland that is of little or no practical use to mankind other than for recreational purposes. They point out that they are using these unoccupied lands so they will not disturb the residents of populated areas. As citizens and taxpayers, they feel that they have a right to use these public lands just as campers and hikers do.

It is apparent that some desert land must be set aside for ORV use. The chief problems are how much land is needed, where it should be located, and how environmental damage can be minimized. One promising trend has been the opening of privately owned ORV sites with established trails that are regularly tended and watered to keep down dust.

In 1980, after four years of hearings, the California Desert Conservation Area Plan was enacted into law and is currently undergoing refinement and legal interpretation. It should eventually provide the framework for an effective desert land-use policy that will offer something for all groups interested in the use, and protection, of the desert.

Figure 16.14 Denudation and erosion of private land by off-road vehicles. This area is near Corona, on the east side of the Los Angeles Basin. (*Howard Wilshire*)

mountain masses are typically steep, rugged, and rocky. The surface at the foot of the mountains is often covered by deposits eroded from the adjacent uplands. If these materials have been transported away from the mountain front, the mountains are typically surrounded by a planed rock surface called a **pediment.** The origin of pediments has been the subject of much controversy, but they are apparently produced by the combined erosional activities of back-wasting, sheet floods following heavy rains, and the lateral shifting of ephemeral streams from the adjacent mountains. The pediments serve as transportation surfaces because their slope angles of 4° or less are just sufficient for floods to carry weathered materials from the adjacent mountains across them. Some pedi-

ments have their solid rock surfaces exposed, but more frequently they are hidden beneath a thin cover of loose rock debris that has been temporarily deposited while being transported across the pediment. The thickness of the rock debris covering the pediment surface tends to increase gradually away from the mountain front until finally, in block-faulted regions like the Great Basin, the pediment slopes steeply downward to form the basement rock of a sediment-filled basin (see Figure 16.16).

The enclosed desert basins into which mountain-derived sediments are transported are termed **bolsons.** Their flat surfaces are often developed upon hundreds or even thousands of feet of debris. Because they contain no drainage outlets, the centers of bolsons

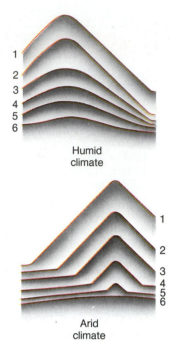

Humid climate

Arid climate

Figure 16.15 Down-wasting processes, which predominate in humid regions, result in a gradual reduction of slope angles through time. Back-wasting processes, which are characteristic of arid regions, cause the slopes to maintain their steepness as they are eroded.

loads enter the relatively flat adjoining pediment surface. Here, the reduction in slope steepness and especially the removal of confining channel walls greatly reduces the stream's speed of flow and allows it to spread out, causing it to deposit most of its sediment load. In origin and general surface configuration, alluvial fans are similar to deltas; but, unlike deltas, they are formed entirely on land, usually along the base of a mountain front. Alluvial fans also form in humid regions on sites such as the foot of a river bluff or terrace. They are best developed in mountainous deserts, however, because of the frequent sudden changes in slope and they are highly visible there because of the lack of vegetation (see Figure 16.17).

Along a linear mountain front, individual alluvial fans may coalesce to form a ramplike alluvial apron called a *piedmont alluvial slope* (or *bajada*). Large fans are favored settlement sites in arid regions because they generally contain near-surface supplies of ground water. In addition, the lower portions of fans consist of fine-textured sediments suitable for cultivation; and their gently sloping surfaces allow for good drainage of water and air (thus reducing frost hazards). A portion of Salt Lake City, Utah, and a number of cities and towns in California's Central Valley have been built on alluvial fans.

become temporary lake basins called **playas** following rainstorms. Playa lakes are usually extensive and very shallow and have boundaries that rapidly shift as water is added by inflow or removed by evaporation and seepage. Most of the time they are completely dry or consist of mudflats that may develop cracked surfaces as their moisture evaporates.

The water of playa lakes is highly saline because salts brought in by the runoff are left behind when the water evaporates. When the playas are dry, the centers of their bolsons may be covered by glistening white precipitated salts (see Figure 16:11). Salts in some localities have accumulated in sufficient quantity to be mined commercially. The mining of nitrates, for example, is a major industry in the Atacama Desert of northern Chile; and the mining of borates has long been carried on in California's Death Valley.

Another common feature of mountainous deserts is the **alluvial fan.** Alluvial fans consist of gently sloping fan-shaped deposits of alluvium. They form where ephemeral mountain streams carrying large sediment

Summary

This chapter deals with a geomorphic process—the wind—and an important surface environment—the desert—that are both associated with areas largely devoid of vegetation. In vegetation-free settings, weathering products generally lie loose on the surface and are easily eroded by even low-energy gradational forces.

The most geographically extensive of the eolian environments are the deserts. Areas that experienced desert conditions in the recent geological past or those located downwind from deserts may also contain wind-produced landforms. Coastlines with sandy beaches are also important eolian settings. Other areas in which wind-produced features are common include the floodplains of streams draining arid and semiarid regions and areas containing fine-textured glacial deposits. Human activities, especially agricultural pursuits, have allowed the wind to erode and redistribute vast quantities of soil.

Wind erosion may act rather evenly to generally lower the surface leaving few distinctive features. More

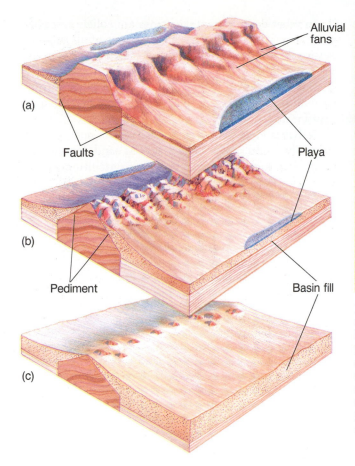

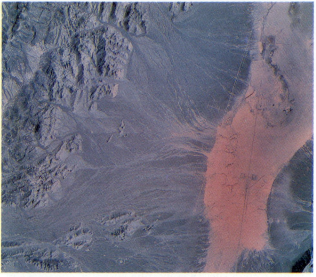

Figure 16.16 *Artwork:* Three stages in the erosion of an upfaulted block (a horst) in an arid climate. As the block is eroded back, the remaining upland segments retain their steep slopes. Interior drainage exists, so sediments are transported only into the adjacent basins, eventually covering most of the surface. *Photo:* Aerial view of residual mountains, a pediment surface covered with coalescing alluvial fans (forming a bajada), and a playa near Las Vegas, Nevada. (*Department of the Interior, U.S. Geological Survey*)

commonly, though, it selectively removes finer or weaker materials, while leaving those of greater resistance. Deflation hollows of various sizes exist in many areas as a result of wind erosion, and many desert regions contain bedrock or desert pavement surfaces from which the finer materials have been selectively winnowed. Wind deposits of sand-sized particles differ considerably from deposits of silt-sized particles. Sand is typically deposited in dunes within collection areas known as ergs. Several major dune forms exist, including the crescent-shaped barchans, longitudinal dunes, and transverse dunes. Dunes not stabilized by vegetation migrate slowly over the surface, mostly within desert regions, in response to prevailing local winds. Silt-sized eolian deposits, commonly termed loess, are typically situated downwind of sand deposits because their smaller size allows them to travel farther. Most loess deposits were originally derived from Pleistocene glacial outwash. At present, most are stabilized by vegetation, and their surfaces have often weathered to

Figure 16.17 Aerial view of a "classic" alluvial fan in California's Death Valley. (*Martin Miller, © JLM Visuals*)

form agriculturally productive soils. Loessial surfaces are predominantly flat and nearly featureless, but the angular nature of the loess particles has allowed streams in some areas to erode steep-walled gullies.

The basic difference between deserts and most humid regions is the absence of a desert vegetation cover. The general lack of surface water also results in a greatly reduced rate of chemical weathering and of soil development. As a result, bedrock features are generally exposed and the topography tends to be angular. Streams in desert regions normally carry water only infrequently, but fluvial geomorphic processes dominate in most arid regions because of the lack of surface vegetative protection. Desert rainstorms, when they do come, can supply copious amounts of precipitation that can produce spectacular and geomorphically effective floods. Many surfaces are covered by shallow washes, or drainage channels, that may be visually enhanced by vegetation patterns. Approximately one-fourth of the world's deserts are ergs, covered by shifting sand; but most are covered with stones or boulders.

Despite the slow rate of landform evolution in arid regions, many deserts are situated in tectonically inactive regions and therefore consist of rather monotonous plains. Flat upland surfaces may likewise have arid or semiarid climates, and they are often bordered by steep-faced escarpments. Mountainous deserts like those in parts of the southwestern United States contain the greatest diversity of features. They are characterized by rocky and rugged mountain masses that tend to remain steep as they are worn back. The mountain masses are typically surrounded by gently sloping, planed pediment surfaces, sometimes partially overlain by alluvial fan deposits. Pediments typically merge downhill with sediment-filled basins of interior drainage called playas.

Review Questions

1. How does the wind erode and transport earth materials that are much denser than air? How does the eolian transport of dust differ from that of sand?

2. In what environmental settings are eolian processes most important? What basic similarities of all these settings enable the wind to be an effective geomorphic agent?

3. Are most desert regions covered with sand? Why or why not? Where are some of the most extensive ergs located?

4. What is loess? Why are loess deposits not normally found in the same places as sand deposits? Why does loess weather into better agricultural soils than sand?

5. Describe the basic differences between the surface appearance of a desert landscape and a humid climate landscape.

6. What is interior drainage? Why is it prevalent only in arid regions?

7. What is a pediment, and what function does it serve in the erosion of a mountainous desert? How does a bolson differ from a pediment?

8. What human activities are most responsible for the process of desertification?

Key Terms

Eolian	Longitudinal dune
Surface creep	Transverse dune
Saltation	Desert
Sandstorm	Exotic river
Abrasion	Interior drainage
Deflation	Back-wasting
Deflation hollows	Pediment
Desert pavement	Bolson
Loess	Playa
Erg	Alluvial fan
Dune	Desertification
Barchan	

CASE STUDY

The Expanding Deserts

The deserts, which already occupy approximately a fourth of the earth's land surface, have in recent decades been increasing in areal extent at an alarming pace. The expansion of desertlike conditions into previously nondesert areas is called **desertification**.

The spread of deserts into nondesert lands has caused a great deal of human suffering, including economic hardship, malnutrition, starvation, and mass migration. The desertification process has been occurring for a long time, but was recently brought forcibly to public attention by the Sahelian drought of 1968–73. (The Sahel is the name given to the vast semiarid region of Africa lying just south of the Sahara.) This drought resulted in the deaths of up to 250,000 people, the loss of millions of livestock, and the southward displacement of hundreds of thousands of people and livestock. More recently, additional major droughts in Ethiopia and its neighboring countries in East Africa have caused hundreds of thousands to die of starvation.

Most regions undergoing desertification lie on the fringes of existing deserts. Included are the areas north and south of the Sahara, the "horn" of East Africa, regions northwest and southeast of the Kalahari, and large portions of the Middle East and southwestern Asia, as well as smaller sections of Australia, eastern Brazil, southern Argentina, the central Andes, northern Mexico, and the western United States (Figure 16.18).

The advance of the desert does not take place in a smooth, continuous manner, but instead occurs in a highly irregular, piecemeal fashion. It may be halted during a period of wet years, only to progress rapidly during a succeeding dry period. Often the less dry

continued on next page

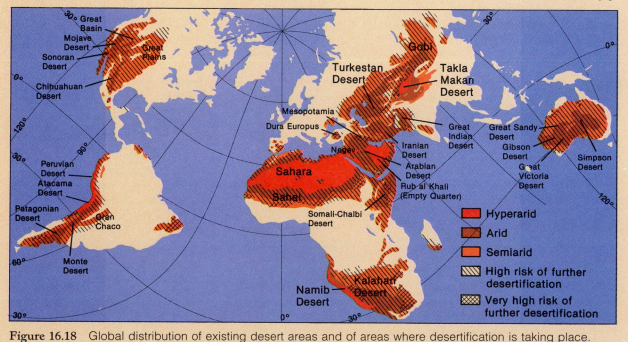

Figure 16.18 Global distribution of existing desert areas and of areas where desertification is taking place.

areas are desertifying the fastest, because they are used more intensively than are drier, less productive areas. Each year some 27,000 square miles (70,000 sq km) of land are turned into desert. This is an area the size of Massachusetts, New Hampshire, and Vermont combined! It has been estimated that the total area threatened by this process covers 14,500,000 square miles (37,600,000 sq km). This is one-fourth of the earth's land surface or an area four times the size of the United States.

Desertification is accomplished primarily through the loss of stabilizing natural vegetation and subsequent accelerated soil erosion by wind and water. In some cases, the loose soil is blown completely away, leaving a stony surface. In other cases, the finer particles may be removed, while the sand-sized particles accumulate to form mobile dune systems.

Even in areas that retain a soil cover, the reduction of vegetation typically results in the loss of the soil's ability to absorb substantial quantities of water. The impact of raindrops on the loose soil tends to transfer fine clay particles into the soil pore spaces, sealing them and producing a largely impermeable surface. Water absorption is greatly reduced, and, as a consequence, runoff is increased, resulting in accelerated erosion rates. The gradual desiccation of the soil caused by its diminished ability to absorb water results in the further loss of vegetation, so that a cycle of progressive surface deterioration is established.

Causes of Desertification

In some regions, desertification is occurring largely because of a trend toward drier climatic conditions. The continued gradual global warming trend following the end of the last Pleistocene glacial stage may have much to do with this process. As the earth warms, the subtropical highs slowly expand in size and shift poleward. This has produced a well documented increase in aridity over the past few thousand years for areas such as North Africa and the southwestern United States. The process may well be accelerated in subsequent decades if an intensified greenhouse effect resulting from human-produced air pollution occurs.

There is little doubt, however, that desertification in most areas results primarily from human activities. The semiarid lands bordering the deserts exist in a delicate ecological balance and have a limited ability to adjust to increased environmental pressures. Most of these areas are located in underdeveloped countries presently experiencing rapid population increases of between 2.5 and 3.5 percent each year. The expanding populations are increasingly pressuring the land to provide food and fuel. In wet periods, the land may be able to meet these demands. During the short-term droughts that are common and expectable phenomena along the desert margins, though, the pressure on the land often far exceeds its diminished supporting capacity, and desertification results.

Four specific activities seem to be the chief contributors to the desertification process. They are overcultivation, overgrazing, firewood gathering, and over-irrigation. The cultivation of crops has expanded into progressively drier regions as population densities have grown. These regions are especially drought-prone, so crop failures are common. Since the raising of most crops necessitates the prior removal of the natural vegetation, crop failures leave extensive tracts of land devoid of a plant cover and susceptible to wind and water erosion.

The raising of livestock is a major economic activity in semiarid lands, where grasses are generally the dominant type of natural vegetation. This is especially true in parts of Africa, where the number of cattle a family owns is a measure of its wealth and status. Livestock numbers have been increasing at about the same rate as the human population, and in many regions they now far exceed the carrying capacity of the land (see Figure 16.19). The consequences of an excessive number of livestock are the reduction of the vegetation cover and the trampling and pulverization of the soil. This is usually followed by the drying of the soil and accelerated erosion.

Firewood is the chief fuel for cooking and heating in many less-developed countries. Expanding populations have led to the removal of woody plants, so many cities and towns, especially in the Sahel region of Africa, are surrounded by large areas completely denuded of trees and shrubs. Journeys to obtain firewood in these areas may take two or three days.

The final major human cause of desertification, soil salinization resulting from over-irrigation, was the subject of a Focus box in Chapter 10. Soil salinization has made especially significant contributions to desertification in the Middle East, Southwest Asia, and North Africa, and is a growing problem in the western United States.

continued on next page

Figure 16.19 Desertification in progress in eastern Africa. The barren land on the right has been overgrazed by cattle. The well-vegetated land in the left distance has been left ungrazed. (© *Breck Kent, JLM Visuals*)

Is Desertification Irreversible?

The extreme seriousness of the desertification problem results from the vast areas of land and the tremendous numbers of people affected as well as from the great difficulty of reversing or even slowing the process. Once the soil has been removed by erosion, only the passage of centuries or millennia will enable new soil to form. In areas where considerable soil still remains, though, a rigorously enforced program of land protection and cover crop planting may make it possible to reverse the present surface deterioration. If the process of desertification is to be halted, the basic rule must be that the intensity of marginal land use must not exceed the land's carrying capacity in the *poorest* years.

Chapter Seventeen

Coastal Processes and Landforms

Outline

Focus Questions

1. What geomorphic processes are chiefly responsible for the production of coastal features?
2. What are the most important types of erosional and depositional coastal landforms?
3. What are the major causes of coastal retreat and advance, and what categories of coastlines have resulted from these coastal changes?

ater is the most geomorphically effective of the gradational tools, and this chapter examines another important category of surface features in which water plays a dominating role: coastal landforms. Coastal landforms are actually polygenetic in origin, since terrestrial, marine, and atmospheric processes all contribute to their development. Because some prior knowledge of these processes is necessary for an understanding of coasts, this subject has been placed in the final chapter of the book.

Coastal processes are often powerful and can perform relatively rapid geomorphic work, but they are highly restricted in areal extent at any given time. Unlike fluvial, glacial, and eolian processes, they affect the land only along its margins. The coastlines of the world, where coastal processes perform all their work, are several hundred thousand miles long, but are generally quite narrow. The global zone of coastal features is somewhat enlarged if the shorelines of large lakes, such as the Great Lakes, are included. Lakeshores are subject to many of the same geomorphic processes as oceanic coasts, but the processes operate on a more limited scale.

The coastal zone, despite its restricted areal extent, is especially critical for human beings. It contains some of the world's highest population densities and provides sites for some of the most intensive human activities. Many of the world's largest cities, including Tokyo, New York, Shanghai, Rio de Janeiro, Los Angeles, and Leningrad, have been built on coasts. The coastal zone is also one of the most dynamic and environmentally sensitive of all natural habitats. As such, major coastal alterations may occur as a result of human actions. Although a thorough discussion of humanity's influence on coastal processes and features is beyond the scope of this text, some aspects of this topic are examined briefly in the Case Study at the end of the chapter.

COASTAL PROCESSES

Coastal processes, like the gradational processes already discussed, can locally erode, transport, or deposit materials, depending upon the existing balance between available energy and the sediment supply. Where a large amount of energy is available and the sediment supply is restricted, erosion is dominant; where sediment is abundant and the available energy is restricted, deposition predominates.

Most coastal sediments undergo multiple episodes of erosion, transportation, and deposition in response to

the shifting balance of forces that influence the coastal environment. In the absence of tectonic uplift or a decline in sea level, though, coastal processes ultimately tend toward erosion because of the inexorable downward force of gravity. A continual net seaward transport of sediments takes place globally, and the deep seafloor is the ultimate repository for sediments derived from the land. Terrestrial sediments must, of course, be transported out to sea through the coastal zone, and coastal processes tend to form a continuous gentle "graded" seaward transportation slope to facilitate this process. As a consequence, steep coasts tend to be eroded back, while low-lying coasts are likely to experience net long-term deposition.

Waves, currents, and tides, the origins and characteristics of which were discussed in Chapter 8, most actively shape coastlines. Of these three oceanic motions, waves generate the fastest water motions and are by far the most powerful and effective geomorphic agent. Their strength results largely from the fact that all the energy they contain is released suddenly as they break, often in contact with the coastline (see Figure 17.1).

Waves and currents are set in motion by the wind. Through its influence on the ocean water, the wind, whose relative weakness as a direct geomorphic agent was discussed in the last chapter, attains considerable additional importance as a shaper of landforms.

The combined motions of waves, currents, and tides are responsible for a variety of coastal gradational

Figure 17.1 Large wave breaking on the California coast near San Francisco. (*Barbara J. O'Donnell/ Biological Photo Service*)

processes. Because they are produced by water movement, these processes are very similar to those performed by streams. Most important is *abrasion,* caused by the impact of materials carried in the water against those on the shore. Sand is the most abundant abrasive material, but breaking waves can sometimes lift even heavy rocks from the seafloor, and large storm waves have been known to hurl rocks for considerable distances. A related erosional process is *attrition*—the wearing down of loose rock materials by impact or rubbing action. Attrition is especially effective shoreward of the wave-break zone because of the ceaseless alternation between the forward motion of wave *swash* and the return flow of wave *backwash.*

Solution is locally important on coasts containing soluble rock types such as limestone, although it affects all rock types to some extent. Finally, *hydraulic action,* or the direct impact of waves on the coast, can be a highly effective erosional mechanism. Pressures generated by the impact of large waves can be enormous, reaching several tons per square foot. Water and/or compressed air can be forced repeatedly into rock fractures, eventually splitting the rocks apart. The greatest effect of hydraulic action, though, is upon unconsolidated shoreline materials. These materials may be removed rapidly when large waves strike the coast.

One of the most obvious characteristics of a sandy beach is the surface smoothness in the zone of wave action. As waves alternately push fine-grained materials up the beach and pull them back, they produce a smooth, gently inclined surface. If waves approach the shoreline at an angle, a down-beach movement of sand grains takes place. This movement, termed **beach drifting,** occurs because the wave swash carries the sand grains up the beach at the oblique angle at which the waves intersect the shore, while the backwash carries them directly downhill in response to gravity (see Figure 17.2). The resulting zigzag movement transports the particles laterally along the beach, and in time may move large quantities of materials for substantial distances.

Although waves normally approach the coastline from a particular direction at any given time, coastal irregularities bend or refract the wave fronts as they approach the shore. **Wave refraction** occurs because shallow water, with a depth of less than half the wavelength (the distance between consecutive waves), exerts a frictional drag on the waves that reduces their rate of landward propagation. If a portion of an incoming wave enters shallow water and is slowed, the axis of the wave is rotated so that it is directed more toward the shallow

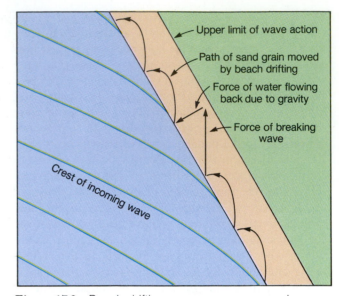

Figure 17.2 Beach drifting occurs as wave swash carries sand grains obliquely up the beach. The return backwash then carries them back toward the ocean directly down the slope of the beach, resulting in a net lateral transport.

water (see Figure 17.3). Along an irregular coastline, this process concentrates wave energy on seaward-extending promontories or *headlands,* while reducing wave energy in coastal indentations such as bays or inlets. The net geomorphic effect of this process is to straighten the coastline by encouraging the erosion of headlands and the deposition of sediment in indentations.

The motion of waves is always toward the coast, and this generates a net shoreward flow of water near the ocean surface. This flow is counterbalanced by a seaward movement of subsurface water, called **undertow.** Undertow serves as the major transporter of fine sediments into deep water.

The effects of eolian and fluvial processes on the coast are also important. The wind forms dunes on many sandy beaches and may erode or redistribute fine-textured coastal sediments that are dry and not stabilized by vegetation. Streams erode valleys that reach the sea to form coastal indentations; if sea levels rise to flood the lower portions of these valleys, extensive estuaries can be produced. Streams that enter the ocean are vital to coastal processes because they supply most of the sediments that are distributed along the coast by waves and currents to form beach deposits.

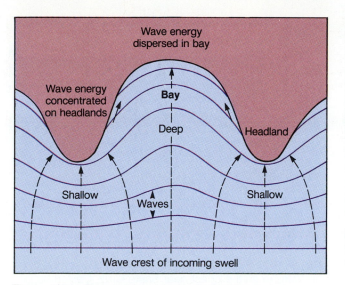

Wave energy
dispersed in bay

Wave energy
concentrated
on headlands

Bay

Deep

Headland

Shallow

Waves

Shallow

Wave crest of incoming swell

Figure 17.3 The process of wave refraction acts to straighten shorelines by concentrating wave erosional energy on headlands. The dashed lines indicate the direction of movement of the wave fronts.

Low-lying coasts are frequently major sites of fluvial deposition and delta formation.

The rates at which coastlines are modified by erosional and depositional processes vary greatly. Modification rates depend primarily on wave size, the strength of the coastal materials, the angle of seaward slope of the coastal surface (which determines how far offshore the waves will break), and the shape of the coastline (which may locally focus or diminish wave energy by refraction).

Just as rivers in flood produce the most rapid fluvial landform alterations, large storm waves produce the most rapid rates of coastal modification. Depending upon local conditions, inland floods can cause either rapid erosion or deposition; similarly, coastal storm waves can either transport sediment onshore, resulting in deposition, or carry it out to sea, resulting in erosion. At any given site, the erosional or depositional energy of the sea is concentrated in the zone of wave impact. Over a 12- or 24-hour period, this zone shifts between high and low tide levels. Areas with large tidal ranges therefore have coastal processes spread over larger vertical and generally larger horizontal distances than do areas with small tidal ranges.

If an erosional tendency exists, erosion will occur most rapidly in unconsolidated coastal materials, such as glacial or alluvial deposits, or beach sand. Individual severe coastal storms, particularly hurricanes and intense winter frontal cyclones, may produce major erosion and even change the configuration of a coastline by eroding and redistributing sediments. Hurricanes crossing North Carolina's Outer Banks, for example, have altered the pattern of tidal channels approximately 30 times in the last 400 years by cutting new channels and filling others. Erosion on a long-term basis can also be relatively rapid along steep coastlines composed of weak sedimentary rock such as chalk or some sandstones. The famed Cliffs of Dover on the southeastern coast of England are being eroded back at rates of up to one mile (1.6 km) per thousand years. Conversely, coasts composed of massive resistant rock such as granite can withstand even intense wave action for long periods with little erosion.

COASTAL LANDFORMS

In this section, features produced directly by coastal marine processes are examined. It should be noted, though, that many coastal features were originally produced by tectonic activity or by the actions of streams, ice, wind, and gravity. These features, if currently situated on the coast, will have been modified to a greater or lesser extent by coastal processes. A common method of categorizing coastal features for purposes of discussion, and the one employed in this section, is by determining whether they are predominantly erosional or depositional in origin.

Erosional Features

Coastal erosion occurs when the near-shore waters have the capacity to transport a greater quantity of sediment than is locally available. If coastal erosion has long been active in an area of consolidated materials, the coast is generally steep and rugged (see Figure 17.4). The existence of such a coast indicates that terrestrial gradational processes such as streams, which erode the surface from the top, have been unable to keep pace with the undercutting action of marine erosion. A steep coast allows large waves to break at or near the shore, concentrating their energy on the land rather than dissipating it farther out to sea.

The most characteristic landform developed along an erosional coast is a **sea cliff** (see Figure 17.5). Sea cliffs are produced by wave undercutting in areas of

Figure 17.4 A cliffed coast with a pocket beach in Oregon. (*David Butler*)

Figure 17.6 Wave erosion has produced a sea arch and stacks on the coast near Santa Cruz, California. (*John S. Shelton*)

strong rock. The undermining of sea cliffs by a combination of wave hydraulic action, abrasion, and solution causes periodic slope failure as a portion of the cliff falls or slides into the ocean, where it is consumed by the waves. Along the California coast, where sea cliffs are well developed, most failures occur as rotational slumps (see Figure 13.7). In other places, they may take the form of rockslides or rockfalls. These mass movements temporarily stabilize the sea cliff, but the process of undermining immediately begins anew.

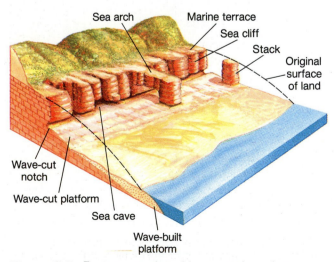

Figure 17.5 Features produced by coastal erosion.

Cliffed coasts are frequently highly irregular in shape. This sometimes reflects differences in the erosional resistance of the rocks comprising the cliff. In other cases it results from differences in the elevation of the land along whose margin the sea cliff has developed. For instance, a ridge cut into by a sea cliff would contain much more rock for the sea to erode than would a valley. For this reason, the ridge would tend to be eroded back more slowly and would form a coastal promontory or *headland,* while the valley would be the site of a coastal indentation. Such indentations, depending upon their sizes and shapes, may be referred to as bays, coves, or inlets.

The process of wave refraction concentrates wave attack on headlands; consequently, the cliffs at their margins are likely to be especially bold and steep. A narrow headland attacked by the sea from opposite sides may eventually be cut through to form a *sea arch.* The collapse of the arch will produce a steep-sided island remnant called a *stack* (see Figure 17.6). These features are common along portions of the west coast of the United States.

Conversely, the coastal indentations between headlands are subject to much less vigorous wave attack. The quieter environment encourages the local deposition of sediments, often derived from the headlands, so that *pocket beaches* may develop even along a coastline that is, as a whole, being worn back by erosion (see Figure 17.4).

Figure 17.7 A wave-cut platform exposed at low tide along the California coast. (*John S. Shelton*)

Coastal erosion on solid rock surfaces is effective only down to the lowest level of wave attack at low tide. As a sea cliff retreats, then, it leaves behind a planed-off rock platform at a level representing the base of erosionally effective wave action. This surface is termed a **wave-cut platform.** A wave-cut platform is essentially a transportation surface between an erosional site—the sea cliff—and a depositional site—the deeper water farther offshore. The platform may be bare, or it may be covered by a thin veneer of rock fragments derived from the sea cliff and in the process of being transported seaward. In form and function, then, a wave-cut platform is analogous to the pediment surface at the foot of a desert mountain range (see Figure 17.7).

Sediment deposition frequently occurs at the normally steeper outer edge of the wave-cut platform. In time, depending upon the steepness of the coast and the amount of available wave energy, these sediments may accumulate to form a **wave-built platform** consisting of mud, sand, gravel, or cobbles. The wave-built platform therefore extends the gradual slope of the continental margin farther seaward.

Depositional Features

Deposition occurs when the coastal waters are supplied with a greater quantity of sediment than they can transport and remove from the coastal zone. These materials are derived from various sources. Some are picked up well offshore and carried to the coastline by large storm waves. Others are eroded from sea cliffs or other coastal features. In most areas, though, a majority of sediments are transported to the coast by streams from inland locations. Waves and currents then sort and spread these sediments along the coastline.

Conditions favor deposition where sediment supplies are plentiful and both onshore and offshore slopes are gentle. Gentle slopes force large, erosionally effective waves to break and expend their energy well away from the shoreline.

The most characteristic feature of a depositional coast, and undoubtedly the best-known coastal landform of all, is the beach. Beaches annually attract millions of tourists to coastal resorts within the United States alone and add billions of dollars to coastal economies (see Figure 17.8). A **beach,** by definition, is a gently sloping shore that is washed by waves or tides; it is normally covered by loose deposits of sand or gravel. Beach sediments often accumulate on a wave-cut platform when conditions favor deposition. The depth of accumulation is highly variable, but is usually quite shallow.

The most familiar beaches are composed of whitish or buff-colored sand-sized quartz grains, but beach deposits can vary considerably in both composition and size. Some portions of the Hawaiian Islands, for example, have beaches of black lava sand. In some areas, shell

Figure 17.8 Sandy beaches annually draw many millions of tourists who spend billions of dollars at North American beach resorts. (*Stefan Schluter/The Image Bank*)

fragments are an important component of beach sand, and on many tropical islands, especially in the Pacific, beaches are formed from crushed coral. Where coastal slopes are fairly steep and wave action is too vigorous for sand-sized deposits to be retained, beaches are typically covered by gravel or cobbles. In some locations, the type of beach materials may change seasonally because of seasonal variations in climatic conditions. Some California beaches, for example, are covered with sand during the tranquil summer months, but winter storms transport the sand seaward, exposing an underlying cobble beach. With the return of summer, the sand is once again carried onshore, where it re-forms the higher summer beach.

The California example illustrates the point that even where beach deposits cover the coast for long periods, these deposits are highly vulnerable to erosion and may be removed quickly under extreme conditions. Beach materials are transient because they are normally fine-textured, are located in a high-energy environment subject to changeable conditions, and are usually not stabilized by vegetation. Material continuously arrives and departs and the beach will remain relatively unchanged only if the quantity of arriving and departing sediments offset one another.

Because beaches are highly sensitive to coastal processes, they normally develop discontinuously along an irregular coastline. Smooth, regular coastlines, conversely, are more likely to be continuously fronted by beaches. The Pacific coast of the United States from California to Washington is predominantly high, rugged, and irregular. As a result, beaches have developed only in more protected sites, especially within bays or coves. Conversely, the Atlantic and Gulf Coast from Long Island to Texas is predominantly low-lying and gently sloping, and beaches have developed along virtually its entire length.

The gentle seaward slope and large sediment supply of a coast along which deposition is occurring allow waves, especially large storm waves, to redistribute the bottom sediments to a considerable distance offshore. Just seaward of the *shoreline,* where land and water meet, there typically exists a trough of relatively deeper water excavated by moderate-sized waves breaking near the shore (see Figure 17.9). Farther seaward, water depths often become shallower again as the bottom rises to form a **sand bar** paralleling the coastline. A sand bar is a largely submerged ridge, usually composed of sand, formed in the waters near shore by currents and larger waves. The top of the bar may be exposed at the surface

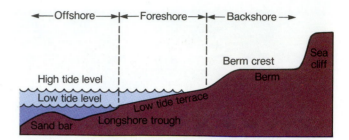

Figure 17.9 Cross-sectional profile of a typical beach.

at low tide, but it is submerged at high tide. While the beach contains an accumulation of sediments thrown forward by normal waves, the sand bar largely represents an accumulation of sediments thrown forward by storm waves. Bars also serve as temporary repositories of sediments eroded from the beach during periods of larger waves, such as those occurring in winter in portions of the middle latitudes.

Continued deposition by storm waves, coupled with a reduction in sea level, may convert bars to **barrier islands** that remain continuously above the ocean surface even at times of high tide (see Figures 17.10 and 17.13). Like the bars from which they are formed, barrier islands are long and narrow and parallel the coastline. They often form in chains along low-lying sandy coasts and are separated from one another by **tidal inlets** through which the sea flows. These inlets enable water levels seaward and landward of the barrier island chain to equalize during storms and as the tides change. The water body between the barrier island chain and the mainland, referred to as a **lagoon,** is generally rather tranquil because it is protected from storms and large waves. The protected waters of coastal marshes and lagoons generally receive a continuous flow of fine sediments and organic materials from rivers entering them from the landward side. As a result, they typically contain a great quantity and variety of aquatic plants and animals and comprise one of the world's most naturally productive and sensitive ecosystems.

Barrier islands vary greatly in length, width, and distance from the coast. Some are situated only a mile (1.6 km) or less offshore, but others are 20 miles (32 km) or more from the mainland, as along the North Carolina coast (see Figure 17.11). Smaller barrier islands are only a few hundred feet wide, but larger ones may exceed a mile in width. These ribbons of land may extend for many miles. Padre Island, off the coast of Texas, is one

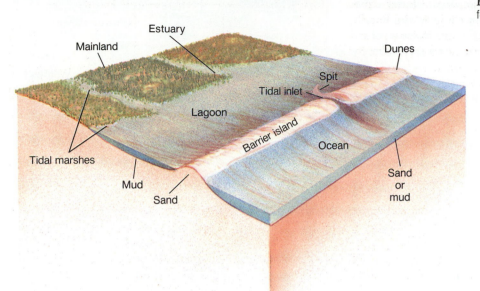

Figure 17.10 Depositional coast features.

of the world's longest, with a length of about 100 miles (160 km). Many barrier islands, as discussed in the Case Study, have been extensively developed as coastal resorts and contain sizable human populations.

Like beaches, bars and barrier islands are highly unstable features and are subject to rapid changes in size, shape, and position. Bars and barrier islands tend to grow in the direction taken by the longshore drift of sediments. If the coastline is irregular, with headlands and bays, they are generally attached at one end to a headland. In this case, the strip of land is a peninsula, rather than an island, and is more appropriately termed a *barrier beach*. Along less rugged, straighter coasts, they are usually entirely separated from the land. A barrier beach attached to the land at one end may extend partially across a bay to form a **spit** ending in open water. Cape Cod, Massachusetts, is an example of a giant coastal spit (see Figure 17.12).

The protected lagoons on the landward sides of barrier beaches are generally tranquil enough that fine silt and clay particles can settle to the bottom. Plants such as marsh grasses that grow in the shallow water near the shore aid in trapping the sediment. This results in a tendency for lagoons to fill gradually with organic matter and fine-textured river sediments. The continued infilling of lagoons by sediments and organic material may eventually convert them to *mudflats,* which are typically submerged only at times of high tide. Coastal mudflats may be diked, flushed of salts, and drained for settlement and agriculture; this has occurred on a large scale on the "polderlands" of the Netherlands. When mudflats are colonized by marsh grasses, they become *salt marshes*. A decline in sea level will convert such marshes to meadows and eventually to dry land.

TYPES OF COASTS

A number of coastal classification systems have been devised, but no one system has proven completely satisfactory. The basic problem is the great variety of processes that affect coasts and the resulting complexity of forms exhibited by the world's coastlines. One very basic division of coasts, which is used as a starting point in some classification systems, is based on whether coasts are retreating (losing land) or advancing (gaining land). This system, on a very general level, is employed in this section.

Retreating Coasts

Coastal retreat caused by the erosion of the land and the deposition of the eroded materials on the seafloor is a slow process. However, another basic process of coastal retreat can operate much more rapidly in the short term. This is submergence caused by a rise in sea level. Submergence is currently causing a relatively rapid retreat of most coastlines because of the post-Pleistocene rise in sea level. The sea level rise results largely from the isostatic rebound of previously glaciated regions in the high latitudes, which displaces ocean water equatorward and increases its depth in the low and middle latitudes. Other related causes of the rise in sea level are the continuing melting of the glaciers and the thermally induced expansion of the ocean water.

In recent decades, coastal submergence rates have been accelerating. Along the east coast of the United States, for instance, the sea level is currently rising at a rate of up to 1 foot (0.3 m) or more per century. Depending upon the steepness of the coast, this generally translates to a rate of horizontal shoreline retreat of 100 to 1000 feet (30 to 300 m) per century. The effects, of course, are most evident along low-lying coasts. It has been speculated that the recent acceleration in the rate

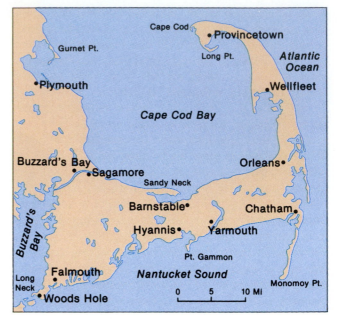

Figure 17.12 Cape Cod, Massachusetts, is a classic example of a large coastal spit. It contains a number of smaller spits, including Long Point and Monomoy Point.

of sea level rise indicates that the widely predicted global warming trend resulting from the human-produced increase in atmospheric carbon dioxide levels has begun (see Chapter 3).

Erosional Coasts

Erosional coasts, as noted earlier, are generally steep, irregular, and rugged. They often contain bold cliffed headlands and intervening indentations, sometimes occupied by pocket beaches. The sea cliffs frequently exhibit a variety of secondary erosional features such as sea caves, arches, and stacks, and they are commonly fronted by wave-cut and wave-built platforms. Beyond the outer platform margins, water depths increase rapidly, so the coastal zone, both seaward and landward of the shoreline, is narrow.

Submergent Coasts

Submergence brings the sea into contact with a land surface that has been shaped by terrestrial processes such as rivers or glaciers. Unless this inherited surface is exceptionally flat, the result is usually a highly irregular coastline.

Figure 17.11 The barrier island chain comprising North Carolina's Outer Banks is situated unusually far from the mainland.

Figure 17.13 A barrier island fronted by sandy beaches and backed by mudflats and a lagoon near Cape Lookout, North Carolina. (*P. Godfrey*)

Coastline development on a surface that has been dominated by fluvial processes produces what is frequently called a **ria coast** (from the Spanish *rio* "river"). Submergence results in the drowning of the lower portions of the river valleys to produce an extremely irregular coastline dominated by numerous bays or estuaries. The higher interfluves, conversely, stand out as broad headlands separating the coastal indentations. Where drowning is more extensive, the remnants of headlands may exist only as strings of aligned islands. With time, erosion of the headlands and deposition within bays straightens the coasts. This process has not proceeded far since the Pleistocene, though, because submergence has been taking place for only about 15,000 years and is an ongoing process.

The east coast of the United States from New York southward to the Carolinas is a prime example of a ria coast. Numerous large embayments, including Delaware Bay, Chesapeake Bay, the Potomac and James River estuaries, and Albemarle and Pamlico Sounds were produced by the drowning of the lower portions of large river valleys (see Figure 17.14).

In high latitude regions where Pleistocene valley glaciers once reached the sea, coastal submergence has resulted in the development of *fjord coasts*. Their characteristics and global distributions are discussed in Chapter 15. Like ria coasts, fjord coasts contain deep indentations. The fjords, however, tend to be much deeper, steeper-walled, and more consistent in width than the drowned river valleys of ria coasts. Both ria

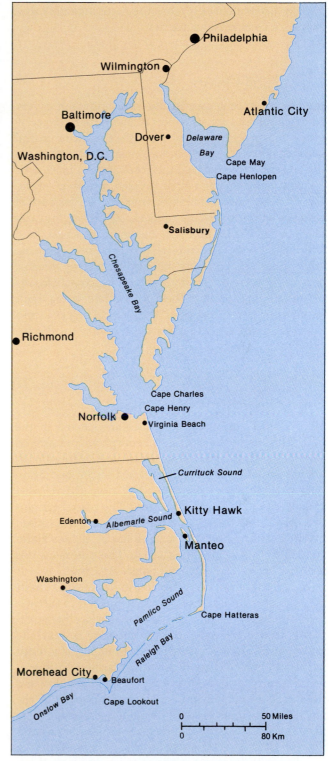

Figure 17.14 The U.S. Mid-Atlantic Coast is a ria coast. Its numerous bays and sounds were produced by the flooding of the lower portions of river valleys by the sea.

and fjord coasts contain protected natural harbors well suited for commercial activities.

Advancing Coasts

Because of recent worldwide rises in ocean levels, advancing coasts are much less common than retreating coasts. They occur only where surface uplift or coastal addition operates quickly enough to more than offset the loss of land from erosional processes and sea level rises.

Advancing coasts can be divided into two categories. First are coasts emerging from the sea because of tectonic uplift or isostatic rebound. A decline in water levels would also produce an emergent coast, and although this is occurring along the shores of many lakes, it is not currently taking place along oceanic coasts. The second category of advancing coasts consists of outbuilding coasts. These coasts are being built outward by either geological or biological processes at a rate sufficiently rapid to allow the coastline to advance seaward.

Emergent Coasts

Most actively emerging coastlines are located in the high latitudes of the northern hemisphere and around the rim of the Pacific. The high latitude locations most notably include the Canadian Shield region and Scandinavia, two areas still rebounding from the released weight of the Pleistocene ice sheets. The coastal margins of the Pacific are rising because of tectonic uplift caused by the overriding of the Pacific Plate by lighter plates containing continental crustal rock. These areas include the west coasts of North and South America.

Emergent coastlines associated with tectonic uplift are relatively straight and regular compared to submergent coastlines because the seafloor surface being exposed in a tectonically uplifted coastline is generally quite smooth. In contrast, surfaces emerging as a result of isostatic rebound are highly irregular because glacial features such as fjords (as in Norway and Labrador) and drumlins (as in Finland) are being exposed. If the emerging coastline is steeply sloping, as is most commonly the case, no offshore bars or barrier islands develop. If it is gently sloping, and especially if it is supplied with large quantities of sediments from rivers or other sources, bars and barrier islands do tend to develop.

Figure 17.15 Multiple marine terraces resulting from sporadic tectonic uplift are evident on the west side of San Clemente Island, California. (*John S. Shelton*)

In general, the most characteristic features of emergent coasts are **marine terraces.** These elevated flat surfaces are the remnants of old wave-cut platforms, wave-built platforms, or beaches that are now located inland well above sea level. Coastal emergence seems in most cases to be sporadic rather than steady, and each terrace represents the position of the shoreline during a previous period of coastal stability. Along portions of the southern California coast, multiple marine terraces extending to elevations as great as 1400 feet (425 m) give the landscape a steplike appearance (see Figure 17.15). Elevated terraces are also common along the shorelines of lakes that formerly contained more water. Because of isostatic rebound, well-developed lake terraces are found adjacent to the northern shores of the Great Lakes. Numerous lake terraces also exist within the western United States, particularly in the Great Basin, because the climate of this region has become warmer and drier since the end of the Pleistocene, and numerous former lakes have either disappeared or diminished greatly in size (see Figure 15.22).

Outbuilding Coasts

Outbuilding coasts are advancing because new material is being deposited on the coastline at a rate rapid enough to counteract processes of erosion and submergence. The types of materials and the modes of deposition vary, forming several distinctive types of outbuilding coasts.

DEPOSITIONAL COASTS   Depositional coasts are experiencing the active deposition of loose sediments. The sediments are transported to the coast largely by rivers and are commonly redistributed by waves and currents. The chief factors favoring the development of this type of coast are an abundant sediment supply and a gently sloping coastal margin. Where wave and current action is sufficiently vigorous to spread these materials along the coastline, depositional coasts are characterized by beaches and by offshore bars and barrier islands.

VOLCANIC COASTS   Volcanic coasts exist where volcanic eruptions have raised new land above the sea. This may occur along the margins of already existing landmasses, such as those around the rim of the Pacific, or it may result in the formation of new volcanic islands. Volcanic coasts are usually very steep, but are typically fairly regular in form. A recent volcanic island is likely to be nearly circular in outline, and a larger volcanic landmass along which several volcanoes have erupted typically has a coastal margin consisting of multiple arcs. After the volcanoes become dormant or extinct, their steep flanks are rapidly eroded and made more irregular by wave action.

CORAL COASTS   Coral coasts are composed of limestone deposited by coral polyps and other reef-building marine organisms. The coastline itself is fronted by a **reef,** which is a ridgelike accumulation of the exoskeletons of large colonies of these small animals. Reefs grow upward through time, since the currently living organisms construct their limestone structures only on the upper and outer portions of the reef. Reef-building organisms require clear, shallow, and warm water in order to survive. As a result, coral coasts occur in the tropics between approximately 30°N and 25°S. In addition, they favor islands rather than mainland coasts containing rivers that discharge large quantities of sediments.

Depending upon their relationship to land, coral reefs can be divided into three types. *Fringing reefs* are attached to the coast, indicating that local land levels have been rising relative to the sea. *Barrier reefs* are separated from the coast by a lagoon that is too deep for coral growth (see Figure 17.16). In form, they are very similar to barrier beaches, and, like barrier beaches, they protect the coast from storms. The largest and best-known barrier reef is the Great Barrier Reef off the east

Figure 17.16   A portion of the Great Barrier Reef off the east coast of Australia. (*Rodman Snead, © JLM Visuals*)

coast of Australia. Most volcanic islands, especially in the western Pacific, are surrounded by barrier reefs. **Atolls,** the final type of reef, differ from the other two in that they are not associated with other landmasses. They typically consist of circular reef growths surrounding a central lagoon. Small low-lying islands of sand derived from crushed coral may surmount high points on the reef to give the complex a "pearl necklace" appearance from the air. It is believed that most or all atolls originated as fringing or barrier reefs surrounding volcanic islands. They later evolved into atolls as the central islands eroded and subsided beneath the ocean while continued upward growth allowed the reefs to remain at or near the ocean surface. Most atolls are located in the western Pacific.

Compound Coasts

Compound coasts add complexity to the variety of coastal types by displaying characteristics of both submergence and emergence. They are common worldwide because of the numerous Pleistocene and Holocene (post-Pleistocene) sea level fluctuations. A good example of a compound coast is the Mid-Atlantic Coast of the United States, where recent drowning of a generally emergent coastline has been taking place. As a consequence, this coast contains both large estuaries, formed by the submergence of lower river valleys, and a well-developed system of barrier islands.

Summary

The coastline, located where air, land, and sea interface, is complexly influenced by a combination of atmospheric, terrestrial, and marine processes. Despite its limited areal extent, this highly dynamic and sensitive physical environment is crucial to human beings. The most powerful and effective geomorphic agent for most coastlines is wave action. Waves can effectively erode shoreline materials through the processes of abrasion, attrition, solution, and hydraulic action. Waves also deposit sediments derived both from the land and from farther offshore along the coast. Currents transport fine-textured sediments along the coastline.

Coasts can be dominated by either erosional or depositional processes. Coasts dominated by long-term erosion are typically steep and rugged. The undercutting action of waves along an elevated coastline characteristically produces sea cliffs, which gradually retreat as they are undermined and collapse. Wave refraction causes erosion to be most active around headlands, while coastal indentations may be the sites of localized deposition. Headland erosion leaves wave-cut platforms that can serve as future sites for the accumulation of beach sediments.

The dominant coastal depositional feature is the beach. Beaches may front mainland shorelines, but often are best developed on offshore barrier islands that are separated from the mainland by coastal lagoons.

Most coastlines are presently retreating. Coastal retreat can be caused either by erosion or by submergence. Erosion typically produces a rugged coast characterized by slowly retreating headlands. Submergent coastlines are geographically widespread because of the nearly global post-Pleistocene rise in sea levels. The nature of the previously dominant geomorphic process acting on a given submergent coastline largely determines its present surface characteristics. Ria coasts, probably the most distinctive of the submergent coastal types, are formed from the partial drowning of a coast dominated by river valleys. They are characterized by numerous bays and estuaries.

Coastlines advance seaward either through tectonic or isostatic uplift or through outbuilding resulting from the addition of materials by geological or biological processes. Large areas in the high latitudes of the northern hemisphere are advancing because of isostatic rebound. Emergent coastlines around the Pacific are rising because of crustal thickening caused by the subduction of the Pacific Plate. Outbuilding coasts include those subject to rapid fluvial or marine deposition, the eruption of volcanic materials, and the growth of reef-building organisms.

Review Questions

1. What processes do waves and currents employ to perform coastal erosion? Briefly describe each process.
2. How are sea cliffs formed? How is their steepness maintained? What feature is produced as a sea cliff is eroded back?
3. What physical conditions favor coastal erosion? What conditions favor coastal deposition?
4. What is the most characteristic feature of a depositional shoreline? Explain how this feature forms. From what sediment sources are its materials derived?
5. What similarities and differences exist between bars and barrier islands? Where and why do these features form? What effects do they have on the characteristics of the mainland coast?
6. What are the two chief mechanisms by which coastlines retreat? Why are more shorelines around the world currently retreating than advancing?
7. What are the two basic causes of present-day coastal advance? In what geographical areas of the world is each occurring?
8. What are marine terraces, and how are they produced? In what part of the United States are marine terraces well developed, and why?
9. What are compound coasts? Why are they common at present?

Key Terms

Beach drifting	Barrier island
Wave refraction	Tidal inlet
Undertow	Lagoon
Sea cliff	Spit
Wave-cut platform	Ria coast
Wave-built platform	Marine terrace
Beach	Reef
Sand bar	Atoll

CASE STUDY

Barrier Beach Migration and Coastal Development

Increases in affluence and leisure time, coupled with the development of efficient modern transportation networks, have resulted in a great increase in tourism within the United States during the past few decades. The chief focus of the nation's multi-billion dollar tourist industry is the string of beach resorts on the more than 300 barrier islands along the Atlantic and Gulf Coasts. Cities such as Atlantic City, New Jersey; Ocean City, Maryland; Virginia Beach, Virginia; Myrtle Beach, South Carolina; and Daytona Beach, West Palm Beach, Fort Lauderdale, and Miami Beach, Florida, are some of the largest and best known of scores of tourist-oriented cities and towns along the East Coast.

Unfortunately, the barrier beaches upon which these cities and towns have been built are unstable and, geologically speaking, short-lived features. Furthermore, the human alteration of these strips of sand has in many places upset the delicate balance of nature and further destabilized the surface. A combination of ongoing natural processes and human influences has reduced the size of the barrier beaches along the Atlantic and Gulf Coasts and has caused them to migrate slowly landward. This raises serious concerns for the future of the billions of dollars worth of real estate on these beaches and for the livelihoods of their hundreds of thousands of human inhabitants.

Causes of Barrier Beach Migration

The basic cause of barrier beach migration is the nearly global rise in ocean levels. Barrier beaches occupy a fixed position relative to the coast just shoreward of the zone of impact of storm waves. If sea levels rise—as they are now doing all along the U.S. Atlantic and Gulf Coasts—this zone shifts landward, resulting in the shoreward migration of the barrier beaches. Because slopes along barrier beach coasts tend to be quite gentle, a small increase in sea level produces a large horizontal migration of both the shoreline and the barrier beaches. For example, the approximate one foot (0.3 m) per century rise in sea level recently experienced along much of the East Coast has produced a landward migration of about 200 feet (60 m) in the shoreline and the offshore barrier beaches.

Barrier beach migration does not occur as a steady movement; rather, it takes place sporadically during major storms. Large storm waves pound at the seaward sides of the barrier beaches, eroding the sand ridges behind the shoreline. If these ridges are eroded through, or if wave levels become high enough to breach them, waves can wash completely over the barrier beach from one side to the other. This *overwash* process causes sand to be removed rapidly from the seaward side of the barrier beach and deposited on the landward side. Through this process, the material comprising the barrier beach slowly rolls over on itself like a moving tank tread (Figure 17.17). Unfortunately, fixed structures are unable to move with the land, and a building originally located well away from the shoreline behind a protective beach ridge may have the shoreline migrate toward it until it is finally undercut by waves and destroyed.

A number of human activities further destabilize barrier beaches, thus hastening their erosion and landward migration. One is the destruction of dune grasses that hold the sand in place and assist in dune formation. These grasses may be either purposefully removed for "aesthetic" reasons or destroyed by trampling or by the use of off-road vehicles. The sand dunes themselves, which form the backbone of barrier beaches, may be flattened to improve the ocean view for residents of homes situated behind them. Of more widespread significance is the upstream damming of rivers that supply sand to the coast. The reservoirs behind the dams trap sediment, greatly reducing the quantity of sand available for barrier beach construction. This factor is believed to play a major role in the rapid erosional losses occurring on barrier islands and mainland beaches in many areas. If the widely anticipated global warming trend caused by the greenhouse effect becomes a reality, increased glacial melting and thermally induced seawater expansion will produce rises in ocean levels even more rapid than those already experienced.

continued on next page

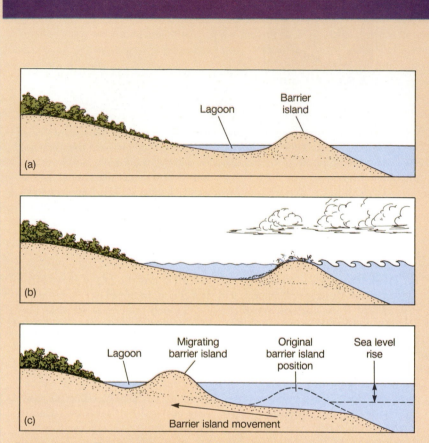

Figure 17.17 The landward migration of barrier island systems, shown here in a time sequence, results largely from the post-Pleistocene rise in sea level. The landward movement actually occurs when storm waves wash sand from the seaward side of the islands and deposit it in the lagoon (center view).

Alternative Solutions

The seriousness and complexity of barrier beach erosion and migration is such that many types of solutions have been tried. One group involves the construction of structures designed either to protect the shoreline from wave attack or to encourage the local deposition of sand. These include *seawalls,* which are rock or wooden barriers built parallel to the coast to absorb and dissipate wave energy and to protect the land behind them. Unfortunately, they do not protect the beach in front of them and often deflect wave energy downward so that they are eventually undermined and destroyed.

Another type of structure is the *groin,* which is a narrow seaward-extending barrier generally built of rock. It is designed to interfere with beach drifting and longshore currents and to encourage the deposition of sand. Series of groins have been constructed in numerous locations along the east coast of the United States. Although they may locally solve the problem of beach erosion, at least temporarily, their interception of sand of-ten causes the accelerated erosion of beaches immediately down the coast. In addition, both seawalls and groins are expensive to build and negatively affect the appearance of the beach and shoreline they are designed to protect.

Another very temporary but aesthetically more pleasing solution is to haul in replacement sand from elsewhere or to pump sand onto the beach from farther offshore. A number of East Coast cities, including Ocean City, in Maryland, and Miami Beach have recently utilized this method of beach restoration. In only a decade or two, and perhaps in much less time if these sites are visited by major storms, this sand will be eroded away. This is especially true where sand pumping has deepened the water just offshore, allowing larger waves to reach the beach.

A different type of solution, favored by various government agencies and many conservation groups, is not to fight the sea, but rather to limit development on barrier beaches and to let nature take its course. In recent years, many barrier beaches along the Atlantic and Gulf Coasts have been set aside as national seashores, national wildlife refuges, and state parks. Coastal changes here are expected and those responsible for these areas adjust to them. This policy is undoubtedly feasible in areas where no major development has occurred, but it is too late for many other areas where enormous investments of money, energy, and materials have already been made.

At present, the future of the developed portions of our barrier beaches is very uncertain. In most cases, development has occurred recently enough that beach migration

continued on next page

has become only an expensive annoyance rather than a major disaster. Technological developments and the expenditure of large sums of money may indefinitely postpone the day of reckoning for some areas, but future coastal storms are certain to cause great destruction in others (see Figure 17.18). In the meantime, increased understanding of coastal processes and greater restraint in coastal development are necessary.

Figure 17.18 An example of a beachfront home destroyed by Hurricane Hugo in Charleston, South Carolina, September 1989.

Appendix A

Metric and Imperial Measurement Equivalencies

Throughout this book, a dual system of numerical values has been used because both our traditional English (or Imperial) system and the metric system are widely employed. Although the metric system is undoubtedly a superior system and is used in most parts of the world, the conversion to metric within the United States is proving to be slower and more difficult than earlier supposed, and it appears that both systems will be in widespread use for some time to come. For convenience, the conversion formulas for those units of measurement especially relevant to the subject of physical geography are provided here.

Length

1 kilometer = 1000 meters = 0.6214 mile = 3281 feet

1 mile = 5280 feet = 1.609 kilometers = 1609.3 meters

1 centimeter = 0.3937 inch

1 inch = 2.540 centimeters

1 micrometer (μm) = 10^{-6} meter = 10^{-4} centimeter = 3.937×10^{-5} inch

Area

1 square kilometer = 0.3861 square mile = 247.1 acres

1 square mile = 2.590 square kilometers = 640 acres

Volume

1 cubic kilometer = 0.2399 cubic mile

1 cubic mile = 4.1684 cubic kilometers

1 cubic meter = 1.308 cubic yards = 61,024 cubic inches

1 cubic yard = 0.7646 cubic meters = 46,656 cubic inches

1 liter = 1.0567 quarts

1 quart = 0.946 liters

Mass

1 metric ton = 1000 kilograms = 2205 pounds

1 short ton = 2000 pounds = 907.2 kilograms

1 kilogram = 1000 grams = 2.2046 pounds

1 pound = 0.4536 kilograms

Speed

1 meter per second = 3.281 feet per second = 3.6 kilometers per hour = 2.237 miles per hour

1 mile per hour = 1.467 feet per second = 1.6093 kilometers per hour = 0.8684 knots

1 kilometer per hour = 0.6214 miles per hour

1 knot = 1.152 miles per hour = 1.853 kilometers per hour

Temperature Scale Conversion Formulas

$°C = \frac{5}{9}(°F - 32°)$

$°F = \frac{9}{5}°C + 32°$

$°K = °C + 273.15°$

Air Pressure

Mean sea level air pressure = 1013.2 millibars = 29.92 inches of mercury = 101.3 kilopascals = 1 atmosphere

Maps and Remote Sensing

This section on maps and remote sensing is placed in the appendix because the topics covered here do not readily fit into the sequence of subjects presented in the main body of the text. Strictly speaking, maps and remote sensing techniques are not in themselves topics in physical geography, because they are human constructs rather than natural phenomena. On the other hand, they are essential tools for the effective geographic analysis of both physical and human phenomena; and a basic understanding of both is of great importance to any geographical study.

MAPS

A map may be defined as a pictorial representation of the geographic locations of selected surface features at a reduced scale. Maps perform two basic functions: (1) they reduce the size of an area of interest so that the entire area can be shown; (2) they simplify spatial distributions by displaying only those phenomena of interest to the map user.

Maps are used for a great variety of purposes by individuals in many different occupations. The subject of maps and map making is one of the few associated more closely with geography than with any other academic field. This close association is a natural one; geographers are interested in the geographic distributions of phenomena, which is precisely what maps are designed to display pictorially. In fact, mapping is the only practical method by which large amounts of data can be presented for many purposes.

Modern maps, such as that shown in Figure B.10, contain a great deal of information. The verbal descrip-

tion of the locations and relationships of all the features shown in this figure would take many pages of text and could never be communicated as effectively as it is on this one map sheet. Nearly all geographical studies employ maps, and maps have rightfully been called the chief research tool of the geographer. By displaying the locational patterns of the phenomena under study, maps often aid in geographic understanding by shedding light on the reasons for the patterns.

Cartography, the art and science of mapmaking, has developed continually over the past several centuries and currently is highly automated and precise. Much of the information contained on modern maps is obtained through the remote sensing techniques discussed later in this appendix section.

Basic Map Information

Maps have been devised to show many different types of spatial information, which can be displayed in a great variety of forms. Before using the map, map users should therefore take the time to acquaint themselves with the types of information contained on the map and the ways it is shown. Among the most basic types of map information are the title, date, location, directional orientation, legend, and scale.

■ *Title.* The map title, which should be prominently displayed, should indicate the area depicted on the map as well as the subject matter of the map. It should be clear and concise, because many people obtain maps, sight unseen, based on their titles.

■ *Date.* The date of the map is important because most types of map information change over time, and the information depicted will likely become outdated. In many cases, the year is sufficient; but in some instances, as with weather maps, information changes so rapidly that the map should be dated to the hour or even to the minute. If the information on the map is significantly older than the date of publication of the map itself, this too should be indicated.

■ *Location.* Many maps indicate locations by the use of numbered parallels and meridians (see Chapter 1), although other locational systems have been devised. A map that shows a small area whose location may not be easily recognizable by its shape or surroundings may contain a small inset map showing the location of the mapped area within its larger geographical setting.

■ *Directional orientation.* The compass orientation of the area shown on a map is commonly indicated by the use of parallels and meridians. Not all maps maintain constant directions everywhere, so these lines, though straight (in a two-dimensional sense) may curve on the map. A north directional arrow is also frequently employed. Most maps are oriented with respect to the four cardinal directions, so that north is toward the top, south toward the bottom, east toward the right, and west toward the left.

■ *Legend.* The legend is the information section of a map. The primary purpose of the legend is to explain the major map properties and symbols. *Symbols* are the devices employed to represent the phenomena depicted on the map. They may be point symbols, such as dots to represent towns; line symbols, such as those representing the roads on a highway map; or area symbols, such as colors to indicate elevations or shading patterns to represent types of land use.

■ *Scale.* The map scale indicates the extent to which earth distances are reduced on a map. Information on scale is especially vital when one of the chief purposes of the map is to indicate distances between places, as on a road map. Unfortunately, for reasons to be explained shortly, the scale of a map showing a large area cannot be made consistent over its entire surface. The scales of maps depicting restricted areas such as states or cities, though, may, for practical purposes, be considered constant.

Map scales are stated by three basic methods, and one or more of these should appear in the map legend. *Verbal scales* are often provided on maps designed for use by people with limited map experience. They fre-

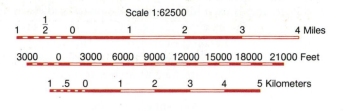

Figure B.1 The map scale information section in a USGS topographic map legend. The scale appears both in representative fraction and in multiple graphic scale formats.

quently appear, for example, on road maps. In this case, the scale is stated in a sentence format, as in the following example:

1 inch equals 8 miles.

In this example, the first figure (1 inch) is the map distance; and the second (8 miles) is the corresponding earth distance.

A *graphic scale* generally takes the form of a bar or line graph on which the map distances corresponding to real earth units are shown. On some maps, multiple bars may be used to show either different units of measure (see Figure B.1) or the scale on different parts of the map. Graphic scales have the advantage of providing a visual indication of the scale and of allowing direct measurements to be taken. Unlike the other two methods of stating scale, they also remain correct if the map is enlarged or reduced in size.

A *representative fraction scale* is stated as a direct ratio between map distances and corresponding earth distances. It may be written in either of the following fashions:

1:63,360 or 1/63,360
(values, of course, vary)

This fraction is a dimensionless quantity that can be thought of as the "degree of reduction" needed to display the portion of the earth's surface depicted on the map surface. In both cases the first number (or fraction numerator) is the map distance, which is always set at a value of one; and the second figure (or fraction denominator) is the corresponding earth distance. A major advantage of the representative fraction (RF) scale is that any units of measure can be employed, as long as the same unit is used for both map distances and earth distances. In the example above, one inch on the map equals 63,360 inches on the earth; one centimeter on the map equals 63,360 centimeters on the earth; and so forth. Because there are 63,360 inches in one statute mile, the RF scale of 1:63,360 is equivalent to the verbal scale "one inch equals one mile."

The scale of a map is decided upon by the cartographer (mapmaker) based largely on the intended use of the map, the amount of information to be shown, and the amount of space available on the map sheet. An unlimited number of scales are possible, but, for general descriptive purposes, maps are often described as being small-scale or large-scale. A *small-scale map* is greatly reduced from reality. This reduction enables the map to show a large surface area, but the map has the disadvantage of being so reduced in size that little surface detail can be shown. Most classroom wall maps are small-scale maps, even though the maps themselves are made large in size for easy viewing from a distance. *Large-scale maps* are less reduced in scale than are small-scale maps. They therefore cannot show as much surface area, but they are capable of showing more detail. An example would be a road map of a city or county.

The surface area depicted on a map changes by the inverse square of the change in linear scale. For example, if one of two equal-sized maps is constructed at a scale three times larger than the other, it will show only one-ninth the total area of the smaller-scale map.

Geography textbooks usually employ numerous maps to show the spatial distributions of the phenomena being studied. Two frequently used methods of presenting this information are the familiar *dot maps,* where each dot represents the presence of a certain number of objects, and *isarithmic maps,* which use lines of equal numerical value. A number of isarithmic maps appear in this book. In either case, the values represented by the dots or lines should be indicated in the title, in the legend, or in both.

Map Properties and Distortion

The great majority of maps are drawn on plane (flat) surfaces, usually consisting of sheets of paper. The planar nature of these maps produces the major problem in cartographic depiction—that of map distortion. Distortion in this context refers to inconsistencies in the spatial properties of different portions of the map. The crucial map properties of size, shape, distance, and direction may differ from one place to another; and it is not possible to eliminate this problem completely.

Map distortion exists because it is geometrically impossible to depict a spherical surface, or any significant part of it, accurately on a plane surface. This dilemma can be illustrated by painting a map of the earth on a large rubber ball, such as a basketball, and then attempting to flatten it out. This obviously cannot be done without considerable stretching, cutting, tearing, or compression. Even pieces of the ball (map) will be curved and cannot be flattened without stretching or tearing. Yet this transformation from a spherical surface to a planar one is exactly what the cartographer must accomplish in producing a flat map. In performing this task, some of the basic map properties of size, shape, distance, and direction must be distorted; and on many maps all of these properties are distorted.

Because map distortion results from the flattening of the earth's curved surface, maps depicting larger surface areas will be more distorted than maps depicting limited areas. Maps of the entire earth, therefore, are subject to the most distortion. On the other hand, large-scale maps generally depict such a small geographical area that distortion can be considered negligible.

The only way to map the earth without distortion is to use a globe, which duplicates the three-dimensional shape of our planet. The geometrical accuracy of globes is their greatest advantage over flat maps. Because globes are not distorted, relative sizes, shapes, distances, and directions can all be shown as they actually exist on the earth.

If globes are accurate and flat maps are distorted, why are flat maps so much more commonly used? The reasons are rather obvious: the handling and use of globes presents numerous practical difficulties. Among the chief problems are the great bulkiness and high cost of globes, difficulties in storing them (see how many will fit in your car's glove compartment!), their small scale and resulting lack of detail, the difficulty of making measurements on a curved surface, and the impossibility of seeing more than half the earth at any one time. Because of these shortcomings, globes are used primarily for display and general reference purposes. They provide an accurate depiction of the earth's surface, but once the correct relationships are mastered, practical work normally proceeds by the use of necessarily distorted flat maps. On the other hand, it is undoubtedly true that globes are not used enough for educational purposes. Many students insufficiently familiar with globes have

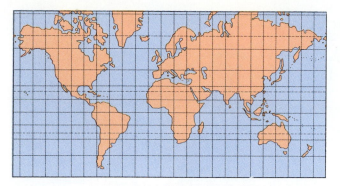

Figure B.2 Mercator projection of the earth. This projection is highly distorted with respect to scale because the relative sizes of the areas depicted increase rapidly in a poleward direction. It has long been used by mariners, however, because a straight line on the map is a line of constant compass bearing.

developed erroneous impressions about the earth's geography by taking literally the views of our planet appearing on flat maps (see Figure B.2).

Although flat map distortion cannot be avoided, it can be manipulated. Maps that have the worst distortion in what are generally considered the less important parts of the earth's surface, such as the polar regions or oceans, can be chosen. Maps whose distortion is confined to certain spatial characteristics, while other characteristics are shown correctly, can also be chosen. The types and amounts of distortion on a map can be estimated by comparing it to a globe or by examining the behavior of the map's parallels and meridians.

Location is a map property of vital importance. All maps, regardless of distortion, should show absolute locations correctly with respect to latitude and longitude (or other referencing system). Two additional basic map properties are the size and shape of the areas depicted. These two properties are strongly influenced by the problem of distortion; in fact, it is impossible for a single flat map to display both an area's size and shape correctly. It is possible, however, to keep either one or the other of these map properties from being distorted.

A map that has the property of consistency in relative size is called an *equal-area map*. All places on such a map are shown in their proper proportional sizes—that is, subject to the same total areal reduction. Proportional sizes can be maintained only by distorting lengths, and therefore shapes; so lengthening in one direction is exactly compensated for by shortening in another. Shape is therefore badly represented on an equal-area map. Equal-area maps are commonly used to show areal distributions of specific phenomena. They are generally used, for example, for dot maps, because the spacing of the dots indicates the density of the subject being represented.

A map that depicts correct shapes is termed a *conformal map*. Actually, only shapes of small areas can be shown correctly on conformal maps. Directions at any point are also correct. Over large distances, shapes and directions remain distorted. Conformality can be achieved only ·by greatly distorting sizes, and high latitude regions usually appear on a much larger scale than low latitude regions. An important use for conformal maps is in determining directions, and they are valuable as navigational charts.

On conformal maps, directions from any point start out correctly, but gradually become distorted with distance. It is impossible to devise a flat map on which all directions remain correct over the entire map surface. However, a third type of map, the *azimuthal map*, retains true directions outward from one central point, regardless of distance.

Many maps are neither equal-area, conformal, nor azimuthal. Instead, they represent some degree of compromise between these basic types. Their distortion is distributed among the various map properties, usually producing a better appearing map. Such maps are commonly used for general reference or classroom purposes.

Map Projections

A geometric figure that cannot be flattened without distortion is said to be undevelopable. The basic principle of mapmaking is to transfer cartographically the earth's undevelopable surface onto a developable surface that can be flattened. A *map projection* of the earth is produced by systematically transferring or projecting the depiction of all or part of the earth's surface onto a developable surface, which is then developed, or flattened. Three particular developable surfaces—planes, cones, and cylinders—are commonly used for map projections (see Figure B.3).

In order to visualize the principle of map projections, imagine that a light bulb has been placed in the center of a transparent globe that has parallels, meridians, and the outlines of the continents drawn on it. Place, in turn, a plane, cone, and cylinder over the globe, as shown in Figure B.3. The shadows of the features on the globe

Figure B.3 The principles of projection for plane, conic, and cylindrical map projections. To envision the projection principle, imagine a light placed at the centers of the globes projecting the outlines of the earth features onto the three surfaces.

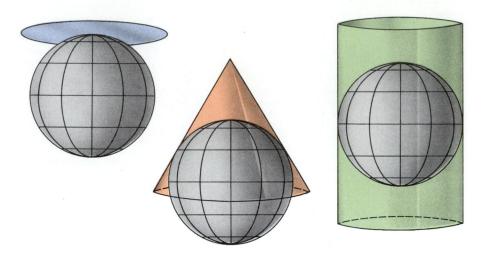

will then be projected onto these surfaces to produce maps that, in the cases of the cone and cylinder, can then be cut and flattened, as shown in Figure B.4.

When map projections are made in this fashion, no distortion of any type occurs at the points or lines where the developable surface rested in contact with the globe. These places are termed *standard points* or *standard lines*. Map projections are often constructed so that standard lines are particular parallels or meridians. Distortion occurs at all other points on the map, increasing in places that lie progressively farther from the globe's surface. Because the projected images expand

outward in all directions after they leave the globe, the locations on the developable surfaces that are located well away from the globe will appear at a larger scale; and angular relationships in these locations will also be distorted.

It is also possible for the developable surface to cut through the globe, rather than lie tangent to it, as shown in Figure B.5. For a plane projection, this will produce a standard line rather than a standard point. For both conic and cylindrical projections, the result will be two standard lines rather than one. Because the amount of contact between the two surfaces has been increased,

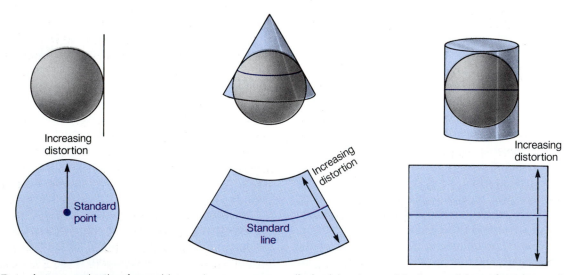

Figure B.4 A map projection formed by a plane, cone, or cylinder lying tangent to the earth's surface has only a single standard point or standard line, with distortion increasing outward from it.

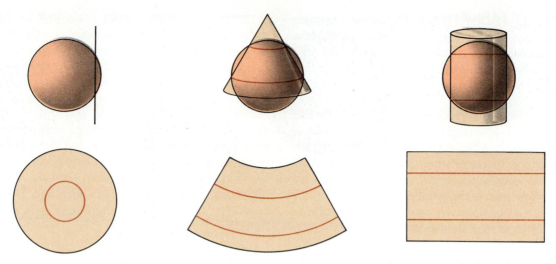

Figure B.5 A map projection formed by a plane cutting through the earth's surface has a circular standard line of no distortion. A map projection formed by a cone or cylinder cutting through the earth's surface has two standard lines, allowing a larger map area to have little or no distortion.

the total map area that has little or no distortion is also increased. The areas of the map that cut inside the globe will now appear at a smaller scale than those at the standard points or lines.

While the preceding paragraphs describe the principles of construction of plane, conic, and cylindrical projections, most modern projections are constructed mathematically and are often altered so as to achieve some desired property. For any projection, however, the transfer of map information must be accomplished in a systematic fashion. Because many types of developable surfaces exist and can be mathematically altered in varying ways, unlimited types of map projections are possible.

Characteristics of Major Map Projections

Most commonly used projections are members of the plane, conic, and cylindrical projection families. Each family has certain basic characteristics, some of which may be regarded as strengths and some as weaknesses. Because map uses are so varied, no projection group or type can be considered best; the type of map projection chosen should depend on the map properties that will allow the map to display its information most clearly and precisely.

PLANE PROJECTIONS A plane projection is a direct projection of a portion of the earth's surface onto a plane surface. The resulting maps are circular in shape

and are capable of showing half of the earth at most. Most plane projections are constructed so that they have a single point of contact with the globe. If so, they are azimuthal (correct in direction) from that point, with distortion increasing symmetrically outward. Many plane projections are centered on the North or South Pole and are termed *polar projections*. They may, however, be centered on any point, such as a city, airport, or park, to show directions and distances to other locations. Any straight line drawn from the standard point to any other point will represent the line of shortest earth distance between the two. Such lines are termed *great circle routes* because they form arcs of great circles on the globe. Plane projections are commonly used for the preparation of navigational charts for ships and aircraft.

CONIC PROJECTIONS Conic projections are generally constructed so that the apex of the cone is situated directly above the North or South Pole. This produces a standard parallel of no distortion in the middle latitudes. Conic projections are therefore well suited for depicting middle latitude areas with long east-west dimensions, such as the United States, Europe, and Australia. If the cone cuts through the globe so as to produce two standard parallels, the width of the zone of little distortion can be expanded. On conic projections, parallels appear as arcs of circles, while meridians appear as straight lines that gradually diverge equatorward.

CYLINDRICAL PROJECTIONS Cylindrical projections are almost always oriented so that the equator forms the standard parallel. As a result, they display the low latitudes most accurately, with scale and distortion increasing poleward. Because only the poles cannot be shown, such projections are frequently used for world maps. On a cylindrical projection, both parallels and meridians appear as straight lines, giving the map a rectangular shape (see Figures 4.6 and 4.7, for example). Unless they have been mathematically adjusted, the spacing of the meridians remains constant in a poleward direction, while that of the parallels increases.

Probably the best-known cylindrical projection is the *Mercator projection,* developed by the Flemish cartographer Gerhardus Mercator in 1569 for use in the construction of navigational charts (see Figure B.2). The unique attribute that has made this projection so valuable and durable is the fact that any straight line is a *rhumb line,* or line of constant compass bearing. In plotting a course, a navigator simply draws a straight line on the map between his current position and his destination and measures the compass orientation of the line. Even though such a course does not normally follow the shortest route, this was a major breakthrough at a time when navigational techniques were very imprecise.

ELLIPTICAL PROJECTIONS Not one of the three basic projection types just discussed is capable of showing the entire earth. In order to meet this need and to avoid the excessive high latitude distortion of cylindrical projections, a group of oval or elliptical projections has been developed. They are currently in widespread use for wall maps and textbooks, including this text (see, for example, Figure 9.8). Elliptical projections are not based on the projection of light, but instead are mathematically derived. Most types contain straight parallels and have meridians that bow outward in opposite directions from a straight central meridian. Distortion is generally worst near the margins, especially in the high latitudes.

INTERRUPTED AND CONDENSED PROJECTIONS A problem shared by most types of projections is that they have only one, or at most two, standard lines where no distortion exists. *Interrupted projections* use several standard lines; consequently, by separating the projection surface in various places (as illustrated in Figure B.6), they increase the amount of surface area that can be accurately depicted. In most cases, map users are more interested in the continents than in the oceans, so the lines of separation are normally made in ocean areas.

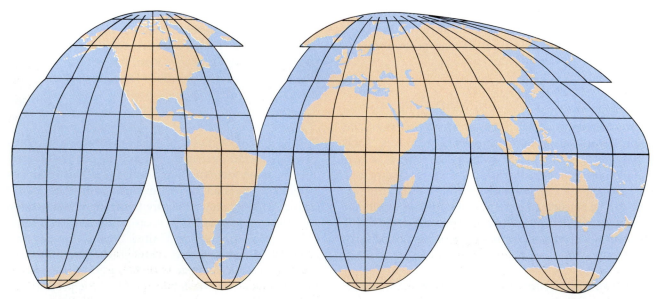

Figure B.6 Goode's interrupted projection of the earth. The separation of the ocean segments allows the more important land segments to be depicted with less distortion.

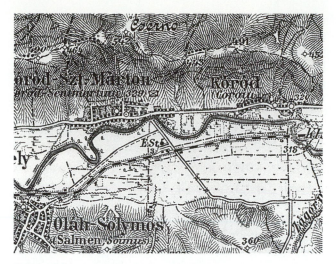

Figure B.7 A portion of a topographic map employing hachures.

On *condensed projections,* portions of the earth's surface are omitted from the map in order to show the remaining areas at a larger scale. Again, the missing portions are usually ocean areas, especially in the Atlantic and the Pacific. Figure 7.20, for example, deletes most of the Atlantic in order to save space.

Topographic Maps

Despite the problem of distortion, maps are intrinsically well suited for displaying horizontal locational information because map sheets are plane surfaces. The earth's surface, however, is far from level. Cartographically indicating the vertical dimensions of surface features therefore presents a major challenge.

Some maps contain no information about the vertical characteristics of the surface, or do not treat the subject systematically. Such a map is called a *planimetric map,* because it presents a plan view or two-dimensional view of the surface. A planimetric map may contain the written names of features such as mountain ranges, may use symbols such as miniature peaks for mountains, or may even provide *spot elevations* of notable points for reference, but does not allow for the systematic determination of elevations over the mapped area as a whole. A good example of a much-used type of planimetric map is a road map.

A map that, in addition to its other data, consistently displays elevational information for the entire area covered is termed a *topographic map.* It contains

three-dimensional information on a two-dimensional surface. Topographic maps can employ several methods to depict the vertical characteristics of the landscape, and two or more methods may be combined for greater effectiveness. The five most common methods involve the use of hachures, color, shading, raised relief, and contour lines.

Hachures

Hachures are short line segments that are drawn directly downslope (see Figure B.7). They appear somewhat like gullies and indicate the direction water would take as it flows downhill. Hachure maps were produced in quantity by several European countries during the first half of the twentieth century, but are no longer in widespread use.

Color

The use of colors to depict elevations is a technique familiar to those who use classroom wall maps. Each color indicates an elevation range that is described in the legend, and a logical gradational sequence of colors is normally employed. The particular color sequence of dark green, light green, orange, red, and brown for increasing elevations is especially common; and many maps use a sequence of shades of blue for increasing ocean depths as well.

Shading

On maps employing shading, the topography is shown by surface brightness differences simulating those resulting from oblique sunlight (see Figure B.8). A realistic and sometimes almost three-dimensional impression can be produced by the skillful use of this technique. Shading is frequently combined with other methods of elevation depiction to enhance the visual impression of topographic relief.

Raised Relief

A *raised relief map* differs from all other types because the map is made three-dimensional, producing a miniaturized model of the landscape it portrays. In most instances, the vertical scale is exaggerated to stress the topographic characteristics of the surface. Raised relief

Figure B.8 The skillful use of shading gives the surface a three-dimensional appearance.

maps are attractive and are popular for displaying surface features in a highly realistic fashion. They have the disadvantage of being expensive, because they must be molded from stiff material (usually plastic); and, like the other methods so far discussed, they do not permit the making of precise elevation measurements.

Contour Lines

Contour lines are lines that connect points of equal elevation above mean sea level. In following a contour line, one remains at the same elevation, while in crossing contour lines, one travels uphill or downhill (see Figure B.9). Contour lines give a less realistic visual impression of the topography than do some of the other methods discussed. However, they are by far the most

Figure B.9 Contour lines can be visualized as the locations in a mapped area that would be intersected by plane surfaces drawn at set elevation levels. In this case, the contour interval is 20 units. (Elevation units are normally feet or meters.) The resulting topographic map view of the twin-peaked island shown in the upper portion of the diagram appears at the bottom of the page.

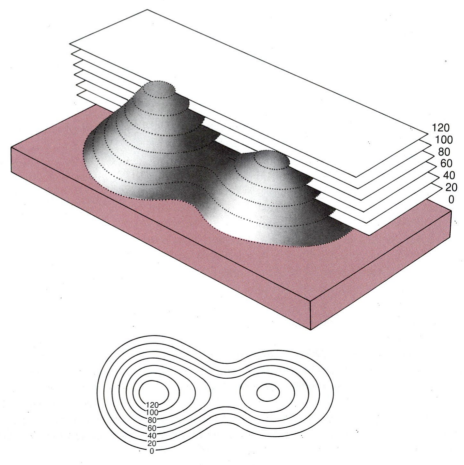

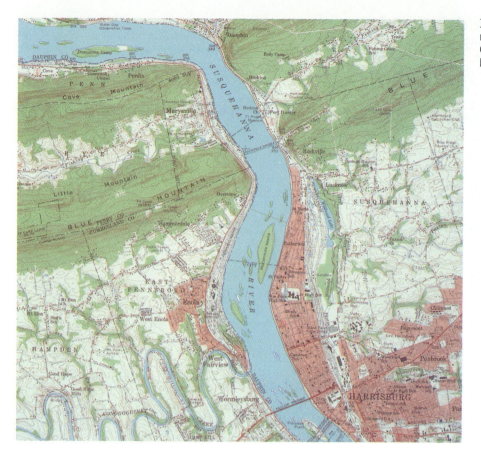

Figure B.10 A portion of a modern USGS topographic map. Contour lines are colored in brown.

useful method of elevation depiction for physical geographers and others engaged in actual fieldwork because they are well suited to large-scale maps and usually permit a reasonably accurate determination of elevations at any point on the map.

Some of the more important characteristics of contour lines are as follows:

1. Each contour line has a specific elevation. The elevation value of a given contour line may be written on it, or the value may have to be determined from the elevations of nearby contour lines.
2. A contour line separates areas with elevations higher than its value from areas with elevations lower than its value.
3. The spacing between contour lines indicates slope steepness. Widely spaced contour lines indicate gentle slopes, while closely spaced contour lines indicate steep slopes.
4. All contour lines eventually form closed figures, although these figures may extend beyond the borders of the map.

5. Contour lines do not meet or cross one another, except in the rare cases of vertical or overhanging topographic features.
6. The bends in contour lines point uphill when crossing streams or other linear lowlands and point downhill when crossing ridges.
7. A closed contour line normally indicates the presence of a high point of land, such as a hilltop. Closed depressions, which are less common, are indicated by special *depression contours,* which have inward-facing (i.e., downhill-extending) hachure marks.

The number and elevational values of the contour lines to be used on a given topographic map are decided by the cartographer. The vertical interval used, called the *contour interval,* is normally constant over the entire map. Some topographic maps, such as those prepared by the U.S. Geological Survey, also use thicker *index contours* with a larger contour interval for faster elevation determination. For example, on a map with a

20-foot contour interval, every fifth line might be thickened to produce a 100-foot index contour interval.

The use of contour lines allows for the exact determination of elevations only on the lines themselves. Elevations of areas between the lines are estimated by the process of interpolation, or proportional spacing. For example, if you wish to estimate the elevation of a house located on a contour map three-fourths of the way between the 320-foot and 340-foot contour lines, you would calculate the number that is three-fourths of the way between 320 and 340. This number is 335, the best estimate of the elevation of the house. Slopes may not be smooth, so there is no guarantee that you will be exactly correct; but, given no evidence to the contrary, a relatively smooth slope is the most reasonable assumption.

Topographic Mapping Agencies

Many countries have governmental agencies responsible for the preparation of topographic maps. In the United States, this agency is the U.S. Geological Survey (USGS). The USGS has been systematically mapping the country since 1879, and the maps it produces are revised periodically. Most USGS map information is now derived from aerial photographs, but the maps are also locationally controlled by carefully surveyed surface sites called *bench marks*.

USGS topographic maps are available for purchase by the public and are frequently used for fieldwork and for lab exercises in many earth science and engineering courses. In addition to elevations, which are indicated by brown-colored contour lines, these maps contain a great deal of other information (see Figure B.10). The larger scale maps, especially, contain a wealth of surface detail pertaining to both natural and cultural features; and over 100 standardized symbols have been developed to identify these features (see Figure B.11).

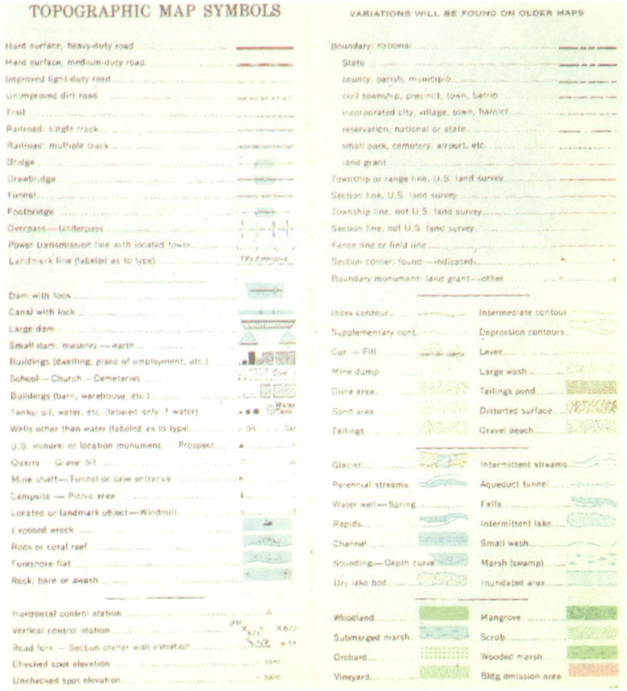

REMOTE SENSING

Every object or substance on the earth's surface absorbs, reflects, and emits energy in the form of electromagnetic radiation. The quantity and wavelengths of the radiation depend on the nature and temperature of the surface materials. In the rapidly developing field of *remote sensing,* information is acquired from a distance by mechanical devices that receive portions of this electromagnetic energy. The information is used to

Figure B.11 USGS standardized topographic map symbols.

increase our knowledge of the nature or characteristics of the radiation sources.

A large number of remote sensing devices are used at the earth's surface for many purposes. Some familiar examples include radios, cameras, binoculars, telescopes, and radar. This section, however, deals with sensing devices that are sent aloft to gather data on the nature of the earth's surface and lower atmosphere. These devices, which operate on a variety of electromagnetic wavelengths, have made possible the accumulation of vast quantities of spatial information.

The oldest remote sensing tool is black and white photography. The earliest aerial photographs were taken in the late 1800s by cameras carried aloft by balloons. With the development of the airplane, it became possible to control the area photographed; and systematic air photo coverage began in the United States and portions of Europe after World War I. This led to the growth of the field of *photogrammetry*—the making of accurate measurements from photographs. Plotting machines that can make accurate topographic maps directly from overlapping series of aerial photographs have been developed (see Figure B.12).

After World War II, color film, which more realistically portrays the surface and provides a greater variety of surface information, gradually increased in

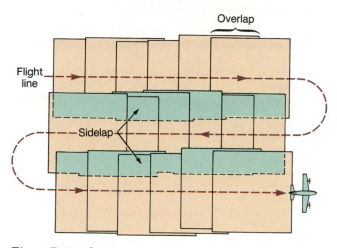

Figure B.12 Overlapping photographs taken from an aircraft flying back and forth over an area. Because surface features are viewed from differing angles on the overlapping portions of adjacent photos, it is possible to determine differing surface elevations directly from the photographs. In a similar manner, our brains construct a single three-dimensional image from the separate two-dimensional views gathered by each of our eyes.

use. About the same time, color infrared film was developed for aerial photography. This film has the advantage of good haze penetration and is highly useful in determining the types and health of vegetation.

In the 1960s, the deployment of orbiting satellites began a revolution in remote sensing technology equivalent to that resulting from the development of the camera and the airplane. From their high altitudes, satellites have greatly extended the field of view of the earth below. In addition, unlike aircraft, satellites can remain aloft and continuously collect and transmit surface data for months and even years. Satellites have been able to acquire data from all parts of the world because they are affected neither by adverse weather conditions nor by international boundaries. They are therefore critically important in gathering military intelligence, and many of their data-collecting capabilities remain classified. At present, hundreds of satellites have been placed in orbit around the earth (see Table B.1 on page 398); and increasingly sophisticated satellites are constantly being deployed. The U.S. government makes aircraft and satellite imagery of all parts of the earth's surface available to the public, and a number of these photos have been employed as illustrations in this book.

Another recent and ongoing revolution with major applications to both cartography and remote sensing lies

in the area of *computer mapping*. Computers can be used to collect, organize, and display huge quantities of data in a tiny fraction of the time that such tasks could be accomplished manually. Modern cartography and remote sensing technology now depend upon the use of computers for preparing maps.

As an example of computer technology, the Soil Conservation Service of the U.S. Department of Agriculture, which compiles soil maps and related data for the nation, is digitizing (converting to computer-readable form) maps of soil data. This information can be processed by computers to produce maps of important soil characteristics, such as soil fertility or erosion potential.

In addition to map preparation, computers are used to store, retrieve, manipulate, and display geographic data for the earth's surface. These computer systems, or *geographic information systems,* are revolutionizing the way geographers work with mapped data. Geographic information systems, which can be considered automated atlases of mapped data, allow geographers to superimpose several maps in the computer in order to identify spatial patterns of earth surface phenomena. For example, maps of soils, geology, and topography can be readily combined to aid in the search for areas suitable for farming or residential development.

Table B.1
Applications of Landsat and Weather Satellites

Atmo-sphere	Agriculture, Forestry, and Range Resources	Cartography and Land Use/Cover	Geology	Water Resources	Oceanography and Marine Resources	Environment
Weather data	Discrimination of vegetative types:	Classification of land uses/cover	Identification of rock types	Determination of water boundaries and surface water area and volume	Determination of turbidity patterns and circulation	Monitoring surface mining and reclamation
Pollution mapping	Crop types	Mapping and map updating	Mapping of major geologic units		Mapping shoreline changes	Mapping and monitoring of water pollution
Cloud types	Timber types Range vegetation	Categorization of land capability	Revising geologic maps	Mapping of floods and floodplains		Determining effects of natural disasters
Water vapor content	Measurement of crop acreage by species	Separation of urban and rural categories	Delineation of unconsolidated rock and soils	Determination of areal extent of snow and snow boundaries	Mapping of shoals and shallow areas	Monitoring environmental effects of human activities (lake eutrophication defoliation, etc.)
Fog	Measurement of timber acreage and volume by species	Preparation of regional plans	Mapping igneous intrusions	Measurement of glacial features	Mapping of ice for shipping	
Haze conditions	Determination of range readiness and biomass	Mapping of transportation networks	Mapping recent volcanic surface deposits	Measurement of sediment and turbidity patterns	Study of eddies and waves	
	Determination of vegetation stress	Mapping of land-water boundaries	Mapping landforms	Determination of water depth		
	Determination of soil associations	Mapping of wetlands	Search for surface guides to mineralization	Delineation of irrigated fields		
	Assessment of grass and forest fire damage		Determination of regional structures	Inventory of lakes and water impoundments		
			Mapping lineaments (fractures)			

Source: The Landsat Story (Washington, D.C.: Earth Resources Satellite Data Application Series, Module U-2. 1980), p. 14.

Appendix C

Weather Map Symbols

Surface weather maps are prepared daily by the National Weather Service in Washington, D.C.; and it is possible to purchase a subscription to a weekly booklet entitled "Daily Weather Maps" from the superintendent of documents. Each map in these booklets depicts the surface weather conditions for the coterminous United States as well as adjacent portions of Canada and Mexico at 7:00 A.M., eastern standard time. An example of one of the maps is shown in Figure 6.8. Also included are smaller maps showing upper air wind-flow patterns, 24-hour high and low temperatures, and precipitation totals for a large number of reporting stations. The ready availability and high quality of these maps have led many individuals interested in the weather as well as numerous educational institutions to subscribe to them.

The weather information contained on the Daily Weather Maps is conveyed by the use of standardized symbols. The two major types of features shown are isobars, which are drawn at four-millibar intervals, and weather fronts, the symbols for which are shown in Table C.1. The positions of high and low pressure centers are also shown.

A large number of individual reporting stations are displayed on the maps, and their information provides the basis for the positions of the isobars and fronts. The following sample report illustrates the format used for displaying reporting station weather data. A translation of the information in the report appears in Figure C.1 on the following pages.

Table C.1
Front Symbols

Symbol	Meaning
▲▲▲	Cold front
●●●	Warm front
▲●▲●	Occluded front
▲▼	Stationary front

Table C.2
Standardized Weather Map Symbols

Information	Meaning (sample station conditions in parentheses)
●	Circle represents station's location. Proportion blackened in represents amount of cloud cover (overcast).
(wind shaft)	Shaft extending from station circle indicates direction from which wind is blowing. Lines on shaft indicate wind speed (northwest wind at 18–22 knots).
32	Current temperature in degrees Fahrenheit (32°F).
30	Current dew point in degrees Fahrenheit (30°F).
3/4	Visibility in miles, if under 10 miles (3/4 mile).
✳✳	Present weather of significance (continuous light snow).
247	Last three digits of air pressure reading to tenths of a millibar, with decimal point omitted (1024.7 mb).
+ 28/	Air pressure change in 3 hours preceding observation, including pressure tendency over the period (pressure has risen steadily a total of 2.8 millibars).
.4	Time precipitation last began or ended (precipitation began as rain between 3 and 4 hours ago).
.45	Amount of precipitation during past 6 hours in inches. If frozen, water equivalent is given (0.45'').
(cirrus symbol)	Type of high cloud, if any (dense cirrus).
(altocumulus symbol)	Type of middle cloud, if any (altocumulus).
(fractostratus symbol)	Type of low cloud, if any (fractostratus).
6	Fraction of sky covered by low- or middle-level clouds (seven-tenths to eight-tenths of sky covered by fractostratus clouds).
2	Height of cloud base or ceiling (ceiling 300–599 feet).

SAMPLE PLOTTED REPORT

399

Figure C.1 Daily Weather Map information. (Abridged from Daily Weather Maps, U.S. Department of Commerce, National Oceanic and Atmospheric Administration, and the Environmental Data Service.)

C_L — Low Clouds

Cloud Abbreviation	C_L	Description (Abridged From W M O Code)
St or Fs-Stratus or Fractostratus	1	Cu of fair weather, little vertical development and seemingly flattened
Ci-Cirrus	2	Cu of considerable development, generally towering, with or without other Cu or Sc bases all at same level
Cs-Cirrostratus	3	Cb with tops lacking clear-cut outlines, but distinctly not cirriform or anvil-shaped; with or without Cu, Sc, or St
Cc-Cirrocumulus	4	Sc formed by spreading out of Cu; Cu often present also
Ac-Altocumulus	5	Sc not formed by spreading out of Cu
As-Altostratus	6	St or Fs or both, but no Fs of bad weather
Sc-Stratocumulus	7	Fs and/or Fc of bad weather (scud)
Ns-Nimbostratus	8	Cu and Sc (not formed by spreading out of Cu) with bases at different levels
Cu or Fc-Cumulus or Fractocumulus	9	Cb having a clearly fibrous (cirriform) top, often anvil-shaped, with or without Cu, Sc, St, or scud
Cb-Cumulonimbus		

C_M — Middle Clouds

C_M	Description (Abridged From W M O Code)
1	Thin As (most of cloud layer semi-transparent)
2	Thick As, greater part sufficiently dense to hide sun (or moon), or Ns
3	Thin Ac, mostly semi-transparent; cloud elements not changing much and at a single level
4	Thin Ac in patches; cloud elements continually changing and/or occurring at more than one level
5	Thin Ac in bands or in a layer gradually spreading over sky and usually thickening as a whole
6	Ac formed by the spreading out of Cu
7	Double-layered Ac, or a thick layer of Ac, not increasing; or Ac with As and/or Ns
8	Ac in the form of Cu-shaped tufts or Ac with turrets
9	Ac of a chaotic sky, usually at different levels; patches of dense Ci are usually present also

C_H — High Clouds

C_H	Description (Abridged From W M O Code)
1	Filaments of Ci, or "mares tails," scattered and not increasing
2	Dense Ci in patches or twisted sheaves, usually not increasing, sometimes like remains of Cb; or towers or tufts
3	Dense Ci, often anvil-shaped, derived from or associated with Cb
4	Ci, often hook-shaped, gradually spreading over the sky and usually thickening as a whole
5	Ci and Cs, often in converging bands, or Cs alone; generally overspreading and growing denser; the continuous layer not reaching 45° altitude
6	Ci and Cs, often in converging bands, or Cs alone, generally overspreading and growing denser; the continuous layer exceeding 45° altitude
7	Veil of Cs covering the entire sky
8	Cs not increasing and not covering entire sky
9	Cc alone or Cc with some Ci or Cs, but the Cc being the main cirriform cloud

h — Height

h	Height in Feet (Rounded Off)	Height in Meters (Approximate)
0	0 - 149	0 - 49
1	150 - 299	50 - 99
2	300 - 599	100 - 199
3	600 - 999	200 - 299
4	1,000 - 1,999	300 - 599
5	2,000 - 3,499	600 - 999
6	3,500 - 4,999	1,000 - 1,499
7	5,000 - 6,499	1,500 - 1,999
8	6,500 - 7,999	2,000 - 2,499
9	At or above 8,000. or no clouds	At or above 2,500. or no clouds

R_t — Time of Precipitation

R_t	Time of Precipitation
0	No Precipitation
1	Less than 1 hour ago
2	1 to 2 hours ago
3	2 to 3 hours ago
4	3 to 4 hours ago
5	4 to 5 hours ago
6	5 to 6 hours ago
7	6 to 12 hours ago
8	More than 12 hours ago
9	Unknown

N — Sky Coverage (Total Amount) / N_h — Sky Coverage (Low And/Or Middle Clouds)

N / N_h	Sky Coverage
0	No clouds
1	Less than one-tenth or one-tenth
2	Two-tenths or three-tenths
3	Four-tenths
4	Five-tenths
5	Six-tenths
6	Seven-tenths or eight-tenths
7	Nine-tenths or overcast with openings
8	Completely overcast
9	Sky obscured

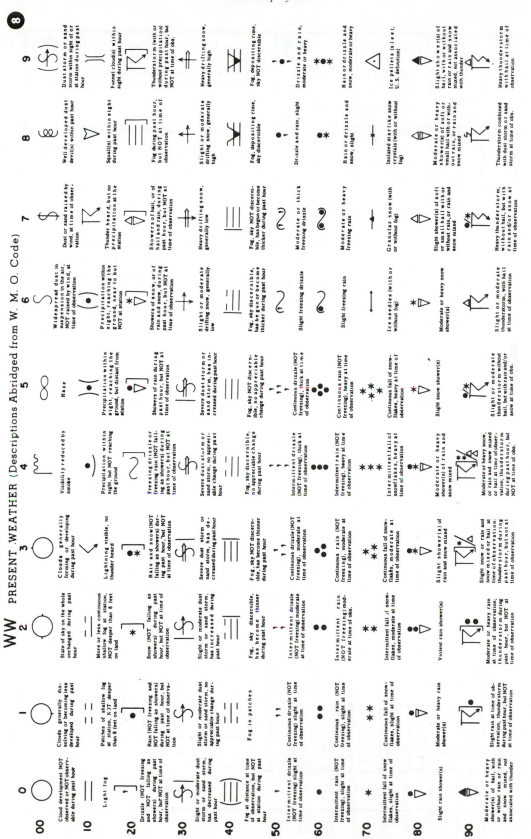

PRESENT WEATHER (Descriptions Abridged from W. M. O. Code)

ff	(MILES) (Statute) Per Hour	⑨ KNOTS	Code Number	ⓐ	BAROMETRIC TENDENCY ⑩
◎	Calm	Calm	0	⁀	Rising, then falling
—	1 - 2	1 - 2	1	╱	Rising, then steady; or rising, then rising more slowly
	3 - 8	3 - 7	2	╱	Rising steadily, or unsteadily
	9 - 14	8 - 12	3	✓	Falling or steady, then rising; or rising, then rising more quickly
	15 - 20	13 - 17	4	—	Steady, same as 3 hours ago
	21 - 25	18 - 22	5	╲	Falling, then rising, same or lower than 3 hours ago
	26 - 31	23 - 27	6	╲	Falling, then steady; or falling, then falling more slowly
	32 - 37	28 - 32	7	╲	Falling steadily, or unsteadily
	38 - 43	33 - 37	8	⁀	Steady or rising, then falling; or falling, then falling more quickly
	44 - 49	38 - 42			
	50 - 54	43 - 47	Code Number	W	PAST WEATHER ⑪
	55 - 60	48 - 52	0		Clear or few clouds
	61 - 66	53 - 57	1		Partly cloudy (scattered) or variable sky
	67 - 71	58 - 62	2		Cloudy (broken) or overcast
	72 - 77	63 - 67	3		Sandstorm or dust-storm, or drifting or blowing snow
	78 - 83	68 - 72	4	≡	Fog, or smoke, or thick dust haze
	84 - 89	73 - 77	5	'	Drizzle
	119 - 123	103 - 107	6	•	Rain
			7	✳	Snow, or rain and snow mixed, or ice pellets (sleet)
			8	▽	Shower(s)
			9	℞	Thunderstorm, with or without precipitation

Barometer now higher than 3 hours ago

Barometer now lower than 3 hours ago

Not Plotted

Glossary

Ablation   The loss of ice from a glacier as a result of melting and sublimation.

Abrasion   Erosion accomplished by the scraping of water-borne particles against other materials.

Acid rain   Rainfall of more-than-usual acidity. Acid rain is believed to result largely from the human release of sulfur dioxide (SO_2) gas, especially from coal-fired electric generating plants, and has proven harmful to plant and animal life as well as to a variety of material substances.

Adiabatic temperature changes   Temperature changes produced within a parcel of air because of the air's expansion or contraction. Adiabatic cooling normally occurs as a parcel of air rises and expands; adiabatic warming normally occurs as air descends and contracts. No energy is actually gained or lost by the air during the process.

Advection   The process of energy transfer through the horizontal movement of a fluid from one place to another. A type of convection.

Advection fog   A fog type that forms when moist air flows over a colder surface that cools it below the dew point.

Air mass   A large body of air with relatively uniform horizontal temperature and moisture characteristics.

Air mass thunderstorm   A thunderstorm that has developed within a single unstable air mass. Air mass thunderstorms typically show a lack of linear organization, and over land areas form most frequently during the afternoon hours.

Air pressure   The force exerted by the air on its surroundings because of the pull of gravity. Air pressure at sea level averages 14.7 pounds per square inch (1.0 kg/cm^2).

Albedo   The short wave energy reflection rate of an object or substance, normally stated in percent.

Alfisol   A soil order containing fine-textured soils with a moderately high cation exchange capacity that are low in humus and have a yellowish-brown A horizon. They are common in forest-grassland transitional environments in the low and middle latitudes.

Alluvial fan   A gently sloping fan-shaped deposit of alluvium that is formed when a stream flowing within a steep, narrow channel suddenly enters an open lowland and loses its capacity to transport its sediment load.

Alluvial soil   A soil formed from alluvium, which is stream-deposited sediment.

Alluvium   An accumulation of stream-deposited sediments, typically found on the surface of a floodplain.

Alpine glacier   A relatively small glacier or glacier system that forms at high elevations in a mountainous region and moves downhill.

Altitude   Distance above sea level; generally used for a location above the earth's surface.

Altitude (solar)   The angular distance in degrees of the sun above the nearest horizon.

Anemometer An instrument, consisting of several rotating cups, that measures wind speeds.

Aneroid barometer A barometer that measures air pressure by means of the force exerted on a collapsible box, called a sylphon cell, from which some of the air has been removed.

Angle of repose The steepest angle that can be maintained on a given slope without slope failure (mass movement) occurring.

Annual A plant that has a life span of one year or less, usually because it is killed by cold or dry weather at the end of the growing season.

Annual temperature range The number of degrees between the warmest and the coldest monthly mean temperatures for the year.

Annual variability of precipitation The average annual percentage deviation of a site's annual precipitation total from the long-term mean total.

Antarctic Circle The most equatorward line of latitude in the southern hemisphere (66 1/2° S) at which the sun remains constantly above the horizon on the date of the southern hemisphere summer solstice, and constantly below the horizon on the date of the southern hemisphere winter solstice.

Anticline An upward fold in one or more rock strata.

Anticyclone A high pressure system, characterized by an outward spiraling wind-flow pattern.

Aphelion The point of the earth's orbit that is farthest from the sun. It occurs about July 4th, when the two bodies are approximately 94.5 million miles apart.

Aquifer A porous and permeable rock or sediment layer that contains large quantities of ground water and permits its ready movement.

Arctic Circle The most equatorward line of latitude in the northern hemisphere (66 1/2° N) at which the sun remains constantly above the horizon on the date of the summer solstice, and constantly below the horizon on the date of the winter solstice.

Arête A narrow, serrated mountain ridge in a region of present or past glaciation. It is formed by the headward growth, and eventual meeting, of cirques from opposite sides of the ridge.

Argon An inert gas comprising almost 1% of the atmosphere.

Arid climate A climate with a potential evaporation rate that greatly exceeds its mean precipitation receipts, so that desert conditions prevail.

Aridisol A soil order containing predominantly shallow, stony, and mineralogically immature soils that have developed in arid environments.

Aspect (of a slope) The compass direction in which the slope faces.

Association See *vegetative association.*

Asthenosphere The plastic upper portion of the mantle. Convection currents in the asthenosphere are believed to produce lithospheric plate motions.

Atmosphere The gaseous envelope surrounding the earth.

Atoll An often ring-shaped reef that forms an island or chain of islands. Atolls are believed to form around oceanic volcanic islands. They persist because of the continued growth of the reef-building organisms after the volcanic island has disappeared due to erosion or submergence.

Autumnal equinox The day that the earth's axial inclination is exactly sideways with respect to the sun, following a period in which one's hemisphere was inclined toward the sun. Marks the official beginning of autumn.

Avalanche The rapid movement of a detached mass of snow and ice down a steep mountain slope.

Average environmental lapse rate The average rate at which the temperature declines with altitude in the troposphere. This value is approximately 3.5F°/1000 feet, or 6.4C°/1000 m.

Axis The line, connecting the North and South Poles, about which the earth rotates.

Azimuthal map A map that retains true directions outward from one central point.

Backwash The return seaward flow of wave swash. It is largely responsible for moving beach sediments into the near-shore water.

Back-wasting Processes of weathering and erosion that result in a mountain mass maintaining its steepness and ruggedness while being reduced in size and areal extent. Back-wasting is characteristic of arid regions and contrasts to the down-wasting processes characterizing humid regions.

Bajada See *piedmont alluvial slope.*

Bar A linear deposit of sand or gravel that is often found within a stream channel at a place where the speed of flow is diminished.

Barchan A relatively small, crescentic sand dune that is oriented with its tapered ends pointing downwind.

Barometer An instrument designed to measure air pressure.

Barrier island An elongated sandy island paralleling the coastline that is formed by the continued deposition of sand by storm waves. This same feature, if connected to the mainland to form a peninsula, is termed a barrier beach.

Basalt A dense, black-colored extrusive igneous rock containing substantial iron and magnesium.

Base level The elevation at any point along a stream's course that provides the stream with a gradient just sufficient for it to transport its sediment load past that point with no erosion or deposition. If a stream is at base level throughout its course, it is considered a graded stream.

Batholith A large mass of intrusive igneous rock, usually granitic in composition. Because of its resistance to erosion, it may form a mountain mass if exposed at the surface.

Beach The gently sloping shore of a body of water that is subject to wave or tide action. It is normally covered by loose deposits of sand or gravel.

Beach drifting The gradual movement of sediments along a beach in a zigzag motion. This occurs as the sediments are alternately pushed up the beach by swash from oblique waves and then carried directly downslope by the wave backwash.

Bed load Sediments that are moved along the bed of a stream by sliding, rolling, or bouncing (saltation).

Big Bang Theory At present, the most widely accepted theory of origin of the universe. It holds that all matter in the universe was initially produced, and hurled outward, by the explosion of an inconceivably tiny and compact entity sometimes referred to as the "cosmic egg."

Biogeography The branch of geography concerned with the global patterns of plants and animals.

Biomass A measure of vegetation density, defined as the mass of vegetation per unit of area.

Biome See *vegetative association.*

Biosphere The sphere of living organisms inhabiting the surface and near-surface portions of the earth.

Bird's-foot delta A delta that contains long fingerlike deltaic projections. It develops when a stream deposits sediments in a quiet, partially enclosed water body that lacks the wave and current action to spread out the deltaic materials.

Bolson An enclosed desert basin into which ephemeral streams transport water and sediments. Following rainfalls, they become the sites of temporary lakes termed playas.

Broadleaf Usually used to describe trees with leaves of considerable width, such as oak, maple, or elm, as opposed to needleleaf trees.

Butte A small, isolated, and steep-sided remnant of a former more elevated topographic surface. Buttes are often left as erosional outliers of retreating escarpments in arid regions.

Caldera A roughly circular volcanic depression caused either by the explosion or collapse of a volcano.

Calorie The amount of heat needed to raise the temperature of one gram of liquid water by one Celsius degree (1.8F°).

Canyon A deep, steep-walled valley of considerable width that is usually carved by a stream that is rapidly downcutting.

Cap rock A resistant surface rock layer that serves as a cap to protect the underlying less resistant strata from erosion. Most cap rock layers exist in arid regions.

Capillary tension The electromagnetic attraction of a thin molecular film of water to solid surfaces, such as soil grains and plant root linings.

Carbon dioxide An atmospheric gas essential to plant photosynthetic processes and largely responsible for the long-wave radiation absorbing capacity of the atmosphere.

Carbonation A chemical weathering process involving the reaction of rock material with carbonic acid (H_2CO_3) produced when carbon dioxide is dissolved in water.

Cartography The art and science of mapmaking.

Catastrophism The widely held view of a few centuries ago that the earth's major surface features were produced in the recent past by cataclysmic events.

Cation exchange capacity (CEC) The quantitative measure of the ability of a given soil to exchange cations with plants.

Cave A natural cavity within or beneath the surface. Caves can be caused by a number of processes, but are especially numerous and important in karst areas, where they are produced by subsurface solution.

Celsius The metric system temperature scale. It is based on the boiling point of water (under standard temperature and pressure conditions) at 100° C, and the freezing point of water at 0° C.

Channelled Scablands A region, located in Washington State, where the surface topography has been produced largely by repeated Pleistocene floods of massive proportions. These floods resulted from the breaching of an ice dam that blocked the outflow of glacial Lake Missoula.

Chaparral The name used in California for a dense cover of woody shrubs and small trees ranging up to 20 feet (6 m) in height. It is a variety of the Mediterranean Woodland and Scrub vegetation subassociation.

Chemical sedimentary rock A sedimentary rock formed by the direct precipitation of chemicals in water.

Chemical weathering Weathering processes involving chemical changes in the rock material that form new substances.

Chinook A warm, dry wind that is adiabatically heated as it descends the eastern slopes of the Rocky Mountains in the winter season.

Chlorofluorocarbon gases (CFCs) A group of synthetic gases that are apparently depleting the ozone layer, causing an increase in the quantity of shortwave radiation reaching the earth's surface.

Cinder cone A conical volcanic hill or low mountain constructed of loose pyroclastic material thrown into the air from a central vent.

Cirques Bowl-shaped mountainside hollows, sculpted by glacial scour and frost action, in which alpine glaciers typically have their heads.

Cirrus From the Latin word meaning "curl of hair," cirrus clouds are the highest of the ten basic cloud types, are composed of ice crystals, and have a wispy appearance.

Clastic sedimentary rock A sedimentary rock composed of particles of preexisting rocks. Based on the sizes of the constituent rock particles, the four major categories are conglomerate, sandstone, siltstone, and shale.

Clay-humus complex A complex combination of tightly bound clay and humus particles that is capable of attracting and exchanging cations needed by vegetation.

Climagraph A graph of monthly mean temperature, precipitation, and sometimes other climatic data for a given reporting station.

Climate The long-term or overall condition of the atmosphere with respect to the weather elements and weather systems.

Climatology The branch of physical science that deals with the study of climate.

Climax community The stable biological community that eventually occupies an ecosystem at the end of a successional sequence. Of all possible communities available to form the ecosystem, it is theoretically the best suited to survive under existing environmental conditions.

Cloud A visible accumulation of minute water droplets or ice crystals, suspended in the air, that have condensed or sublimed around dust particles because the air has cooled below the dew point.

Coal An organic sedimentary rock composed primarily of carbon and formed from the burial, compaction, and lithification of plant remains in a poorly drained environment.

Cold front A frontal boundary along which an advancing mass of cold air is underrunning relatively lighter warm air, forcing it to rise.

Cold ice glacier A dry-base glacier that is frozen fast to the underlying bedrock surface. Glacial movement must therefore take place solely by internal deformation.

Colloids Very fine clay particles that are capable of remaining suspended indefinitely in water and are soft and gelatinous in nature.

Composite volcano A volcano constructed of alternating layers of pyroclastic materials and lava. Most large volcanoes are of this type.

Condensation The phase change of a substance from a vapor to a liquid.

Condensation nuclei Atmospheric particulates that serve as the collection centers for water molecules, enabling clouds and ultimately precipitation to form.

Conduction The transfer of thermal energy by physical contact between two objects or substances having different temperatures.

Conformal map A map that has the property of showing the correct shapes of the areas it depicts.

Conglomerate A clastic sedimentary rock comprised largely of rounded gravel or larger-sized clasts.

Conifer A tree that produces its seeds within cones.

Continental arctic air mass A very cold, dry air mass that has formed over a snow-and-ice covered surface.

Continental climate Climatic characteristics typical of a large landmass, especially including relatively large daily and annual temperature ranges and low amounts of precipitation, fog, cloudiness, and humidity.

Continental crust The relatively light, discontinuous layer of granitic rock that comprises the uppermost portion of the crust and forms the continental masses.

Continental drift The somewhat outdated term for the movements of the lithospheric plates as a result of convection currents in the asthenosphere.

Continental glacier A glacier that forms at high latitudes and covers an extensive land area. Continental glaciers currently cover Antarctica and interior Greenland, and existed during the Pleistocene in northern North America and northwestern Eurasia.

Continental polar air mass A relatively cold, dry air mass that has formed over a high latitude landmass.

Continental tropical air mass A relatively hot, dry air mass that has formed over a low latitude landmass.

Contour lines Lines of constant elevation on a topographic map that are generally numbered from mean sea level.

Convection The process of energy transfer through the physical movement of a fluid from one place to another. On the earth, the chief fluids involved are air and water.

Convectional lifting The rising of air because it is warmer, and therefore lighter, than the surrounding air.

Core The innermost layer of the earth, believed to be primarily metallic in composition. The core consists of two segments—a solid inner core and a liquid outer core.

Coriolis effect The apparent deflection of moving fluids, especially winds and ocean currents, on the earth's surface, occurring as a result of the earth's rotation. The direction of deflection is to the right in the northern hemisphere and to the left in the southern hemisphere.

Cosmic radiation Charged atomic particles that enter the earth's atmosphere at nearly the speed of light and are capable of destroying or mutating plant and animal cells.

Creep The very slow downhill movement of soil or rock material over a period of years. Creep is most important in humid climate regions, and involves the entire mass of near-surface weathered material on most slopes.

Crest (of a flood) The peak flow of a flood, measured in terms of either discharge or water level.

Crevasse A large crack or fissure in the near-surface portion of a glacier, caused by the brittle fracturing of the ice as it passes over a convexity in its bed.

Critical erosion velocity The minimum velocity of flow that will enable a stream to erode streambed particles of a given size.

Crust The relatively thin "skin" of rigid low-density rock that forms the outermost lithospheric layer of the earth. The crust varies in thickness from 3 to 40 miles (5-70 km) and consists of two segments, the continental and the oceanic crust.

Cumulonimbus The cloud type responsible for thunderstorms, hail, and tornadoes; probably the world's leading precipitation producing cloud.

Cumulus From the Latin word meaning "pile" or "heap," cumulus clouds are the most common of the ten basic cloud types. They have a puffy appearance, with flat bases and uneven sides and tops.

Cyclone A low pressure system or storm center, characterized by an inward spiraling wind-flow pattern.

Cyclonic lifting Atmospheric lifting associated with the convergence of air into a low pressure system.

Daily mean temperature The average temperature of the day, calculated by adding the daily high and low temperatures and dividing by two.

Daily temperature range The number of degrees between the high and the low temperatures for the calendar day.

Daylight saving time A variation of standard time in which the time is set ahead by one hour in order to achieve greater correlation between the normal human activity period and the daylight period.

Debris avalanche The rapid movement of a mixture of soil, rock, and vegetation down a preexisting mountainside hollow. Movement is nearly always triggered by heavy rainfall; it typically begins as a slide and becomes a flow.

Deciduous A leafy plant that loses all its foliage during a given season, usually as a mechanism for coping with cold or dry conditions.

Deep ocean current An ocean current flowing at considerable depths that is produced by density differences resulting from variations in temperature and salinity from the surrounding water. It often displays significant vertical as well as horizontal components of motion.

Deflation The erosion of loose surficial material directly by the force of the wind.

Deflation hollow A shallow, often roughly circular depression of variable dimensions that has been formed by deflation.

Delta An often fan-shaped deposit of alluvium formed when a stream enters a standing body of water and loses its ability to transport its sediment load.

Dendritic stream pattern A treelike drainage pattern that develops on homogeneous surfaces that exert no significant structural control on stream location. It is the most widespread of the drainage pattern types.

Desert A region of substantial size that is largely or entirely devoid of vegetation, usually as a result of aridity.

Desert pavement A desert surface of close-fitting pebbles or stones that protects the underlying surface from erosion. It results largely from the eolian removal of fine-textured particles.

Dew Water droplets that have condensed on surface objects because the overlying air has cooled below the dew point.

Dew point The temperature at which a given parcel of air would become saturated with water vapor.

Diastrophism Solid-state movements of the earth's crust, including folding and faulting, that result from tectonic forces.

Differential erosion Erosion occurring at differing rates within a given area because of the uneven operation of gradational forces.

Dike A sheet of basalt that was injected as magma into a nearly vertical preexisting rock fracture.

Dike ridge A wall-like basaltic ridge formed by the erosional exposure of a dike injected into weaker sedimentary strata.

Dip The angle at which a rock stratum or any planar feature is inclined from the horizontal.

Discharge The volume of stream flow past a given point during a specified time span, normally measured in cubic feet per second (cfs) or in cubic meters per second (cms).

Doldrums Another name for the equatorial belt of variable winds and calms.

Doline See *sinkhole*.

Dome A rounded anticlinal structure. Smaller domes are usually formed by the surfaceward movement of material such as magma or salt.

Doppler radar A type of radar that holds great promise for tornado identification and prediction. It can remotely measure and display the speed of winds moving toward or away from the radar site, allowing for the detection of developing rotary motions in thunderstorms.

Drainage basin The total surface area drained by a given stream system.

Drift See *glacial drift*.

Drizzle Liquid precipitation droplets with diameters less than 0.02″ (0.5 mm).

Drumlin A smooth, elongated hill composed of glacial till. They generally occur in groups or "swarms" and are oriented so that that their tapered ends point in the direction the ice was moving.

Dune A mound or ridge of wind-deposited sand. Dunes assume a variety of forms and, unless stabilized by vegetation, tend to migrate slowly over the surface because of wind action.

Duricrust   A hard, nearly impermeable surface or near-surface layer, typically formed by the precipitation of salts in an enclosed basin in an arid climate.

Dust   The collective term for all liquid and solid particulates suspended in the atmosphere.

Earth flow   The slow flowage of a mass of nearly saturated soil down a moderate-to-steep slope.

Earth system   The earth and its atmosphere, treated as a unit.

Easterly wave   A north-south-oriented low pressure trough in the trade wind belt that drifts westward and may have the potential to develop into a hurricane.

Ecosystem   A functioning biological entity consisting of all the organisms in a biological community, as well as the environment in which and with which they interact.

Elevation   The distance above sea level of a point on the earth's surface.

El Niño   The anomalous warm ocean current that sometimes flows eastward across the southern Pacific to reach the coast of South America. It has recently been discovered to have a major impact on global climate patterns.

Eluviation   The removal of solid or dissolved materials from a soil horizon by downward percolating water.

Entisol   A soil order containing poorly developed soils, mostly of recent origin, that display little or no profile development.

Eolian   The term used for geomorphic processes proformed or products produced by the wind.

Ephemerals   Small, short-lived desert plants that pass through their life cycles quickly following a substantial rainfall.

Ephemeral stream   A stream that flows only during wet periods when it is supplied by surface or soil water.

Epiphyte   An "air plant" that grows on other plants, but derives its moisture and nutrients from the air and is therefore not parasitic. They are especially diverse and abundant in the tropical rainforests.

Equal area map   A map that has the property of showing the correct proportional sizes of the areas it depicts.

Equator   The line of fastest rotation on the earth. It is a full circumference, located midway between the North and South Poles, and is the starting (0°) line of latitude.

Equatorial belt of variable winds and calms   The zone of light and variable winds associated with the Intertropical Convergence Zone.

Equatorial lows   Weak low pressure centers that develop along the ITCZ and drift westward, attended by extensive cloudiness and rainfall.

Erg   A sandy desert.

Erosion   The removal of earth material from a given site, normally by the action of water, ice, or wind.

Estuary   A semi-enclosed water body located either in a drowned river valley or a submerged tectonic trough.

Evaporation   The phase change of a substance from a liquid to a vapor.

Evapotranspiration   The total water evaporated from both the earth's surface and from vegetation.

Evergreen   A leafy plant that retains its foliage throughout the year.

Exfoliation   The separation of rock into sheets paralleling the surface as a result of unloading and chemical weathering.

Exotic stream   A large perennial stream, flowing through an arid region, that has its headwaters in a humid region from which it has derived most of its water.

Extrusive igneous rock   Igneous rock that was extruded onto the earth's surface in a liquid state (as lava). Subsequent rapid cooling allowed little or no time for mineral crystals to form.

Fahrenheit   The temperature scale most commonly used by the general public in the United States. It is based on the boiling point of water (at standard temperature and pressure) at 212° F, and the freezing point of water at 32° F.

Fault escarpment (or Fault scarp)   Ideally, a linear, smooth-sided slope consisting of the exposed surface of the higher side of a fault.

Faulting   The fracturing of rock, followed by the movement of the two sides of the fracture relative to one another.

Fault line   The trace of a fault on the earth's surface.

Fennoscandian ice sheet   The continental ice sheet that covered most of northwestern Eurasia during the glacial stages of the Pleistocene.

Fjord   A glacial trough that has been partially occupied by seawater to form a deep oceanic coastal inlet.

Flood   The inundation of the land along the banks of a stream that occurs when the stream's discharge exceeds its carrying capacity.

Flood basalt   A huge outpouring of fluid lava from a fissure that can cover an extensive region to form a lava plain or plateau.

Floodplain   A broad belt of low, flat land adjacent to a stream that is formed by fluvial erosional and depositional processes, and is periodically inundated during floods.

Floodplain   A broad, flat-floored valley covered by alluvium and subject to flooding at times of high water.

Fluvial   An adjective used to identify stream-related processes and features.

Fog   A visible accumulation of minute water droplets or ice crystals suspended in air immediately overlying the surface. Fog normally forms as the air is cooled below the dew point, forcing the condensation or sublimation of water vapor on dust particles.

Folding   The bending of rock layers subjected to tectonic stresses.

Forest   An area in which trees are spaced sufficiently close together that their foliage overlaps to form a more or less continuous canopy.

Front   See *weather front.*

Frontal cyclone   A low pressure system that has developed on a weather front. Most frontal cyclones have an associated warm front and cold front. They are major precipitation producers for the middle and high latitudes.

Frontal lifting   The lifting of relatively light warm air by denser cold air along a weather front.

Frontal thunderstorm   A thunderstorm that has developed in association with a weather front. Frontal thunderstorms typically develop in a linear pattern and are most frequently associated with cold fronts.

Frontal wave   A poleward bend in a weather front, normally associated with an area of weak low pressure. Frontal waves often develop into frontal cyclones.

Frost   Ice crystals that have sublimed on surface objects because the overlying air has cooled below the dew point.

Frost wedging   The process of rock breakup resulting from the growth of ice crystals within rock fractures or hollows.

Galeria   Strip growths of forest in the moist soils along riverbanks in savanna grassland regions.

Geography   The study of the distributions and interrelationships of earth phenomena.

Geomorphology   The systematic study of landforms, including their origin, characteristics, and distribution.

Geothermal heat   Heat from the earth's interior, believed to result largely from the radioactive decay of uranium and thorium. It serves as the energy source for plate tectonic processes.

Glacial drift   A general term for all deposits of glacial origin.

Glacial lacustrine plain   A plain covered by fine-textured lake sediments. These sediments were deposited at a time when the surface was covered by a lake whose drainage outlet was blocked by a glacier.

Glacial trough   A deep and generally U-shaped valley, usually of fluvial origin, that was subsequently occupied and modified by a valley glacier.

Glacier   A mass of freshwater ice, formed on land, that is or has been in motion as a result of gravity and/or internal pressure.

Glacio-fluvial material   Rock material that was partially sorted by meltwater after being deposited by a glacier. It therefore displays some stratification.

Global radiation balance   The long-term global balance between incoming shortwave insolation and outgoing long-wave terrestrial radiation.

Gorge   A narrow, deep, and steep-walled valley that is usually carved by a stream that is rapidly downcutting.

Graben   A block of land bounded by two normal faults that has been downdropped relative to the blocks on either side.

Gradational forces   Solar powered forces that, in conjunction with the force of gravity, act to lower the earth's surface. The three major "tools" of gradation are water, ice, and wind.

Graded stream   A stream that has adjusted its flow parameters to the quantity and texture of its sediment load, so that it is neither eroding nor depositing materials (at least on a long-term basis) on its bed.

Granite   A low-density intrusive igneous rock, consisting primarily of quartz, feldspar, and mica, that comprises much of the continental crust.

Grass   A large and biologically diverse family of plants with bladelike opposing leaves connected to a jointed stem by a sheathlike attachment. The cereal grains that form the basic foodstuffs of humans and livestock are grasses.

Greenhouse effect   The heat-trapping ability of the atmosphere, especially of the gases carbon dioxide and water vapor. The process is analogous to the heat-trapping ability of the glass in a greenhouse.

Greenwich meridian   The starting line of longitude; also known as the prime meridian.

Ground water   Gravitational water that completely fills the pore spaces or other voids in the soil or rock below the water table.

Growing season   The period of time during the year when weather conditions continuously favor the growth of vegetation. In the United States, it is generally considered as the period between the last killing frost of spring and the first killing frost of fall.

Hail   A precipitation form consisting of relatively large, rounded particles of ice that typically exhibit internal layering. They form as strong updrafts and downdrafts in a thunderstorm carry ice pellets alternately above and below the freezing level.

Halophytes   Salt-tolerant plants that normally occupy lowland basins of interior drainage in arid regions.

Hanging valley   A small valley that enters a larger valley high up on its flank, and therefore "hangs" above the larger valley. This markedly ungraded situation generally results from either glaciation or fault displacement.

Haystack hill   A rounded hill, somewhat resembling a haystack, that remains on a karst surface as an erosional remnant of a former higher topographic level.

Heat of condensation   The quantity of heat energy, totaling approximately 590 calories per gram of H_2O, that is released when water vapor condenses to liquid water. It is the direct power source for many atmospheric processes.

Heat of fusion   The quantity of heat energy, totaling approximately 80 calories per gram of H_2O, that is released when liquid water freezes.

High   See *high pressure system.*

High pressure system   An area of higher barometric pressure than the surrounding areas, characterized by an anticyclonic wind flow pattern and typically associated with fair weather.

Histosol   A soil order consisting of dark-colored organic soils of poorly drained areas. Most histosols are highly acidic and low in fertility.

Horn   A sharp-crested, pyramidal peak, representing the remnant rock mass between three or more coalescing cirques.

Horst   A raised block of land bounded by two normal faults.

Human geography   The branch of geography concerned with the spatial aspects of earth phenomena that are produced or are primarily controlled by humans.

Humidity   A term referring to the water vapor content of the air. Several types of humidity measurements exist.

Humus   Finely divided, brownish-black, decomposed organic matter that is a major supplier of plant nutrients.

Hurricane   A tropical cyclonic storm that forms over water and has sustained winds of at least 75 miles per hour (33.5 m/sec). Hurricanes are the world's most destructive storm type.

Hydration   A chemical weathering process involving the absorption, or adhesion, of water molecules to the molecules of a mineral.

Hydraulic action   The direct removal of loose materials by the friction and turbulence generated by flowing water in a stream or by oceanic wave action.

Hydrologic cycle   The continuous cycle of phase change and movement of H_2O between the earth's surface and the atmosphere.

Hydrolysis   A chemical weathering process consisting of a permanent chemical combination of minerals with water.

Hydrosphere   The aqueous envelope of the earth, including water in the atmosphere, at the earth's surface, and in the near-surface layers of the ground.

Hygrophyte   A plant especially adapted to grow in a very moist or waterlogged surface environment.

Ice-scoured plain   An extensive, lake-strewn, low relief surface that was subjected to glacial scour by a continental ice sheet. Ice-scoured plains presently occupy much of interior eastern Canada and Scandinavia.

Iceberg   An islandlike mass of freshwater ice produced by the breakup of a glacier as it enters the ocean or other large standing water body.

Igneous rock   A rock that has formed directly from the solidification of magma or lava resulting from tectonic activity.

Illuviation   The deposition, usually in a lower soil horizon, of solid or dissolved materials that were removed (eluviated) from another horizon.

Inceptisol   A soil order containing chemically immature soils with weak profile development.

Infiltration capacity   The maximum rate at which a given surface can absorb water. It is determined primarily by the texture, degree of compaction, and existing water content of the surface materials.

Insolation   A shortening of the term "incoming solar radiation," referring to the total quantity of solar radiation reaching the earth system.

Interception   The process by which vegetation intercepts precipitation before it reaches the ground. This process becomes ineffective once the vegetation has been fully wetted.

Interfluve   The generally well-drained higher ground between streams.

Interior drainage   Drainage that does not reach the ocean, but instead enters an enclosed basin where it is absorbed into the ground or evaporates. Interior drainage is characteristic of most arid regions.

International Date Line   This line is based on the 180th meridian and is located in the central Pacific. Time varies by a full 24 hours on either side, with the west side being a day later than the east side.

Intertropical Convergence Zone (ITCZ)   The zone of low barometric pressure and wind convergence centered near the equator, caused primarily by the prevailing high temperatures.

Intrusive igneous rock   Igneous rock that has formed inside the earth. The slowness of the solidification process usually allows visible mineral crystals to form.

Island arc   An arc-shaped group of islands paralleling, at a distance of several hundred miles, a submarine trench. It is formed when magma from the melting subducted plate that produced the trench has been erupted in sufficient mass to protrude above sea level.

Isobars   Lines connecting places having the same barometric pressure.

Isostatic rebound   The buoyant uplift of a portion of continental crust that had formerly been depressed by the weight of a continental glacier.

Isostatic uplift   The broad, gentle upwarping of a region that has become buoyant because of its reduction in mass as a result of erosion or deglaciation.

Isotherms   Lines connecting places having the same temperature.

Jet stream   Relatively narrow bands of strong westerly winds centered in the upper troposphere above the middle latitudes and the subtropics of both hemispheres. They act as steering currents to direct the movements of surface weather systems.

Jungle   A nearly impenetrable growth of low growing plants that is common in warm, moist areas where sunlight can reach the near-surface vegetation layers.

Karst   Topography developed largely by the differential solution and subsequent erosion of highly soluble rocks, especially limestone and dolomite.

Karst plain   A limestone plain covered with large numbers of sinkholes and sometimes other solution features.

Kelvin temperature scale   A temperature scale with its starting temperature (0° K) at absolute zero—the temperature at which all molecular motion ceases. Like the Celsius scale, it employs 100 units between the freezing and boiling points of water (273° K and 373° K, respectively).

Kettle   A pit or depression within a till plain formed by the weight of a detached block of glacial ice. Kettles may be occupied by small lakes.

Laccolith   A dome-shaped igneous intrusion of relatively acidic magma that is injected into a sequence of near-surface sedimentary strata.

Lagoon A shallow, protected body of salt water located between a barrier island chain and the mainland. It may consist of an extensive mudflat at the time of low tide.

Lake A landlocked body of water with a horizontal surface level.

Land and sea breezes A diurnally reversing coastal wind-flow pattern established in some areas by land and water temperature differences. Sea breezes are generated during the day, and land breezes at night.

Landforms Individual earth surface features.

Lapse rate See *average environmental lapse rate.*

Latent heat Energy absorbed by melting ice or evaporating water. This energy is released as sensible heat when the reverse phase change processes of freezing or condensation occur.

Laterite A hard, iron-rich subsurface layer that sometimes develops in humid tropical regions following the removal of vegetation.

Latitude Angular distance in degrees north or south of the equator.

Laurentide ice sheet The continental ice sheet that covered most of Canada and the northern United States during the glacial stages of the Pleistocene.

Lava Molten rock material formed by the expulsion of magma onto the earth's surface.

Lava plain (or lava plateau) An extensive, gently rolling plain or plateau surface formed by consecutive fissure outpourings of basaltic lava.

Leaching The removal of dissolved materials and fine-textured particles from the upper soil by downward percolating soil water. These materials may be deposited deeper within the soil, or they may be removed completely from the soil.

Leap year A year containing 366 days (including February 29).

Leeward coast A coast with a prevailing land-to-water wind flow.

Liana Any of a variety of climbing plants that root in the ground. Large woody lianas are characteristic of the tropical rainforest.

Limestone An organic sedimentary rock composed of calcium carbonate ($CaCO_3$) and formed in quiet tropical seafloor environments either by direct chemical precipitation or through the accumulation of the remains of marine organisms.

Lithosphere The solid portion of the earth, composed largely of rock and metal. More recently, this term has also been used to refer to the rigid uppermost portion of the mantle and the overlying crust that forms the lithospheric plates.

Lithospheric plate A discrete segment of the earth's crust and uppermost mantle that moves as an entity when transported by convection currents in the mantle.

Loam A soil of intermediate texture that generally contains substantial proportions of sand, silt, and clay.

Local time The time of day based on the east-west position of the sun at a given locality.

Loess Silt-sized eolian deposits that are typically buff colored and are composed of a well-sorted mixture of angular quartz and calcite ($CaCO_3$) particles.

Longitude Angular distance in degrees east or west of the prime meridian, stated to values of 180° E or W.

Longitudinal dune A long ridge of sand that is typically aligned parallel to the direction of the prevailing winds in an arid region.

Low See *low pressure system.*

Low pressure system An area of lower barometric pressure than the surrounding areas, characterized by a cyclonic wind-flow pattern and typically associated with cloudiness and precipitation.

Magma Molten rock material inside the earth. If it reaches the surface in a molten state, it is termed lava.

Magnetic North Pole The northern locus of the earth's magnetic field and the place to which a compass needle points. It is located in the Canadian Arctic.

Magnetic South Pole The southern locus of the earth's magnetic field, located near the coast of Antarctica.

Mantle The most extensive of the internal layers of the earth, occupying an intermediate position between the core and the crust. The mantle is approximately 1800 miles (2900 km) thick and is apparently composed primarily of peridotite rock.

Map A pictorial representation of the geographic location of selected surface features at a reduced scale.

Map scale   The ratio between the separation of places on a map and on the earth's surface.

Marine terrace   An elevated surface of low relief that represents the remains of an old wave-cut or wave-built platform. These features occur in areas where the land has been uplifted or sea level has dropped.

Maritime climate   Climatic characteristics typical of the area over or near a large water body, especially including relatively small daily and annual temperature ranges and high quantities of atmospheric moisture.

Maritime polar air mass   A relatively cold, moist air mass that has formed over a high latitude water body.

Maritime tropical air mass   A relatively warm, moist air mass that has formed over a water body in the low latitudes.

Mass wasting   The downslope movement of surficial material under the direct influence of gravity.

Meander   A broad looping bend in the course of a stream occupying a floodplain. Meanders typically display a distinctive cycle of formation, growth, and disappearance.

Mercurial barometer   A barometer that measures air pressure by the height of the mercury column in a glass tube from which the air has been removed.

Meridian   Another term for a line of longitude.

Mesa   An isolated, steep-sided and flat-topped remnant of a previous more elevated topographic surface. Mesas are often left as erosional outliers of retreating escarpments in arid regions.

Mesopause   The boundary between the mesosphere and the thermosphere. At this level, temperatures stop falling with increasing altitude.

Mesophyte   A plant adapted to an environment with a moderate water supply that is available throughout the year. It is intermediate in characteristics between a xerophyte and a hygrophyte.

Mesosphere   The thermal layer of the atmosphere between the stratosphere and the thermosphere, in which temperatures fall with increasing altitude.

Metamorphic rock   Rock subjected to solid-state alteration from its original form by heat, pressure, and/or chemical activity. Each igneous and sedimentary rock type has one or more metamorphic rock equivalents.

Meteorology   The branch of physical science that systematically studies the weather.

Milky Way galaxy   The organized aggregation of stars in which our solar system is located.

Mineral   A naturally occurring, solid, inorganic compound that has a specific chemical composition and usually exhibits a crystalline structure that results from a distinctive atomic arrangement of its constituent elements.

Minute   An angular distance of 1/60th of a degree; used for accurately stating latitude and longitude.

Mollisol   A soil order containing soils with a thick, dark-colored, humus-rich A horizon and a high cation exchange capacity. Most mollisols are highly fertile and agriculturally productive and are associated with major grain producing regions.

Monsoon winds   A seasonally reversing regional-scale wind-flow pattern caused by land and water temperature differences. It is especially well developed in southern and eastern Asia and in parts of North Africa.

Monthly mean temperature   The average temperature for the month, calculated by averaging the daily mean temperature values for each day of the month.

Moraine   A glacial depositional feature composed of till. Several types of moraines exist, depending on their sites of deposition. These include lateral, medial, terminal, recessional, and ground moraines.

Mudflow   The perceptible and often rapid downhill flow of mud, usually mixed with rocks and other debris.

Muskeg   A swampy area within the Northern Coniferous Forest that is covered with mosses, swamp grasses, shrubs, and scattered hygrophytic trees such as larches.

Mutation   A sudden and inheritable alteration in the characteristics of a plant or animal caused by a molecular rearrangement of its genetic material.

Natural levees   Low and often broad ridges of stream deposits that parallel each side of a stream channel within a floodplain. They are formed by deposition resulting from the sudden increase in friction as a flooding stream overtops its banks.

Neap tides   Tides with exceptionally small vertical ranges, occurring approximately twice a month, caused by the perpendicular alignment of the moon and sun with respect to the earth.

Needleleaf Used to describe trees with needle-shaped leaves, such as pine, spruce, or fir, as opposed to broadleaf trees.

Nimbus A Latin term meaning "rain cloud," used in conjunction with the two cloud types responsible for most of the world's precipitation—nimbostratus and cumulonimbus.

Nitrogen A relatively inactive gas that comprises approximately 78 percent of the lower atmosphere.

Normal fault A fault in which the vertical displacement exceeds the horizontal and an expansionary component is present.

North Pole The northern end of the earth's axis, located at 90° N.

Oasis A desert locality where the water table intersects the surface, resulting in the growth of significant vegetation.

Oblate ellipsoid A three-dimensional ellipse that is flattened at the poles. Earth assumes this shape because the centrifugal force of rotation bulges out the equatorial regions.

Occluded front A frontal boundary separating, at the surface, two relatively cold air masses, with a warmer air mass aloft. It is formed by the overtaking of a warm front by a cold front.

Ocean The interconnected body of salt water that dominates the earth's surface. It is commonly divided into segments named the Pacific, Atlantic, Indian, and Arctic Oceans.

Ocean current The continuous movement of an ocean water stream of similar temperature and density.

Oceanic crust The thin, continuous layer of dense basaltic rock that comprises the lowermost portion of the crust and forms the oceanic basins.

Organic sedimentary rock Sedimentary rock produced from the remains of plants and animals. Most are composed of carbonate minerals, especially limestone ($CaCO_3$).

Orographic lifting The lifting of air over a topographic barrier.

Outwash Glacio-fluvial materials that have been transported and deposited beyond the edge of the glacier by meltwaters.

Outwash plain A relatively extensive and flat glacial outwash surface deposited by multiple meltwater streams issuing from the margin of a glacier.

Overland flow A nonchannelized, sheetlike flow of water that is common on agricultural fields and near drainage divides.

Oxbow lake An arcuate lake that occupies an abandoned stream meander.

Oxidation A chemical weathering process involving the chemical combination of minerals with oxygen that has often first been dissolved in water.

Oxisol The deepest and most highly chemically weathered of the soil types. They are quite limited in natural fertility and have developed on ancient surfaces within the humid tropics.

Oxygen The second most abundant gas in the atmosphere, necessary to animal respiratory processes, and derived from carbon dioxide by means of plant photosynthesis.

Ozone A triatomic form of oxygen (O_3), occurring in very small quantities in the atmosphere, that helps to protect the earth's surface from cosmic ray bombardment.

Ozone layer The layer of the upper atmosphere, extending from about 10 to 40 miles (15–65km) altitude, in which ozone is most abundant.

Parallelism The idea that the earth's axis is constantly oriented in the same direction with respect to the plane of the ecliptic.

Parallels Another term for lines of latitude.

Parent material The solid inorganic and organic material from which the soil develops. The inorganic parent material may be either residual (originally derived from that location) or transported (derived from elsewhere and deposited in that location).

Ped A soil structural aggregate formed by the tendency of individual soil particles to adhere to one another. Different soil types tend to develop peds of differing sizes and shapes.

Pediment A gently sloping bedrock erosional surface that is commonly found at the foot of a mountain mass in an arid region.

Pedologist A soil scientist.

Perennial A plant that has an expectable life span of at least several years. It is therefore not normally killed by adverse weather conditions at the end of the growing season, although it may experience a period of dormancy.

Perennial stream   A stream that flows continuously because it has a dependable source of ground water.

Perihelion   The closest approach between the earth and sun. It occurs about January 3, when the two bodies are approximately 91.5 million miles apart.

Permafrost   The condition of perennially frozen subsoil and bedrock that underlies large areas in arctic, subarctic, and alpine environments.

Permeability   The ability of the soil or rock to transmit the flow of water.

Photosynthesis   The formation of carbohydrates in the chlorophyll-containing tissues of plants exposed to light.

Physical geography   The branch of geography concerned with the spatial aspects of natural earth phenomena, particularly with weather and climate, natural vegetation, soils, and landforms.

Physical weathering   Weathering processes that reduce the size of rock masses without altering their chemical composition.

Piedmont alluvial slope   A ramplike alluvial apron, typically found along the base of a desert mountain range, that is formed by the coalescence of numerous alluvial fans. Also called a bajada.

Place   A specific location on the earth's surface, containing a unique assemblage of spatial attributes.

Plain   An area of restricted local relief that has a predominantly flat-to-gently rolling surface.

Plane of the ecliptic   The angular plane on which the earth's orbit around the sun is located.

Planimetric map   A map that provides systematic information about only the horizontal characteristics of the areas it depicts.

Plant community   A plant assemblage whose members occupy a specific environmental setting of restricted size.

Plate   See *lithospheric plate*.

Plateau   A relatively level upland surface that is frequently bounded on at least one side by an abrupt descent to lower elevations. Also termed a tableland.

Plate tectonics   Encompasses the processes involved in the formation, movement, and destruction of lithospheric plates.

Playa   A shallow desert lake, often of considerable areal extent, that temporarily occupies the center of a bolson following a rainfall.

Pleistocene   The geologic epoch, extending from about 2,000,000 B.P. to 10,000 B.P., during which continental and alpine glaciers alternately advanced and receded over large portions of the world, especially in North America and Eurasia.

Pluton   A body of intrusive igneous rock that has generally solidified well beneath the surface.

Pocket beach   A small, crescent-shaped beach located on a coastline that is, as a whole, being worn back by vigorous wave attack. These beaches are common along the U.S. West Coast, and often develop in areas of relatively weak rock.

Point bar   A streambed bar of sand or gravel that develops on the inside of a meander. It is caused by the diminished flow at this point.

Polar easterlies   The belts of easterly winds blowing from the polar high pressure centers of each hemisphere toward the subpolar lows.

Polar front   The major weather front of the middle latitudes, separating air masses of subtropical and subpolar origins. It is a zone along which frontal cyclones frequently form.

Polar front zone   The zone of highly variable winds associated with the subpolar low pressure belts in each hemisphere.

Polar highs   High pressure caps centered near the North and South Poles, resulting largely from cold surface temperatures.

Polar zones of variable winds and calms   The zones of light and variable winds associated with the centers of the polar highs.

Pore spaces   The voids between the solid soil particles that contain soil liquids and gases.

Porosity   The volume of soil pore spaces divided by the total volume of the soil.

Potential evapotranspiration   The total quantity of water that would evaporate and transpire from a given site if the available moisture supply were unrestricted.

Prairie   Tall grasslands generally associated with subhumid portions of the middle latitudes. Most prairie grasslands have been destroyed by human agricultural activities.

Precipitation   Water in liquid or solid form falling through the atmosphere toward the earth's surface. The major forms are rain, snow, sleet, hail, and drizzle.

Pressure gradient   The rate of change in air pressure per unit of distance. When uncompensated for by other factors, it causes the wind to blow.

Prime meridian   The starting line of longitude. A separation line between the eastern and western hemispheres.

Radar   An instrument used to determine the approximate size, shape, and intensity of areas of precipitation. It emits pulses of microwave energy that are reflected from the precipitation and displayed on a screen.

Radiation   The process of energy transmission by electromagnetic waves.

Radiation fog   A fog type that commonly forms over land areas on calm, clear nights because surface radiation chills the overlying air below the dew point.

Radiosonde   A weather instrument package that is sent aloft by means of a hydrogen- or helium-filled balloon and automatically radios back upper atmosphere data on temperature, air pressure, and humidity.

Rain gage   A funnel-shaped, tubular instrument used for the collection and measurement of precipitation.

Rainshadow effect   The reduction of precipitation by the blocking effect of a topographic barrier.

Recumbent folds   Folds that have been compressed so tightly that the fold structures have closed in on one another and collapsed.

Reef   A ridgelike accumulation of the exoskeletons of large colonies of coral polyps or other reef-building marine organisms. The reef itself may be entirely beneath the water or may protrude above the water surface to form a coastal shelf or island. A fringing reef is attached to the mainland, while a barrier reef is separated from the mainland by a lagoon.

Region   A portion of the earth that displays relative similarity in one or more attributes.

Regional metamorphism   Large-scale metamorphism that occurs in conjunction with tectonic activity; especially found in the root zones of mountain systems.

Regolith   Weathered rock material, commonly located between unweathered bedrock (below) and soil (above).

Relative humidity   The percentage ratio of the amount of water vapor actually in the air to the maximum amount of water vapor the air could hold at that temperature and pressure.

Relief (topographic)   Differences in elevation within an area. The topographic relief of a mapped area is computed by taking the elevational difference between the highest and lowest point.

Remote sensing   The collection of information at a distance from a location by means of mechanical devices that receive emitted or reflected electromagnetic radiation.

Reversal of topography   Occurs when intense folding or faulting causes rock strata to overturn, so that older strata overlie younger strata.

Reverse fault   A fault in which the vertical displacement exceeds the horizontal and a compressional component is present, with some overlapping of the two sides.

Revolution   The yearly orbital motion of the earth around the sun.

Ria coast   A highly irregular erosional coast dominated by the presence of numerous bays or estuaries that were produced by the gradual submergence of river valleys. The Middle Atlantic coast of the United States is a classic example.

Richter scale   The most commonly used scale for measuring earthquake magnitudes. Each number on the scale has ten times the earthquake wave amplitude of the preceding number.

Rift valley   A large-scale graben, sometimes formed during an early stage in the separation of two continental plate segments.

Riparian   Riverbank vegetation that often flourishes in the moist soils along streams in arid and semiarid regions.

Roche moutonée   A bedrock outcrop in glaciated terrain that exhibits a smoothly rounded up-ice side and an irregularly plucked down-ice side.

Rock   A solid aggregate of minerals. Different rocks are defined on the basis of their proportional mineral composition. Rocks may also consist of organic materials.

Rock cycle   The continuous cycle of rock formation, erosion, deposition, and reconstitution.

Rockfall   The falling of an individual rock or small rock mass down a nearly vertical cliff.

Rockslide   The sliding of a mass of detached rock down a moderate-to-steep slope.

Rotation   The spinning of the earth on its axis. This motion is the basis of our calendar day.

Saline water intrusion   The intrusion of salt water into a freshwater aquifer, often as the result of the excessive withdrawal of well water.

Salinization   The increase in soil salinity, usually in poorly drained arid areas, that occurs as a consequence of irrigation for a prolonged period. The salt is derived either from the deeper soil layers or from the irrigation water itself.

Saltation   An asymmetrical bouncing motion exhibited by relatively coarse-textured sediments being transported either by a stream or by the wind.

Sand bar   A largely or entirely submerged ridge composed of sand that is formed in the near-shore oceanic waters by currents and larger waves.

Sandstone   A clastic sedimentary rock comprised largely of sand-sized clasts that are usually composed of silica.

Sandstorm   A stinging mass of windblown sand grains extending to only a few feet above the ground.

Savanna   A tropical grassland that contains scattered trees or shrubs. Savannas are dominant in forest/grassland transitional areas in regions with alternating wet and dry seasons.

Sclerophyll   A plant with small, hard, thick, leathery leaves that serve to minimize transpirational losses.

Scrub   A dense growth of shrubs and small trees, typical of many tropical and subtropical areas with alternating wet and dry seasons.

Sea cliff   A cliff formed along or just behind a shoreline by wave undercutting in areas of competent (strong) materials.

Seafloor spreading   The process of formation of new oceanic crust by the extrusion of lava along divergent lithospheric plate boundaries.

Sea ice   Ice that forms on the ocean surface from the freezing of seawater. Most exists in the Arctic Ocean and near the coast of Antarctica.

Second   An angular distance of 1/60th of a minute, or 1/3600th of a degree.

Sedimentary rock   Rock that has formed from the lithification of sediments and typically accumulates in layers (strata).

Sediments   Solid materials of either organic or inorganic origin that were transported to their depositional sites by water, ice, wind, or gravity. The lithification of sediments forms sedimentary rocks.

Seismograph   An instrument used to measure earthquake shock waves.

Semiarid climate   A climate with a potential evaporation rate that slightly exceeds its mean precipitation total, so that it is moderately dry.

7th Approximation   A nickname for the U.S. Comprehensive Soil Classification System, the system currently in official use in the United States.

Shale   The most abundant sedimentary rock type, composed of clay-sized particles and often containing fossils. Most shales formed from muds deposited on the deep ocean floor.

Shield   An ancient block of continental crust that has been tectonically stable for a lengthy geologic time span.

Shield volcano   A broad, gently sloping volcano constructed primarily of basalt derived from fluid lava.

Silicate minerals   A mineral family consisting primarily of silicon and oxygen. It is the dominant mineral group of the crust and mantle.

Sill   A relatively thin sheet of magma injected more or less horizontally between near-surface sedimentary rock strata.

Sinkhole   A rounded depression formed by subsurface solution, especially in regions underlain by limestone. Sinkholes can develop either by gradual subsidence or by the sudden collapse of the surface. Also called a doline.

Sinking creek   A stream that enters a swallow hole in a karst region and continues its course beneath the surface.

Sleet   A precipitation form consisting of small ice pellets that have formed by the freezing of descending raindrops.

Slump   The intermittent and relatively slow movement of a mass of earth or rock along a curved slip-plane, or landslide boundary surface.

Soil   A loose mixture of weathered rock material, organic matter, water, and air that can support plant growth. Most undisturbed soils tend to develop distinctive layers (horizons) that parallel the surface and differ in their physical, chemical, and biological characteristics.

Soil horizons   Soil layers, roughly paralleling the surface, that differ in their physical, chemical, and biological characteristics as a result of their formation under differing environmental conditions.

Soil pH   The acidity or alkalinity of the soil, determined by its supply of exchangeable H+ ions.

Soil profile   Vertical slices of the soil in which the various horizons are exposed for analysis.

Soil texture   The size of the individual soil particles. Different texture categories are determined by the soil's proportions of sand, silt, and clay-sized particles.

Soil water   Water that adheres in a molecular film to the soil particle surfaces and partially occupies the pore spaces between the soil particles.

Solar day   The period of time needed for the sun to make one apparent 360° circuit of the sky.

Solar noon   The exact time that the sun reaches its zenith (high point in the sky) at any given locality.

Solar system   The sun and the other astronomical bodies, notably the nine planets, that revolve around the sun because of its gravitational attraction.

Source region   A large land or water body over which air masses form and from which they derive their temperature and moisture characteristics.

South Pole   The southern end of the earth's axis, located at 90° S.

Specific heat   The quantity of heat energy that must be absorbed or released by a particular substance in order to undergo a given temperature change.

Specific humidity   The ratio of the mass of water vapor in a parcel of air to the total mass of the air, including the water vapor. Usually stated in grams per kilogram.

Spit   A typically hook-shaped barrier beach that is attached to the mainland at one end and often extends partially across a bay on the other end.

Spodosol   A soil order containing acidic soils with a light-colored eluvial E horizon and an illuvial B horizon. They are typically associated with a subarctic climate and with taiga vegetation.

Spring   The site of an outflow of subsurface water.

Spring equinox   See *vernal equinox.*

Spring tides   Exceptionally high and low tides, occurring approximately twice a month, caused by the parallel alignment of the moon and sun with the earth.

Squall line   A line of heavy showers or thunderstorms associated with strong, gusty winds. Squall lines frequently develop in advance of strong cold fronts, in which case they are known as pre-frontal squall lines.

Stable air   Air with a vertical temperature profile that causes it to resist any lifting influences.

Stack   A small, steep-sided island located near the shoreline that is an erosional remnant of a retreating sea cliff.

Stalactite   An icicle-like travertine feature that hangs downward from the roof of a cave.

Stalagmite   The upward-growing counterpart of a stalactite, formed of travertine and found on the floor of a cave in a limestone region.

Standard time system   The currently used time system for most of the world. Time is based on the east-west position of the sun at meridians with longitudes divisible by 15, and changes by whole-hour increments as time zones are crossed.

Stationary front   A relatively stationary frontal boundary between two differing air masses. Such a front eventually will either dissipate or will begin to move as a warm front or cold front.

Steppe   Short grasslands generally associated with semiarid climates in the low and middle latitudes. Grass heights are usually less than 2 feet (0.6 m).

Storm surge   The very high ocean water levels that accompany the landfall of a hurricane.

Strata   Horizontal layers, usually of sedimentary rocks.

Stratopause   The boundary between the stratosphere and the mesosphere. At this level, temperatures stop rising with increasing altitude.

Stratosphere   The thermal layer of the atmosphere above the troposphere, in which temperatures increase with altitude.

Stratus   From the Latin word meaning "layer," stratus clouds are low, dense, gray clouds that usually cover the entire sky.

Stream   A channelized body of water that flows downhill in response to gravity.

Stream terrace   A terrace consisting of the remnant of a former floodplain surface that was developed at a topographic level higher than the present stream valley level.

Striations   Sets of parallel bedrock scratches caused by the passage of glacial ice containing embedded rock fragments.

Strike   The compass orientation of a horizontal line in the plane of an inclined rock stratum or outcrop.

Structure (of soil)   The characteristic shape, size, and organization of the soil aggregates or peds.

Subduction   The angular descent of the margin of an oceanic lithospheric plate into the asthenosphere, where it is eventually destroyed. Subduction results from a collision between two plates, with the denser plate being subducted.

Sublimation   The direct phase change of water vapor to ice, or vice-versa. Approximately 670 calories of energy are involved for each gram of H_2O that sublimes.

Subpolar jet   A jet stream that is produced by temperature contrasts between the low and high latitudes. Its position is closely related to the surface position of the polar front.

Subpolar lows   Zones of rising air and low pressure centered in the subpolar latitudes of both hemispheres, caused primarily by the earth's rotation.

Subsidence   The vertical settling of the surface as a result of the removal of subsurface support by mining, solution, or other processes.

Subsoil   The soil horizon that underlies the topsoil. Also referred to as the B horizon, it usually contains little organic matter and is frequently an illuvial horizon.

Subtropical belts of variable winds and calms   The narrow belts of light and variable winds associated with the centers of the northern and southern hemisphere subtropical highs; also referred to as the "horse latitudes."

Subtropical highs   Zones of subsiding air and high pressure centered in the subtropics of both hemispheres and caused primarily by the earth's rotation.

Subtropical jet   A jet stream that is centered above the surface positions of the subtropical highs.

Summer solstice   The day that one's hemisphere reaches its maximum inclination toward the sun. The sun is directly overhead at a latitude of 23 1/2°. Marks the official beginning of summer.

Surface creep   The slow transport of windblown surface materials by rolling or sliding movements. This process accounts for about 20 percent of all eolian surface sediment transport.

Suspended load   The portion of the sediment load of a stream that consists of fine-textured solid particles transported within the flow and kept from settling by turbulence.

Suturing   The fusion of two lithospheric plates into a single plate following their collision.

Swamp   A poorly drained area, frequently covered with hygrophytic vegetation.

Swash   The landward flow of water resulting from the momentum of a breaking wave. Before returning as backwash, this water typically transports loose sediments to higher portions of the beach.

Syncline   A downward fold in one or more rock strata.

System   A set of interrelated components through which energy flows to produce orderly changes. A closed system has a limited supply of energy from an internal source. An open system has an unlimited supply of energy from an external source.

Taiga   The Russian name for the Northern Coniferous Forest, the most poleward of the forest subassociations.

Talus cone   A conical accumulation of angular rocks resting against a cliff or mountainside, derived largely from the process of frost wedging.

Tectonic forces   Landform-producing forces originating inside the earth, primarily as a result of the motions of tectonic plates.

Temperature   A measure of the thermal energy contained by an object or substance. This energy level is related to the rate of motion of its constituent molecules.

Temperature inversion   An increase in temperature with altitude in the troposphere. This situation represents a reversal of the normal decline in temperatures with altitude.

Terrace   A steplike surface that is relatively flat and linear. It is bounded on one side by a slope to a lower level and on the other by a slope to a higher elevation.

Terranes   Small pieces of continental crust, often of uncertain origin, that may be swept up by the movements of lithospheric plates and accreted onto the margin of a continent.

Terrestrial radiation   The longwave electromagnetic radiation emitted by the earth system, which we sense as heat.

Thermal high   A high pressure system produced by the compaction of air as a result of cold surface temperatures.

Thermal low   A low pressure system produced by the expansion of air as a result of hot surface temperatures.

Thermosphere   The uppermost of the four thermal layers of the atmosphere, in which temperatures rise with increasing altitude.

Throughflow   The lateral flow of water near, but slightly below, the soil surface. It results from a temporary condition of saturation in the lower soil.

Thrust fault   A fault in which extreme compression causes one side to be thrust at a low angle over the other, often for a considerable distance.

Thunderstorm   A storm associated with cumulonimbus clouds, thunder, and lightning. It is the world's most common storm type, and is often associated with heavy rains and strong local winds.

Tidal inlet   An oceanic inlet between barrier islands that connects the coastal lagoon to the open sea.

Tides   The rhythmic rise and fall of the ocean surface level produced by the gravitational influence of the moon and sun.

Till   Rock material deposited directly by glacial ice, and therefore displaying a lack of sorting.

Till plain   An extensive low relief surface covered by a thick accumulation of ground moraine. Much of the Midwest United States and the North European Plain are till plains.

Topographic map   A map that provides systematic information on surface elevations for the areas it depicts by means of colors, hachures, contour lines, or some other method.

Topography   The combination of landform characteristics and distributions within a region.

Topsoil   The uppermost mineral horizon of the soil; often referred to as the A horizon.

Tornado   A tubular vortex of whirling air surrounding a central core of extremely low pressure, which extends downward from the base of a cumulonimbus cloud.

Tower karst   Steep-sided remnant limestone peaks of mountainous proportions.

Trade winds   The major wind belt of the tropics, blowing from the subtropical high pressure centers of each hemisphere toward the Intertropical Convergence Zone. Prevailing wind directions are northeasterly in the northern hemisphere and southeasterly in the southern hemisphere.

Transcurrent fault   A fault in which the two sides are predominantly horizontally offset. Also called a strike-slip fault.

Translocation   The transfer of dissolved materials and fine-textured particles from one place to another in the soil as a result of soil water transport.

Transpiration   The evaporation from plant surfaces of water previously absorbed through the plant root systems.

Transverse dune   A linear sand dune with an undulating crest line that is elongated perpendicular to the prevailing wind direction.

Travertine   Calcium carbonate ($CaCO_3$) deposits generally found in caves. They assume a wide variety of forms, including stalactites, stalagmites, columns, and ribbons.

Tree   A large woody plant normally containing a single primary stem or trunk that develops many branches.

Trellis stream pattern   Consists of a parallel pattern of major streams, which are joined at nearly right angles by much smaller and shorter tributaries. This pattern typically develops in regions containing alternating bands of resistant and non-resistant sedimentary strata.

Tropic of Cancer   Located at 23 1/2° N, it is the northernmost line of latitude that the sun's vertical rays reach during the course of the year.

Tropic of Capricorn   Located at 23 1/2° S, it is the southernmost line of latitude that the sun's vertical rays reach during the course of the year.

Tropical depression   A weak closed low pressure center that has formed over a tropical water body and has maximum sustained winds of less than 40 miles per hour (18 m/sec). It may have the potential to develop into a tropical storm and, eventually, into a hurricane.

Tropical storm A storm that has developed over a tropical water body and has maximum sustained winds of between 40 and 74 miles per hour (18–33.5 m/sec). Tropical storms are given names, are capable of causing destruction if they strike land, and have the potential to strengthen into hurricanes.

Tropopause The boundary between the troposphere and the stratosphere. At this level, temperatures stop falling with increasing altitude.

Tropophyte A plant adapted to alternating favorable and unfavorable climatic conditions for growth. Most tropophytes grow in areas with alternating warm and cold or wet and dry seasons.

Troposphere The lowermost of the four thermal layers of the atmosphere, in which temperatures normally decline with increasing altitude. Most weather phenomena exist in this layer.

Turbulence The chaotic eddying or swirling motions imparted to wind or flowing water by the effects of friction.

Typhoon The term used for a hurricane in the western North Pacific.

Ultisol A soil order consisting of highly chemically weathered soils of limited natural fertility that have developed in warm, moist climates.

Undertow The seaward movement of subsurface water near the shoreline. It occurs as a counterbalance to the landward flow of water carried by surface waves.

Uniformitarianism The geologic principle stating that the earth's surface features have evolved from the long-continued application of currently operative geomorphic processes.

Unloading A physical weathering process involving rock breakup resulting from the gradual decrease in gravitational pressure as erosion removes the overlying material.

Unstable air Air with a vertical temperature profile that produces a tendency to rise, especially if the air is given an initial lift.

Upwelling The flow of deep ocean water to the surface.

Urban heat island The islandlike area of relatively warm temperatures associated with a city. The heat island results both from the nature of the city surface materials and from human activities that release heat.

Valley train A linear train of glacio-fluvial materials deposited by a meltwater stream issuing from the margin of a glacier.

Vegetative association A global-scale community of plants that exists in relative equilibrium with its environment. Also referred to as a biome.

Vegetative succession The gradual change in the composition of a plant community in response to changing environmental conditions following a disturbance such as a fire.

Vernal equinox The day that the earth's axial inclination is exactly sideways with respect to the sun, following a period in which one's hemisphere was inclined away from the sun. Marks the official beginning of spring.

Vertisol A soil order consisting of soils that contain large amounts of clay that swells when wetted and shrinks when dried. This results in a churning action that inhibits profile development.

Volcanic neck A solid spire of hardened lava that was in the vent of a volcano at the end of its final eruption.

Volcanism Processes involving the transfer of molten rock material either from one subsurface location to another, or its expulsion onto the surface.

Volcano A hill or mountain constructed of materials from the interior of the earth that have been ejected under pressure from a vent.

Warm front A frontal boundary along which an advancing mass of warm air is overrunning and displacing relatively denser cold air.

Warm ice glacier A wet-base glacier that has an underlying film of meltwater acting as a lubricant to aid slippage over its bedrock base.

Water table The upper boundary of the zone of permanently saturated soil or rock; the boundary between soil water and ground water.

Water vapor Water (H_2O) in the invisible gaseous state. It comprises an average of about 1.4 percent of the atmosphere and is the source of moisture for clouds and precipitation.

Wave A rhythmic vertical motion of the near-surface portion of any water body, usually generated by the wind. The water molecules move in circular or elliptical paths, with little or no net forward motion.

Wave-built platform A platform of unconsolidated marine sediments that is generally deposited in the near-shore waters immediately seaward of a wave-cut platform.

Wave-cut platform A rock platform that has been planed off at the base of erosionally effective wave action along an eroding coast.

Wave refraction The bending of the angle of approach of incoming waves. Refraction occurs as the waves obliquely approach the shoreline and encounter a shallow bottom that slows and rotates them more directly toward shallow water.

Weather The short-term condition of the atmosphere with respect to temperature, air pressure, wind, humidity, clouds, and precipitation.

Weather front A boundary zone along which two or sometimes three air masses of differing temperature and moisture characteristics meet, often producing an extensive area of cloud cover and precipitation.

Weathering The combined action of physical and chemical processes that disintegrate and decompose bedrock. An essential preliminary for soil formation and gradation.

Westerlies The major wind belts of the middle latitudes, blowing from the subtropical high pressure centers of each hemisphere toward the subpolar lows.

Wilting point The soil moisture level at which plants begin to wilt. Wilting occurs because the remaining soil water is so tightly bound to the soil particles that the plants are unable to extract it.

Windward coast A coast with a prevailing water-to-land wind flow.

Winter solstice The day that one's hemisphere reaches its maximum inclination away from the sun. The sun is directly overhead at a latitude of 23 1/2° in the opposite hemisphere. Marks the official beginning of winter.

Woodland An area dominated by trees that are too widely spaced to produce the closed vegetative canopy of a forest.

Xerophyte A plant especially adapted to grow in a region deficient in water.

Zone of aeration The zone above the water table, where pore spaces or other openings are jointly occupied by soil water and gases.

Zone of saturation The zone below the water table, where pore spaces or other openings are completely occupied by ground water.

Index

James River, 377
Jet streams, 68–70
Joint/fracture formation, 283–84

Kalahari Desert, 196, 197, 354, 358, 365
Kamchatka Peninsula, 141, 247, 272
Karakoram Mountains, glaciers and, 325
Karst plains, 295
Karst topography, 292–93
    cause and global distribution of,
        293–94
    features of, 294–98
Kelvin (Absolute) scale, 45
Kettles, 341
Kilimanjaro, Mount, 274
Krakatau Volcano, 274, 279–81

Labrador Current, 81, 165
Laccoliths/laccolithic domes, 276
Lacustrine plains, 342
Lagoons, 374
Lake Ontario Plain, 341
Lakes, 156–57
    breeze, lake, 65, 67
    glacial, 158, 340, 342
    origins of, 157–58, 340, 342
    oxbow, 315
    playa, 362
    rising water levels, 159
    saline, 157, 362
    snowstorms, lake-effect, 95–96
Land and water distribution, temperature
    and, 44–45
Land breeze, 67
Landforms
    classification of, 234–38
    coastal, 369–79, 381–83
    desert, 197–98, 352, 356–62
    diastrophic, 259–65, 267–69
    fluvial (rivers/streams), 155, 158,
        160–63, 303–6, 308–19
    glacial, 51, 318–19, 325–44
    volcanic, 242–43, 269–77
LaSal Mountains, 276
Latent heat, 38
Lateral moraines, 330–31
Laterite, 225
Latitude(s), 13–14
    high, 13, 100–105, 107, 109–14,
        141–44
    low, 13, 125–32
    middle, 13, 100–105, 107, 109–14,
        132–41
    temperature and, 146
Laurentide ice sheet, 336–37, 339, 343–44
Lava, 242–43, 274–75
Leaching, 208

clearing of forests and soil, 202–3,
    224–26
Leap year, 7
Leeward side, 46, 88, 146
Levees, natural, 315–16
Lianas, 188
Limestone, 244–45
Lithosphere, 3, 234, 240
Lithospheric plates, 250
Loam, 206
Local/regional winds, 65, 67–68, 72–73
Localized storms, of the high and middle
    latitudes, 110–14
Local time, 15
Loess, 351, 355–56
Longitude, 13–15
Longitudinal dunes, 354
Los Angeles Basin, 33
Low Latitude Dry Climate, 129–32
Low latitudes, 13
    climates of, 125–32
Lows
    equatorial, 116
    thermal, 53–55
    traveling low-pressure systems
        (cyclones), 64–65, 105, 107,
        109–10

Magma, 242, 252–53, 269, 272, 274, 276
Magnetic north and south, 12
Mammoth Cave, 296
Mantle, of the earth, 240
Maquis, 191
Mariana Trench, 163
Marine Climate, 136–39
Marine terraces, 378
Maritime air masses, 98
Maritime climate, 44
Marshes, salt, 375
Mass wasting, 287–92
Matterhorn, 332
Mauna Kea, 273
Mauna Loa, 273
Mayon, Mount, 274
Mazama, Mount, 274
Mead, Lake, 158
Meanders, 314–15
Medial moraines, 331
Mediterranean Basin, 190–91
Mediterranean Sea, 254, 272
Mediterranean Woodland and Scrub,
    190–91
Melting, 75–76
    of glaciers, 51, 318–19, 326, 343–44,
        355
Mercurial barometers, 56
Mercury scale (air pressure), 56
Meridians, 14–15
Mesas, 359

Mesopause, 31
Mesophytes, 180
Mesosphere, 30–31
Metamorphic rocks, 246–48
Meteorology, 31
Methane, 27
Michigan, Lake, 159
Mid-Atlantic Ridge, 249–51, 272
Middle latitudes, 13
    climates of, 132–41
    localized storms of, 110–14
    secondary circulation systems of,
        100–105, 107, 109–14
Mid Latitude Deciduous and Mixed Forest,
    191–92
Mid Latitude Dry Climate, 135–36
Mid Latitude Evergreen Forest, 192–93
Milky Way galaxy, 7
Millibar scale, 56
Minerals, 241–42
Minutes, 14
Mississippi River, 159, 161, 342, 355
    channel modification of, 311, 321–23
    delta of, 316, 317, 321–23
    soils of, 219, 226
Mississippi Valley, 222, 230, 314, 321, 355
    earthquakes and, 268
Missoula, Lake, 346–47
Missouri River, 342, 350, 355
Mojave Desert, 197
Mollisols, 219
Monsoons, 67–68, 72–73
Monthly mean temperature, 41
Moon, tides and the, 168–69
Moraines, 329–31, 341
Mountains
    in deserts, 359–62
    erosional, 312–13
Mountain vegetation, 199
Mudflats, 375
Mudflows, 291
Multiple faulting, 267–68
Mutations, of natural vegetation, 176

NASA (National Aeronautics and Space
    Administration), 27
National Weather Service, 300
Natural bridges, 297
Natural levees, 315–16
Natural tunnels, 297
Nazca Plate, 251
Needleleaf trees, 185
Niagara Peninsula, 341
Niger River, 317, 357
Night and day, lengths of, 21
Nile River, 161, 219, 317, 357
Nile Valley, 212
Nitrogen, 25
Nonvariable gases, 25